WHAT YOU SHOULD CONSIDER IN . . .

Trigonometry with Applications

■ **Comprehensive and well organized content** gives students excellent preparation for more advanced courses. (See Contents, pages iii–vii.)

■ **Real-life applications** throughout the text and exercises show students how trigonometry can be used to solve problems in daily life and in careers. (See pages 18–19, 176–177, 198–199.) Special *Application* sections appear at the end of each chapter. (See pages 186–187.)

■ **Appropriate emphasis on skills and theory** provides students with a balanced course in trigonometry. (See pages 53–55, 89–91, 108–112, 142–143.)

■ **Helpful worked-out examples** reinforce concepts and prepare students for the exercises. (See pages 132–133, 234–235.)

■ **An abundance of exercises** provides practice with skills, applications, and theory at three levels of difficulty. (See pages 56–58, 139–141, 160–161.) *Challenge* exercises provide extra motivation for very capable students. (See page 23.) *Extra Practice* is included for each chapter. (See pages 368–381.)

■ **Using calculators and computers** to help students learn and apply trigonometry is discussed throughout. (See pages 2, 68, 173, 293.) *Computer Investigations* following each chapter show how a computer can be used to solve a variety of mathematical and applied problems. (See page 119.)

■ **Frequent reviews and tests,** including cumulative and mixed review, make it easy to monitor student progress. (See pages 58, 122, 182–183, 274, 344–355.) The *Algebra and Geometry Review* section offers review of prerequisite skills. (See pages 382–389.)

Supplementary materials: The **Teacher's Manual with Solutions** provides a pacing chart, teaching suggestions, and complete, worked-out solutions for all exercises. **Tests** (duplicating masters) and a **Resource Book** containing tests, practice, and activities on blackline masters are also available.

TRIGONOMETRY
with Applications

John A. Graham
Robert H. Sorgenfrey

TEACHER CONSULTANTS

Patricia Barnette

Linda R. Hunter

William P. Larsen

Gerald R. Rutz

Stephen F. Woodruff

HOUGHTON MIFFLIN COMPANY / BOSTON

Atlanta   Dallas   Geneva, Ill.   Palo Alto   Princeton   Toronto

AUTHORS

John A. Graham, Mathematics Teacher, Buckingham Browne and Nichols School, Cambridge, Massachusetts.

Robert H. Sorgenfrey, formerly Professor of Mathematics, University of California, Los Angeles.

TEACHER CONSULTANTS

Patricia Brady Barnette, Mathematics Teacher, Eau Gallie High School, Melbourne, Florida.

Linda R. Hunter, Mathematics Department Chair, Douglas MacArthur High School, San Antonio, Texas.

William P. Larsen, Mathematics Instructor, Grand Junction High School, Grand Junction, Colorado.

Gerald R. Rutz, Mathematics Department Chairman, Arcadia High School, Phoenix, Arizona.

Stephen F. Woodruff, Director of Mathematics, Stoughton High School, Stoughton, Massachusetts.

ISBN 0-395-46141-3

DEFGHIJ-D-987654321

Contents

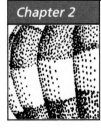

Chapter 6

VECTORS IN THE PLANE   **190**

Chapter 7

COMPLEX NUMBERS   **228**

Chapter 8

INFINITE SERIES   **280**

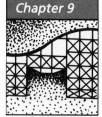

Chapter 9

VECTORS IN SPACE   **318**

Lines and Planes

USING TECHNOLOGY IN THIS COURSE

Scientific calculators, graphing calculators, and computers are particularly useful tools for exploring concepts and solving problems in trigonometry. There are so many ways of using these technological tools throughout the course, it is not possible to indicate them all. However, to highlight the importance of technology, a variety of calculator and computer uses have been suggested in the textbook. (See, for example, pages 11, 44-45, 68, 90-91, 111-112, 173, 237-238, 293, 296, 307.) In addition, there are special Computer Investigation sections that show how a computer can be used to solve mathematical and applied problems (pages 39, 79, 119, 149, 184-185, 224-225, 270-271, 315, 341).

The *Resource Book* accompanying this textbook includes exploratory activities that can be done with a graphing calculator or computer graphing software such as *Precalculus Plotter Plus*. Selected activities from *Precalculus Computer Activities* can also be used to explore trigonometry. A chart correlating these activities to the textbook and suggestions for using technology in the course are included in the *Teacher's Manual with Solutions*.

List of Symbols

THE GREEK ALPHABET

A	α	Alpha	H	η	Eta	N	ν	Nu	T	τ	Tau
B	β	Beta	Θ	θ	Theta	Ξ	ξ	Xi	Y	υ	Upsilon
Γ	γ	Gamma	I	ι	Iota	O	Φ	Omicron	Φ	ϕ	Phi
Δ	δ	Delta	K	κ	Kappa	Π	π	Pi	X	χ	Chi
E	ϵ	Epsilon	Λ	λ	Lambda	P	ρ	Rho	Ψ	ψ	Psi
Z	ζ	Zeta	M	μ	Mu	Σ	σ	Sigma	Ω	ω	Omega

METRIC UNITS OF MEASURE

Length:	cm	centimeter	Area:	cm^2	square centimeter
	m	meter		m^2	square meter
	km	kilometer		km^2	square kilometer
Speed:	m/s	meters per second	Mass:	g	gram
	km/h	kilometers per hour		kg	kilogram
	rad/s	radians per second			
Force:	N	newton	Work and Energy:	J	joule
				kW • h	kilowatt-hour

1

Trigonometric Functions

The photograph shows the horsehead, a region of dark dust that hides a vast region of brightly glowing interstellar material in the constellation of Orion. The bright star to the left of the horsehead is Zeta Orionis. Astronomy was one of the major interests of early mathematicians.

GEOMETRIC BEGINNINGS

Angles and Degree Measure

Objective To generalize geometric angles and to find their degree measures.

One of the first accomplishments of trigonometry was comparing the distance between Earth and the moon with the distance between Earth and the sun. The discovery that made this accomplishment possible was that if the angle E in Figure 1-1 is known, then the ratio of the distances EM and ES can be found since it does not depend on the size of right triangle EMS.

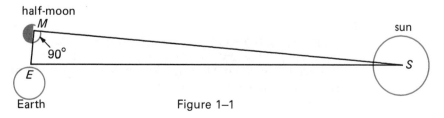

Figure 1–1

In fact, for any right triangle, if an acute angle is known, the ratios of the lengths of the sides of the triangle can be found. This discovery by Hipparchus of Nicaea in about 140 B.C. marked the beginning of trigonometry.

Since angles play an important role in trigo- nometry, we shall start with a discussion of them. Recall from geometry that an **angle** consists of two rays having the same endpoint. The rays are the **sides** of the angle, and their common endpoint is the **vertex** of the angle. The angle shown in Figure 1-2 can be named ∠ *ABC*, ∠ *B*, or simply θ. Greek letters, such as θ, listed on page ix, are often used to name angles.

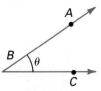

Figure 1–2

An angle is often measured in **degrees**. A right angle has measure 90 degrees, which we write 90°. We denote the measure of an angle θ by $m(\theta)$, or, if the context is clear, by θ alone. For example, if ϕ is one half of a right angle, we could write $m(\phi) = 45°$, or simply $\phi = 45°$. Each degree can be divided into decimal parts (one fourth of a right angle has measure 22.5°) or into minutes and seconds.

One *minute* (1′) is $\dfrac{1}{60}$ of 1°, and one *second* (1″) is $\dfrac{1}{60}$ of 1′.

At one time, the degree-minute-second system was almost universally used, but now decimal degrees are preferred for most applications except cartography. Most scientific calculators operate with decimal degrees, but can convert from one system to another.

To convert from decimal degrees to degrees and minutes without a calculator, keep the whole number of degrees and multiply the fractional part by 60. For example, a calculator display of 38.4 degrees can be written as 38°(0.4 × 60)′ = 38°24′.

Example 1

Express 51.724° in the degree-minute-second system. Give the answer to the nearest second.

Solution

51.724° = 51°(0.724 × 60)′
= 51°43.44′ = 51°43′(0.44 × 60)″ = 51°43′26.4″
To the nearest second, 51.724° = 51°43′26″. ∎

To convert from the degree-minute-second system to decimal degrees, begin by dividing the number of seconds by 60.

Example 2

Express 51°32′30″ in decimal degrees to the nearest 0.0001°.

Solution

First convert seconds to minutes. Then convert minutes to degrees.
$$51°32'30'' = 51°\left(32 + \frac{30}{60}\right)' = 51°32.5' = \left(51 + \frac{32.5}{60}\right)° = 51.541\overline{6}°,$$
which can be rounded to 51.5417°. (The bar over the 6 indicates a repeating decimal.) ∎

Most angles you met in geometry had measures between 0° and 180°. In trigonometry, however, we shall use angles of arbitrary measures, positive and negative. To do this, we regard each angle as being generated by a rotation, about the vertex, of one side, the **initial side**, onto the other side, the **terminal side**. If the rotation is counterclockwise, we regard the angle as positive and give it positive measure. If the rotation is clockwise, the angle and its measure are negative. Curved arrows, such as those shown in Figure 1-3, are used to indicate **directed angles**. In connection with directed angles, one degree is $\frac{1}{360}$ of one full counterclockwise revolution and 360° is one counterclockwise revolution.

Figure 1–3

An angle is in **standard position** (with respect to an *xy*-coordinate system) if its vertex is at the origin and its initial side coincides with the positive *x*-axis. Some examples are shown in Figure 1-4.

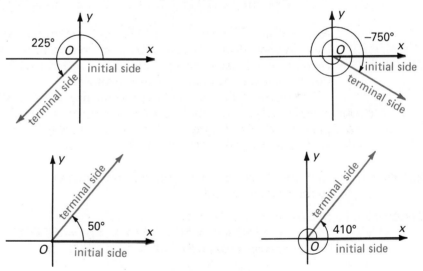

Figure 1–4

Notice that the 50° angle and the 410° angle begin and end at the same place. Angles such as these are said to be **coterminal** because their terminal sides coincide when they are in standard position. Figure 1-5 at the top of the next page shows angles coterminal with a 150° angle. The measures of *all* angles coterminal with that angle are given by 150° + $n \cdot 360°$, n an integer.

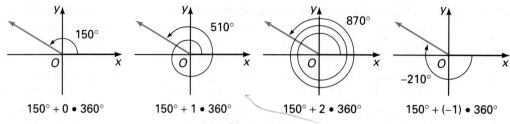

Figure 1–5

In Figure 1-5, n takes on the values 0, 1, 2, and -1. Every angle is coterminal with some angle θ such that $0° \leq \theta < 360°$.

Example 3   Find an angle θ, $0° \leq \theta < 360°$, that is coterminal with (a) a 1325° angle and (b) a $-400°$ angle.

Solution   (a) Subtract integral multiples of 360° until the difference is between 0° and 360°.
$$1325 - 3 \cdot 360 = 1325 - 1080 = 245$$
Hence, $1325° = 245° + 3 \cdot 360°$ and $\theta = 245°$.
(b) Add integral multiples of 360° until the sum is between 0° and 360°.
$$-400 + 2 \cdot 360 = -400 + 720 = 320$$
Hence, $-400° = 320° - 2 \cdot 360°$ and $\theta = 320°$.   ∎

EXERCISES 1-1

Estimate the degree measure of angle θ.

A 1.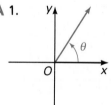

2.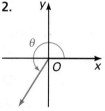

3.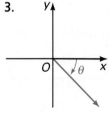

4.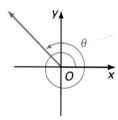

5.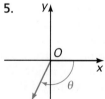

6.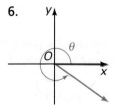

7.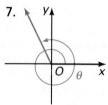

8.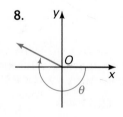

Draw an angle in standard position having the given measure.

9. 45°   10. 180°   11. 270°   12. 135°

13. 225°   14. −90°   15. −135°   16. 120°

17. −240°   18. −60°   19. 405°   20. 450°

Find the degree measure of each of the following angles.

21. $\frac{3}{4}$ of a counterclockwise revolution   22. $\frac{2}{3}$ of a counterclockwise revolution

23. $\frac{3}{8}$ of a clockwise revolution   24. $\frac{1}{6}$ of a clockwise revolution

25. $\frac{2}{5}$ of a clockwise revolution   26. $\frac{3}{10}$ of a counterclockwise revolution

27. $1\frac{2}{3}$ of a counterclockwise revolution   28. $2\frac{1}{4}$ of a counterclockwise revolution

Express in degrees, minutes, and seconds. Give answers to the nearest second.

29. 23.75°   30. 62.5°   31. 15.43°

32. 41.16°   33. 24.687°   34. 54.321°

Express in decimal degrees. Give answers to the nearest 0.0001°.

35. 9°45′   36. 82°15′   37. 67°22′30″

38. 27°15′45″   39. 0°0′27″   40. 42°42′42″

Find two angles, one positive and one negative, that are coterminal with and different from the given angle.

41. 90°   42. 180°   43. 270°

44. 60°   45. −240°   46. −120°

Find an angle between 0° and 360° that is coterminal with the given angle.

47. 460°   48. −300°   49. −60°

50. 500°   51. −120°   52. −180°

B 53. 750°   54. 1000°   55. 821.7°

56. −521.3°   57. −803.25°   58. 1206.37°

Two positive angles are **complementary** if the sum of their measures is 90°. Find the degree-minute-second measure of the complement of each of the following angles. (Hint: 90° = 89°59′60″.)

59. 48°17′40″   60. 16°43′22″

Two positive angles are **supplementary** if the sum of their measures is 180°. Find the degree-minute-second measure of the supplement of each angle.

61. 72°18′40″   62. 109°7′42″

63. A machine part rotates through an angle of 150° every second. How many revolutions does it make in one minute?

64. A long-playing record makes $33\frac{1}{3}$ revolutions per minute. Through what angle does it rotate in one second?

The following are measures of angles in standard position. In each case, how many terminal sides will there be as k takes on the values 1, 2, 3, . . . ? (Hint: In Exercise 65, find the smallest positive integral k such that $20° + k \cdot 60° = 20° + n \cdot 360°$ where n is an integer.)

65. $20° + k \cdot 60°$ 66. $40° + k \cdot 120°$ 67. $10° + k \cdot 75°$ 68. $25° + k \cdot 210°$

THE TRIGONOMETRIC FUNCTIONS

1-2 Trigonometric Functions of Acute Angles

Objective: To study the trigonometric functions of acute angles and to find relationships between them.

Recall that a **function** from a set D to a set R can be thought of as a rule that assigns to each member of D, the **domain**, a unique member of R, the **range.** In algebra, the domain of a function is usually a set of numbers. In this section we shall define several important functions whose domains are sets of acute angles.

Let θ be an acute angle: $0° < \theta < 90°$. Place θ in standard position and choose a point $P(x, y)$ other than O on the terminal side of θ. Let r be the distance OP, that is, $r = \sqrt{x^2 + y^2}$ by the Pythagorean theorem. (Figure 1-6.)

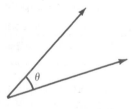
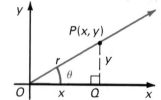

Figure 1–6

Although the numbers x, y, and r certainly depend on the choice of P, the ratios $\dfrac{y}{r}$, $\dfrac{x}{r}$, and $\dfrac{y}{x}$ do not. (See Exercise 53.) Since these ratios depend only on θ, they can be used to define these **trigonometric functions** of θ:

> The **sine** of θ, written $\sin \theta$, is $\dfrac{y}{r}$.
>
> The **cosine** of θ, written $\cos \theta$, is $\dfrac{x}{r}$.
>
> The **tangent** of θ, written $\tan \theta$, is $\dfrac{y}{x}$. $x \neq 0$

Example 1

When the acute angle ϕ is placed in standard position, its terminal side passes through $P(5, 12)$. Find $\sin \phi$, $\cos \phi$, and $\tan \phi$ to four decimal places.

Solution

Here $x = 5$, $y = 12$, and by the distance formula,
$$r = \sqrt{x^2 + y^2} = \sqrt{5^2 + 12^2} = \sqrt{169} = 13.$$

$$\sin \phi = \tfrac{12}{13} \qquad \cos \phi = \tfrac{5}{13} \qquad \tan \phi = \tfrac{12}{5}$$
$$= 0.9231 \qquad\quad = 0.3846 \qquad\quad = 2.4000 \quad \blacksquare$$

Figure 1-7 suggests how the definitions of $\sin \theta$, $\cos \theta$, and $\tan \theta$ can be restated to apply to an acute angle of a right triangle.

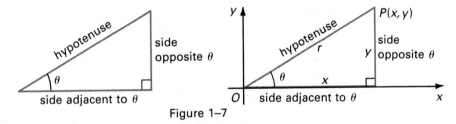

Figure 1–7

> $$\sin \theta = \frac{\text{length of the side opposite } \theta}{\text{length of the hypotenuse}} \qquad \frac{y}{r}$$
>
> $$\cos \theta = \frac{\text{length of the side adjacent to } \theta}{\text{length of the hypotenuse}} \qquad \frac{x}{r}$$
>
> $$\tan \theta = \frac{\text{length of the side opposite } \theta}{\text{length of the side adjacent to } \theta} \qquad \frac{y}{x}$$

Numerical values of the trigonometric functions of most angles cannot be given exactly. However, exact values of the sine, cosine, and tangent of 30°, 45°, and 60° can be read from the triangles in Figure 1-8. These values in simplest radical form have been entered into the following table. For example, $\tan 30° = \dfrac{1}{\sqrt{3}} = \dfrac{\sqrt{3}}{3}$.

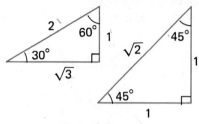

θ	$\sin \theta$	$\cos \theta$	$\tan \theta$
30°	$\dfrac{1}{2}$	$\dfrac{\sqrt{3}}{2}$	$\dfrac{\sqrt{3}}{3}$
45°	$\dfrac{\sqrt{2}}{2}$	$\dfrac{\sqrt{2}}{2}$	1
60°	$\dfrac{\sqrt{3}}{2}$	$\dfrac{1}{2}$	$\sqrt{3}$

Figure1–8

Example 2 Find the lengths a and c in the right triangle shown below. Leave answers in simplest radical form.

Solution To find a, choose the tangent. To find c, choose the cosine.

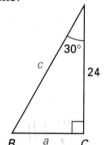

$$\tan 30° = \frac{a}{24} \qquad \cos 30° = \frac{24}{c}$$

$$\frac{\sqrt{3}}{3} = \frac{a}{24} \qquad \frac{\sqrt{3}}{2} = \frac{24}{c}$$

$$3a = 24\sqrt{3} \qquad \sqrt{3}\, c = 48$$

$$a = 8\sqrt{3}$$

$$c = \frac{48}{\sqrt{3}} = 16\sqrt{3} \quad \blacksquare$$

There are three other trigonometric functions defined for acute angles θ. Let x, y, and r be as in the definitions on page 7.

The **cotangent** of θ, written cot θ, is $\dfrac{x}{y}$. $\quad y \neq 0$

The **secant** of θ, written sec θ, is $\dfrac{r}{x}$. $\quad x \neq 0$

The **cosecant** of θ, written csc θ, is $\dfrac{r}{y}$. $\quad y \neq 0$

Notice the following *reciprocal relationships*:

$$\csc \theta = \frac{1}{\sin \theta} \qquad \sec \theta = \frac{1}{\cos \theta} \qquad \cot \theta = \frac{1}{\tan \theta}$$

Example 3 If α is an acute angle and $\cos \alpha = \dfrac{2}{3}$, find the other five trigonometric functions of α.

Solution Since α is acute, it can be made an angle of a right triangle, as shown. Let the length of the hypotenuse be 3. Then the length of the side adjacent to α must be 2 since $\cos \alpha = \dfrac{2}{3}$.

To find a, use the Pythagorean theorem:
$a^2 + 2^2 = 3^2;\ a^2 = 3^2 - 2^2 = 5;\ a = \sqrt{5}.$
From the triangle:

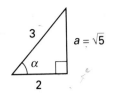

$$\sin \alpha = \frac{\sqrt{5}}{3} \qquad \cos \alpha = \frac{2}{3} \qquad \tan \alpha = \frac{\sqrt{5}}{2}$$

From the reciprocal relations or the figure:

$$\csc \alpha = \frac{3}{\sqrt{5}} = \frac{3\sqrt{5}}{5}, \qquad \sec \alpha = \frac{3}{2}, \qquad \cot \alpha = \frac{2}{\sqrt{5}} = \frac{2\sqrt{5}}{5} \quad \blacksquare$$

In Figure 1-9, $\angle A$ and $\angle B$ are **complementary**: $\angle A + \angle B = 90°$. The trigonometric functions of A and B are related as follows.

$$\sin A = \frac{a}{c} = \cos B \qquad \cos A = \frac{b}{c} = \sin B$$

$$\tan A = \frac{a}{b} = \cot B \qquad \cot A = \frac{b}{a} = \tan B$$

$$\sec A = \frac{c}{b} = \csc B \qquad \csc A = \frac{c}{a} = \sec B$$

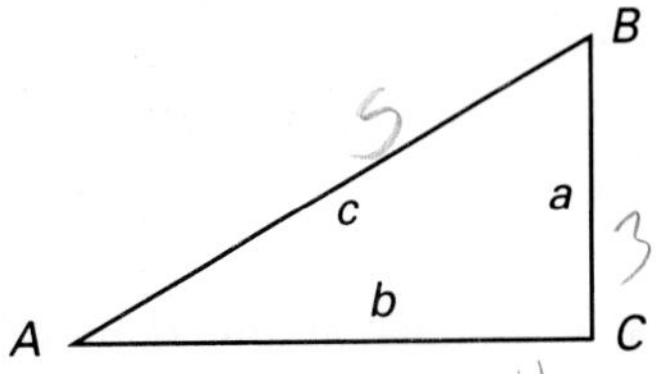

Figure 1-9

The sine and cosine are called **cofunctions.** Other pairs of cofunctions are the tangent and cotangent and the secant and cosecant.

> Any trigonometric function of an acute angle is equal to the cofunction of the complement of the angle.

Example 4 Find the acute angle β if $\sin \beta = \dfrac{1}{\sec 40°}.$

Solution $\sin \beta = \dfrac{1}{\sec 40°} = \cos 40°$ since $\sec \theta = \dfrac{1}{\cos \theta}.$ Since sine and cosine are cofunctions, β and $40°$ are complementary. Therefore, $\beta = 50°.$ $\blacksquare$

EXERCISES 1-2

When the acute angle θ is placed in standard position, its terminal side passes through the given point. Find $\sin \theta$, $\cos \theta$, and $\tan \theta$ to four decimal places. Use a calculator or the table of square roots on page 407 as necessary.

A 1. $(3, 4)$ 2. $(12, 5)$ 3. $(24, 7)$ 4. $(20, 21)$ 5. $(2, 1)$ 6. $(3, 2)$

7. $(10, 15)$ 8. $(4, 12)$ 9. $(1, \sqrt{3})$ 10. $(\sqrt{5}, 2)$ 11. $(\sqrt{21}, 2)$ 12. $(3, \sqrt{7})$

One function of the acute angle α is given. Find the other five trigonometric functions of α. Leave answers in simplest radical form.

13. $\cos \alpha = \dfrac{4}{5}$

14. $\sin \alpha = \dfrac{5}{13}$

15. $\tan \alpha = \dfrac{21}{20}$

16. $\cot \alpha = \dfrac{24}{7}$

17. $\sec \alpha = \dfrac{7}{3}$

18. $\csc \alpha = \dfrac{9}{7}$

19. $\cot \alpha = \dfrac{2\sqrt{10}}{3}$

20. $\tan \alpha = \dfrac{4\sqrt{2}}{7}$

21. $\csc \alpha = 2$

22. $\cos \alpha = \dfrac{1}{4}$

23. $\sin \alpha = \dfrac{2}{5}$

24. $\csc \alpha = \sqrt{5}$

When the acute angle ϕ is placed in standard position, its terminal side passes through the given point. Find ϕ without using a calculator or tables.

25. $(3, 3)$

26. $\left(3, 3\sqrt{3}\right)$

27. $\left(\sqrt{3}, 1\right)$

28. $\left(\sqrt{2}, \sqrt{2}\right)$

The triangle at the right establishes the notation used in Exercises 29–34. Find the lengths and the angle measures that are not given. Leave answers in simplest radical form.

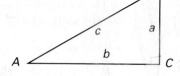

29. $\angle A = 60°$; $c = 12$

30. $c = 18$; $\angle B = 45°$

31. $\angle B = 45°$; $b = 4$

32. $\angle A = 30°$; $b = 6$

33. $a = \sqrt{3}$; $b = 1$

34. $a = 3$; $c = 6$

Find the exact value of the acute angle ϕ.

35. $\sin \phi = \cos 25°$

36. $\cot \phi = \tan 75°$

37. $\csc \phi = \dfrac{1}{\sin 10°}$

38. $\sec \phi = \dfrac{1}{\sin 80°}$

39. $\sin \phi \sec 70° = 1$

40. $\tan 50° \cot \phi = 1$

In Chapter 3, we shall prove the following formulas.

$\sin (a + b) = \sin a \cos b + \cos a \sin b \qquad \cos (a + b) = \cos a \cos b - \sin a \sin b$

$\sin (a - b) = \sin a \cos b - \cos a \sin b \qquad \cos (a - b) = \cos a \cos b + \sin a \sin b$

Use these formulas to find exact values of the following in simplest radical form. (Hint: In Exercise 41, $15° = 45° - 30°$.)

B 41. $\sin 15°$

42. $\sin 75°$

43. $\cos 15°$

44. $\cos 75°$

Use the results of Exercises 41–44 to find exact values of the following in simplest radical form.

45. $\sec 15°$

46. $\sec 75°$

47. $\csc 15°$

48. $\csc 75°$

In Exercises 49–52, find the length *x*. Leave answers in simplest radical form.

49.

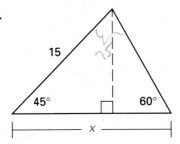

50.

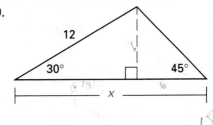

51.

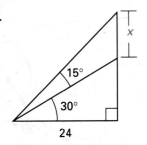

52.

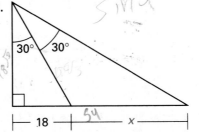

C 53. $P(x, y)$ and $P'(x', y')$ are two points on the terminal side of an acute angle θ in standard position. Let $OP = r$ and $OP' = r'$. Use similar triangles to show that $\dfrac{y}{r} = \dfrac{y'}{r'}$, $\dfrac{x}{r} = \dfrac{x'}{r'}$, and $\dfrac{y}{x} = \dfrac{y'}{x'}$. (This shows that using point P' to define the trigonometric functions of θ will yield the same values as using point P.)

1-3   ## Calculating with Trigonometric Functions

Objective: To find values of trigonometric functions and to use them in calculations.

It is not possible to give exact trigonometric function values for most angles (30°, 45°, and 60° are exceptions). You can, however, obtain good approximations by using a scientific calculator or a table of values.

When you use a calculator, be sure to study its instruction manual to see how to find the trigonometric functions of an angle and how to find an angle when a function of it is given. If angles are measured in degrees, be sure that the calculator is set in the *degree* mode (rather than *RAD* or *grad*).

Even if you have access to a scientific calculator, it is a good idea to learn how to use tables. Table 1-1 on the following page is a simplified version of Table 1, page 390, Table 2, page 398, and Table 3, page 403, and all three are used in the same way.

To find the value of a function of an angle θ:
1. If $0° \leq \theta \leq 45°$, find the function name in the *top* row and read *down* until opposite the degree measure of θ at the extreme *left*. For example, cos 29.8° ≈ 0.8678.
2. If $45° \leq \theta \leq 90°$, find the function name in the *bottom* row and read *up* until opposite the degree measure of θ at the extreme *right*. For example, sin 65.7° ≈ 0.9114.

We need only a table of values for angles between 0° and 45°, since sin θ and cos θ, tan θ and cot θ, and sec θ and csc θ are cofunctions.

To find θ given one of its function values, essentially reverse the process just described. For example, if tan θ ≈ 0.4533, then θ is about 24.4°.

θ Degrees	θ Radians	sin θ	cos θ	tan θ	cot θ	sec θ	csc θ		
24.0	.4189	.4067	.9135	.4452	2.246	1.095	2.459	1.1519	66.0
24.1	.4206	.4083	.9128	.4473	2.236	1.095	2.449	1.1502	65.9
24.2	.4224	.4099	.9121	.4494	2.225	1.096	2.439	1.1484	65.8
24.3	.4241	.4115	.9114	.4515	2.215	1.097	2.430	1.1467	65.7
24.4	.4259	.4131	.9107	.4536	2.204	1.098	2.421	1.1449	65.6
29.5	.5149	.4924	.8704	.5658	1.767	1.149	2.031	1.0559	60.5
29.6	.5166	.4939	.8695	.5681	1.760	1.150	2.025	1.0542	60.4
29.7	.5184	.4955	.8686	.5704	1.753	1.151	2.018	1.0524	60.3
29.8	.5201	.4970	.8678	.5727	1.746	1.152	2.012	1.0507	60.2
29.9	.5219	.4985	.8669	.5750	1.739	1.154	2.006	1.0489	60.1
30.0	.5236	.5000	.8660	.5774	1.732	1.155	2.000	1.0472	60.0
		cos θ	sin θ	cot θ	tan θ	csc θ	sec θ	θ Radians	θ Degrees

TABLE 1-1

Example 1

A flagpole is braced by a wire 8.7 m long extending from the top of the pole to the ground. If the brace makes a 58° angle with the ground, how tall is the pole?

Solution

From the figure, $\sin 58° = \dfrac{x}{8.7}$.

Hence, $x = 8.7 \sin 58°$.
From Table 1, sin 58° ≈ 0.8480.
$x ≈ 8.7 \times 0.8480 = 7.3776$
Therefore, the flagpole is 7.4 m tall. ∎

Example 2

A certain stretch of a cog railway rises 271 m while it covers a horizontal distance of 865 m. Assuming the track to be straight, what angle does it make with the horizontal?

Solution

From the figure, $\tan \theta = \dfrac{271}{865} ≈ 0.3133$.

From Table 1, $\theta ≈ 17.4°$ to the nearest 0.1°. ∎

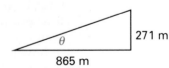

Most measurements, such as those in Examples 1 and 2, are approximations as are most values obtained from tables or calculators. Nevertheless, it is customary to use " = ", rather than "≈" when working with them, and we shall do so from now on.

The more significant digits there are in a measurement, the more *accurate* it is. A **significant digit** is any nonzero digit or any zero whose purpose is not just to place the decimal point. In the following examples, the significant digits are printed in red.

$$3004, 0.0205, 430.0, 0.00120$$

When you multiply or divide measurements, round the answer to the least number of significant digits in any of the numbers.

In Example 1, we rounded to two significant digits because the measurement 8.7 m was given to that accuracy. The following is a guide to corresponding accuracies of length and angle measurements.

A length measured to   an angle measured to
  1 significant digit ⎤   ⎧ 10°
  2 significant digits ⎥ corresponds to ⎨ 1°
  3 significant digits ⎥   ⎪ 0.1° or 10'
  4 significant digits ⎦   ⎩ 0.01° or 1'

In Example 2, the lengths were accurate to three significant digits, so we may use the answer $\theta = 17.4°$.

It is not clear which of the zeros in a numeral such as 37,000 are significant. For our purposes in this text we shall regard all zeros to the right of a nonzero digit to be significant.

The significant digits of a number x can be seen by using scientific notation. To write x in **scientific notation**, express it in the form $a \times 10^n$, where a is a terminating decimal such that $1 \le a < 10$, and n is an integer. The digits in a are the significant digits of x. The examples that follow illustrate how this notation shows which digits are significant.

$1430.0 = 1.4300 \times 10^3$		5 significant digits
$0.0040 = 4.0 \times 10^{-3}$		2 significant digits
$400 = 4.00 \times 10^2$		3 significant digits
$0.04 = 4 \times 10^{-2}$		1 significant digit

When you add or subtract measurements, round the answer to the least number of decimal places in any of the numbers.

For example, 3.571 cm + 4.3 cm = 7.9 cm.

Figure 1-10 shows a hot-air balloon followed by its pickup truck. To see the balloon, the driver's line of sight must be raised, or elevated, an angle θ above the horizontal: θ is called the **angle of elevation** of the balloon. Similarly, ϕ is called the **angle of depression** of the truck from the balloon. Notice that the angle of elevation and the angle of depression are equal, since they are alternate interior angles.

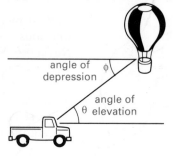

Figure 1-10

Example 3   When the angle of elevation of the sun is 53°, a tree casts a shadow 6.5 m long. How tall is the tree?

Solution   Let x be height of the tree.
From the figure:

$$\tan 53° = \frac{x}{6.5}$$

$$x = 6.5 \tan 53°$$

From Table 1, $\tan 53° = 1.327$.

$$x = 6.5 \times 1.327 = 8.6255.$$

Therefore, the tree is 8.6 m tall. ■

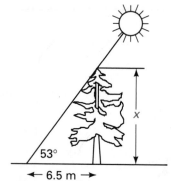

In view of the accuracy of the given data, it would be misleading to give the answer in Example 3 to more than two significant digits.

EXERCISES 1-3

Express in decimal form.

A 1. 3.10×10^{-2} 2. 6.06×10^{-1} 3. 4.00×10^{-3}

   4. 9.020×10^{2} 5. 2.100×10^{1} 6. 6.30×10^{-4}

In Exercises 7–14, (a) express in scientific notation and (b) state the number of significant digits.

   7. 106.0   8. 0.0021   9. 0.0340   10. 40.10

   11. 102.30   12. 0.0005   13. 321   14. 0.321

Use Table 1 to find the following to four decimal places.

   15. $\cos 60°$ 16. $\sin 73°$ 17. $\tan 41°$ 18. $\cot 78°$

Use a calculator, Table 1, or Table 2 to find the following to four significant digits.

19. sin 13.7°   **20.** cos 42.5°   **21.** csc 68.6°   **22.** sec 27.2°

23. cot 18°20′   **24.** tan 71°40′   **25.** cos 23°30′   **26.** sin 80°30′

Use Table 1 to find the measure of the acute angle θ to the nearest degree.

27. sin θ = 0.2950   **28.** cos θ = 0.6401   **29.** cot θ = 0.4101   **30.** tan θ = 4.000

Use a calculator or Table 1 to find the measure of the acute angle θ to the nearest 0.1°.

31. cos θ = 0.9300   **32.** sin $\theta \doteq$ 0.2006   **33.** tan θ = 1.575

34. cot θ = 0.4043   **35.** sec θ = 3.000   **36.** csc θ = 1.692

36.2

Find x or θ in each right triangle.

37.
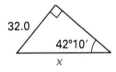
27.0   66.5° x

tan66.5 = 27.0 / Y

38.
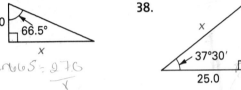
x 37°30′   25.0

39.

x 28°40′   54.0

40.

32.0   42°10′ x

41.
23.0   56.0 θ

42.
130   90.0 θ

43. A ladder 4.20 m long is leaning against a building. The foot of the ladder is 1.75 m from the building. Find the angle the ladder makes with the ground and the distance it reaches up the building.

44. What is the angle of elevation of the sun when a building 38.6 m tall casts a shadow 24.5 m long?

45. From the top of a vertical lakeside tower 106 m high, the angle of depression of a rowboat is 32.5°. How far is the boat from the bottom of the tower?

46. A ramp is to be built to a loading dock 2.10 m high. How long must the ramp be if the angle it makes with the ground is to be 20.0°?

The figure shows a light ray passing from a medium where the speed of light is u into a medium where the speed is v; α is the **angle of incidence** and β is the **angle of refraction**. Snell's law of optics states that $\dfrac{\sin \alpha}{\sin \beta} = \dfrac{u}{v}$. Use this fact in Exercises 47–49.

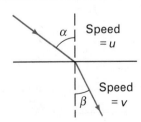
α Speed = u

β Speed = v

B 47. Suppose that a light ray passes from air, where its speed is 3.00×10^8 m/s into water. If its angles of incidence and refraction are 48.2° and 34.1°, respectively, find the speed of light in water.

48. A light ray passes from glass, where its speed is 1.88×10^8 m/s, into air, where its speed is 3.00×10^8 m/s. Find its angle of refraction if its angle of incidence is 32.5°.

C 49. The figure shows a light ray passing through a flat pane of glass surrounded by air. Use Snell's law to show that $\gamma = \alpha$ and therefore, the light ray does not change direction.

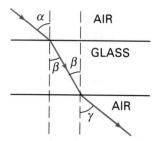

50. Refer to the figure at the right.
    (a) Use similar triangles to show that the value of x in the figure is $\dfrac{\sqrt{5} - 1}{2}$.

    (b) Use (a) to find $\cos 72°$ in simplest radical form.

    (c) By approximating $\sqrt{5}$, convert the answer in (b) to decimal form and compare it with $\cos 72°$ as obtained from a calculator or table.

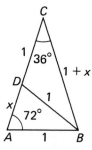

51. (a) Use Exercise 50(a) to find $\cos 36°$ in simplest radical form.
    (b) Repeat Exercise 50(c) for 36°.

52. In the figure, $\overline{AB}$ is a chord of a circle with center O. In the first known trigonometric table, Hipparchus of Nicaea essentially gave values of the ratio $\dfrac{AB}{AO}$ as a function of the angle θ. Give an equivalent expression for this ratio using the sine function.

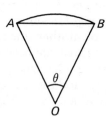

Self Quiz

1. Find θ in the figure at the left below to the nearest 0.1°.

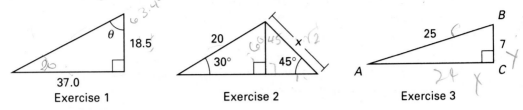

   | 37.0 |
   Exercise 1

   30°   45°
   Exercise 2

   A   *B*   25   7   *C*
   Exercise 3

2. Find the exact value of x in the middle figure above.

3. Find the six trigonometric functions of $\angle A$ in the figure at the right above.

4. Find angles between 0° and 360° that are coterminal with   (a) 550°
   (b) $-140°$.

5. Find the acute angle α if $\cos \alpha \csc 65° = 1$.

6. Express in scientific notation: (a) 0.00650   (b) 304.00

7. (a) Write 1°32′24″ in decimal degrees.   (b) 32.67° = 32°?′?″.

1-4   Solving Right Triangles

Objective: To find the lengths of the sides and measures of the angles of right triangles.

To **solve a triangle,** you find the lengths of all its sides and the measures of all its angles. As you may have concluded from previous sections, you can solve a *right* triangle if you are given the length of one side and either the length of another side or the measure of an acute angle. The examples that follow illustrate how to treat the cases that can arise.

   (In dealing with triangles, it is customary to denote each angle by a capital letter and the side opposite that angle by the corresponding lower-case letter as shown in Figure 1-11.)

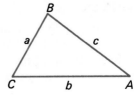

Figure 1-11

Example 1   Solve $\triangle ABC$ given that $\angle C = 90°$, $b = 38.0$, and $c = 62.7$.

Solution   Draw a figure and label the known parts. Drawing your figure to scale will help you avoid errors.
   To find $\angle A$, you can use

$$\cos A = \frac{b}{c} = \frac{38.0}{62.7} = 0.6061.$$

From this, $\angle A = 52.7°$.

   To find $\angle B$, you can use the fact that the acute angles of a right triangle are complementary. Thus,

$\angle B = 90° - \angle A = 90° - 52.7° = 37.3°$.

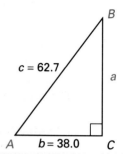

   To find side a, you can use the Pythagorean theorem ($a^2 = c^2 - b^2$) or the equation $\tan A = \dfrac{a}{38.0}$ to obtain $a = 49.9$. (Since the former method involves only given values, using it will help prevent errors.)   ∎

Example 2   Triangle ABC is isosceles with base $\overline{AB}$. Solve $\triangle ABC$ if $c = 26.4$ and $\angle A = 68.5°$.

Solution   Draw and label a figure. The bisector, $\overline{CD}$, of vertex angle C divides $\triangle ABC$ into two *right* triangles. In $\triangle ADC$, $\angle A = 68.5°$ and

$AD = \dfrac{1}{2} \times 26.4 = 13.2$. Therefore,

$$\cos 68.5° = \frac{AD}{AC} = \frac{13.2}{b}$$

$$b = \frac{13.2}{\cos 68.5°}$$

$$= \frac{13.2}{0.3665} = 36.016$$

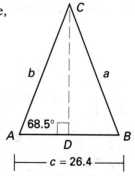

Hence, $a = b = 36.0$.
Since $\angle A + \angle B + \angle C = 180°$ and $\angle B = \angle A = 68.5°$,
$\angle C = 180° - (68.5° + 68.5°) = 43.0°$.
Therefore, the parts of $\triangle ABC$ are $a = b = 36.0$, $c = 26.4$,
$\angle A = \angle B = 68.5°$, and $\angle C = 43.0°$.   ∎

Example 3   A television camera in a blimp is focused on a football field with angle of depression 25.8°. A range finder on the blimp determines that the field is 925 m away. How high up is the blimp?

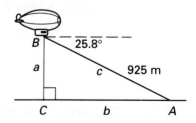

Solution   Alternate interior angles of parallel lines cut by a transversal are congruent. Therefore, $\angle A = 25.8°$. Since $\sin A = \dfrac{BC}{AB}$,

$$BC = AB \sin A$$
$$= 925 \sin 25.8°$$
$$= 925 \times 0.4352$$
$$= 402.56.$$

Therefore, the blimp is 403 m high up. ■

Sometimes a problem may require you to consider two right triangles.

Example 4   A research ship finds that the angle of elevation of a volcanic island peak is 22.8°. After the ship has moved 1250 m closer (approximated to the nearest 10 m), the angle of elevation is 30.5°. Find the height of the peak above sea level.

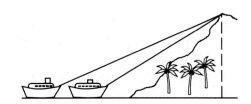

Solution   Draw and label a figure.

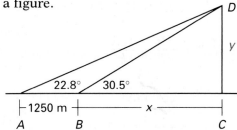

From $\triangle BCD$:   From $\triangle ACD$:

$$\cot 30.5° = \frac{x}{y} \qquad\qquad \cot 22.8° = \frac{x + 1250}{y}$$
$$x = y \cot 30.5° \qquad\qquad y \cot 22.8° = x + 1250$$
$$x = 1.698y \qquad\qquad\quad 2.379y = x + 1250$$

Substitute $1.698y$ for x in $2.379y = x + 1250$.

$$2.379y = 1.698y + 1250$$
$$0.681y = 1250$$
$$y = \frac{1250}{0.681} = 1835.5$$

The peak is 1840 m above sea level (to the nearest 10 m). ■

EXERCISES 1-4

In Exercises 1–12, $\angle C = 90°$. Solve $\triangle ACB$.

A 1. $c = 252$; $\angle A = 31.5°$ 2. $c = 42.6$; $\angle A = 58°20'$

3. $a = 35.6$; $\angle B = 64°40'$ 4. $b = 305$; $\angle A = 38.6°$

5. $a = 0.47$; $\angle A = 72°$ 6. $b = 0.26$; $\angle B = 41°$

7. $a = 3.60$; $b = 2.50$ 8. $a = 52.0$; $b = 38.0$

9. $b = 46.0$; $c = 63.0$ 10. $b = 3.20$; $c = 5.50$

In Exercises 11 and 12, give exact values of measures.

11. $b = 3\sqrt{3}$; $c = 3\sqrt{6}$ 12. $a = 5\sqrt{6}$; $c = 10\sqrt{2}$

In Exercises 13–16, solve isosceles $\triangle ABC$, having $\overline{AB}$ as base.

13. $a = 27$; $\angle A = 43°$ 14. $a = 14$; $c = 22$

15. $a = 16.8$; $\angle C = 76.4°$ 16. $c = 47.6$; $\angle C = 52.4°$

17. Find $\angle B$ in the figure if $\angle A = 52°40'$, $AD = 38.1$, and $BC = 70.0$.

18. Find $\angle A$ in the figure if $\angle B = 37.2°$, $BC = 17.8$, and $AD = 11.2$.

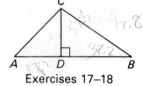

Exercises 17–18

19. What is the angle of elevation of the sun when a pole 32.0 feet tall casts a shadow 45.0 feet long?

20. The levels in a parking garage are 12.5 feet apart, and a ramp from one level to the next level is 138 feet long. What angle does the ramp make with the horizontal?

21. A vertically-directed searchlight shines a spot of light on the bottom of the cloud cover over an airport. The angle of elevation of the spot from an observation post 975 m from the searchlight is 63.5°. How high is the cloud cover?

22. In one minute a plane descending at a constant angle of depression of 12.4° travels 1600 m along its line of flight. How much altitude has it lost?

23. A beam bracing a wall makes an angle of 59° with the ground and reaches 18 feet up the wall. How long is the beam?

24. A ladder leans against a wall and makes an angle of 65°10' with the ground. How long is the ladder if its foot is 2.10 m from the bottom of the wall?

B 25. The figure gives the cross-sectional plan for a roof. Given the specifications shown, find the height allowed for the skylight.

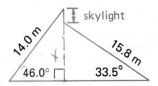

26. In the given diagram, rods BC and AC are joined at C so that as C moves in a clockwise circle, the piston attached at A moves back and forth inside the cylinder. Suppose that rod AC has length 40 cm. If C moves from the position shown, where $\angle ABC = 90°$ and $\angle C = 50°$, to a position in line with $\overline{AB}$, how far must the piston move?

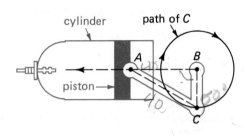

27. The arms of a compass are each 12.5 cm long and their ends are 6.50 cm apart. Find the angle between them.

28. The opposite corners of a small rectangular park are joined by diagonal paths, each 250 m long. Find the dimensions of the park if the paths intersect at a 20.0° angle.

29. A right triangle having a 32.5° angle is inscribed in a circle with radius 42.5 cm. How long are the legs of the triangle? (Hint: A right angle inscribed in a circle subtends a semicircular arc.)

30. The side of the beach shelter shown is made up of four 30°–60° – 90° triangles. Find the exact length x.

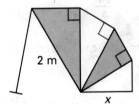

Exercises 31 and 32 refer to the crane pictured below. The boom (or arm) of the crane is 65.0 m long and can be raised or lowered by varying the angle β. The distance from the center of the load to the center line of the tower is called the radius.

31. At what angle β (to the nearest tenth of a degree) should the boom be set if the radius is to be 62.5 m?

32. Find the minimum possible radius (to the nearest tenth of a meter) if the largest angle β at which the boom can be set is 70.0°.

33. A pendulum 40 cm long is moved 26° from the vertical. How much is the lower end of the pendulum raised?

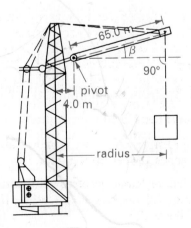

34. In square *ABCD*, *P* is the midpoint of $\overline{AB}$. Find $\angle\ CPD$ to the nearest 0.1°.

35. In its original form, the Great Pyramid at Giza had a square base of about 230 m on each side. The angle of elevation (represented by $\angle\ PRO$ in the given diagram) measured about 51°50′. Find the original height of the Great Pyramid.

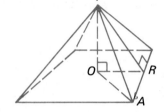

36. Use the information given in Exercise 35 to find the angle of elevation represented by $\angle\ PAO$.

In Exercises 37 and 38, express *x* in terms of α, β, and *y*.

37.

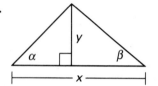

38.

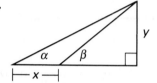

39. Kim and Luis are in line with a 50-foot flagpole but on opposite sides of it. The angles of elevation of the top of the pole from their positions are 45° and 33°. How far apart are they? (See Exercise 37.)

40. Two planes flying at 4500 m above sea level are directly east of a control tower 1200 m above sea level. How far apart are the planes if their angles of elevation from the tower are 21.2° and 39.6°?

C 41. In the figure for Exercise 38, express *y* in terms of α, β, and *x*.

42. In the figure for Exercise 37, express *y* in terms of α, β, and *x*.

43. A plane is in line with a 1200 m landing strip. The pilot measures the angles of depression of the ends of the strip to be 15.8° and 11.5°. How high up is the plane and how far is it horizontally from the nearer end of the strip?

44. Two observers 1500 m apart on a straight road measure the angles of elevation of a helicopter hovering over the road between them. If these angles are 32.5° and 51.0°, how high is the helicopter and how far is it from the nearer observer?

45. Fire lookouts at *A* and *B*, 2.42 km apart, spot a fire at *F*. How far is the fire from *A* and from *B* if $\angle\ FAB = 44.5°$ and $\angle\ FBA = 50.5°$?

46. Point *P* lies in the horizontal line through the bases of two telephone poles of equal height. The angles of elevation from point *P* to the tops of the poles are α and β, as shown. If *P* is *c* meters from the nearer pole and the poles are *x* meters apart, show that $x = c\left(\dfrac{\tan \alpha}{\tan \beta} - 1\right)$.

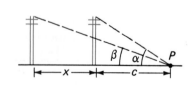

47. An 8-foot pole and an 18-foot pole are braced by two guy wires, each extending from the bottom of one pole to the top of the other. How far apart are the poles if the wires cross at right angles?

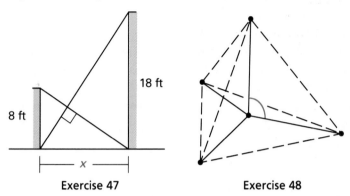

Exercise 47   Exercise 48

48. In a molecule of carbon tetrachloride, the four chlorine atoms are at the vertices of a regular tetrahedron, with the carbon atom at the center. Find the angle between two of the carbon-chlorine bonds (shown in red in the figure).

Challenge

A weight is suspended from a string that is attached to two vertical poles of unequal height. The weight can slide along the string and is allowed to come to a rest position at point B as shown in the diagram. A law of physics tells us that the two angles (labeled θ) between the string and a vertical line through B are equal. If the poles are b meters apart and if the string is c meters long, show that $\sin \theta = \dfrac{b}{c}$ (and hence the angle θ is independent of the relative heights of the poles).

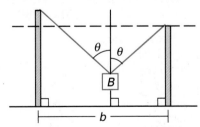

1-5   Trigonometric Functions of Arbitrary Angles

Objective: To find values of trigonometric functions of arbitrary angles.

In Section 1-2, we defined the trigonometric functions for acute angles. We now define them for angles of any measure.

Given an angle θ, place it in standard position. Choose a point $P(x, y)$ (other than O) on the terminal side of θ and let $r = OP$ (Figure 1-12).

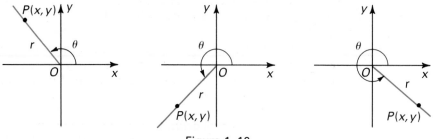

Figure 1–12

Then we make the following definitions.

$$\sin \theta = \frac{y}{r} \qquad \cos \theta = \frac{x}{r} \qquad \tan \theta = \frac{y}{x} \text{ if } x \neq 0$$

$$\csc \theta = \frac{r}{y} \text{ if } y \neq 0 \qquad \sec \theta = \frac{r}{x} \text{ if } x \neq 0 \qquad \cot \theta = \frac{x}{y} \text{ if } y \neq 0$$

Keep in mind that r is positive while x and y may be positive or negative.

Example 1   When an angle θ is placed in standard position its terminal side passes through $P(6, -3)$. Find the values of the six trigonometric functions of θ. Leave answers in simplest radical form.

Solution   With, $x = 6$ and $y = -3$, $r = \sqrt{x^2 + y^2} = \sqrt{36 + 9} = \sqrt{45} = 3\sqrt{5}$.

$$\sin \theta = \frac{y}{r} = \frac{-3}{3\sqrt{5}} = -\frac{\sqrt{5}}{5} \qquad\qquad \csc \theta = \frac{r}{y} = \frac{3\sqrt{5}}{-3} = -\sqrt{5}$$

$$\cos \theta = \frac{x}{r} = \frac{6}{3\sqrt{5}} = \frac{2\sqrt{5}}{5} \qquad\qquad \sec \theta = \frac{r}{x} = \frac{3\sqrt{5}}{6} = \frac{\sqrt{5}}{2}$$

$$\tan \theta = \frac{y}{x} = \frac{-3}{6} = -\frac{1}{2} \qquad\qquad \cot \theta = \frac{x}{y} = \frac{6}{-3} = -2 \quad \blacksquare$$

Recall that the coordinate axes divide the plane into four regions called **quadrants,** numbered as indicated in Figure 1-13. An angle θ is **in a** particular **quadrant** if its terminal side lies in that quadrant when θ is in standard position. Thus a 240° angle is in Quadrant III, or is a third-quadrant angle. Angles such as those measuring 90°, 180°, and 270°, whose terminal sides lie along an axis, are called **quadrantal angles.**

QUADRANT II	QUADRANT I
$x < 0$ $y > 0$	$x > 0$ $y > 0$
$x < 0$ $y < 0$	$x > 0$ $y < 0$
QUADRANT III	QUADRANT IV

Figure 1–13

In the definitions of the trigonometric functions, r is positive, so the signs of the functions of θ are determined by the signs of x and y. These depend only on the quadrant of θ, and therefore we have the following table of algebraic signs.

θ in Quadrant	$\sin \theta$ $\csc \theta$	$\cos \theta$ $\sec \theta$	$\tan \theta$ $\cot \theta$
I	+	+	+
II	+	−	−
III	−	−	+
IV	−	+	−

TABLE 1-2

Although the definitions on page 24 did not use right triangles, our work will be simplified if we introduce the so-called *reference triangle.* Given any nonquadrantal angle θ, let P, a point other than the origin, be a point on the terminal side of θ and let Q be the foot of the perpendicular from P to the <u>x-axis</u> (Figure 1-14).

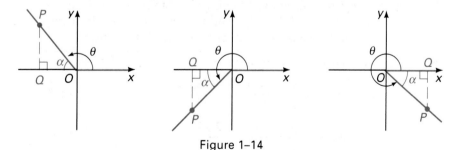

Figure 1–14

Right triangle *PQO* is the **reference triangle** of θ, and the acute angle $\alpha = \angle POQ$ is the **reference angle** of θ. Since $\sin \theta = \dfrac{y}{r}$ and $\sin \alpha = \dfrac{|y|}{r}$, we have the first of the following statements. (The notation "±" means "+ or −.") The correct sign is determined by the quadrant of θ.

$$\sin \theta = \pm \sin \alpha \qquad \cos \theta = \pm \cos \alpha \qquad \tan \theta = \pm \tan \alpha$$
$$\csc \theta = \pm \csc \alpha \qquad \sec \theta = \pm \sec \alpha \qquad \cot \theta = \pm \cot \alpha$$

Example 2

Find tan 304° and cos 304° to four decimal places.

Solution

A sketch may be helpful. In this case,

$$\alpha = 360° - 304° = 56°.$$

Since 304° is a fourth-quadrant angle,

$$\tan 304° = -\tan 56° = -1.483$$
$$\cos 304° = \cos 56° = 0.5592 \quad \blacksquare$$

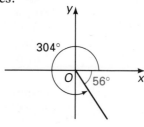

Example 3

Find θ to the nearest 0.1° given that 0° < θ < 270° and sin θ = −0.6450.

Solution

Since 0° < θ < 270°, θ is in Quadrant I, II, or III. Because sin θ < 0, θ must be in Quadrant III, and θ and α are related as in the sketch. Moreover,

$$\sin \alpha = -\sin \theta = 0.6450.$$

Therefore, α = 40.2°, and

$$\theta = 180° + \alpha = 180° + 40.2° = 220.2°. \quad \blacksquare$$

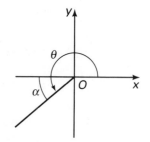

The sine and cosine are defined for all angles, but the other functions are undefined for certain quadrantal angles. For example, tan 90° is undefined. (See Exercises 39–42.)

In the definitions on page 7, $r^2 = x^2 + y^2$. Thus, for any angle θ,

$$(\cos \theta)^2 + (\sin \theta)^2 = \left(\frac{x}{r}\right)^2 + \left(\frac{y}{r}\right)^2 = \frac{x^2 + y^2}{r^2} = 1$$

We usually write $(\cos \theta)^2$ as $\cos^2 \theta$ and $(\sin \theta)^2$ as $\sin^2 \theta$, so we have:

$$\cos^2 \theta + \sin^2 \theta = 1$$

If $\cos \theta \neq 0$, $\dfrac{\sin \theta}{\cos \theta} = \dfrac{\dfrac{y}{r}}{\dfrac{x}{r}} = \dfrac{y}{x} = \tan \theta$. Therefore:

$$\tan \theta = \frac{\sin \theta}{\cos \theta} \quad \text{if } \cos \theta \neq 0$$

The boxed equations above are called *identities* because they are true for all values of θ for which both sides are defined. In Chapter 3, you will see many other trigonometric identities.

Example 4 If $90° < \theta < 360°$ and $\sin \theta = \frac{2}{3}$, find $\cos \theta$ and $\tan \theta$.

Solution Because $90° < \theta < 360°$, θ must be in Quadrant II, III, or IV. Since $\sin \theta > 0$, θ must be in Quadrant II and $\cos \theta$ must be negative. From the figure at the right, $\cos \theta = \dfrac{-\sqrt{5}}{3}$ and $\tan \theta = -\dfrac{2}{\sqrt{5}} = -\dfrac{2\sqrt{5}}{5}$. ∎

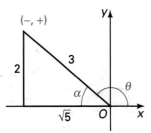

We could also solve Example 4 as follows.

Since $\cos^2 \theta + \left(\dfrac{2}{3}\right)^2 = 1$, $\cos^2 \theta = 1 - \dfrac{4}{9} = \dfrac{5}{9}$ and $\cos \theta = \dfrac{-\sqrt{5}}{3}$.

Using $\tan \theta = \dfrac{\sin \theta}{\cos \theta}$, we have $\tan \theta = \dfrac{\dfrac{2}{3}}{\dfrac{-\sqrt{5}}{3}} = \dfrac{2}{-\sqrt{5}} = -\dfrac{2\sqrt{5}}{5}$.

EXERCISES 1-5

Name the quadrant of angle θ.

A 1. $\sin \theta > 0$; $\cos \theta > 0$ 2. $\sin \theta < 0$; $\cos \theta > 0$

3. $\sin \theta < 0$; $\cos \theta < 0$ 4. $\sin \theta > 0$; $\tan \theta > 0$

5. $\cos \theta > 0$; $\tan \theta < 0$ 6. $\cos \theta < 0$; $\tan \theta < 0$

Use reference angles to find the value of each of the following.

7. $\sin 235°$ 8. $\cos 148°$ 9. $\tan 313°$ 10. $\cot(-160°)$

11. $\sec(-231.5°)$ 12. $\csc 505.5°$ 13. $\cos 416°20'$ 14. $\sin(-205°40')$

Use reference angles to give the exact value of the six trigonometric functions of each angle. Leave answers in simplest radical form.

15. 240°

16. 150°

17. 225°

18. 315°

19. −240°

20. −60°

21. 480°

22. 690°

In Exercises 23–30, the terminal side of an angle θ in standard position passes through the given point. (a) Find $\sin \theta$ and $\cos \theta$. (b) Find θ to the nearest 0.1°. (Assume that $0° < \theta < 360°$.)

23. $(-6, 8)$

24. $(5, -12)$

25. $(-8, -15)$

26. $(-20, 21)$

27. $\left(\sqrt{5}, -2\right)$

28. $\left(-3, \sqrt{7}\right)$

29. $\left(\sqrt{3}, -\sqrt{6}\right)$

30. $\left(-5, -\sqrt{11}\right)$

In Exercises 31–38, one of $\sin \theta$, $\cos \theta$, or $\tan \theta$ is given. Find the other two. Leave answers in simplest radical form.

B 31. $\cos \theta = \dfrac{4}{5}$; θ in Quadrant IV

32. $\tan \theta = -\dfrac{4}{3}$; θ in Quadrant II

33. $\sin \theta = -\dfrac{1}{3}$; θ in Quadrant III

34. $\cos \theta = \dfrac{3}{4}$; θ in Quadrant IV

35. $\tan \theta = \dfrac{1}{2}$; $90° < \theta < 360°$

36. $\sin \theta = -\dfrac{3}{4}$; $0° < \theta < 270°$

37. $\cos \theta = \dfrac{1}{4}$; $0° < \theta < 270°$

38. $\cos \theta = -\dfrac{2}{5}$; $-90° < \theta < 180°$

Copy and complete the following table. If a certain function is undefined for a certain angle, indicate this with a —.

	θ	$\sin \theta$	$\cos \theta$	$\tan \theta$	$\cot \theta$	$\sec \theta$	$\csc \theta$
39.	0°	?	?	?	?	?	?
40.	90°	?	?	?	?	?	?
41.	180°	?	?	?	?	?	?
42.	270°	?	?	?	?	?	?

43. Let α be the reference angle of θ, where $0° < \theta < 360°$. Express θ in terms of α if the quadrant of θ is (a) I, (b) II, (c) III, and (d) IV.

44. Repeat Exercise 43 for the case $-360° < \theta < 0°$.

Copy and complete the following table.

Quadrant of θ	I	II	III	IV
45. Quadrant of $-\theta$	?	III	?	?
46. Quadrant of $90° - \theta$	I	?	?	?
47. Quadrant of $180° - \theta$	?	?	?	III
48. Quadrant of $270° - \theta$	?	?	?	IV

Find all angles θ, $0° \leq \theta < 360°$, that make each statement true.

C 49. $\cos \theta = 0$

50. $\sin \theta = 0$

51. $\tan \theta = 0$

52. $\cos \theta = 1$

53. $\sin \theta = 1$

54. $\sin \theta = -1$

55. $\cos \theta = -1$

56. $\tan \theta = 1$

57. $\sin \theta = \sin 50°$

58. $\cos \theta = \cos 100°$

59. $\cos \theta = \cos (-40°)$

60. $\sin \theta = \sin (-110°)$

Challenge

The diagram is an illustration of one of the first steam engines. Steam pressure forced the beam to pivot up and down at point X. The beam, in turn, moved the shaft. Gear **A** then revolved in a path around a Gear **B** of equal diameter, causing Gear **B** to rotate and turn the attached wheel. Suppose the beam moved from Y to Z, sweeping out an angle of 40°. If YZ equals the diameter of the path of Gear **A** around Gear **B**, and if XY has length 203 cm, find the diameter of Gear **B**.

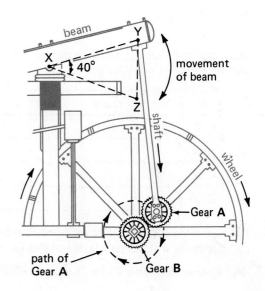

1-6 Radian Measure

Objective: To understand and use radian measure.

In advanced mathematics, angles are usually measured in *radians* rather than degrees. To define the radian measure of an angle θ, let c be a circle of radius r centered at the vertex of θ (Figure 1-15). Let s be the length of the arc of c intercepted by θ. Take s to be positive if θ is a positive angle, negative if θ is negative. Then:

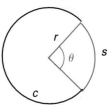

Figure 1–15

> The **radian measure** of θ is $\dfrac{s}{r}$.

Because of properties of similar figures (Figure 1-16), the ratio $\dfrac{s}{r}$ does not depend on the size of the circle P but rather only on θ.

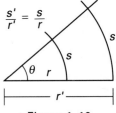

$$\frac{s'}{r'} = \frac{s}{r}$$

Figure 1–16

We can express the definition given above by the formula

$$\theta = \frac{s}{r},$$

in which θ stands for the measure of angle θ. Thus, if θ is a counterclockwise angle intercepting an arc that is 2 radii long, then $s = 2r$ and $\theta = \dfrac{2r}{r}$, or $\theta = 2$. Notice that no units are used with radian measure.

When no unit of angle measure is specified, you can assume that radian measure is meant. For example, the statement $\theta = 2$ means that the measure of angle θ is 2 radians, not 2 degrees.

A 180° angle intercepts a semicircular arc. Since the circumference of a circle of radius r is $2\pi r$, a semicircle of radius r has length πr. Thus, the radian measure of 180° angle is $\dfrac{\pi r}{r}$, or π. This leads to:

> $$180° = \pi \text{ radians}$$

which we can use to change from one system of measure to the other. In particular, we have

$$1° = \frac{\pi}{180} \text{ radians and } 1 \text{ radian} = \frac{180°}{\pi} \approx 57.3°.$$

Example 1   Convert to radian measure.
(a) 60°   (b) 225°   (c) 143.6°

Solution   (a) $60° = 60 \cdot \dfrac{\pi}{180} = \dfrac{\pi}{3}$

(b) $225° = 225 \cdot \dfrac{\pi}{180} = \dfrac{5\pi}{4}$

(c) $143.6° = 143.6 \cdot \dfrac{\pi}{180} = \dfrac{143.6 \times 3.1416}{180} \approx 2.506$ ■

It is customary to leave the radian measures of angles that are multiples of 30° and 45° in terms of π. This was done in Example 1(a) and in 1(b).

Example 2   Convert the following radian measures to degrees.

(a) $\dfrac{\pi}{4}$ (b) $\dfrac{2\pi}{3}$ (c) 1.508

Solution   (a) $\dfrac{\pi}{4} = \left(\dfrac{\pi}{4} \cdot \dfrac{180}{\pi} \right)^{\circ} = 45°$

(b) $\dfrac{2\pi}{3} = \left(\dfrac{2\pi}{3} \cdot \dfrac{180}{\pi} \right)^{\circ} = 120°$

(c) $1.508 = \left(1.508 \cdot \dfrac{180}{\pi} \right)^{\circ} \approx \left(\dfrac{1.508 \times 180}{3.1416} \right)^{\circ} \approx 86.40°$ ■

The following table gives degree and radian measures of some frequently-occurring angles.

Degree measure	0°	30°	45°	60°	90°	120°	135°	150°	180°
Radian measure	0	$\dfrac{\pi}{6}$	$\dfrac{\pi}{4}$	$\dfrac{\pi}{3}$	$\dfrac{\pi}{2}$	$\dfrac{2\pi}{3}$	$\dfrac{3\pi}{4}$	$\dfrac{5\pi}{6}$	π

You will soon learn the exact values of the trigonometric functions of the angles in the table above. To find approximate function values for other angles measured in radians, you can use either a calculator set in the RAD mode or Table 3, page 403.

Example 3   Find: (a) $\cos \dfrac{\pi}{2}$ (b) $\tan \dfrac{2\pi}{3}$ (c) $\sin 1.150$

Solution   (a) $\cos \dfrac{\pi}{2} = \cos 90° = 0$

(b) $\tan \dfrac{2\pi}{3} = \tan 120° = -\sqrt{3}$

(c) $\sin 1.150 = 0.9128$ ■

Radian measure provides us with the simple formulas for circular-arc length and circular-sector area given in Figure 1-17.

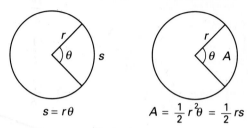

$$s = r\theta \qquad\qquad A = \frac{1}{2}r^2\theta = \frac{1}{2}rs$$

Figure 1–17

The formula $s = r\theta$ is equivalent to the definition of radian measure. To derive the area formulas, we use the proportion

$$\frac{\text{area of sector}}{\text{area of circle}} = \frac{\text{measure of central angle of sector}}{\text{measure of central angle of circle}}.$$

Then,

$$\frac{A}{\pi r^2} = \frac{\theta}{2\pi}.$$

Solving for A gives $A = \pi r^2 \cdot \dfrac{\theta}{2\pi} = \dfrac{1}{2}r^2\theta$. Using $\theta = \dfrac{s}{r}$, we have

$$A = \frac{1}{2}r^2 \cdot \frac{s}{r} = \frac{1}{2}rs.$$

Example 4   A central angle of a circle of radius 6.25 cm measures 84.0°. Find (a) the length of the arc it intercepts and (b) the area of the sector it determines.

Solution   To use the length and area formulas above, θ must be in radians.

$$r = 6.25, \text{ and } \theta = 84° = 84 \cdot \frac{\pi}{180} = \frac{7\pi}{15} = 1.466$$

(a) $s = r\theta = (6.25)(1.466) = 9.1625$. Hence, $s = 9.16$ cm.

(b) $A = \dfrac{1}{2}r^2\theta = \dfrac{1}{2}(6.25)^2(1.466) = 28.638$. Hence, $A = 28.6$ cm². ■

EXERCISES 1-6

Convert to radian measure. Leave answers in terms of π.

A 1. 210°   2. 240°   3. 300°   4. $-120°$

   5. $-90°$ 6. 315°   7. 36°   8. 495°

   9. 105°   10. 24°   11. 570°   12. 144°

Convert to degree measure.

13. $\dfrac{5\pi}{4}$

14. $\dfrac{5\pi}{3}$

15. $\dfrac{3\pi}{2}$

16. $\dfrac{9\pi}{4}$

17. $-\dfrac{2\pi}{3}$

18. $-\dfrac{3\pi}{4}$

19. $\dfrac{11\pi}{6}$

20. $\dfrac{7\pi}{6}$

21. 5π

22. $\dfrac{9\pi}{2}$

23. $\dfrac{13\pi}{6}$

24. $\dfrac{19\pi}{3}$

Convert to radian measure. Give answers to four significant digits.

25. $43.25°$

26. $15.50°$

27. $100.00°$

28. $155.20°$

Convert to degree measure. Give answers to the nearest $0.01°$.

29. 0.8000

30. 0.5236

31. 2.381

32. 1.275

Give exact values of each of the following.

33. $\cos\dfrac{3\pi}{4}$

34. $\sin\dfrac{7\pi}{6}$

35. $\tan\dfrac{\pi}{3}$

36. $\cos\dfrac{5\pi}{3}$

Give each of the following to four significant digits.

37. $\sin 1.000$

38. $\cos 1.375$

39. $\tan 0.2500$

40. $\sin 0.6667$

In Exercise 41–50, r, θ, s and A are as shown in the figure at the right. In each exercise, two of the four measures are given; find the other two. If θ is asked for, give its measure in both radians and degrees.

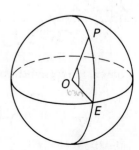

41. $r = 3.00$ cm; $\theta = 2.00$

42. $r = 40.0$ m; $\theta = 0.500$

43. $r = 60.0$ m; $s = 90.0$ m

44. $r = 1.20$ cm; $s = 3.00$ cm

B 45. $s = 2.40$ m; $\theta = 0.800$

46. $\theta = 0.375$; $s = 1.20$ m

47. $r = 6.00$ cm; $A = 24.0$ cm^2

48. $r = 1.25$ cm; $A = 0.750$ cm^2

49. $A = 1.08$ m^2; $\theta = 1.5$

50. $A = 0.250$ cm^2; $\theta = 2.00$

The latitude of point P shown in the drawing of Earth at the right is the measure of $\angle EOP$. The radius of Earth is about 6367 km.

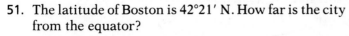
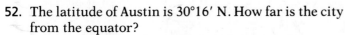

51. The latitude of Boston is $42°21'$ N. How far is the city from the equator?

52. The latitude of Austin is $30°16'$ N. How far is the city from the equator?

53. A clock pendulum one meter long makes a back-and-forth round trip every two seconds. If the greatest angle the pendulum makes with the vertical is 12°, how many kilometers does its bottom end travel in one day?

54. One **nautical mile** is the distance corresponding to one minute of Earth's circumference. How many kilometers is this?

55. About 200 B.C., a Greek mathematician named Eratosthenes found that when the sun's rays (assumed to be parallel) struck Earth at a certain time, they made an angle of about 7.2° with a vertical pole in the city of Alexandria. At the same time, the rays were exactly vertical at the city of Syene, 800 km to the south, since the bottom of a deep well there was illuminated. From these data, estimate the circumference of Earth both in km and in miles.

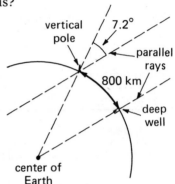

The figure at the right shows pulleys of radii 16 cm and 24 cm with a belt passing tightly over them.

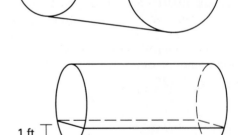

56. Through how many radians does the larger pulley turn when the smaller one makes a complete revolution?

57. How long is the belt if the centers of the pulleys are 64 cm apart?

58. A cylindrical tank 4 feet in diameter is lying on its side as shown at the right. What percent of the tank is filled when it contains water to a depth of one foot?

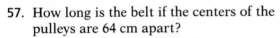

Self Quiz

1-4 | 1-5 | 1-6

1. The terminal side of an angle θ in standard position contains $P(9, -12)$. Find the exact values of the six trigonometric functions of θ.

2. Solve $\triangle PQR$ in which $\angle Q = 90°$, $\angle P = 56.5°$, and $p = 42.6$.

3. Find (a) the length of arc AB and (b) the area of sector AOB. Leave answers in terms of π.

4. If $\sin \theta = -\dfrac{24}{25}$ and $-90° < \theta < 90°$, find $\cos \theta$ and $\tan \theta$.

5. Find (a) $\cos \dfrac{5\pi}{3}$ and (b) $\tan \dfrac{3\pi}{4}$.

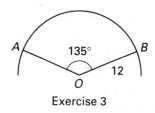

Exercise 3

ADDITIONAL PROBLEMS

1. The legs of a step ladder are each 2.25 m long and the angle between them is 35.6°. How far apart are the bottoms of the legs?

2. The terminal side of an angle ϕ in standard position passes through the point $\left(-\sqrt{7}, 3\right)$. Find the exact values of the six trigonometric functions of ϕ.

3. Find the acute angle α such that $\sin 75° \sec \alpha = 1$.

4. Express (a) 406.0 and (b) 0.0002600 in scientific notation.

5. A pulley rotates through 210° every second. Through how many revolutions does it turn in one hour?

**Exercises 6 and 7 refer to the figure at the right.
Leave answers in terms of π.**

6. Find the length s.

7. Find the area A.

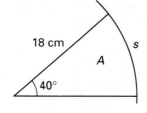

8. Express 12.635° in the degree-minute-second system.

9. Express 67.5° in radians to the nearest 0.001 radian.

10. The diagonal of a rectangle is 462 cm long and makes an angle of 37.2° with one of the sides. Find the dimensions of the rectangle.

11. Find the exact values of the six trigonometric functions of the acute angle ϕ given that $\cot \phi = \dfrac{20}{21}$.

12. Express (a) 4.6002×10^3 and (b) 7.60×10^{-2} as decimals.

13. If α is acute and $\tan \alpha = 0.75$, find the cosine of the supplement of α.

14. Express (a) 7°29′42″ and (b) 66°39′36″ in the decimal-degree system.

15. Find the exact value of *x* in the following figure.

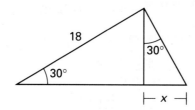

16. Express 285° in radians (in terms of π).

17. The legs of a right triangle are 27.6 m and 32.5 m long. Find the angles of the triangle.

18. Find two angles, one positive and one negative, that are coterminal with and different from a 200° angle.

19. Express *x* in terms of α, β, and *d*.

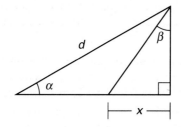

20. Two sailboats are in line with the bottom of a tower 40.0 m tall and are on the same side of it. Their angles of depression from the top of the tower are 10.3° and 12.8°. How far apart are the boats?

CHAPTER SUMMARY

1. In trigonometry, an angle often is regarded as being generated by a rotation about its vertex of one side, the initial side, into the other, the terminal side. The angle is positive (negative) if the rotation is counterclockwise (clockwise). Angles can be measured in decimal degrees or in degrees, minutes, and seconds.

2. An angle is in standard position when its vertex is at the origin and its initial side coincides with the positive *x*-axis. Two angles are co-terminal if their terminal sides coincide when the angles are in standard position.

3. To define the trigonometric functions of an angle θ, place θ in standard position and let $P(x, y)$ be a point (other than the origin O) on its

terminal side. Let $r = OP$. Then:

$$\sin \theta = \frac{y}{r} \qquad \cos \theta = \frac{x}{r} \qquad \tan \theta = \frac{y}{x} \ (x \neq 0)$$

If θ is an acute angle of a right triangle, then:

$$\sin \theta = \frac{\text{length of side opposite } \theta}{\text{length of hypotenuse}}$$

$$\cos \theta = \frac{\text{length of side adjacent to } \theta}{\text{length of hypotenuse}}$$

$$\tan \theta = \frac{\text{length of side opposite } \theta}{\text{length of side adjacent to } \theta}$$

The functions csc θ, sec θ, and cot θ are the reciprocals of sin θ, cos θ, and tan θ, respectively.

4. Any trigonometric function of an acute angle is equal to the cofunction of the complement of the angle. Exact values of the trigonometric functions of multiples of 30° and 45° can be found with the help of 30°-60°-90° and 45°-45°-90° triangles. Values of functions of all angles can be approximated by calculators or tables.

5. The number of significant digits in a measurement indicates its accuracy. A significant digit is any nonzero digit or any zero whose purpose is not just to place the decimal point. Scientific notation can be used to specify digits that are significant.

6. Trigonometric functions of acute angles can be used to solve right triangles.

7. Angles can be measured in radians as well as degrees. The two systems of measurement are related by the formula 180° = π radians. Arc length and sector area of a circle of radius r are given by the formulas $s = r\theta$ and $A = \dfrac{1}{2} r^2\theta$, respectively, where θ is the radian measure of the central angle.

CHAPTER TEST

1-1   1. (a) Express 67.845° in the degree-minute-second system.
         (b) Express 16°21′54″ in the decimal-degree system.

    2. Give two angles, one positive and one negative, that are co-terminal with and different from 575°.

1-2   3. Angle ϕ is acute and $\tan \phi = \dfrac{12}{5}$. Find the exact values of the other five trigonometric functions of ϕ.

4. Find *x* and *y* in the following figure. Give answers in simplest radical form.

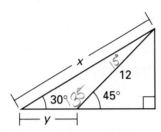

1-3   5. Find the angle of elevation of the sun when a 125 foot flagpole casts a shadow 218 feet long.

6. Express in scientific notation: (a) 126.30   (b) 0.00410.

1-4   7. The base of an isosceles triangle is 23.8 cm long and the vertex angle measures 55.0°. How long are the other sides?

8. How long is the shadow of a 38 meter tall monument when the angle of elevation of the sun is 63°?

1-5   9. When θ is in standard position, its terminal side passes through the point $(20, -21)$. Find the exact values of the six trigonometric functions of θ.

10. Find $\cos \alpha$ and $\tan \alpha$ if $\sin \alpha = -\dfrac{1}{3}$ and $90° < \alpha < 270°$.

1-6   11. (a) Convert 67.5° to radian measure.   (b) Convert $\dfrac{21\pi}{4}$ to degree measure.

12. Find (a) the perimeter and (b) the area of the shaded region shown at the right. Leave answers in terms of π.

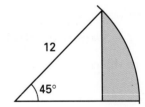

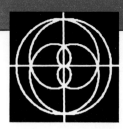

Linear Interpolation

The diagram at the right shows the graph of a function f on the interval $a \le x \le b$ and $\overline{AB}$, the graph of a *linear* function g. Since the slope of $\overline{AM}$ equals the slope of $\overline{AB}$,

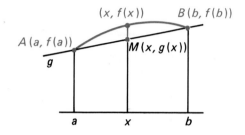

$$\frac{g(x) - f(a)}{x - a} = \frac{f(b) - f(a)}{b - a}$$

This proportion enables us to approximate, for example, cos 35°53′. Substitute $a = 35°50′$, $b = 36°00′$, $f(a) = \cos 35°50′ = 0.8107$, and $f(b) = \cos 36°00′ = 0.8090$ into the proportion.

$$\frac{g(x) - 0.8107}{35°53′ - 35°50′} = \frac{0.8090 - 0.8107}{36°00′ - 35°50′}$$

Solve the proportion for $g(x)$, to get $g(x) = 0.8102$, approximately cos 35°53′.

If we write the proportion as

$$g(x) = (x - a)\left(\frac{f(b) - f(a)}{b - a}\right) + f(a),$$

then we can find $g(x)$ when a, b, $f(a)$, $f(b)$, and x are known. This method of approximating $f(x)$ by finding $g(x)$ is called *linear interpolation*.

EXERCISES

1. Write a computer program to accept a, b, $f(a)$, $f(b)$, and x as input and print $g(x)$ as output.
2. Use the program in Exercise 1 to approximate cos for values of x from 35° to 36° every ten minutes. Compare the output with the values found from Table 2.

Choosing a Course

Inez enters a rowing-running contest, as shown. A trophy is awarded to the contestant that can row from P to Q, and then run from Q to R in the least time. What course $\theta = \angle SPQ$ should Inez steer so as to win if she can row at 6 mi/h and can run at 8 mi/h?

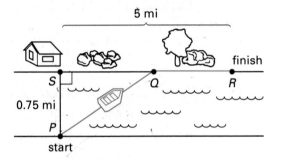

Since distance equals rate $\times$ time, the time it takes to row from P to Q is $\dfrac{PQ}{6}$ and the time it takes to run from Q to R is $\dfrac{5 - SQ}{8}$. To win, Inez wants to minimize T:

$$\frac{PQ}{6} + \frac{5 - SQ}{8}$$

To choose a course θ, Inez needs to write T in terms of θ. From right triangle SPQ

$$PQ = \frac{0.75}{\cos \theta} \qquad\qquad SQ = 0.75 \tan \theta.$$

Therefore,

$$T = \frac{0.75}{6 \cos \theta} + \frac{5 - 0.75 \tan \theta}{8}.$$

To see what θ gives a minimum value of T, Inez wrote and ran a computer program. Part of the output is shown in the following table.

θ	T	θ	T	θ	T	θ	T
30	.7152	40	.7095	50	.7077	60	.7126
31	.7145	41	.7091	51	.7079	61	.7137
32	.7138	42	.7088	52	.7080	62	.7149
33	.7132	43	.7085	53	.7083	63	.7163
34	.7125	44	.7082	54	.7086	64	.7179
35	.7120	45	.7080	55	.7090	65	.7197
36	.7114	46	.7079	56	.7075	66	.7218
37	.7109	47	.7078	57	.7101	67	.7241
38	.7104	48	.7077	58	.7109	68	.7266
39	.7099	49	.7077	59	.7117	69	.7296

From the table, she should choose a course θ such that $48° \leq \theta \leq 50°$.

EXERCISES

1. A large cluster of rocks is 0.75 mi down river from S. Given the diagram, should Inez steer a course to the left or the right of the rock cluster?
2. A large oak tree is 1.3 mi down the river from S. Given the diagram, should Inez steer a course to the left or the right of the tree?
3. Suppose that she runs every day before the contest and increases her running speed to 9.5 mi/h. Write the new function for T in terms of θ.
4. Given the function from Exercise 3, what value or values of θ will minimize T?

2 Circular Functions, Graphs, and Inverses

The photograph shows hot-air balloons beginning their ascent at a fiesta in the Southwest. Inverse trigonometric functions can be used to express angles of elevation as functions of time.

CIRCULAR FUNCTIONS AND CIRCULAR MOTION

2-1 Circular Functions

Objective: To define and use the six circular functions.

The equation $y = 1.25 \sin (0.24x)$ describes the shape of an ocean wave (height y and horizontal distance x in meters). The equation $V = 115\sqrt{2} \cos (120\pi t)$ gives the line voltage V of ordinary alternating current (t in seconds). In neither case are angles involved. These examples suggest that we need to define $\sin t$ and $\cos t$, where t represents a *real number*.

To make these definitions, let C be the unit circle, $x^2 + y^2 = 1$, A be $(1, 0)$, and t be any real number (Figure 2-1). Start at A and measure $|t|$ units around C in a counter-clockwise direction if $t \geq 0$ and in a clock-wise direction if $t < 0$, arriving at $P_t(x, y)$. The sine and cosine of t are then defined by the equations on the following page.

Figure 2-1

$$\sin t = y \text{ and } \cos t = x$$

These functions are often called **circular functions.** The circular functions **tangent, cotangent, secant,** and **cosecant** are defined as follows.

$$\tan t = \frac{\sin t}{\cos t} \text{ if } \cos t \neq 0 \qquad \cot t = \frac{\cos t}{\sin t} \text{ if } \sin t \neq 0$$

$$\sec t = \frac{1}{\cos t} \text{ if } \cos t \neq 0 \qquad \csc t = \frac{1}{\sin t} \text{ if } \sin t \neq 0$$

There is a close relationship between the circular functions and the corresponding trigonometric functions. For each real number t, there is a unique angle θ_t in standard position having radian measure t. Since the radius of C is 1, the terminal side of θ_t passes through P_t (Figure 2-2). Therefore:

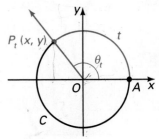

Figure 2-2

$$\sin \theta_t = \frac{y}{1} = y = \sin t$$

$$\cos \theta_t = \frac{x}{1} = x = \cos t.$$

In general, a circular function of a number t is equal to the corresponding trigonometric function of any angle having radian measure t. Because of this fact, you can find approximate values of circular functions by using a calculator set in the RAD mode or Table 1. Table 2-1 gives exact values of circular functions of some frequently-occurring numbers.

$t =$	0	$\dfrac{\pi}{6}$	$\dfrac{\pi}{4}$	$\dfrac{\pi}{3}$	$\dfrac{\pi}{2}$
$\sin t$	0	$\dfrac{1}{2}$	$\dfrac{\sqrt{2}}{2}$	$\dfrac{\sqrt{3}}{2}$	1
$\cos t$	1	$\dfrac{\sqrt{3}}{2}$	$\dfrac{\sqrt{2}}{2}$	$\dfrac{1}{2}$	0
$\tan t$	0	$\dfrac{\sqrt{3}}{3}$	1	$\sqrt{3}$	—

Table 2-1

Example 1   Find the exact values of $\sin \dfrac{5\pi}{3}$, $\cos \dfrac{5\pi}{3}$, $\tan \dfrac{5\pi}{3}$.

Solution

The angle having radian measure $\dfrac{5\pi}{3}$ is in Quadrant IV, and its refer-

ence angle measures $2\pi - \dfrac{5\pi}{3} = \dfrac{\pi}{3}$ radians. (See the unit-circle dia-

gram at the right.) Therefore:

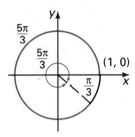

$$\sin\frac{5\pi}{3} = -\sin\frac{\pi}{3} = -\frac{\sqrt{3}}{2}$$

$$\cos\frac{5\pi}{3} = \cos\frac{\pi}{3} = \frac{1}{2}$$

$$\tan\frac{5\pi}{3} = -\tan\frac{\pi}{3} = -\sqrt{3} \quad \blacksquare$$

Example 2 Find $\cos 4.0$ to two significant digits.

Solution 1 A scientific calculator set in the RAD mode gives $\cos 4.0 = -0.6536$, or, to two significant digits, $\cos 4.0 = -0.65$.

Solution 2 The angle having radian measure 4.0 is in Quadrant III, and its reference angle has radian measure

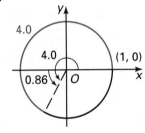

$$4.0 - \pi \approx 4.0 - 3.14 = 0.86.$$

From Table 3, $\cos 0.86 = 0.6524$. Therefore, to two significant digits, $\cos 4.0 = -\cos 0.86 = -0.65.$ $\blacksquare$

Example 3 Find the number t if $\tan t = -1.24$ and $0 < t < \pi$.

Solution From either a calculator (using the "inverse" and "tangent" keys) or a table (read "backwards") find that $\tan 0.89 \approx 1.24$. Therefore, the angle θ_t having radian measure t has a reference angle measuring 0.89 radians. Now θ_t is in Quadrant II, because $\tan t < 0$ and $0 < t < \pi$. Hence, $t = \pi - 0.89 \approx 3.14 - 0.89 = 2.25.$ $\blacksquare$

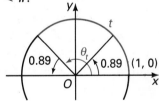

Because the unit circle has the equation $x^2 + y^2 = 1$, and the definitions of the circular functions specify that $x = \cos t$ and $y = \sin t$,

$$\cos^2 t + \sin^2 t = 1.$$

This is the same identity proved on page 26 for trigonometric functions.

Example 4 Find $\cos t$ and $\tan t$ if $\sin t = -0.36$ and $5 < t < 6$.

Solution
$$\cos^2 t + \sin^2 t = \cos^2 t + (-0.36)^2 = 1$$
$$\cos^2 t = 1 - (-0.36)^2 \approx 1 - 0.130 = 0.870$$

(Solution continued next page)

$$\cos t = \pm \sqrt{0.870} \approx \pm 0.93$$

Since $\dfrac{3\pi}{2} < 5 < t < 6 < 2\pi$, $\cos t > 0$. Hence, $\cos t = 0.93$.

$$\tan t = \frac{\sin t}{\cos t} = \frac{-0.36}{0.93} = -0.39 \quad \blacksquare$$

It is customary to refer to circular functions as trigonometric functions (because of the close relationships between them), and we shall often do so from now on. It will be clear from the context whether we are talking about functions of numbers or functions of angles.

EXERCISES 2-1

Find the exact values, if they exist, of sin t, cos t, and tan t for the given values of t.

A 1. $\dfrac{2\pi}{3}$　　　　2. $\dfrac{5\pi}{4}$　　　　3. $\dfrac{7\pi}{6}$　　　　4. $-\dfrac{\pi}{6}$

5. $-\dfrac{4\pi}{3}$　　　　6. 3π　　　　7. $\dfrac{3\pi}{2}$　　　　8. $\dfrac{5\pi}{6}$

9. $\dfrac{11\pi}{4}$　　　　10. $-\dfrac{11\pi}{6}$　　　　11. $\dfrac{14\pi}{3}$　　　　12. $\dfrac{19\pi}{6}$

13-20. Find the exact values, if they exist, of cot t, sec t, and csc t for the values of t given in Exercises 1-8.

Find each value to four significant digits.

21. cos 1.20　　　　　　22. sin 0.65　　　　　　23. tan 1.00

24. cos 1.02　　　　　　25. sin 2.16　　　　　　26. tan 3.50

27. tan (−4.00)　　　　28. cos (−0.76)　　　　29. cot 4.00

30. sec 5.00　　　　　　31. csc 7.00　　　　　　32. cot (−2.00)

Find the number t to two decimal places.

33. $\sin t = 0.65;\ -\dfrac{\pi}{2} < t < \dfrac{\pi}{2}$　　　　34. $\cos t = 0.26;\ 0 < t < \pi$

35. $\tan t = -1.23;\ 0 < t < \pi$　　　　36. $\sin t = -0.40;\ -\dfrac{\pi}{2} < t < \dfrac{\pi}{2}$

37. $\cos t = -0.75;\ \pi < t < 2\pi$　　　　38. $\tan t = 2.36;\ \pi < t < 2\pi$

B 39. $\cot t = -1.95;\ 0 < t < \pi$　　　　40. $\sec t = 3.00;\ 0 < t < \pi$

In Exercises 41-44, one of sin *t* or cos *t* is given. Find the other one and tan *t*. Use $\cos^2 t + \sin^2 t = 1$.

41. $\sin t = 0.72$ and $0 < t < 1$.

42. $\cos t = -0.68$ and $2 < t < 3$.

43. $\cos t = -0.25$ and $4 < t < 5$.

44. $\sin t = 0.32$ and $-4 < t < -3$.

In Exercises 45-49, $0 < t < \dfrac{\pi}{2}$. Refer to the figure at the right.

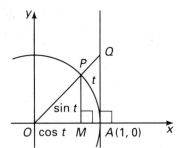

45. Show that tan *t* = *AQ*.
    (Hint: $\triangle OAQ$ is similar to $\triangle OMP$.)

46. Show that sec *t* = *OQ*.

C 47. Show that $\sin t < t < \tan t$. (Hint: Compare the areas of $\triangle OAP$, sector *AOP*, and $\triangle OAQ$.)

48. Show that $1 < \dfrac{t}{\sin t} < \dfrac{1}{\cos t}$.

49. When *t* is very small, sin *t* is close to 0 and cos *t* is close to 1. What is $\dfrac{t}{\sin t}$ close to? (The result is important in calculus.)

50. The unit circle *C*, with center initially at *O*, rolls to the left along the line $y = -1$ as shown in the figure at the left below. After *C* has rolled a distance *t*, the point *P* of *C*, initially at (1, 0) has moved to P'. Find the coordinates of P' in terms of *t*. (Hint: $\overset{\frown}{AP'}$ has length *t*.)

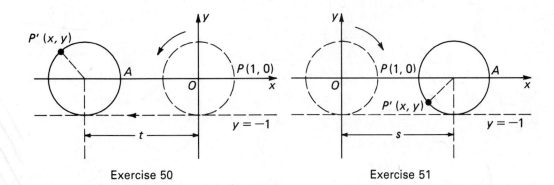

Exercise 50   Exercise 51

51. Repeat Exercise 50, but with *C* rolling to the right.

2-2

Circles and Circular Motion

Objective: To find the position and speed of a point moving on a circle.

Figures 2-3 and 2-4 show a circle C of radius r centered at O; M is the point $(r, 0)$; $P(x, y)$ is an arbitrary point of C; and θ is the measure of $\angle MOP$. It follows from the definitions of $\sin \theta$ and $\cos \theta$ that

$$x = r \cos \theta \text{ and } y = r \sin \theta.$$

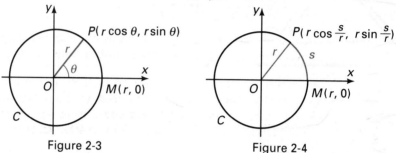

Figure 2-3 Figure 2-4

If θ is measured in radians and s is the length of $\overset{\frown}{MP}$, then $\theta = \dfrac{s}{r}$, and we have

$$x = r \cos \frac{s}{r} \text{ and } y = r \sin \frac{s}{r}.$$

Example 1 A particle starts at $(5, 0)$ and moves counterclockwise a distance 18.0 around the circle $x^2 + y^2 = 25$. Find the final position P of the particle. Give the answer to three significant digits.

Solution Draw a sketch and mark the points $M(5, 0)$ and $P(?, ?)$ on the circle $x^2 + y^2 = 25$. Let θ be the positive angle MOP. Since $\overset{\frown}{MP}$ has

length $s = 18.0$, $\theta = \dfrac{s}{r} = \dfrac{18.0}{5} = 3.60$.

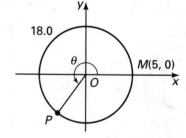

Therefore:

$x = r \cos \theta$

$\quad = 5 \cos 3.60 = -4.48$

$y = r \sin \theta$

$\quad = 5 \sin 3.60 = -2.21$

Therefore, the final position of the particle is $P(-4.48, -2.21)$. ∎

Suppose now that a point P is moving with constant speed around a circle of radius r and center O (Figure 2-5). We call this **uniform circular motion.** If in time t, the radius $\overline{OP}$ sweeps out an angle θ, then the

angular speed ω of P is given by $\omega = \dfrac{\theta}{t}$. The

linear speed v of P is, of course, $v = \dfrac{s}{t}$, where s is the distance P travels in t seconds. That is,

angular speed $\omega = \dfrac{\theta}{t}$ and linear speed $v = \dfrac{s}{t}$.

If θ is measured in radians, then $s = r\theta$, and

$$v = \frac{r\theta}{t} = r\frac{\theta}{t}. \text{ Thus, } v = r\omega.$$

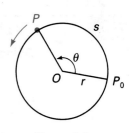

Figure 2-5

Angular speeds are given in various units; for example: revolutions per minute (rpm), degrees per hour, and radians per second.

Example 2   All points on Earth (except the poles) describe uniform circular motion with the same angular speed. (a) Find this angular speed in radians/min. (b) Find the linear speed in miles per minute of a point P having latitude 30°. (See page 33.) Use 4000 miles as the radius of Earth.

Solution   (a) $\omega = 1$ rev/day $= 2\pi$ rad/day

$$= \frac{2\pi}{24 \times 60} \text{ rad/min}$$

$$= \frac{\pi}{720} \text{ rad/min} \approx 4.36 \times 10^{-3} \text{ rad/min}$$

(b) The radius of the circle on which P moves is

$r = 4000 \cos 30°$

$= 4000 \cdot \dfrac{\sqrt{3}}{2}$

$= 2000\sqrt{3}$

Hence, $v = r\omega$

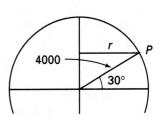

$$= 2000\sqrt{3} \cdot \frac{\pi}{720}$$

$$= \frac{25\pi\sqrt{3}}{9} \text{ miles/min}$$

$v = 15.1$ miles/min ■

We can find the position of a particle moving with angular speed ω by using the formulas $x = r \cos \theta$ and $y = r \sin \theta$. Referring to Figure 2-6, suppose that the particle is at P_0 at time $t = 0$ and at $P(x, y)$ at time t. Since $\phi = \omega t$, $\theta = \phi + \theta_0 = \omega t + \theta_0$. Therefore:

$$x = r \cos (\omega t + \theta_0), \qquad y = r \sin (\omega t + \theta_0)$$

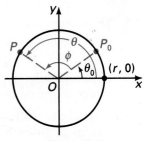

Figure 2-6

Example 3   A particle starts at the point $(-2, 2)$ and travels in a counterclockwise direction around the circle $x^2 + y^2 = 8$ with angular speed $63°/\text{min}$. Where is it 5 min later?

Solution   We plan to use the formulas just derived. $r = \sqrt{8} = 2\sqrt{2}$, $\omega = 63°/\text{min} = 63 \cdot \dfrac{\pi}{180} = \dfrac{7\pi}{20}$ rad/min. Since P_0 is $(-2, 2)$, $\theta_0 = \dfrac{3\pi}{4}$ and $t = 5$. Therefore:

$$x = r \cos(\omega t + \theta_0) \qquad\qquad y = r \sin(\omega t + \theta_0)$$

$$= 2\sqrt{2} \cos\left(\frac{7\pi}{20} \cdot 5 + \frac{3\pi}{4}\right) \qquad = 2\sqrt{2} \sin \frac{5\pi}{2}$$

$$= 2\sqrt{2} \cos\left(\frac{7\pi}{4} + \frac{3\pi}{4}\right) \qquad\qquad = 2\sqrt{2}$$

$$= 2\sqrt{2} \cos \frac{5\pi}{2}$$

$$= 0$$

Hence, the particle is at the point $\left(0, 2\sqrt{2}\right)$. ∎

EXERCISES 2-2

A particle moves with linear speed v and angular speed ω around a circle of radius r. Find the unknown quantity. Leave answers in terms of π.

A 1. $r = 6$ cm; $\omega = 8$ rad/s; $v = $ __?__ cm/s

2. $r = 10$ cm; $v = 15$ cm/s; $\omega = $ __?__ rad/s

3. $r = 2.0$ m; $v = 2.4$ m/s; $\omega = $ __?__ rad/min

4. $r = 3.6$ m; $\omega = 20$ rpm; $v = $ __?__ m/min

5. $r = 7.2$ cm; $\omega = 120°/\text{min}$; $v = $ __?__ cm/min

6. $r = 1.5$ m; $v = 900$ m/h; $\omega = $ __?__ rpm

7. $v = 200\pi$ ft/min; $\omega = 20$ rpm; $r = $ __?__ ft

8. $\omega = 270°/s$; $v = 30\pi$ cm/s; $r = $ __?__ cm

9. The linear speed of a point 10.8 cm from the center of a phonograph record is 12π cm/s. Find the angular speed of the phonograph record in revolutions per minute.

10. Find the linear speed to the nearest cm/s of a point 15 cm from the center of the record of Exercise 9.

11. A helicopter has just landed and its rotor is idling at 90 rpm. How fast, to the nearest m/s, are the tips of its 4.5-meter-long rotor blades moving?

12. A Ferris wheel 40 feet in diameter makes one revolution every two minutes. What is the speed of a seat on the rim of the wheel?

Exercises 13 and 14 refer to a fan belt joining a pulley 16 cm in diameter to one that is 9 cm in diameter. The smaller pulley is rotating at 4000 rpm.

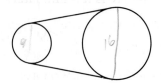

13. Find the speed of a point of the fan belt.

14. How fast is the larger pulley rotating?

In Exercises 15-32, a particle starts at point *M* and travels for *s* cm or with angular speed *ω* counterclockwise around the circle $x^2 + y^2 = r^2$. Find the coordinates of the final position of the particle. Give exact answers if possible. Otherwise, give answers to three significant digits.

15. $M = (6, 0); r = 6; s = 4\pi$

16. $M = (8, 0); r = 8; s = 14\pi$

17. $M = (9, 0); r = 9; s = 48\pi$

18. $M = (12, 0); r = 12; s = 26\pi$

19. $M = (4, 0); r = 4; s = 3$

20. $M = (3, 0); r = 3; s = 4$

21. $M = (4, 0); r = 4; \omega = \dfrac{\pi}{6}$ rad/s; $t = 5$ s

22. $M = (6, 0); r = 6; \omega = \dfrac{4\pi}{30}$ rad/min; $t = 10$ min

23. $M = (10, 0); r = 10; \omega = 21°/\text{min}; t = 15$ min

24. $M = (15, 0); r = 15; \omega = 84°/\text{h}; t = 5$ h

25. $M = (7, 0); r = 7; \omega = 15$ rpm; $t = 45$ s

26. $M = (8, 0); r = 8; \omega = 35$ rpm; $t = 20$ s

B 27. $M = (0, -2); r = 2; \omega = \dfrac{5\pi}{18}$ rad/min; $t = 3$ min

28. $M = (-3, 3); r = 3\sqrt{2}; \omega = \dfrac{3\pi}{20}$ rad/s; $t = 5$ s

29. $M = \left(-2, 2\sqrt{3}\right); r = 4; \omega = 1000°/\text{h}; t = 45$ min

30. $M = \left(1, -\sqrt{3}\right); r = 2; \omega = 630°/\text{min}; t = 40$ s

31. $M = (-4, -4); r = 4\sqrt{2}; \omega = 3.4$ rpm; $t = 1$ min 15 s

32. $M = \left(3\sqrt{3}, 3\right); r = 6, \omega = 10$ rpm; $t = 50$ s

33. Find the linear speed (due to the rotation of Earth) of London, which has latitude 51°30′ N.

34. Find the linear speed of your school. Use a reasonable degree-minute estimation of latitude taken from an atlas.

The bicycle referred to in Exercises 35 and 36 has wheels 26 inches in diameter.

35. Find the angular speed of the wheels (in rpm) when the bicycle is traveling at 15 mph. (One mile = 5280 feet)

36. How fast (in mph) is the bicycle traveling when its wheels are revolving at 150 rpm?

A particle in uniform circular motion is held in orbit by **centripetal force**, a force directed toward the center of the orbit. The magnitude of this force is given by $F = \dfrac{mv^2}{r}$, where m is mass, v is linear speed, and r is the radius of the orbit.

37. Express the magnitude of the force in terms of angular speed, ω.

38. If the mass and radius are held constant, how will the centripetal force change when either linear speed or angular speed are doubled?

C 39. What is the angular speed in radians per second of the hour hand of a clock? If the hand is 6 cm long, how fast is its tip moving in millimeters per second?

40. A car's wheel (with tire) is 26 inches in diameter and its center is 12 inches above the road, as shown at the left below. If the car is traveling at 60 feet per second, how fast is the highest point of the tire moving relative to the ground?

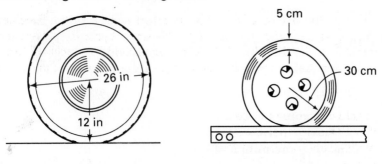

Exercise 40   Exercise 41

41. A wheel of a train traveling at 90 km/h is 30 cm in radius and has a 5 cm flange that extends below the top of the rail, as indicated at the right above. How fast is the lowest point of the flange moving and in what direction?

42. A circle of radius a begins with its center at the point $(0, a)$ and rolls to the right along the x-axis with angular velocity ω. A point on the circumference of this circle starts at the origin and is at position P at time t. Find equations for the coordinates of P at time t. (Hint: OA = the length of $\overset{\frown}{PA} = a\theta$ where θ is in radians.)

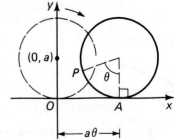

GRAPHS OF CIRCULAR FUNCTIONS

2-3   Periodicity and Other Properties

Objective: To use periodicity and other properties as an aid in graphing functions.

On page 44 we defined cos t and sin t to be, respectively, the first and second coordinates of the point P_t on the unit circle C. Because C has circumference 2π, the points P_t and $P_{t+2\pi}$ coincide (Figure 2-7). Therefore, for all real numbers t:

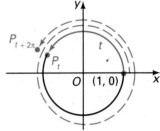

$$\cos (t + 2\pi) = \cos t \qquad \sin (t + 2\pi) = \sin t.$$

Equivalently, in terms of the variable x:

$$\cos (x + 2\pi) = \cos x \quad \sin (x + 2\pi) = \sin x.$$

Figure 2-7

We say that cos x and sin x have *period* 2π.

In general, a function f is **periodic** if for some positive constant p,

$$f(x + p) = f(x)$$

for all x in the domain of f. The smallest such p is the **fundamental period**, or simply ***the period***, of f. The word cycle is sometimes used for period, as in the phrase "cycles per second." A **cycle of the graph** of a function having period p is a part of the graph on any interval of length p.

Example 1

One cycle of the graph of a function f having period 2 is shown. Graph f in the interval $-3 \le x \le 5$.

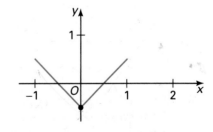

Solution

Since the given part of the graph covers one period, repeat it to the right until you reach 5 and to the left until you reach -3. ∎

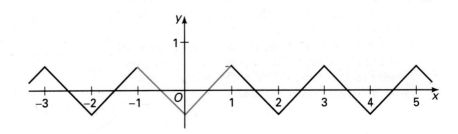

Figure 2-8 shows that

$$\cos(-t) = x = \cos t$$

and

$$\sin(-t) = -y = -\sin t.$$

We say that the cosine is an *even function* and that the sine is an *odd function*. In general,

f is an **even function** if $f(-x) = f(x)$

and

f is an **odd function** if $f(-x) = -f(x)$

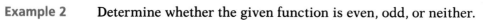

Figure 2-8

for every x in the domain of f. Unlike an integer, a function need not be even or odd; it can be neither.

Example 2

Determine whether the given function is even, odd, or neither.

(a) $f(x) = \dfrac{x}{x^2 + 1}$ (b) $g(x) = x^2 - 4x$ (c) $h(x) = x \sin x$

Solution

(a) $f(-x) = \dfrac{-x}{(-x)^2 + 1} = -\dfrac{x}{x^2 + 1} = -f(x)$. Hence, f is odd.

(b) Since $g(-x) = (-x)^2 - 4(-x) = x^2 + 4x$ and $g(x) = x^2 - 4x$, g is not even. Since $g(-x) = x^2 + 4x$ and $-g(x) = -x^2 + 4x$, g is not odd. Hence, g is neither even nor odd.

(c) $h(-x) = (-x)\sin(-x) = (-x)(-\sin x) = x \sin x = h(x)$. Hence, h is even. ∎

Figure 2-9 illustrates the effects of evenness and of oddness on the graphs of functions.

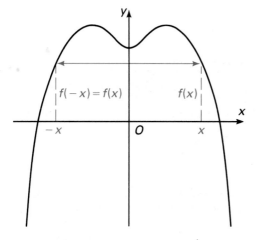

f even: graph is symmetric with respect to the y-axis

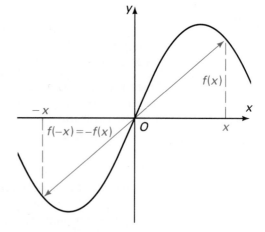

f odd: graph is symmetric with respect to the origin.

Figure 2-9

Example 3   Part of the graph of a function
f is shown at the right. Graph
f in the interval $-3 \le x \le 3$,
given that f is (a) even, (b)
odd.

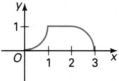

Solution   (a) Reflect the given part of the
graph in the *y*-axis.

(b) Reflect the given part
of the graph in the origin.   ■

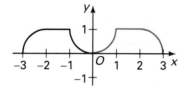   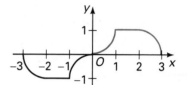

Figure 2-10 shows the graph of the function $f(x) = \dfrac{x-1}{x-2}$. Notice that

when x is close to 2, the denominator of $\dfrac{x-1}{x-2}$ is close to 0, so that $|f(x)|$

is large. Therefore, as x approaches 2, the point $P(x, f(x))$ moves on the
graph in such a way that (a) P moves away from the origin and (b) the
distance between P and the vertical line $L: x = 2$ becomes arbitrarily
small. We say that L is an *asymptote* of the graph.

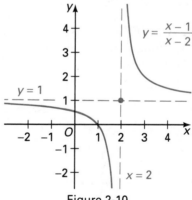

Figure 2-10

In general:

A line L is an **asymptote** of a graph G if the distance between L
and a point P moving on G becomes arbitrarily small as P
moves away from the origin.

The horizontal line $y = 1$ is another asymptote of the graph of $f(x) =$
$\dfrac{x-1}{x-2}$ because when $|x|$ is very large the numerator and denominator are
almost equal and therefore have quotient about 1.

EXERCISES 2-3

Give the period of the periodic function whose graph is shown. Is this function even, odd, or neither?

A 1.

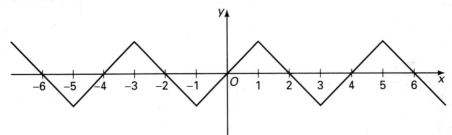

2.

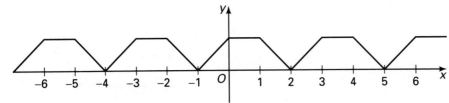

3.

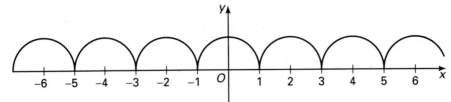

4.

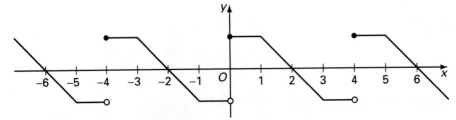

Part of the graph of a function f having the stated period p is given. Graph f in the indicated interval and state whether f is even, odd, or neither.

5.

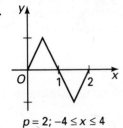

$p = 2; -4 \le x \le 4$

6.

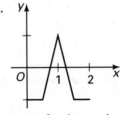

$p = 2; -4 \le x \le 4$

7.

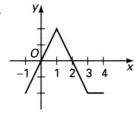

$p = 5; -10 \le x \le 10$

8.

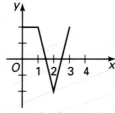

$p = 3; -6 \le x \le 6$

9.

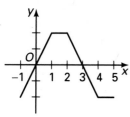

$p = 6; -6 \le x \le 6$

10.

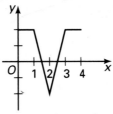

$p = 4; -8 \le x \le 8$

Part of the graph of a function f is given. Graph f in the interval $-3 \le x \le 3$ assuming that f is (a) even, (b) odd.

11.

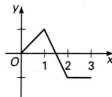

12.

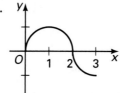

13.

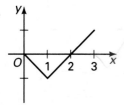

14.

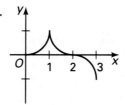

Determine whether the function f is even, odd, or neither.

15. $f(x) = x^3 - 4x$

16. $f(x) = x^4 - 4x$

17. $f(x) = \dfrac{x^2}{x + 1}$

18. $f(x) = \dfrac{x^2 + 1}{x}$

19. $f(x) = x^2\sqrt{x^2 - 9}$

20. $f(x) = x^2 - \sqrt{1 + x^2}$

B 21. $f(x) = x^2 \sin x$

22. $f(x) = x \cos x$

23. $f(x) = \sin x + \cos x$

24. $f(x) = \sin x \cos x$

Find the vertical and horizontal asymptotes (if any) of the graph of f.

25. $f(x) = \dfrac{x + 2}{x + 1}$

26. $f(x) = \dfrac{1}{x - 1}$

27. $f(x) = \dfrac{2x + 1}{x}$

28. $f(x) = \dfrac{4x}{2x - 1}$

29. Show that the product of two odd functions is even.

30. Show that the product of an even function and an odd function is odd.

In Exercises 31-34, let f be an arbitrary function defined for all real numbers.

31. Let $g(x) = f(x^2)$. Is g even, odd, or neither? Explain.

32. Let $h(x) = f(|x|)$. Is h even, odd, or neither? Explain.

C 33. Let $E(x) = \dfrac{f(x) + f(-x)}{2}$. Show that E is even.

34. Let $O(x) = \dfrac{f(x) - f(-x)}{2}$. Show that O is odd.

35. Show that any function defined for all real numbers is the sum of an even function and an odd function. (Hint: Use Exercises 33 and 34.)

36. Express $f(x) = x^3 - 5x^2 + 4x + 3$ as the sum of an even function and an odd function.

Self Quiz

| 2-1 | 2-2 | 2-3 |

1. Find the exact value of (a) $\tan \dfrac{-5\pi}{3}$ and (b) t if $\sec t = -2$ and $0 \le t \le \pi$.

2. The figure at the left below shows one cycle of the graph of a function f having period 4.   (a) Graph f in the interval $-4 \le x \le 8$.   (b) Is f even, odd, or neither?

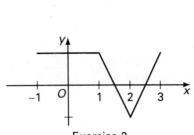

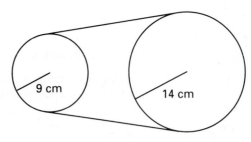

Exercise 2   Exercise 3

3. The figure at the right above shows two pulleys connected by a belt that is moving at 630 m/min. Find the angular speed of each pulley.

4. Find the vertical and horizontal asymptotes of the graph of $g(x) = \dfrac{2x + 1}{x - 2}$.

5. A particle starts at (4, 0) and moves with angular speed $33\frac{1}{3}$ rpm counterclockwise around the circle $x^2 + y^2 = 16$. Where is it after 105 s? Give an exact answer.

6. Find $\cos t$ and $\tan t$ given that $\sin t = -0.60$ and $-\dfrac{\pi}{2} < t < \dfrac{\pi}{2}$.

2-4   Graphs of the Trigonometric Functions

Objective: To graph the six trigonometric functions.

In order to graph $y = \sin x$ and $y = \cos x$, we first make a table of values.

x	0	$\dfrac{\pi}{6}$	$\dfrac{\pi}{4}$	$\dfrac{\pi}{3}$	$\dfrac{\pi}{2}$	$\dfrac{2\pi}{3}$	$\dfrac{3\pi}{4}$	$\dfrac{5\pi}{6}$	π
	0.00	0.52	0.79	1.05	1.57	2.09	2.36	2.62	3.14
$\sin x$	0.00	0.50	0.71	0.87	1.00	0.87	0.71	0.50	0.00
$\cos x$	1.00	0.87	0.71	0.50	0.00	-0.50	-0.71	-0.87	-1.00

We next plot $(x, \sin x)$ and $(x, \cos x)$ and join the resulting points with smooth curves to obtain the partial graphs in Figures 2-11 and 2-12.

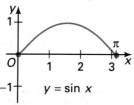

Figure 2-11

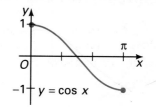

Figure 2-12

Since $\sin x$ is odd and $\cos x$ is even, we reflect these partial graphs in the origin and the y-axis, respectively, to get Figures 2-13 and 2-14.

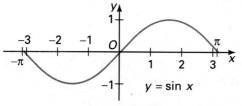

Figure 2-13

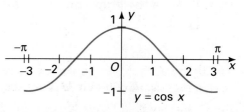

Figure 2-14

We now have the graphs of $y = \sin x$ and $y = \cos x$ on intervals of length 2π and can use periodicity to obtain the final graphs below.

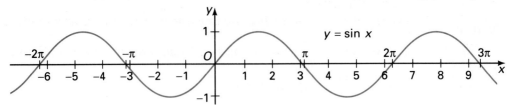

Figure 2-15

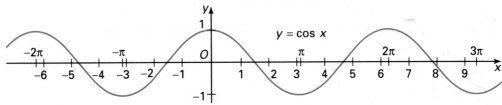

Figure 2-16

Of course, these graphs extend without limit to the right and to the left because both the sine and the cosine functions have the set of all real numbers as domain.

Example 1   Graph $y = |\cos x|$ in the interval $-2\pi \le x \le 3\pi$.

Solution   If $\cos x \ge 0$, $|\cos x| = \cos x$. If $\cos x < 0$, $|\cos x| = -\cos x$. Therefore, reflect in the x-axis the part of the graph of $y = \cos x$ that lies below the x-axis (shown as a dashed curve). ■

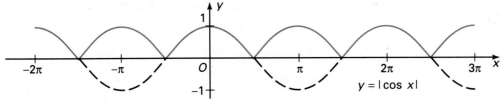

$y = |\cos x|$

Before graphing $y = \tan x$, let us derive a few properties of the tangent.

1. The function $\tan x$ is odd, since

$$\tan(-x) = \frac{\sin(-x)}{\cos(-x)} = \frac{-\sin x}{\cos x} = -\tan x.$$

2. The period of $\tan x$ is π. Referring to the unit circle in Figure 2-17, we see:

$$\sin(t + \pi) = -y = -\sin t$$
$$\cos(t + \pi) = -x = -\cos t$$

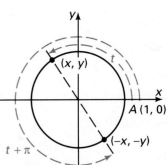

Figure 2-17

Since $\tan (x + \pi) = \dfrac{\sin (x + \pi)}{\cos (x + \pi)} = \dfrac{-\sin x}{-\cos x} = \tan x,$

$\tan x$ has period π.

3. The graph of $y = \tan x$ has the line $x = \dfrac{\pi}{2}$ as a vertical asymptote. If x is very close to $\dfrac{\pi}{2}$, $\sin x$ is close to 1 and $\cos x$ is close to 0. Hence, $|\tan x| = \left| \dfrac{\sin x}{\cos x} \right|$ is very large.

We start the graph of $y = \tan x$ by using a short table of values and property 3 above to obtain Figure 2-18.

x	$\tan x$
0	0
$\dfrac{\pi}{6} = 0.52$	0.58
$\dfrac{\pi}{4} = 0.79$	1.00
$\dfrac{\pi}{3} = 1.05$	1.73
$\dfrac{\pi}{2} = 1.57$	—

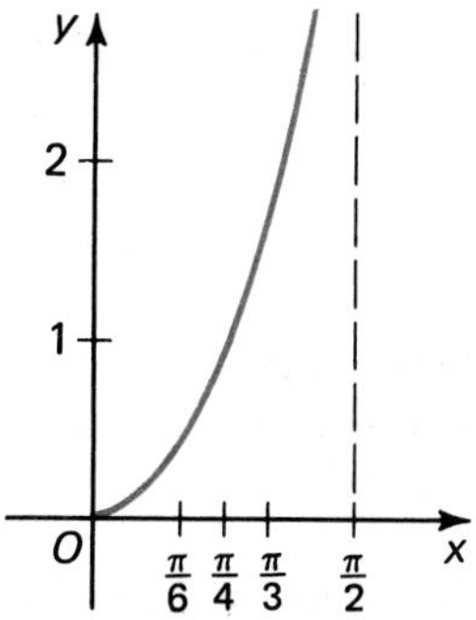

Figure 2–18

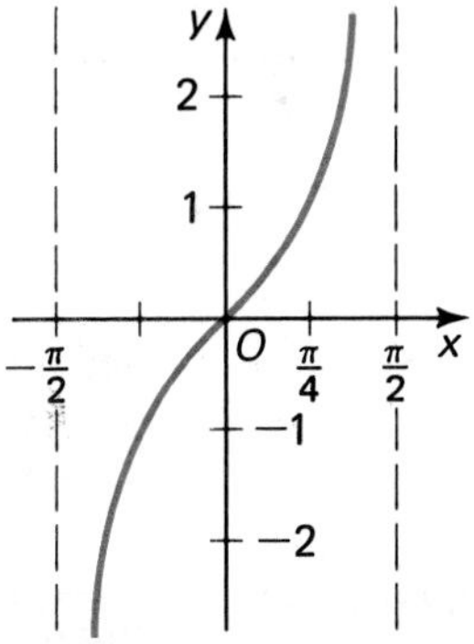

Figure 2–19

Next we use the fact that $\tan x$ is odd and reflect this part of the graph in the origin to obtain Figure 2-19. Lastly we use the fact that $\tan x$ has period π to obtain the final result shown in Figure 2-20.

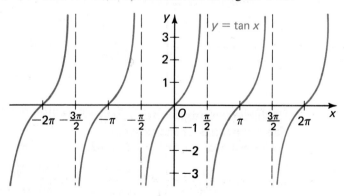

Figure 2–20

Using methods like those above, we obtain the graphs of the cotangent, secant, and cosecant functions. They are shown in red in Figures 2-21, 2-22, and 2-23, with the graphs of their reciprocal functions.

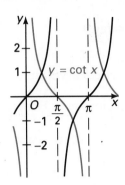

Figure 2-21

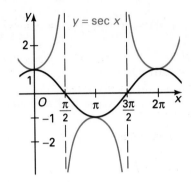

Figure 2-22

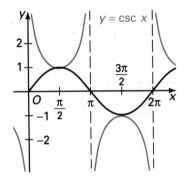

Figure 2-23

Example 2   Show that $f(x) = \tan \frac{1}{2}x$ has period 2π and sketch one cycle.

Solution   To show that f has period 2π, show that $f(x + 2\pi) = f(x)$.

$$f(x + 2\pi) = \tan \frac{1}{2}(x + 2\pi)$$

$$= \tan \left(\frac{1}{2}x + \pi \right)$$

$$= \tan \frac{1}{2}x = f(x)$$

The graph is shown at the right.

Notice that the graph of $f(x) = \tan \frac{1}{2}x$

has vertical asymptotes at $x = \pi$ and

$x = -\pi$, since $\tan \frac{1}{2}(\pi)$ and $\tan \frac{1}{2}(-\pi)$ are undefined.   ■

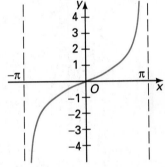

EXERCISES 2-4

Graph the given equation on the specified interval. Do as little point-by-point plotting as possible.

A 1. $y = -\sin x;\ -\pi \le x \le 2\pi$

2. $y = -\cos x;\ -\pi \le x \le 2\pi$

3. $y = |\sin x|;\ -2\pi \le x \le 3\pi$

4. $y = |\tan x|;\ -2\pi \le x \le 2\pi$

5. $y = \cos(x + \pi);\ -\pi \le x \le 2\pi$

6. $y = \sin(x + \pi);\ -\pi \le x \le 2\pi$

7. $y = \tan |x|;\ -\pi \le x \le 2\pi$

8. $y = \sin |x|;\ -\pi \le x \le 2\pi$

Show that $f(x)$ has period p and sketch one cycle of its graph.

9. $f(x) = \cos 2x;\ p = \pi$

10. $f(x) = \sin 2x;\ p = \pi$

11. $f(x) = \sin\frac{1}{2}x; \; p = 4\pi$

12. $f(x) = \cos 4x; \; p = \frac{\pi}{2}$

13. $f(x) = \sin \pi x; \; p = 2$

14. $f(x) = \cos 2\pi x; \; p = 1$

Is the function $f(x)$ even, odd, or neither? Justify your answer.

15. $f(x) = \cot x$

16. $f(x) = \sec x$

17. $f(x) = \sin x - \cos x$

18. $f(x) = 2 \sin x \cos x$

19. $f(x) = \cos^2 x - \sin^2 x$

20. $f(x) = \sin x + \sin^2 x$

Graph each equation in the interval $-2\pi < x < 2\pi$.

B 21. $y = \sin\left(x + \frac{\pi}{2}\right)$

22. $y = \cos\left(x - \frac{\pi}{2}\right)$

23. What do Exercises 21 and 22 suggest about the relationship between the graphs of $y = \sin x$ and $y = \cos x$?

Graph each equation in the specified interval.

Example $x = \cot y; \; 0 < y < \pi$.

Solution   Mentally graph $y = \cot x, \; 0 < x < \pi$. Then interchange the names of the axes. Finally revolve through $180°$ about the line $y = x$ so that the new y-axis will point up and the new x-axis will point to the right.

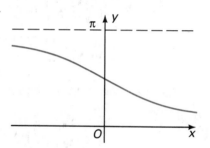

24. $x = \sin y; \; -\frac{\pi}{2} \le y \le \frac{\pi}{2}$

25. $x = \cos y; \; 0 \le y \le \pi$

26. $x = \tan y; \; -\frac{\pi}{2} < y < \frac{\pi}{2}$

C 27. Graph $y = \sin^2 x$ on the interval $-\pi \le x \le 2\pi$. (Hint: In Chapter 5, we shall show that the graph is a displaced sine curve of period π.)

28. Graph $y = \cos^2 x$ on the interval $-\pi \le x \le 2\pi$. (See the hint for Exercise 27.)

29. Graph $y = \sin x, \; y = x,$ and $y = \tan x, \; 0 \le x < \frac{\pi}{2}$ in the same coordinate plane. (Note that the figure illustrates the inequality $\sin x < x < \tan x$ on the interval $0 < x < \frac{\pi}{2}$.)

30. Graph $y = \cos x, \; -\frac{\pi}{2} \le x \le \frac{\pi}{2}$ and $y = 1 - \frac{x^2}{2}, \; -\sqrt{2} \le x \le \sqrt{2}$ in the same coordinate plane. (The answer illustrates that the polynomial $1 - \frac{x^2}{2}$ is a good approximation to $\cos x$ near $x = 0$.)

INVERSE FUNCTIONS

2-5	**Inverse Functions and Relations**

Objective: To find and graph inverses of functions and relations.

You may have learned in previous mathematics courses that a **relation** is any set of ordered pairs. The set of all first members of the ordered pairs of a relation is its **domain** and the set of all second members is its **range**.

Example 1

Graph the relation $K = \{(x, y): (x - 3)^2 + y^2 = 1\}$ and find its domain and range.

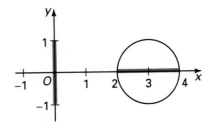

Solution

The graph of K is the circle of radius 1 centered at $(3, 0)$. The domain of K is the interval $2 \le x \le 4$ and its range is the interval $-1 \le y \le 1$. ∎

A **function** can be defined as a relation in which no two ordered pairs have the same first members. The relation K of Example 1 is *not* a function. For one thing, K contains both the pairs $(3, 1)$ and $(3, -1)$.

The **inverse**, R^{-1}, of relation R is obtained by interchanging the members of the ordered pairs of R. (R^{-1} is read "R inverse". The superscript -1 is *not* an exponent.) For example, the inverse of the relation K of Example 1 is $K^{-1} = \{(y, x): (x - 3)^2 + y^2 = 1\}$. Since we usually use x to denote the first member of an ordered pair and y the second, we interchange x and y *throughout* and write

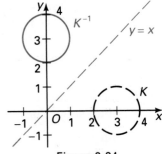

$$K^{-1} = \{(x, y): (y - 3)^2 + x^2 = 1\}$$
$$= \{(x, y): x^2 + (y - 3)^2 = 1\}.$$

Figure 2-24

The graph of K^{-1} is shown in Figure 2-24 along with the graph of K (shown dashed). Notice that the graph of the inverse R^{-1} of any relation R can be obtained by reflecting the graph of R in the line $y = x$.

We shall mainly be concerned with relations that are functions. The inverse of a function is, of course, a relation but need not be a function.

Example 2

Graph each function and its inverse in the same coordinate plane. Is the inverse a function?   (a) $g(x) = x^2 + 1$ (b) $h(x) = x^3 + 1$

Solution

(a) The graph of g is a parabola with vertex $(0, 1)$ and opening upward. To obtain the graph of g^{-1} (shown in red) reflect the graph of g in the line $y = x$. Because it contains the points $(2, 1)$ and $(2, -1)$, g^{-1} *is not a function.*

(b) The graph of h is the graph of $y = x^3$ raised up one unit. The graph of h^{-1} is shown in red. Because no vertical line intersects the graph of h^{-1} in more than one point, h^{-1} *is a function.* ■

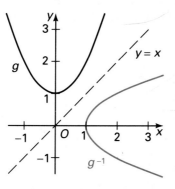

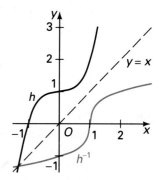

The test used in Example 2(b) is called the **vertical-line test**. If no two ordered pairs of a function f have the same *second* member, then no two ordered pairs of f^{-1} have the same *first* member, and f^{-1} is a function. Such a function f is said to be **one-to-one**. This property can be detected by the **horizontal-line test**: A function is one-to-one if no horizontal line intersects the graph of the function more than once.

Let f be a one-to-one function. Then f^{-1} also is a function. It follows from the definition of f^{-1} that:

$$y = f^{-1}(x) \text{ if and only if } x = f(y)$$

From this we have:

$$f(f^{-1}(x)) = x \text{ for all } x \text{ in the domain of } f^{-1}.$$
$$f^{-1}(f(x)) = x \text{ for all } x \text{ in the domain of } f.$$

These equations can be used to check certain results.

Example 3 If $f(x) = \dfrac{2x + 3}{x - 1}$, find $f^{-1}(x)$ and check the result.

Solution Let $y = f^{-1}(x)$. Then $x = f(y)$.

That is, $x = \dfrac{2y + 3}{y - 1}$.

Solve for y: $x(y - 1) = 2y + 3$

(Solution continued next page.)

$$xy - x = 2y + 3$$
$$xy - 2y = x + 3$$
$$(x - 2)y = x + 3$$
$$y = \frac{x + 3}{x - 2}$$

Therefore: $\qquad f^{-1}(x) = \dfrac{x + 3}{x - 2}$

$$Check: f(f^{-1}(x)) = f\left(\frac{x + 3}{x - 2}\right) = \frac{2\left(\dfrac{x + 3}{x - 2}\right) + 3}{\dfrac{x + 3}{x - 2} - 1}$$

$$= \frac{2(x + 3) + 3(x - 2)}{(x + 3) - (x - 2)} = \frac{5x}{5} = x \quad \checkmark$$

Showing that $f^{-1}(f(x)) = x$ is left for you. ■

EXERCISES 2-5

In Exercises 1-4, the graph of a relation R is given. (a) Find the domain and range of R. (b) Is R a function? (c) Is R^{-1} a function?

A 1. 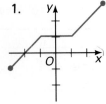 2. 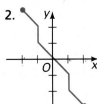 3. 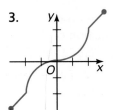 4.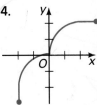

In Exercises 5-8, a relation R is given. (a) Graph R and R^{-1} in the same coordinate plane. (b) Is R a function? (c) Is R^{-1} a function?

5. $R = \{(x, y): 2x + 3y = 6\}$ 6. $R = \{(x, y): x^2 + (y - 2)^2 = 4\}$

7. $R = \{(x, y): x = 2|y|\}$ 8. $R = \{(x, y): x = y^2\}$

In Exercises 9 and 10, f and g are one-to-one functions.

9. (a) If $f(5) = 2$, find $f^{-1}(2)$. (b) Find $f(f^{-1}(5))$. (c) Find $f^{-1}(f(4))$.

10. (a) If $g^{-1}(7) = 3$, find $g(3)$. (b) Find $g^{-1}(g(2))$. (c) Find $g(g^{-1}(0))$.

11. Let f be a function. Explain why the domain of f^{-1} is the range of f and why the range of f^{-1} is the domain of f.

12. If f is a one-to-one function, need f^{-1} be one-to-one? Explain.

In Exercises 13-16, find $f^{-1}(x)$ and graph f and f^{-1} in the same coordinate plane.

13. $f(x) = 2x - 3$ 14. $f(x) = \dfrac{x + 5}{2}$ 15. $f(x) = x^3$ 16. $f(x) = \dfrac{12}{x}$

In Exercises 17-22, find $f^{-1}(x)$ and check the result.

17. $f(x) = \dfrac{x - 2}{x}$

18. $f(x) = \dfrac{2x + 1}{x - 3}$

B 19. $f(x) = \sqrt{x - 1}$

20. $f(x) = \sqrt{4 - x}$

21. $f(x) = (8 - x)^3$

22. $f(x) = \sqrt[3]{x + 1}$

23. What do you think would happen if you applied the method used in Example 3 to a function that is not one-to-one?

24. Show that the line $y = x$ is the perpendicular bisector of the segment joining (a, b) and (b, a). (Hint: One way to do this is to show that an arbitrary point on the line is equidistant from the two points.)

In Exercises 25-28, f is the linear fractional function defined by $f(x) = \dfrac{ax + b}{cx + d}$ where $ad - bc \neq 0$.

C 25. Find $f^{-1}(x)$.   26. Verify that $f(f^{-1}(x)) = x$ and $f^{-1}(f(x)) = x$.

27. (a) Find $f(f(x))$. (b) Find a condition on a, b, c, and d to make f be its own inverse.

28. Show that f is a constant function if $ad - bc = 0$.

2-6   ## Inverse Trigonometric Functions

Objective: To define, evaluate, and use inverse trigonometric functions.

The function $\cos x$ is not one-to-one. For example, $\cos x = \dfrac{1}{2}$ if x has any of the values $\dfrac{\pi}{3}$, $-\dfrac{\pi}{3}$, or any number differing from one of these by an integral multiple of 2π. (See Figure 2-25).

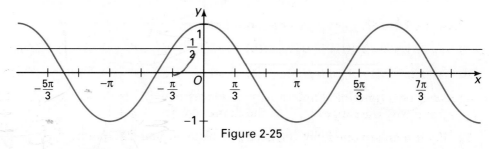

Figure 2-25

Therefore, $\cos^{-1} x$, the inverse of $\cos x$, is a relation that is not a function.

Since we prefer to work with functions, we restrict the domain of $\cos x$ to the interval $0 \le x \le \pi$ and call the resulting function $\text{Cos } x$. Its graph, shown in Figure 2-26, illustrates that $\text{Cos } x$ is one-to-one and therefore has an inverse function, $\text{Cos}^{-1} x$. Now, $y = \text{Cos}^{-1} x$ if and only if $\text{Cos } y = x$ (recall the boxed statement on page 65). Therefore, we can define the **inverse cosine** function in terms of cosine by

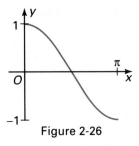

Figure 2-26

$$y = \text{Cos}^{-1} x \text{ if and only if } \cos y = x \text{ and } 0 \le y \le \pi.$$

Example 1   Evaluate: (a) $\text{Cos}^{-1} \dfrac{\sqrt{3}}{2}$ (b) $\text{Cos}^{-1}(-0.66)$

Solution   Except for notation, these are problems of the type you solved in Section 2-1. (See Example 3, page 45.)

(a) $\text{Cos}^{-1} \dfrac{\sqrt{3}}{2}$ is the number between 0 and π whose cosine is $\dfrac{\sqrt{3}}{2}$.

Hence, $\text{Cos}^{-1} \dfrac{\sqrt{3}}{2} = \dfrac{\pi}{6}$. (Recall Table 2-1, page 44.)

(b) $\text{Cos}^{-1}(-0.66)$ is the number between 0 and π whose cosine is -0.66. To find it, we can use either a calculator or Table 3.

From a calculator set in the RAD mode, $\text{Cos}^{-1}(-0.66) = 2.2916.$

From Table 3, $\text{Cos}^{-1} 0.66 = 0.85.$ Hence, $\text{Cos}^{-1}(-0.66) = \pi - 0.85 = 3.14 - 0.85 = 2.29.$ ■

To obtain the graph of $y = \text{Cos}^{-1} x$, shown in Figure 2-27, reflect the graph of $y = \text{Cos } x$ in the line $y = x$.

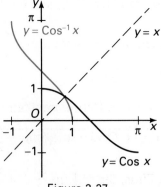

Figure 2-27

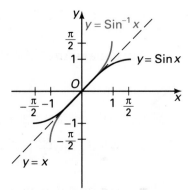

Figure 2-28

In defining the **inverse sine** function, we proceed as before except that to obtain a one-to-one function Sin x we restrict the domain of sin x to the interval $-\frac{\pi}{2} \leq x \leq \frac{\pi}{2}$. Therefore,

$$y = \text{Sin}^{-1} x \text{ if and only if } \sin y = x \text{ and } -\frac{\pi}{2} \leq y \leq \frac{\pi}{2}.$$

Figure 2-28 shows the graphs of $y = \text{Sin } x$ and $y = \text{Sin}^{-1} x$.

The **inverse tangent** and **inverse cotangent** functions are defined as follows.

$$y = \text{Tan}^{-1} x \text{ if and only if } \tan y = x \text{ and } -\frac{\pi}{2} < y < \frac{\pi}{2}.$$
$$y = \text{Cot}^{-1} x \text{ if and only if } \cot y = x \text{ and } 0 < y < \pi.$$

The graphs of $y = \text{Tan}^{-1} x$ and $y = \text{Cot}^{-1} x$ are shown in Figure 2-29.

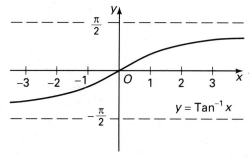
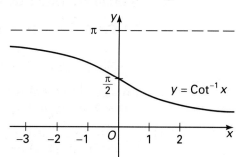

Figure 2-29

The functions $\text{Sec}^{-1} x$ and $\text{Csc}^{-1} x$ are discussed on page 73.

Example 2 Evaluate (a) $\text{Tan}^{-1}\left(-\sqrt{3}\right)$, (b) $\tan\left(\text{Tan}^{-1} 2\right)$, and (c) $\text{Sin}^{-1}\left(\sin \frac{2\pi}{3}\right)$.

Solution (a) Let $y = \text{Tan}^{-1}\left(-\sqrt{3}\right)$. Then, $\tan y = -\sqrt{3}$ and $-\frac{\pi}{2} < y < \frac{\pi}{2}$.

Hence, $y = -\frac{\pi}{3}$, and $\text{Tan}^{-1}\left(-\sqrt{3}\right) = -\frac{\pi}{3}$.

(b) Let $v = \text{Tan}^{-1} 2$. Then, $\tan v = 2$. Hence, $\tan\left(\text{Tan}^{-1} 2\right) = 2$.

(c) $\text{Sin}^{-1}\left(\sin \frac{2\pi}{3}\right) = \text{Sin}^{-1}\left(\frac{\sqrt{3}}{2}\right)$. Find y such that $-\frac{\pi}{2} \leq y \leq \frac{\pi}{2}$

and such that $\sin y = \frac{\sqrt{3}}{2}$. Thus, $\text{Sin}^{-1}\left(\sin \frac{2\pi}{3}\right) = \frac{\pi}{3}$. ∎

It sometimes is convenient to think of values of inverse trigonometric functions as measures of angles. For example, you can think of $\text{Tan}^{-1} x$ as (the radian measure of) the angle whose tangent is x (and which lies between $-\dfrac{\pi}{2}$ and $\dfrac{\pi}{2}$). Example 2(b) would then become "Find the tangent of the angle whose tangent is 2."

Example 3   Evaluate $\cos\left(\text{Sin}^{-1}\left(-\dfrac{2}{3}\right)\right)$.

Solution   Make a sketch showing $\theta = \text{Sin}^{-1}\left(-\dfrac{2}{3}\right)$ as the angle between 0

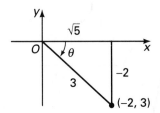

and $-\dfrac{\pi}{2}$ whose sine is $-\dfrac{2}{3}$. Read from the sketch the fact that

$$\cos\left(\text{Sin}^{-1}\left(-\dfrac{2}{3}\right)\right) = \cos\theta = \dfrac{\sqrt{5}}{3}. \quad \blacksquare$$

In some applications, values of inverse functions are given in degrees. For example, one might give the value of $\text{Cos}^{-1}\left(-\dfrac{1}{2}\right)$ as 120°.

Note: Some writers prefer to work with the relation $\cos^{-1} x$ (called a **multiple-valued function**) and call $\text{Cos}^{-1} x$ the **principal value** of $\cos^{-1} x$. Other writers use the notations $\cos^{-1} x$ or arccos x (read "*arc cosine*") for what we have called $\text{Cos}^{-1} x$.

Example 4   A lightship L, 5 mi from port P, a buoy B, 4 mi from port P, and port P itself form a right triangle as shown. A ship S initially 10 mi from port, sails at 12 mi/h along the line that contains B and P.
(a) Express $\theta = \angle PLS$ as a function of time.
(b) Find θ after the ship has sailed 4.8 mi.

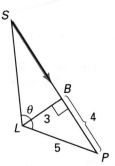

Solution   (a) From the diagram and right triangle trigonometry, $BL = 3$ and $\angle PLB = 53.1°$. Furthermore, after t hours, the ship has sailed $12t$ mi. Now there are two cases to consider. When $t = \dfrac{1}{2}$, the ship reaches the buoy (since initially the ship is 6 mi from the buoy). When $t = \dfrac{5}{6}$, the ship reaches port.

Case 1 The buoy is between the ship and port when

$$0 \le t \le \frac{1}{2}, \theta = 53.1° + \angle BLS, \text{ and } SB = 6 - 12t.$$

Case 2 The ship is between the buoy and port when

$$\frac{1}{2} \le t \le \frac{5}{6}, \theta = 53.1° - \angle BLS, \text{ and } SB = 12t - 6.$$

Hence:

$$\theta = \begin{cases} 53.1° + \text{Tan}^{-1}\left(\dfrac{6 - 12t}{3}\right) = 53.1° + \text{Tan}^{-1}(2 - 4t) & \text{if } 0 \le t \le \dfrac{1}{2} \\ 53.1° - \text{Tan}^{-1}\left(\dfrac{12t - 6}{3}\right) = 53.1° - \text{Tan}^{-1}(4t - 2) & \text{if } \dfrac{1}{2} \le t \le \dfrac{5}{6} \end{cases}$$

Notice that the two formulas agree when $t = \dfrac{1}{2}$.

(b) The ship sails 4.8 mi in $\dfrac{4.8}{12}$ hours. Since $\dfrac{4.8}{12} < \dfrac{1}{2}$, use the first formula for θ.

$$\theta = 53.1° + \text{Tan}^{-1}(2 - 4t) = 53.1° + \text{Tan}^{-1}(0.4)$$
$$= 53.1° + 21.8° = 74.9° \quad \blacksquare$$

The function for θ in Example 4 is called a **piecewise function**, a function defined by different formulas on different intervals of the domain.

EXERCISES 2-6

Evaluate. Give exact values.

A 1. $\text{Cos}^{-1} 0$ 2. $\text{Cos}^{-1}(-1)$ 3. $\text{Cos}^{-1} 1$

4. $\text{Sin}^{-1} 0$ 5. $\text{Sin}^{-1} 1$ 6. $\text{Sin}^{-1}(-1)$

7. $\text{Tan}^{-1} 0$ 8. $\text{Tan}^{-1} 1$ 9. $\text{Cot}^{-1}(-1)$

10. $\text{Sin}^{-1} \dfrac{1}{2}$ 11. $\text{Cos}^{-1} \dfrac{\sqrt{3}}{2}$ 12. $\text{Cot}^{-1}(-\sqrt{3})$

13. $\text{Tan}^{-1} \dfrac{\sqrt{3}}{3}$ 14. $\text{Cos}^{-1}\left(-\dfrac{1}{2}\right)$ 15. $\text{Sin}^{-1}\left(-\dfrac{\sqrt{3}}{2}\right)$

Evaluate. Give answers to four decimal places if you use a calculator, to two decimal places if you use Table 3.

16. $\text{Sin}^{-1} 0.621$ 17. $\text{Cos}^{-1} 0.219$ 18. $\text{Tan}^{-1} 1.21$

19. $\text{Sin}^{-1}\, 0.214$

20. $\text{Cos}^{-1}\, (-0.170)$

21. $\text{Tan}^{-1}\, (-2.650)$

22. $\text{Sin}^{-1}\, (-0.932)$

23. $\text{Cot}^{-1}\, 0.235$

24. $\text{Cot}^{-1}\, (-1.560)$

Evaluate. Give exact answers.

25. $\cos\, (\text{Cos}^{-1}\, 0.4)$

26. $\sin\, (\text{Sin}^{-1}\, (-0.7))$

27. $\cot\, (\text{Cot}^{-1}\, 6.0)$

28. $\sin\, \left(\text{Cos}^{-1}\, \dfrac{4}{5} \right)$

29. $\cos\, \left(\text{Sin}^{-1}\, \dfrac{3}{5} \right)$

30. $\tan\, (\text{Cot}^{-1}\, 2)$

31. $\cot\, \left(\text{Tan}^{-1}\, \dfrac{1}{3} \right)$

32. $\cos\, \left(\text{Sin}^{-1}\, \left(-\dfrac{1}{3} \right) \right)$

33. $\sin\, \left(\text{Cos}^{-1}\, \left(-\dfrac{3}{4} \right) \right)$

34. $\cot\, (\text{Tan}^{-1}\, (-3))$

35. State the domain and range of (a) $\text{Cos}^{-1}\, x$ and (b) $\text{Sin}^{-1}\, x$.

Evaluate. Give exact answers.

36. $\text{Cos}^{-1}\, \left(\cos \dfrac{\pi}{3} \right)$

37. $\text{Sin}^{-1}\, \left(\sin \dfrac{3\pi}{4} \right)$

38. $\text{Tan}^{-1}\, \left(\tan \dfrac{\pi}{4} \right)$

39. $\text{Cos}^{-1}\, \left(\cos \left(-\dfrac{3\pi}{2} \right) \right)$

40. State the domain and range of (a) $\text{Tan}^{-1}\, x$ and (b) $\text{Cot}^{-1}\, x$.

Express each of the following without using trigonometric or inverse trigonometric functions.

Example $\tan\, (\text{Cos}^{-1}\, u)$

Solution   Let $v = \text{Cos}^{-1}\, u$. Then $\cos v = u$. Now use $\sin^2 v + \cos^2 v = 1$:
$\sin v = \pm\sqrt{1 - \cos^2 v} = \sqrt{1 - u^2}$. (Use the $+$ sign, because $0 \le v \le \pi$ and therefore $\sin v > 0$.)

Therefore, $\tan\, (\text{Cos}^{-1}\, u) = \tan v = \dfrac{\sin v}{\cos v} = \dfrac{\sqrt{1 - u^2}}{u}$.

(Problems of this kind can also be done by using a sketch.)   ∎

B 41. $\sin\, (\text{Cos}^{-1}\, u)$ 42. $\cos\, (\text{Sin}^{-1}\, u)$ 43. $\cot\, (\text{Tan}^{-1}\, u)$

44. $\sin\, (\text{Tan}^{-1}\, u)$ 45. $\cos\, (\text{Tan}^{-1}\, u)$ 46. $\tan\, (\text{Sin}^{-1}\, u)$

47. Refer to Example 4. (a) Find θ after the ship has sailed 2.1 mi, 4 mi, and 7.5 mi. (b) After how much time (to the nearest tenth of an hour) will θ be $10°$?

48. An observer is standing 110 m from the spot where a hot-air balloon begins its ascent. If the balloon ascends vertically at the rate of 40 m/min, (a) express the balloon's angle of elevation as a function of time, (b) find the angle of elevation after 4 min, 5 min, and 6 min, and (c) find how many minutes it will take for the angle of elevation to be 60°.

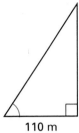

110 m

49. An observer spots a hot-air balloon when it is 1000 m away from him and has an altitude of 420 m. The balloon descends vertically at the rate of 1.5 m/s. (a) Express the balloon's angle of elevation as a function of time. (b) After how many seconds will that angle be 20°? (c) Find the angle of elevation after 4 s, 12 s, 45 s. (d) Find the angle of elevation after 5 min. Explain.

50. A camera at C, 40 ft from a road, films an automobile at A traveling at 30.8 ft/s. Initially, the automobile is 500 ft down the road from B.
(a) Express the angle θ between $\overline{AC}$ and $\overline{BC}$ as a function of time.
(b) What is θ after 0 s, 1 s, 2 s, and 4 s?   (c) What will θ be 2 s after the automobile has passed point B?

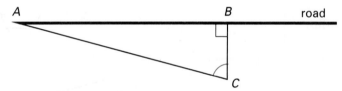

51. An observer standing on a 300 m tall cliff watches a hot-air balloon begin its ascent from a point 160 m from the base of the cliff. The balloon ascends at the rate of 44 m/min.   (a) Express the balloon's angle of depression and subsequent angle of elevation as a function of time.   (b) At what angle will the observer spot the balloon after 21 min, 1h?

We define the inverse secant and inverse cosecant functions for $|x| \geq 1$ by:

$$\text{Sec}^{-1} x = \text{Cos}^{-1}\left(\frac{1}{x}\right) \text{ and } \text{Csc}^{-1} x = \text{Sin}^{-1}\left(\frac{1}{x}\right)$$

52. Justify the definition of $\text{Sec}^{-1} x$ by giving a reason for each step below.

$$y = \text{Cos}^{-1}\left(\frac{1}{x}\right)$$

$$\cos y = \frac{1}{x}$$

$$\sec y = x$$

$$y = \text{Sec}^{-1}x$$

53. Justify the definition of $\text{Csc}^{-1} x$ as was done in Exercise 52.

C 54. Graph $y = \text{Sec}^{-1} x$. (Note that $0 < y < \pi$.)

55. Graph $y = \text{Csc}^{-1} x$. $\left(\text{Note that } -\frac{\pi}{2} < y < \frac{\pi}{2}. \right)$

56. Show that if $0 \le x \le 1$, then $\text{Sin}^{-1} x + \text{Cos}^{-1} x = \frac{\pi}{2}$. (Hint: Let $\theta = \text{Sin}^{-1} x$ and $\phi = \text{Cos}^{-1} x$. Then $\sin \theta = x = \cos \phi$. Now use properties of cofunctions (page 9).)

57. Show that if $x > 0$, then $\text{Tan}^{-1} x + \text{Cot}^{-1} x = \frac{\pi}{2}$. (See the hint for Exercise 56.)

1. Show that $\text{Sin}^{-1} x = \text{Tan}^{-1} \left[\dfrac{x}{\sqrt{1 - x^2}} \right]$, $-1 < x < 1$.

(Hint: Use the fact that $x = \sin t$ for some t, $-\dfrac{\pi}{2} < t < \dfrac{\pi}{2}$, and $\sin^2 t + \cos^2 t = 1$.)

2. Draw the graph of $y = \text{Tan}^{-1} (\tan x)$ for $0 \le x \le 2\pi$.

Self Quiz

2-4 | 2-5 | 2-6

1. Let $R = \{(x, y): x + 2y = 4\}$. (a) Graph R and R^{-1} in the same coordinate plane. (b) Is R a function? Is R^{-1} a function?

2. Evaluate $\sin \left(\text{Cos}^{-1} \left(-\dfrac{2}{3} \right) \right)$.

3. Is f even, odd, or neither? (a) $f(x) = x \tan x$ (b) $f(x) = x(\cos x - \sin x)$.

4. Show that $f(x) = \tan \pi x$ has period 1 and sketch one cycle of its graph.

5. Express $\tan (\text{Cos}^{-1} u)$ without using trigonometric or inverse trigonometric functions.

6. Find $f^{-1}(x)$ if $f(x) = \dfrac{x + 3}{2x}$. Check your result.

ADDITIONAL PROBLEMS

In Exercises 1-4, give exact values.

1. Find $\sin t$, $\cos t$, and $\tan t$ if:

   (a) $t = \dfrac{5\pi}{6}$ (b) $t = \dfrac{7\pi}{4}$ (c) $t = 7\pi$

2. Find $\tan t$ if $\cos t = -\dfrac{3}{5}$ and $0 \le t \le \pi$.

3. Find: (a) $\text{Cos}^{-1}\left(-\dfrac{1}{2}\right)$ (b) $\text{Tan}^{-1}\sqrt{3}$ (c) $\text{Sin}^{-1}\left(-\dfrac{\sqrt{2}}{2}\right)$

4. Find: (a) $\text{Sin}^{-1}(1)$ (b) $\text{Cos}^{-1}(1)$ (c) $\text{Tan}^{-1}(1)$
   (d) $\text{Sin}^{-1} 0$ (e) $\text{Cos}^{-1} 0$ (f) $\text{Tan}^{-1} 0$
   (g) $\text{Sin}^{-1}(-1)$ (h) $\text{Cos}^{-1}(-1)$ (i) $\text{Tan}^{-1}(-1)$

In Exercises 5-7 use a calculator or Table 3. If you use a calculator, give your answer to four significant digits.

5. Find: (a) $\cos 2.1$ (b) $\tan 0.65$ (c) $\csc 1.95$

6. Find t if $\sin t = 0.54$ and $|t| < \dfrac{\pi}{2}$.

7. Find: (a) $\text{Sin}^{-1} 0.25$ (b) $\text{Cos}^{-1}(-0.56)$ (c) $\text{Tan}^{-1}(-1.55)$

8. The graph of a periodic function g is shown at the left below.
   (a) What is the period of g?   (b) Is g even, odd, or neither?

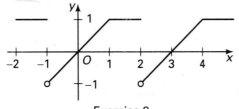
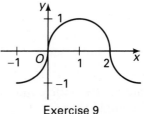

             Exercise 8   Exercise 9

9. One cycle of the graph of a periodic function f is shown at the right above. (a) Graph f in the interval $-4 \le x \le 8$. (b) Is f even, odd, or neither?

10. A particle started at the point $(5, 0)$ and moved 12 units counter-clockwise around the circle $x^2 + y^2 = 25$. What point did it reach? (Use a calculator or Table 3.)

11. Pulleys 10 cm and 24 cm in diameter are connected by a belt. The larger pulley has angular speed 300 rpm. Find the linear speed of the belt and the angular speed of the smaller pulley.

12. Graph $y = \sqrt{1 - \cos^2 x}$ over $-\pi \leq x \leq 2\pi$. (Hint: $\sqrt{a^2} = |a|$.)

13. Show that $f(x) = \sin x \cos x$ has period π.

14. Determine whether the function g is even, odd, or neither.
    (a) $g(x) = |x| \tan x$ \qquad (b) $g(x) = x^3 \sin x$

15. Graph the relations $R = \{(x, y): y = 1 + |x|\}$ and R^{-1} in the same coordinate plane. Which, if either, is a function?

16. Find $f^{-1}(x)$ if $f(x) = \sqrt[3]{x - 8}$. Check your result.

17. Find the exact value of $\tan \left(\text{Cos}^{-1} \dfrac{2}{3} \right)$.

18. Express $\cot (\text{Sin}^{-1} u)$ without using trigonometric or inverse trigonometric functions.

19. Let $f(x) = mx + b$, where $m \neq 0$. (a) Find $f^{-1}(x)$. (b) If $f^{-1} = f$, what form does f have?

20. A camera in a blimp begins to film a football game when the blimp has an altitude of 1000 ft and is a horizontal distance of 4000 ft from the field. The blimp approaches the field at the rate of 50 ft/s and, at the same time, descends at the rate of 12 ft/s. (a) Express the field's angle of depression as a function of time. (b) Find the angle of depression after 30 s, 45 s, and 1 min.

CHAPTER SUMMARY

1. For many applications it is necessary to have the sine, cosine, tangent, cotangent, secant, and cosecant defined for real numbers. When this is done, they are called circular functions.

2. The coordinates of P on the circle $x^2 + y^2 = r^2$ are $(r \cos \theta, r \sin \theta)$, where θ is the angle that $\overrightarrow{OP}$ makes with the positive x-axis. If P is moving with constant linear speed v and angular speed ω, then $v = r\omega$.

3. Properties of periodicity and symmetry are useful in graphing the trigonometric functions. The period of a function f is the smallest positive number p such that $f(x + p) = f(x)$ for all x in the domain of f. For $\sin x$, $\cos x$, $\csc x$, and $\sec x$, $p = 2\pi$; for $\tan x$ and $\cot x$, $p = \pi$. The functions $\cos x$ and $\sec x$ are even—they satisfy $f(-x) = f(x)$—and their graphs are symmetric with respect to the y-axis. The other four functions are odd, they satisfy $f(-x) = -f(x)$, and their graphs are symmetric with respect to the origin.

4. A function f is one-to-one if $f(x_1) \neq f(x_2)$ whenever $x_1 \neq x_2$. The inverse of a one-to-one function f is the function f^{-1} such that

$$y = f^{-1}(x) \quad \text{if and only if} \quad f(y) = x.$$

The graph of $y = f^{-1}(x)$ is obtained by reflecting the graph of $y = f(x)$ in the line $y = x$.

5. By restricting the domain of $\cos x$ to the interval $0 \leq x \leq \pi$, a one-to-one function $\text{Cos } x$ is obtained. The inverse cosine function is then defined by

$$y = \text{Cos}^{-1}x \quad \text{if and only if} \quad \cos y = x \text{ and } 0 \leq y \leq \pi.$$

Similarly:

$$y = \text{Sin}^{-1}x \quad \text{if and only if} \quad \sin y = x \text{ and } -\frac{\pi}{2} \leq y \leq \frac{\pi}{2}$$

$$y = \text{Tan}^{-1}x \quad \text{if and only if} \quad \tan y = x \text{ and } -\frac{\pi}{2} < y < \frac{\pi}{2}$$

$$y = \text{Cot}^{-1}x \quad \text{if and only if} \quad \cot y = x \text{ and } 0 < y < \pi$$

The functions $\text{Sec}^{-1} x$ and $\text{Csc}^{-1} x$ are seldom used.

CHAPTER TEST

2-1　**1.** Find the exact values of $\sin t$, $\cos t$, and $\tan t$ if:

(a) $t = \dfrac{5\pi}{4}$　　(b) $t = \dfrac{2\pi}{3}$　　(c) $\dfrac{11\pi}{6}$

2. Use $\cos^2 t + \sin^2 t = 1$ to find $\cos t$ and $\tan t$ if $\sin t = -\dfrac{1}{3}$ and $-1.5 < t < 0$.

2-2　**3.** Find the speed in km/h of a point on a rotating 45 rpm record 8.00 cm from its center.

2-3　**4.** Determine whether f is even, odd, or neither.

(a) $f(x) = \dfrac{x}{\sqrt{1 - x^2}}$

(b) $f(x) = x^3 \tan x$

5. Find the vertical and horizontal asymptotes of the graph of $y = \dfrac{x - 3}{2x - 1}$.

2-4　**6.** Show that the period of $g(x) = \sin 4\pi x$ is $\dfrac{1}{2}$.

7. Graph over the interval $-2\pi < x < 2\pi$.
   (a) $y = \csc x$ (b) $y = |\csc x|$ (c) $y = \csc |x|$

2-5

8. Graph the relations $R = \{(x, y): y^2 = x - 1\}$ and R^{-1} in the same coordinate plane. Which, if either, is a function?

9. Find $f^{-1}(x)$ given that $f(x) = \dfrac{x}{2x + 3}$. Check your result.

2-6

10. Find the exact value of $\tan\left(\operatorname{Cos}^{-1}\left(-\dfrac{1}{4}\right)\right)$.

11. Express $\tan(\operatorname{Cot}^{-1} u)$ without using trigonometric or inverse trigonometric functions.

Analyzing Models

The figure that accompanies Example 4 of Section 2-6 (page 70) is a *pictorial model* of the problem stated there. By looking at the figure, we can imagine ∠ PLS changing as the ship sails past the buoy. The equation

$$\theta = 53.1° + \text{Tan}^{-1}(2 - 4t) \qquad 0 \le t \le 0.83$$

is a *trigonometric model* that can be used to study ∠ PLS as it changes. The following computer-generated table also models θ as time passes.

t	θ	t	θ
.00	116.72	.45	65.04
.05	114.34	.50	54.20
.10	111.54	.55	41.54
.15	107.59	.60	31.48
.20	103.48	.65	22.66
.25	98.51	.70	15.19
.30	92.50	.75	8.06
.35	83.91	.80	3.09
.40	75.09	.83	.27

EXERCISES

1. (a) From the figure on page 70, does it seem that θ decreases over time?
   (b) Is your answer to Exercise 1 confirmed by the table?
2. (a) Approximate θ for $t = .375$ by averaging the values of θ when $t = .35$ and $t = .40$. (b) Use a calculator to evaluate the function above for $t = .375$. (Give θ to two decimal places.) (c) Find the absolute value of the difference between the answers to (a) and (b).
3. If the mathematical model for the problem is correct, then the value of θ for $t = .375$ should be less than the value of θ for $t = .35$. Is it?

Sinusoidal Data

Often a set of data suggests that a certain phenomenon is periodic in nature and that the variation of a particular quantity is sinusoidal in relation to the variation of another quantity.

One example of observed sinusoidal variation is the relationship between the numbers of hours of daylight in each day and the time of year. The following table gives such data for the area near Boston, Massachusetts for the calendar year 1988. The data is found in *Albert's Almanac* by Richard Albert (Needham, Mass. WCVB-TV, 1987).

t	h	t	h	t	h	t	h
7	9.23	98	12.98	189	15.15	280	11.52
14	9.37	105	13.32	196	15.00	287	11.20
21	9.55	112	13.62	203	14.82	294	10.88
28	9.80	119	13.98	210	14.60	301	10.58
35	10.08	126	14.20	217	14.37	308	10.27
42	10.35	133	14.50	224	14.07	315	9.97
49	10.67	140	14.75	231	13.78	322	9.72
56	11.00	147	14.93	238	13.45	329	9.48
63	11.32	154	15.10	245	13.12	336	9.30
70	11.68	161	15.22	252	12.82	343	9.20
77	12.00	168	15.28	259	12.50	350	9.10
84	12.33	175	15.28	266	12.17	357	9.08
91	12.67	182	15.27	273	11.85	364	9.10

The graph of this data is shown below. The horizontal axis gives the day of the year and the vertical axis gives the number of hours of daylight on that day.

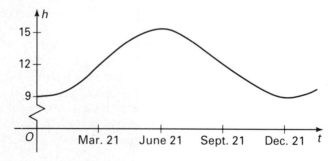

Note that it does not pass through (0, 0) or (0, 1) (as do the sine and cosine graphs, respectively), and its central horizontal axis is not the *x*-axis, but its shape resembles a sine curve.

This is a concrete example of a function whose domain consists of real numbers rather than angle measures.

EXERCISES

1. Approximate the range of values of *h*, the number of hours of daylight in a day.
2. Using the table and the result of Exercise 1, write an equation for the central axis of the graph.
3. What is the period of the function whose graph is given on the preceding page?
4. Describe how the number of hours of daylight changes from March 21 to September 21.
5. Approximately how many hours of daylight are there on the 100th day of the year?
6. Approximately how many hours of daylight can we expect on April 11, 1991? November 25, 1991?

CUMULATIVE REVIEW Chapters 1-2

Chapter 1

1. Give the degree measure of an angle between 0° and 360° that is coterminal with an angle of (a) −112°30′ and (b) 505.25°.

2. Find sin θ and cos θ to four decimal places.

3. Suppose θ is an acute angle in standard position. If the terminal side of θ passes through $(\sqrt{3}, 3)$, find θ without using tables or a calculator.

4. Use tables or a calculator to find each of the following to four significant digits.
   (a) sin 13.5° (b) cos 62°20′
   (c) θ if cos θ = 0.8910 (d) θ if sin θ = 0.9436

5. Express csc 30° · cos 45° + tan 60° in simplest radical form.

6. Find all of the missing sides and angles in the figure if θ = 32°30′, AC = 170, and BD = 120.

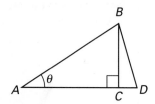

7. Two fire towers, A and B, are 10 miles apart. Rangers in them spot a fire at C. If $\angle CAB$ = 42° and $\angle CBA$ = 68°, how far is the fire from tower A and how far is the fire from tower B? (Hint: Draw the altitude $\overline{CD}$ from C to $\overline{AB}$.)

8. The terminal side of an angle in standard position passes through $(-3, \sqrt{7})$. Find all six trigonometric functions of the angle.

9. (a) In which quadrants is the secant positive?
   (b) In which quadrants is the cotangent negative?

10. List all undefined trigonometric functions for specific angles θ, 0° ≤ θ ≤ 180°. (You should have six answers.)

11. Find tan θ if cos θ = −0.4281 and 360° < θ < 540°.

12. Convert (a) 0°, (b) 45°, and (c) 280° to radian measure.

13. Convert (a) $\dfrac{3\pi}{2}$, (b) $\dfrac{7\pi}{6}$, and (c) 3π to degree measure.

14. Find the length of an arc of a circle of radius 25 cm that is intercepted by a central angle of 105°.

Chapter 2

1. Find each of the following to two significant digits.
   (a) tan 2.8   (b) sin (-1.7)

2. When a car travels at 88 km/h, a point on the edge of its 28 cm radius wheel has a linear speed of 88 km/h (or 2444 cm/s). Find its angular speed.

3. (a) What is the angular speed of the second hand of a watch?
   (b) If the second hand is 1 cm long, how far does the tip travel in 1 minute?

4. Part of the graph of a function f is given. Graph f in the interval $-5 \le x \le 5$ if f is (a) even and (b) odd.

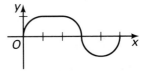

5. Determine whether each function is even, odd, or neither.
   (a) $f(x) = -x^5 + 3x^3 - x$ (b) $f(x) = \tan x \sin x$

6. Given the function $f(x) = |\tan x|$, state whether it is even or odd, and sketch the graph of the function for $-2\pi \le x \le 2\pi$.

7. Graph $y = \sin\left(x + \dfrac{\pi}{2}\right)$, $0 \le x \le 2\pi$.

8. Graph $y = |\sec x|$, $-\pi \le x \le \pi$.

9. Evaluate: (a) $\text{Tan}^{-1}\left(\dfrac{\sqrt{3}}{3}\right)$ (b) $\cot\left(\text{Sin}^{-1} 0.6\right)$

   (c) $\text{Cos}^{-1}\left(\cos\dfrac{\pi}{3}\right)$ (d) $\text{Sin}^{-1}\left(\sin\left(-\dfrac{7}{6}\pi\right)\right)$

10. State the domain and range of the function $y = \text{Tan}^{-1} x$ and draw its graph.

11. A ship that stands 42 ft above its water line approaches a drawbridge. The bridge stands 18 ft above the water and has draw spans 66 ft long.
    (a) What is the minimum angle through which the draw span must move to enable the ship to pass under it?
    (b) If the span is raised at 0.75°/s, how long will it take to reach the angle found in part (a).

3 Properties of Trigonometric Functions

A medium such as water tends to bend light rays and create illusion. Trigonometric equations are useful in the study of light rays and optics as the Application section on pages 120 and 121 illustrates.

TRIGONOMETRIC IDENTITIES

Simplifying Trigonometric Expressions

Objective: To simplify trigonometric expressions.

In algebra it is often helpful to simplify an expression by means of a known **identity**, an equation true for all values of the variables for which both sides are defined. For example, by virtue of the identities

$$x^2 - y^2 = (x - y)(x + y), \text{ and } \frac{ab}{ac} = \frac{b}{c}, \text{ where } a \neq 0 \text{ and } c \neq 0,$$

we can write $\dfrac{1 - x^2}{1 - x} = \dfrac{(1 - x)(1 + x)}{1 - x} = 1 + x$ where $x \neq 1$.

In trigonometry we can make use of algebraic identities to simplify expressions, but we can also use trigonometric identities. For example,

$$\frac{\sin^2 x}{\sin x \cos x} = \frac{\sin x}{\cos x} = \tan x, \text{ where } \sin x \neq 0 \text{ and } \cos x \neq 0.$$

Here, we used the identity $\tan x = \dfrac{\sin x}{\cos x}$. It is listed on the next page along with some other identities that you have already seen.

$$\tan x = \frac{\sin x}{\cos x} \quad (1) \qquad \cot x = \frac{\cos x}{\sin x} \quad (2)$$

$$\sec x = \frac{1}{\cos x} \quad (3) \qquad \csc x = \frac{1}{\sin x} \quad (4)$$

$$\tan x = \frac{1}{\cot x} \quad (5) \qquad \cot x = \frac{1}{\tan x} \quad (6)$$

Notice that in identities (1)–(6) we omitted stating what values are restricted from the domains. It is customary to do this when the restrictions are clear. Identities (3)–(6) are called *reciprocal identities*.

The following identity is also one that you have seen already.

$$\sin^2 x + \cos^2 x = 1 \qquad (7)$$

Using this identity and identities (1)–(6), we can derive the following:

$$1 + \tan^2 x = \sec^2 x \qquad (8)$$
$$1 + \cot^2 x = \csc^2 x \qquad (9)$$

Identities (7)–(9) are called the **Pythagorean identities** since they are closely related to the Pythagorean theorem. (See Exercises 39 and 40.)

The following examples illustrate how identities can be used to write trigonometric expressions in terms of a single function.

Example 1 Express $\csc^2 x + \cot^2 x$ in terms of $\sin x$.

Solution Choose available identities to make substitutions. Using identities (4) and (2), we have:

$$\csc^2 x + \cot^2 x = \left(\frac{1}{\sin x}\right)^2 + \left(\frac{\cos x}{\sin x}\right)^2 = \frac{1}{\sin^2 x} + \frac{\cos^2 x}{\sin^2 x}$$

$$= \frac{1 + \cos^2 x}{\sin^2 x}$$

$$= \frac{1 + (1 - \sin^2 x)}{\sin^2 x} \text{ by (7)}$$

$$= \frac{2 - \sin^2 x}{\sin^2 x}, \text{ or } \frac{2}{\sin^2 x} - 1 \quad \blacksquare$$

Example 2 Express $1 - \sin x \cos x \tan x$ in terms of $\cos x$.

Solution By identity (1), we have:

(Solution continued next page.)

$$1 - \sin x \cos x \tan x = 1 - \sin x \cos x \frac{\sin x}{\cos x}$$

$$= 1 - \sin^2 x$$

$$= \cos^2 x \quad \text{by identity (7)} \quad \blacksquare$$

Example 3   Express $\dfrac{\tan \theta}{1 + \sec \theta} + \dfrac{1}{\tan \theta}$ in terms of a single trigonometric function.

Solution   Using $(1 + \sec \theta)(\tan \theta)$ as a common denominator and adding, we have:

$$\frac{\tan \theta}{1 + \sec \theta} + \frac{1}{\tan \theta} = \frac{\tan^2 \theta + 1 + \sec \theta}{(\tan \theta)(1 + \sec \theta)}$$

$$= \frac{\sec^2 \theta + \sec \theta}{(\tan \theta)(1 + \sec \theta)} \quad \text{by identity (8)}$$

$$= \frac{(\sec \theta)(\sec \theta + 1)}{(\tan \theta)(1 + \sec \theta)} \quad \text{by factoring the numerator}$$

$$= \frac{\sec \theta}{\tan \theta}$$

$$= \frac{\dfrac{1}{\cos \theta}}{\dfrac{\sin \theta}{\cos \theta}} \quad \text{by identities (3) and (1)}$$

$$= \frac{1}{\cos \theta} \cdot \frac{\cos \theta}{\sin \theta} = \frac{1}{\sin \theta}, \text{ or } \csc \theta \quad \blacksquare$$

EXERCISES 3-1

Express the following in terms of sin x. Give your answers in simplest form.

A 1. $\tan^2 x$ **2.** $\cot^2 x$ **3.** $\sec x \tan x$ **4.** $\dfrac{\cot x}{\sec x}$

5. $\dfrac{\csc x}{1 + \cot^2 x}$ **6.** $\dfrac{\sec^2 x - \tan^2 x}{\csc x}$ **7.** $(\cos x)(\tan x + \cot x)$ **8.** $\dfrac{1 + \cot x}{\sin x + \cos x}$

Express the following in terms of cos x. Give your answers in simplest form.

9. $1 + \tan^2 x$ **10.** $\sec^2 x - 1$

11. $(\sec x)(\cos x + \sin^2 x \sec x)$ **12.** $(1 + \cot x)(1 - \cot x)$

13. $\dfrac{\sin^2 x}{1 + \cos x}$ **14.** $(\csc^2 x)(1 - \cos x)$

15. $(\csc x)(\csc x + \cot x)$ **16.** $\dfrac{\sec x + 1}{\sin^2 x \sec x}$

Express each of the following in terms of a single trigonometric function.

17. $\dfrac{\csc x - \sin x}{\cos x}$

18. $\sec x \csc x - \tan x$

19. $\dfrac{\sin t \cos t}{1 - \cos^2 t}$

20. $\dfrac{1 + \tan^2 t}{\csc t \sec t}$

21. $\dfrac{(\tan \theta)(1 + \cot^2 \theta)}{\cot \theta}$

22. $\dfrac{\sin x + \tan x}{1 + \sec x}$

23. $(\sin x)(\cos x + \sin x \tan x)$

24. $(\csc x)(\sec x - \cos x)$

25. $(\csc x)(1 - \cos x)(1 + \cos x)$

26. $\dfrac{(\sec x - \tan x)(\sec x + \tan x)}{\cos x}$

B 27. $\dfrac{\cos \theta}{1 + \sin \theta} + \tan \theta$

28. $\dfrac{\sec \theta + \csc \theta}{1 + \tan \theta}$

29. $\dfrac{\sec y}{\sin y} - \dfrac{\sec y}{\csc y}$

30. $\dfrac{\sin x + \tan x}{(\tan x)(\csc x + \cot x)}$

31. $1 - \dfrac{\cos^2 x}{1 + \sin x}$

32. $\dfrac{(\tan \theta + 1)^2}{\sec \theta} - \sec \theta$

33. $\dfrac{\sec x - \cos x}{\sin x}$

34. $\dfrac{\cot^2 x}{\csc x + 1} + 1$

35. $\dfrac{\sec \theta - \cos \theta}{\sin^2 \theta \sec^2 \theta}$

36. $\dfrac{\tan \theta + \cot \theta}{\sec^2 \theta}$

37. $\dfrac{1 - \tan^4 x}{\cos^2 x - \sin^2 x}$

38. $\dfrac{1}{2}\left(\dfrac{\sin x}{1 - \cos x} + \dfrac{1 - \cos x}{\sin x}\right)$

Use identity (7) and any of identities (1)–(6) on page 86 to derive the following.

39. $1 + \tan^2 x = \sec^2 x$

40. $1 + \cot^2 x = \csc^2 x$

Show that each of the following expressions equals either 0 or 1.

C 41. $\dfrac{1 + \sec x}{\sec x - 1} + \dfrac{1 + \cos x}{\cos x - 1}$

42. $\dfrac{(\sec^2 x)(1 + \csc x) - (\tan x)(\sec x + \tan x)}{(\csc x)(1 + \sin x)}$

43. $\dfrac{\sec x}{1 - \cos x} - \dfrac{\sec x + 1}{\sin^2 x}$

44. $\dfrac{\tan x}{\tan x + \sin x} - \dfrac{1 - \cos x}{\sin^2 x}$

45. $\dfrac{\csc \theta}{1 + \sec \theta} - \dfrac{\cot \theta}{1 + \cos \theta}$

46. $\dfrac{\sin \theta + \cos \theta - 1}{\sin \theta - \cos \theta + 1} - \dfrac{\cos \theta}{\sin \theta + 1}$

3-2 Proving Identities

Objective: To prove trigonometric identities involving a single variable.

In this section we will use the identities presented in Section 3-1 to prove other identities. The procedures for doing this can be best illustrated by examples.

Example 1 Prove that $\sec x - \cos x = \sin x \tan x$.

Solution

$$\sec x - \cos x = \frac{1}{\cos x} - \cos x \quad \text{by identity (3)}$$

$$= \frac{1 - \cos^2 x}{\cos x} \quad \text{using subtraction}$$

$$= \frac{\sin^2 x}{\cos x} \quad \text{by identity (7)}$$

$$= \sin x \, \frac{\sin x}{\cos x} \quad \text{using algebra}$$

$$= \sin x \tan x \quad \text{by identity (1)} \quad \blacksquare$$

Notice that in Example 1 we first selected just one side of the given equation. Then we used known identities to transform it until it was shown to be equivalent to the other side of the equation. When proving an identity, we must be careful to consider each side of the given equation as an independent expression. That is, we must not assume that the two sides are equal until we prove them to be.

Proving that an equation is an identity requires practice. Sometimes it will take more than one attempt to succeed in completing the proof.

Example 2 Prove that $\dfrac{\sin x}{\csc x} + \dfrac{\cos x}{\sec x} = \sec^2 x - \tan^2 x$.

Solution Starting with the left side of the given equality, we have:

$$\frac{\sin x}{\csc x} + \frac{\cos x}{\sec x} = \frac{\sin x}{\frac{1}{\sin x}} + \frac{\cos x}{\frac{1}{\cos x}} = \sin^2 x + \cos^2 x = 1$$

Since this last expression cannot be simplified further, we work separately with the expression on the right.

$$\sec^2 x - \tan^2 x = (1 + \tan^2 x) - \tan^2 x \quad \text{by identity (8)}$$
$$= 1$$

Because both sides have independently been shown to equal the same expression, they are equal to each other. Hence,

$$\frac{\sin x}{\csc x} + \frac{\cos x}{\sec x} = \sec^2 x - \tan^2 x. \quad \blacksquare$$

The following may be helpful in proving identities:

Strategies for Proving Identities

1. If one side of the identity contains two or more rational expressions, try combining over a common denominator.
2. Try converting one side of the identity to an expression involving only the functions sin x and cos x.
3. If one side of the identity is given in factored form, try multiplying out the factored expression.
4. If one side of the identity can be factored, try factoring that side. In particular, keep in mind identities such as
$a^2 - b^2 = (a - b)(a + b)$ and $a^2 + 2ab + b^2 = (a + b)^2$.
5. Work with the more complicated side of the given equation first.
6. The presence of squared functions (for example, $\sin^2 x$) should suggest using one of the Pythagorean identities.
7. Remember that the known identities can be used in different forms. For example, the identity
$\sin^2 x + \cos^2 x = 1$ can be written $1 - \cos^2 x = \sin^2 x$.
8. As a last resort, try simplifying each side of the equation to the same expression.

The next example illustrates the use of some of these strategies.

Example 3 Prove that $1 + \tan^2 \theta + \csc^2 \theta = \sec^2 \theta \csc^2 \theta$.

Solution Begin by using strategy 5.

$$1 + \tan^2 \theta + \csc^2 \theta = \sec^2 \theta + \csc^2 \theta \quad \text{identity (8)}$$

$$= \frac{1}{\cos^2 \theta} + \frac{1}{\sin^2 \theta} \quad \text{identities (3) and (4)}$$

$$= \frac{\sin^2 \theta + \cos^2 \theta}{\cos^2 \theta \sin^2 \theta} \quad \text{strategy 1}$$

$$= \frac{1}{\cos^2 \theta \sin^2 \theta} \quad \text{identity (7)}$$

$$= \frac{1}{\cos^2 \theta} \cdot \frac{1}{\sin^2 \theta} \quad \frac{1}{ab} = \frac{1}{a} \cdot \frac{1}{b}$$

$$= \sec^2 \theta \csc^2 \theta \quad \text{identities (3) and (4)} \quad \blacksquare$$

Calculator and computer graphics can be used to help decide whether a given trigonometric equation is an identity. For example, could we prove that $\cos 2x = \cos x + \cos x$ for all values of x? The answer is NO since, when $x = 0$, $\cos 0 = 1$ and $\cos 0 + \cos 0 = 2$. Figure 3-1 illustrates a graphical solution. In that figure, the graph of $y = \cos 2x$ is shown together with the graph of $y = \cos x + \cos x$. From the figure, $\cos 0 \neq \cos 0 + \cos 0$. For almost all values of x, $\cos 2x \neq \cos x + \cos x$.

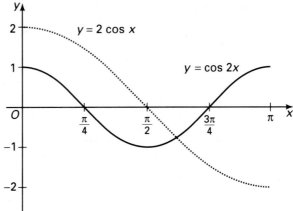

Figure 3-1

Figure 3-2 shows the graph of $y = \cos 2x$ together with that of $y = \cos^2 x - \sin^2 x$ (raised a bit for purposes of clarity). Notice that the graphs would appear to coincide for all points plotted. In Section 3-4, you will see a trigonometric proof that $\cos 2x = \cos^2 x - \sin^2 x$ is an identity.

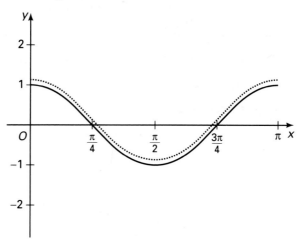

Figure 3-2

EXERCISES 3-2

Prove each identity.

A 1. $\dfrac{x^2 - 1}{x} - \dfrac{2x - 1}{2} = \dfrac{1}{2} - \dfrac{1}{x}$

2. $\dfrac{1}{x - 1} - \dfrac{1}{x + 1} = \dfrac{2}{x^2 - 1}$

3. $\dfrac{1}{1 + y} - \dfrac{y^2}{1 + y} = 1 - y$

4. $\dfrac{1 + t^2}{1 - t^2} - \dfrac{t}{1 - t} = \dfrac{1}{1 + t}$

5. $\dfrac{1}{x}(x - 1)^2 = x - 2 + \dfrac{1}{x}$

6. $\dfrac{x^2 - 2x + 1}{x^2 - 1} = \dfrac{x - 1}{x + 1}$

7. $\dfrac{x^2 + y^2}{x^4 - y^4} \cdot \left(\dfrac{1}{y} - \dfrac{1}{x}\right) = \dfrac{1}{xy(x + y)}$

8. $x(x + 1) - y(y + 1) = (x + 1 + y)(x - y)$

In Exercises 9–16, is the given equation an identity? Give an algebraic or graphical justification of your answer.

9. $\sin 2x = \sin x + \sin x$

10. $\sin (x - \pi) = -\sin x$

11. $\dfrac{\csc x}{\sin x} + \dfrac{\sec x}{\cos x} = 1$

12. $\cos \left(\dfrac{\pi}{4} + x\right) = \dfrac{\sqrt{2}}{2} \cos x + \dfrac{\sqrt{2}}{2} \sin x$

13. $\cos^2 x - \sin^2 x = 1 - 2 \sin^2 x$

14. $-\sin (-x) = \sin x$

15. $(\sin x + \cos x)^2 = 1$

16. $\sin (x + \pi) = \sin (x - \pi)$

Prove each identity.

17. $(\sec^2 x)(1 - \cos^2 x) = \tan^2 x$

18. $\dfrac{\sin^2 x + \cos^2 x}{\tan x} = \cot x$

19. $(\cos x)(\sec x - \cos x) = \sin^2 x$

20. $(\sin x)(\csc x + \sin x \sec^2 x) = \sec^2 x$

21. $\cos \theta + \sin \theta \tan \theta = \sec \theta$

22. $(\sec \theta)(\csc \theta - \cot \theta \cos \theta) = \tan \theta$

23. $\dfrac{1}{1 + \tan^2 x} + \dfrac{1}{1 + \cot^2 x} = 1$

24. $\dfrac{\sec^2 x - 1}{\csc^2 x - 1} = \tan^4 x$

25. $\sec^2 x + \csc^2 x = \sec^2 x \csc^2 x$

26. $\csc t - \sin t = \cos t \cot t$

27. $(\tan x + \sin x)(1 - \cos x) = \sin^2 x \tan x$

28. $(\cot x - \cos x)(\csc x + 1) = \cos x \cot^2 x$

29. $(1 - \cos \theta)(\csc \theta + \cot \theta) = \sin \theta$

30. $(\sec \theta + 1)(\csc \theta - \cot \theta) = \tan \theta$

31. $\cos \theta + \dfrac{\sin \theta}{\cot \theta} = \dfrac{\cot \theta + \tan \theta}{\csc \theta}$

32. $\dfrac{\cos x}{1 - \sin x} - \dfrac{\cos x}{1 + \sin x} = 2 \tan x$

33. $\dfrac{1}{\sec t - 1} + \dfrac{1}{\sec t + 1} = 2 \cot t \csc t$

34. $\dfrac{(\sin x + \cos x)^2}{\sin x} = \csc x + 2 \cos x$

35. $(\cos x)(1 + \tan x)^2 = \sec x + 2 \sin x$

36. $\tan \theta + \cot \theta = \sec \theta \csc \theta$

37. $\dfrac{\sin A}{\sec A - 1} - \dfrac{\sin A}{\sec A + 1} = 2 \cos A \cot A$

B 38. $\dfrac{1 + \sec x}{\tan x} + \dfrac{\tan x}{1 + \sec x} = 2 \csc x$

39. $\dfrac{\cos x}{1 + \sin x} + \dfrac{1 + \sin x}{\cos x} = 2 \sec x$

40. $\dfrac{\sec \theta}{\sec \theta - 1} - \dfrac{\sec \theta + 1}{\tan^2 \theta} = 1$

41. $\dfrac{1}{1 - \sin x} = \dfrac{\cot x}{\cot x - \cos x}$

42. $\dfrac{\sec \theta - 1}{1 - \cos \theta} = \sec \theta$

43. $\dfrac{\csc x - 1}{\cot x} + \dfrac{\cot x}{\csc x + 1} = \dfrac{2 \cos x}{1 + \sin x}$

44. $\dfrac{\sec x - 1}{\tan x} + \dfrac{\tan x}{\sec x + 1} = \dfrac{2 \sin x}{1 + \cos x}$

45. $\dfrac{1 - \cos x}{\sin x} = \dfrac{\sin x}{1 + \cos x}$ $\left(\text{Hint: Show that } \dfrac{1 - \cos x}{\sin x} - \dfrac{\sin x}{1 + \cos x} = 0.\right)$

46. $\dfrac{\tan t}{1 + \sec t} = \dfrac{\sec t - 1}{\tan t}$ (Hint: See the hint for Exercise 45.)

47. $\dfrac{\sec x - \tan x}{\cos x} - \dfrac{\cos x}{\sec x + \tan x} = \dfrac{\sin^2 x}{1 + \sin x}$

48. $(\cot \theta + \tan \theta)^2 = \csc^2 \theta \sec^2 \theta$

49. $\dfrac{\cos \theta}{\sec \theta - \tan \theta} + \dfrac{\sin \theta \cos \theta}{\sec \theta + \tan \theta} = \cos^2 \theta + 2 \sin \theta$

50. $(\sec \theta - \cos \theta)^2 = \tan^2 \theta - \sin^2 \theta$

51. $\dfrac{\cos \theta}{1 - \tan \theta} + \dfrac{\sin \theta}{1 - \cot \theta} = \sin \theta + \cos \theta$

52. $\dfrac{\tan^2 x}{1 - \cos x} = \sec x + \sec^2 x$ **53.** $\dfrac{1 + \sin x}{1 - \sin x} = (\sec x + \tan x)^2$

C 54. $\dfrac{\sec x + \tan x}{\sec x - \tan x} = \dfrac{1 + 2 \sin x + \sin^2 x}{\cos^2 x}$

55. $(\sin x + \csc x)^2 + (\cos x - \sec x)^2 = \sec^2 x \csc^2 x + 1$

56. $(1 + \tan t)^2 + (1 + \cot t)^2 = (\sec t + \csc t)^2$

57. Use the given unit-circle diagram to give a geometric proof of the identity in Exercise 56 for an acute angle t. (Hint: Show that $AN = \tan t$, $AO = \sec t$, $CM = \cot t$, and $CO = \csc t$.)

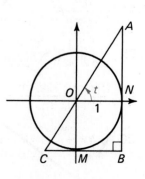

| 3-3 | **Sum and Difference Formulas for Sine and Cosine** |

Objective: To derive and apply addition formulas for sine and cosine.

If a and b are two real numbers and we know their sines and cosines, then we can find $\sin (a + b)$, $\sin (a - b)$, $\cos (a + b)$, and $\cos (a - b)$.

We start by deriving a formula for $\cos (a - b)$. To do this, we first assume that $0 \le a - b < 2\pi$. (The general result follows from this as is shown in Exercise 39 on page 98.) On the unit circle starting at (1, 0), draw arcs of lengths a and b. Let the endpoints of these arcs be P and Q, respectively. (See Figure 3-3.) The length of arc PQ is therefore $a - b$. On another unit circle starting at $A(1, 0)$, draw arc RA of length $a - b$. (See Figure 3-4.)

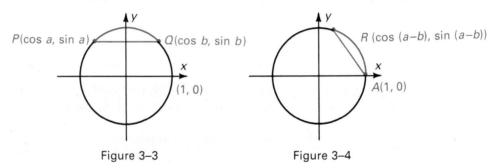

Figure 3–3   Figure 3–4

Since $\overparen{PQ}$ and $\overparen{RA}$ are equal, chords $\overline{PQ}$ and $\overline{RA}$ are equal. Hence,

$$(PQ)^2 = (RA)^2.$$

Now, if d is the distance between two points (x_1, y_1) and (x_2, y_2), then the distance formula tells us that $d^2 = (x_1 - x_2)^2 + (y_1 - y_2)^2$. Therefore,

$$
\begin{aligned}
(PQ)^2 &= (\cos a - \cos b)^2 + (\sin a - \sin b)^2 \\
&= \cos^2 a + \sin^2 a + \cos^2 b + \sin^2 b - 2 \cos a \cos b - 2 \sin a \sin b \\
&= 2 - 2(\cos a \cos b + \sin a \sin b)
\end{aligned}
$$

and

$$
\begin{aligned}
(RA)^2 &= (\cos (a - b) - 1)^2 + (\sin (a - b) - 0)^2 \\
&= \cos^2 (a - b) + \sin^2 (a - b) - 2 \cos (a - b) + 1 \\
&= 2 - 2 \cos (a - b)
\end{aligned}
$$

Since $(PQ)^2 = (RA)^2$, $2 - 2 \cos (a - b) = 2 - 2(\cos a \cos b + \sin a \sin b)$. Therefore,

$$\cos (a - b) = \cos a \cos b + \sin a \sin b \qquad (10)$$

Example 1   Find $\cos 15°$ in simplest radical form.

Solution Express 15° as a difference of two angles whose sines and cosines we know. For example, 15° = 45° − 30°. By identity (10), we have:

$$\cos 15° = \cos (45° - 30°)$$
$$= \cos 45° \cos 30° + \sin 45° \sin 30°$$
$$= \frac{\sqrt{2}}{2} \cdot \frac{\sqrt{3}}{2} + \frac{\sqrt{2}}{2} \cdot \frac{1}{2} = \frac{\sqrt{6} + \sqrt{2}}{4} \quad \blacksquare$$

We can derive a formula for cos (a + b) by using the fact that a + b = a − (−b), and applying identity (10).

$$\cos (a + b) = \cos (a - (-b)) = \cos a \cos (-b) + \sin a \sin (-b)$$

Using the fact that sine is an odd function and cosine is an even function, we obtain the formula for cos (a + b).

$$\cos (a + b) = \cos a \cos b - \sin a \sin b \qquad (11)$$

To derive the formulas for sin (a + b) and sin (a − b), we first need the following identities, which are proved in Exercises 25 and 26 on page 97.

$$\cos \left(\frac{\pi}{2} - x \right) = \sin x \qquad (12)$$
$$\sin \left(\frac{\pi}{2} - x \right) = \cos x \qquad (13)$$

To derive a formula for sin (a + b), apply identity (12).

$$\sin (a + b) = \cos \left[\frac{\pi}{2} - (a + b) \right] \qquad (12)$$
$$= \cos \left[\left(\frac{\pi}{2} - a \right) - b \right]$$
$$= \cos \left(\frac{\pi}{2} - a \right) \cos b + \sin \left(\frac{\pi}{2} - a \right) \sin b \qquad (10)$$
$$= \sin a \cos b + \cos a \sin b \qquad (12), (13)$$

Therefore,

$$\sin (a + b) = \sin a \cos b + \cos a \sin b \qquad (14)$$

The following formula for sin (a − b) can be derived using identity (14). (See Exercise 31 on page 97.)

$$\sin (a - b) = \sin a \cos b - \cos a \sin b \qquad (15)$$

Example 2   Find the exact value of $\sin \dfrac{13\pi}{12}$.

Solution   Express $\dfrac{13\pi}{12}$ as a sum of numbers whose sines and cosines we know. One possible choice is

$$\frac{13\pi}{12} = \frac{5\pi}{6} + \frac{\pi}{4}.$$

(You may be able to discover this choice more easily by converting to degree measure: $195° = 150° + 45°$.) Using identity (14), we have:

$$\sin \frac{13\pi}{12} = \sin \frac{5\pi}{6} \cos \frac{\pi}{4} + \cos \frac{5\pi}{6} \sin \frac{\pi}{4}$$

$$= \frac{1}{2} \cdot \frac{\sqrt{2}}{2} + \left(-\frac{\sqrt{3}}{2}\right) \cdot \frac{\sqrt{2}}{2} = \frac{\sqrt{2} - \sqrt{6}}{4} \quad \blacksquare$$

Example 3   Find $\sin (a + b)$ if $\sin a = \dfrac{3}{5}$, $\sin b = \dfrac{5}{13}$, $0 < a < \dfrac{\pi}{2}$, and $\dfrac{\pi}{2} < b < \pi$.

Solution   To use identity (14), we first use identity (7) to find $\cos a$ and $\cos b$.

$$\left(\frac{3}{5}\right)^2 + \cos^2 a = 1 \text{ so } \cos a = \frac{4}{5} \text{ since } 0 < a < \frac{\pi}{2}.$$

Similarly, $\cos b = -\dfrac{12}{13}$. Applying identity (14), we have

$$\sin (a + b) = \frac{3}{5}\left(-\frac{12}{13}\right) + \frac{4}{5}\left(\frac{5}{13}\right) = -\frac{16}{65} \quad \blacksquare$$

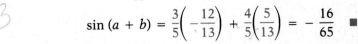

EXERCISES 3-3

Find the exact value of each trigonometric function.

A 1. $\sin 105°$ **2.** $\cos 165°$ **3.** $\sin 15°$ **4.** $\cos 345°$

5. $\cos 75°$ **6.** $\sin 165°$ **7.** $\cos (-75°)$ **8.** $\sin (-75°)$

9. $\cos \left(-\dfrac{\pi}{12}\right)$ **10.** $\sin \dfrac{5\pi}{12}$ **11.** $\cos \dfrac{13\pi}{12}$ **12.** $\cos \left(-\dfrac{5\pi}{12}\right)$

Find each of the following if $\sin r = \dfrac{4}{5}$, $\sin s = -\dfrac{12}{13}$, $0 < r < \dfrac{\pi}{2}$, and $\dfrac{3\pi}{2} < s < 2\pi$.

13. $\cos (r + s)$

14. $\sin (r + s)$

15. $\sin (r - s)$

16. $\cos (r - s)$

Find each of the following if $\cos s = -\dfrac{3}{5}$, $\cos t = -\dfrac{15}{17}$, $\dfrac{\pi}{2} < s < \pi$, and $\pi < t < \dfrac{3\pi}{2}$.

17. $\sin (s - t)$

18. $\cos (s - t)$

19. $\cos (s + t)$

20. $\sin (s + t)$

21. $\sin \left(\dfrac{\pi}{3} - s \right)$

22. $\cos \left(\dfrac{\pi}{3} + s \right)$

Prove each identity.

23. $\cos \left(\dfrac{3\pi}{2} - x \right) + \cos \left(\dfrac{3\pi}{2} + x \right) = 0$

24. $\sin (\pi + x) + \sin (\pi - x) = 0$

25. Use identity (10) on page 94 to prove that $\cos \left(\dfrac{\pi}{2} - x \right) = \sin x$.

26. Use identity (12) on page 95 to prove that $\sin \left(\dfrac{\pi}{2} - x \right) = \cos x$.

B 27. Illustrate the identity in Exercise 23 with a unit-circle diagram in which x is a number between 0 and $\dfrac{\pi}{2}$.

28. Illustrate the identity in Exercise 24 on a unit-circle diagram in which x is a number between $\dfrac{\pi}{2}$ and π.

29. Draw a unit-circle diagram illustrating angles of measure $45° - x$ and $45° + x$ for an arbitrarily chosen positive angle x of measure less than $45°$. From the diagram, guess a relationship between $\sin (45° - x)$ and $\cos (45° + x)$. Prove this relationship using the formulas of this section.

30. Repeat Exercise 29 for angles of measure $135° - x$ and $135° + x$. Does the relationship that you guessed hold for an arbitrary angle x (for example, one whose measure is not necessarily less than $45°$)?

31. Use identity (14) to prove that $\sin (a - b) = \sin a \cos b - \cos a \sin b$.

32. Prove that $\sec \left(\dfrac{\pi}{2} - x \right) = \csc x$.

Exercises 33 and 34: The identities given are useful in transforming a product of sines or cosines into a sum (and occasionally vice versa). Prove each identity.

33. $\cos a \cos b = \dfrac{1}{2}[\cos (a - b) + \cos (a + b)]$

(Hint: Use identities (10) and (11).)

34. $\sin a \sin b = \dfrac{1}{2}[\cos (a - b) - \cos (a + b)]$

35. Express $\sin a \cos b$ in terms of $\sin (a + b)$ and $\sin (a - b)$.
(Hint: Use identities (14) and (15).)

36. (a) Express $\sin b \cos a$ in terms of $\sin (a + b)$ and $\sin (a - b)$.
(b) Use the result of (a) to show that

$$\sin \frac{x}{2} \cos nx = \frac{1}{2}\left[\sin \left(\left[n + \frac{1}{2}\right]x\right) - \sin \left(\left[n - \frac{1}{2}\right]x\right)\right].$$

C 37. Prove: $\dfrac{1}{2} + \cos x + \cos 2x + \cos 3x + \cdots + \cos nx =$

$$\frac{\sin \left(\left[n + \dfrac{1}{2}\right]x\right)}{2 \sin \left(\dfrac{x}{2}\right)}, \text{ where } n \text{ is a positive integer and } x \neq 2\pi n.$$

38. Prove: $\sin x + \sin 2x + \cdots + \sin nx =$

$$\frac{\cos \left(\dfrac{x}{2}\right) - \cos \left(\left[n + \dfrac{1}{2}\right]x\right)}{2 \sin \left(\dfrac{x}{2}\right)}$$

(Hint: Use the method suggested in Exercise 37 and the result of Exercise 34.)

39. Prove identity (10) for all real numbers s and t. (Hint: There is an integer k such that $0 \le s - (t + 2k\pi) < 2\pi$. Note that $\cos (s - t) = \cos (s - t - 2k\pi)$.)

Prove each of the following identities.

40. $\dfrac{\sin (\alpha - \beta)}{\sin \beta} + \dfrac{\cos (\alpha - \beta)}{\cos \beta} = \dfrac{\sin \alpha}{\sin \beta \cos \beta}$

41. $\dfrac{\cos (\alpha + \beta)}{\sin \beta} + \dfrac{\sin (\alpha + \beta)}{\cos \beta} = \dfrac{\cos \alpha}{\sin \beta \cos \beta}$

Evaluate each expression.

42. $\sin\left(\mathrm{Cos}^{-1}\left(\dfrac{3}{5}\right) + \mathrm{Cos}^{-1}\left(\dfrac{5}{13}\right)\right)$

43. $\cos\left(\mathrm{Sin}^{-1}\left(\dfrac{15}{17}\right) - \mathrm{Sin}^{-1}\left(\dfrac{4}{5}\right)\right)$

44. $\sin\left(\mathrm{Sin}^{-1}\left(-\dfrac{3}{5}\right) - \mathrm{Cos}^{-1}\left(\dfrac{12}{13}\right)\right)$

45. $\cos\left(\mathrm{Cos}^{-1}\left(-\dfrac{5}{13}\right) + \mathrm{Cos}^{-1}\left(\dfrac{24}{25}\right)\right)$

46. $\cos\left(\mathrm{Tan}^{-1}\left(\dfrac{3}{4}\right) + \mathrm{Tan}^{-1}\left(\dfrac{7}{24}\right)\right)$

47. $\sin\left(\mathrm{Tan}^{-1}(-1) + \mathrm{Tan}^{-1}\sqrt{3}\right)$

Exercise 48 will prove identity (14) on page 95 directly for α and β such that $0° < \alpha + \beta < 90°$.

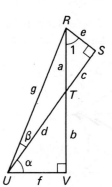

48. (a) Show that $\angle\,1 = \alpha$.
    (b) Show that $ab = cd$ by using similar triangles.
    (c) Use triangle RST to find expressions for $\sin\alpha$ and $\cos\alpha$.
    (d) Find expressions for $\sin\beta$ and $\cos\beta$.
    (e) Find an expression for $\sin(\alpha + \beta)$.
    (f) Use steps (c) and (d) to give an expression for $\sin\alpha\cos\beta + \cos\alpha\sin\beta$.
    (g) Show that the expression in step (f) is equivalent to the one in step (e). (Hint: Use step (b).)

Self Quiz

$\boxed{\textbf{3-1}}\quad\boxed{\textbf{3-2}}\quad\boxed{\textbf{3-3}}$

Write in terms of a single trigonometric function.

1. $1 - \dfrac{\cos^2 x}{1 + \sin x}$

2. $\dfrac{(\sec^2 x - 1)}{\sin^2 x} - 1$

Prove each identity.

3. $\dfrac{1 + \tan x}{\sin x + \cos x} = \sec x$

4. $\dfrac{\sin x}{1 + \cos x} + \cot x = \csc x$

If $\sin r = \dfrac{4}{5}$, $\sin s = \dfrac{5}{13}$, $0 < r < \dfrac{\pi}{2}$, and $\dfrac{\pi}{2} < s < \pi$, find:

5. $\sin(r + s)$

6. $\cos(r + s)$

7. Prove the identity: $\sin\left(\dfrac{\pi}{6} + x\right) + \sin\left(\dfrac{\pi}{6} - x\right) = \cos x$

3-4

Double- and Half-Angle Formulas for Sine and Cosine

Objective: To derive and apply double- and half-angle formulas for sine and cosine.

The trigonometric addition formulas presented in Section 3-3 can be used to derive formulas for $\sin 2x$ and $\cos 2x$.

$$\sin 2x = \sin (x + x) = \sin x \cos x + \cos x \sin x$$
$$\cos 2x = \cos (x + x) = \cos x \cos x - \sin x \sin x$$

Therefore,

$$\sin 2x = 2 \sin x \cos x \qquad (16)$$
$$\cos 2x = \cos^2 x - \sin^2 x \qquad (17)$$

Using the identity, $\sin^2 x + \cos^2 x = 1$, we can write identity (17) in two other forms.

$$\cos 2x = 1 - 2 \sin^2 x \qquad (17a)$$
$$\cos 2x = 2 \cos^2 x - 1 \qquad (17b)$$

Example 1 Prove that $\tan x = \dfrac{\sin 2x}{\cos 2x + 1}$.

Solution We start with the right side, since it is the more complicated one. Using identities (16) and (17b), we have:

$$\frac{\sin 2x}{\cos 2x + 1} = \frac{2 \sin x \cos x}{(2 \cos^2 x - 1) + 1}$$

$$= \frac{2 \sin x \cos x}{2 \cos^2 x} = \frac{\sin x}{\cos x} = \tan x \quad \blacksquare$$

The choice of identity (17), (17a), or (17b) is dictated by convenience. In Example 1, our desire to produce a denominator consisting of a single term prompted us to use identity (17b).

Example 2 Prove that $\sin 3x = 3 \sin x - 4 \sin^3 x$.

Solution Note that $\sin 3x = \sin (2x + x)$.

$$\sin 3x = \sin 2x \cos x + \cos 2x \sin x \qquad \text{by identity (14)}$$
$$= (2 \sin x \cos x)\cos x + (1 - 2 \sin^2 x) \sin x \qquad \text{by (16) and (17a)}$$
$$= 2 \sin x \cos^2 x + \sin x - 2 \sin^3 x$$

$$= (2 \sin x)(1 - \sin^2 x) + \sin x - 2 \sin^3 x \quad \text{by (7)}$$
$$= 2 \sin x - 2 \sin^3 x + \sin x - 2 \sin^3 x$$
$$\sin 3x = 3 \sin x - 4 \sin^3 x \quad \blacksquare$$

To derive a formula for $\cos \dfrac{x}{2}$, we use identity (17b) as follows.

$$\cos 2\left(\frac{x}{2}\right) = 2 \cos^2 \left(\frac{x}{2}\right) - 1$$

Thus,
$$2 \cos^2 \left(\frac{x}{2}\right) = 1 + \cos x$$

$$\cos^2 \left(\frac{x}{2}\right) = \frac{1 + \cos x}{2}$$

Therefore,

$$\cos \frac{x}{2} = \pm\sqrt{\frac{1 + \cos x}{2}} \qquad (18)$$

We can choose the proper sign once we know the quadrant of $\dfrac{x}{2}$.

Similarly, we can use identity (17a) to obtain a formula for $\sin \dfrac{x}{2}$.

$$\sin \frac{x}{2} = \pm\sqrt{\frac{1 - \cos x}{2}} \qquad (19)$$

Example 3 Find $\cos \dfrac{\theta}{2}$ if $\cos \theta = -\dfrac{31}{81}$ and $180° < \theta < 360°$.

Solution Since $180° < \theta < 360°$, then $90° < \dfrac{\theta}{2} < 180°$. Thus, $\cos \dfrac{\theta}{2}$ is negative.

$$\cos \frac{\theta}{2} = -\sqrt{\frac{1 + \cos \theta}{2}}$$

$$= -\sqrt{\frac{1 + \left(-\dfrac{31}{81}\right)}{2}} = -\sqrt{\frac{\dfrac{50}{81}}{2}} = -\sqrt{\frac{25}{81}} = -\frac{5}{9} \quad \blacksquare$$

In Example 1 of Section 3-3, we found cos 15° by using the difference formula for cosine. An alternative method is identity (18).

Example 4   Find the exact value of $\cos 15°$.

Solution   Use identity (18) with $\dfrac{x}{2} = 15°$ and thus, with $x = 30°$.

$$\cos 15° = +\sqrt{\frac{1 + \cos 30°}{2}}$$

$$= \sqrt{\frac{1 + \dfrac{\sqrt{3}}{2}}{2}} = \sqrt{\frac{2 + \sqrt{3}}{4}} = \frac{\sqrt{2 + \sqrt{3}}}{2} \quad\blacksquare$$

To see that the answer in Example 4 is equivalent to the answer we found in Example 1 of Section 3-3, note that $\dfrac{2 + \sqrt{3}}{4} = \left(\dfrac{\sqrt{6} + \sqrt{2}}{4}\right)^2$.

EXERCISES 3-4

For each angle θ satisfying the given condition in the given quadrant, find
(a) sin 2θ and (b) cos 2θ.

A 1. $\sin \theta = \dfrac{4}{5}$; I

2. $\cos \theta = \dfrac{12}{13}$; I

3. $\cos \theta = -\dfrac{5}{13}$; II

4. $\sin \theta = -\dfrac{3}{5}$; III

5. $\sin \theta = -\dfrac{3}{4}$; III

6. $\cos \theta = \dfrac{2}{3}$; IV

7. $\cos \theta = -\dfrac{1}{3}$; II

8. $\sin \theta = -\dfrac{1}{4}$; IV

9. $\cos \theta = \dfrac{1}{5}$; IV

Exercises 10–18: For the angles in Exercises 1–9, find cos 4θ.

In Exercises 19–24, use a half-angle formula to evaluate each expression.

19. $\sin 15°$

20. $\cos 75°$ $\dfrac{\sqrt{2 - \sqrt{3}}}{2}$

21. $\cos 67.5°$

22. $\sin 157.5°$

23. $\sin 75°$

24. $\sin 67.5°$

25. $\cos \dfrac{x}{2}$, if $\cos x = -\dfrac{8}{25}$ and $0 < x < \pi$

26. $\sin \dfrac{x}{2}$, if $\cos x = -\dfrac{31}{49}$ and $\pi < x < 2\pi$

27. $\sin \dfrac{x}{2}$, if $\sin x = \dfrac{24}{25}$ and $0 < x < \dfrac{\pi}{2}$

28. $\cos \dfrac{x}{2}$, if $\sin x = -\dfrac{4\sqrt{5}}{9}$ and $\dfrac{3\pi}{2} < x < 2\pi$

29. Derive identity (19) from identity (17a).

Prove each of the following identities.

B 30. $\csc x - 2 \sin x = \csc x \cos 2x$

31. $\dfrac{\sin 2x}{1 - \cos 2x} = \cot x$

32. $\dfrac{1 - \cos 2\theta}{\cos 2\theta + 1} = \tan^2 \theta$

33. $\dfrac{1 - \cos 2\theta}{\sin \theta \sin 2\theta} = \sec \theta$

34. $1 - \tan^4 x = \dfrac{\cos 2x}{\cos^4 x}$

35. $\tan x + \tan y = \dfrac{\sin (x + y)}{\cos x \cos y}$

36. $(\sin \theta + \cos \theta)^2 = 1 + \sin 2\theta$

37. $\cos^4 \theta - \sin^4 \theta = \cos 2\theta$

38. $\cot \theta - \tan \theta = 2 \cot 2\theta \quad \left(\text{Hint: } \cot 2\theta = \dfrac{\cos 2\theta}{\sin 2\theta} \right)$

39. $\cot \theta + \tan \theta = 2 \csc 2\theta$

40. $\sin 4x = 4 \sin x \cos^3 x - 4 \sin^3 x \cos x$

41. $\cos 4x = 8 \cos^4 x - 8 \cos^2 x + 1$

42. $\cot 2\theta = \dfrac{1}{2} \left(\cot \theta - \dfrac{1}{\cot \theta} \right)$

43. $\tan \dfrac{x}{2} = \pm \sqrt{\dfrac{1 - \cos x}{1 + \cos x}}$

C 44. $\dfrac{\cos 2x + \sin^4 x}{\sin 2x} = \dfrac{1}{2} \cot x \cos^2 x$

45. $\sec 2x = \dfrac{\sec^2 x}{2 - \sec^2 x}$

46. $\csc^2 2x = \dfrac{\csc^4 x}{4(\csc^2 x - 1)}$

47. $\dfrac{\cos x + \sin x}{\cos x - \sin x} = \dfrac{1 + \sin 2x}{\cos 2x}$

Exercise 48 outlines a way to derive the formula for $\cos \dfrac{\theta}{2}$ for an acute angle θ. It is thought to be similar to the method used by Hipparchus of Nicaea (page 1) to construct the first known table of trigonometric functions in intervals of 7.5°.

48. (a) Explain why $\angle PAB = \dfrac{\theta}{2}$.

(b) Explain why $\cos \angle PAB = \dfrac{1 + \cos \theta}{PA}$.

(c) Use the distance formula to find PA in simplest radical form.

(d) Substitute the value for PA found in (c) in the equation in (b) to get the formula $\cos \dfrac{\theta}{2} = \sqrt{\dfrac{1 + \cos \theta}{2}}$.

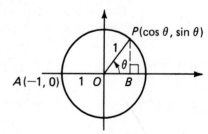

49. Explain how one could compute the values of $\cos \theta$ and $\sin \theta$ for θ in intervals of 7.5° between 0° and 90°, using only:

1. the half-angle formulas
2. the values of sine and cosine of 30° and 45°
3. the identities $\sin (90° - \theta) = \cos \theta$ and $\cos (90° - \theta) = \sin \theta$

Formulas for Tangent

Objective: To derive and apply tangent formulas.

We can derive the sum and difference formulas for the tangent function from the identity

$$\tan \theta = \frac{\sin \theta}{\cos \theta}.$$

We start by deriving a formula for $\tan (a + b)$.

$$\tan (a + b) = \frac{\sin (a + b)}{\cos (a + b)} = \frac{\sin a \cos b + \cos a \sin b}{\cos a \cos b - \sin a \sin b}$$

In order to express the right side of this equation in terms of $\tan a$ and $\tan b$, we divide both the numerator and the denominator by $\cos a \cos b$.

$$\tan (a + b) = \frac{\dfrac{\sin a \cos b}{\cos a \cos b} + \dfrac{\cos a \sin b}{\cos a \cos b}}{\dfrac{\cos a \cos b}{\cos a \cos b} - \dfrac{\sin a \sin b}{\cos a \cos b}}$$

Therefore:

$$\tan (a + b) = \frac{\tan a + \tan b}{1 - \tan a \tan b} \qquad (20)$$

for all a and b such that $\cos a \neq 0$, $\cos b \neq 0$, and $\cos (a + b) \neq 0$.

The following formula for $\tan (a - b)$ can be deduced from identity (20) and the fact that the tangent function is an odd function. (See Exercise 17.)

$$\tan (a - b) = \frac{\tan a - \tan b}{1 + \tan a \tan b} \qquad (21)$$

Example 1 If $\sin x = -\dfrac{12}{13}$ and $\dfrac{3\pi}{2} < x < 2\pi$, find $\tan \left(x + \dfrac{\pi}{4} \right)$ and $\tan \left(x - \dfrac{\pi}{4} \right)$.

Solution First, find $\tan x$. Since $\sin x = -\dfrac{12}{13}$, and x is in the fourth quadrant, $\cos x = \dfrac{5}{13}$. Hence, $\tan x = -\dfrac{12}{5}$. Then apply identity (20).

$$\tan\left(x + \frac{\pi}{4}\right) = \frac{-\dfrac{12}{5} + \tan\dfrac{\pi}{4}}{1 - \left(-\dfrac{12}{5}\right)\tan\dfrac{\pi}{4}} = \frac{1 - \dfrac{12}{5}}{1 + \dfrac{12}{5}} = -\frac{7}{17}$$

$$\tan\left(x - \frac{\pi}{4}\right) = \frac{-\dfrac{12}{5} - \tan\dfrac{\pi}{4}}{1 + \left(-\dfrac{12}{5}\right)\tan\dfrac{\pi}{4}} = \frac{-\dfrac{12}{5} - 1}{1 + \left(-\dfrac{12}{5}\right)} = \frac{17}{7} \quad\blacksquare$$

The following double-angle formula for the tangent can also be deduced from identity (20). (See Exercise 18 on page 107.)

$$\tan 2x = \frac{2\tan x}{1 - \tan^2 x} \qquad (22)$$

To derive a half-angle formula for the tangent, recall that in Example 1 of Section 3-4, we showed that

$$\tan x = \frac{\sin 2x}{\cos 2x + 1}.$$

If we replace x in this identity with $\dfrac{x}{2}$ we obtain:

$$\tan\left(\frac{x}{2}\right) = \frac{\sin 2\left(\dfrac{x}{2}\right)}{\cos 2\left(\dfrac{x}{2}\right) + 1}$$

Therefore,

$$\tan\frac{x}{2} = \frac{\sin x}{1 + \cos x} \qquad (23a)$$

We can obtain an alternative formula for $\tan\dfrac{x}{2}$ from identity (23a).

$$\begin{aligned}
\tan\frac{x}{2} &= \frac{\sin x}{1 + \cos x} \cdot \frac{1 - \cos x}{1 - \cos x} \\[2mm]
&= \frac{(\sin x)(1 - \cos x)}{1 - \cos^2 x} \\[2mm]
&= \frac{(\sin x)(1 - \cos x)}{\sin^2 x} \qquad \text{by identity (7)}
\end{aligned}$$

Therefore,

$$\tan \frac{x}{2} = \frac{1 - \cos x}{\sin x} \qquad \text{(23b)}$$

A third formula for $\tan \frac{x}{2}$ is as follows. (See Exercise 43 on page 103.)

$$\tan \frac{x}{2} = \pm \sqrt{\frac{1 - \cos x}{1 + \cos x}} \qquad \text{(24)}$$

Example 2   If $\tan \theta = \frac{4}{3}$, find $\tan \frac{3\theta}{2}$.

Solution   Apply a half-angle formula to find $\tan \frac{\theta}{2}$.

Then apply the addition formula. From the right triangle shown at the right,

$\sin \theta = \frac{4}{5}$ and $\cos \theta = \frac{3}{5}$.

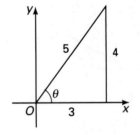

By identity (23a), $\tan \dfrac{\theta}{2} = \dfrac{\dfrac{4}{5}}{1 + \dfrac{3}{5}} = \dfrac{1}{2}.$

Thus, $\tan \dfrac{3\theta}{2} = \dfrac{\tan \theta + \tan \dfrac{\theta}{2}}{1 - \tan \theta \tan \dfrac{\theta}{2}} = \dfrac{\dfrac{4}{3} + \dfrac{1}{2}}{1 - \left(\dfrac{4}{3}\right)\left(\dfrac{1}{2}\right)} = \dfrac{11}{2}. \blacksquare$

EXERCISES 3-5

Find the exact value of each expression.

A **1.** $\tan 75°$ **2.** $\tan 165°$ **3.** $\tan(-15°)$ **4.** $\tan 67.5°$

Given that $\tan x = \frac{1}{3}$ and $\tan y = \frac{6}{5}$, find each of the following.

5. $\tan(x + y)$ **6.** $\tan(x - y)$ **7.** $\tan 2x$ **8.** $\tan 4x$

Given that $\sin x = -\dfrac{4}{5}$ and $\pi < x < \dfrac{3\pi}{2}$, find each of the following.

9. $\tan \dfrac{x}{2}$ **10.** $\tan 2x$ **11.** $\tan 3x$ **12.** $\tan \left(x + \dfrac{x}{2} \right)$

In Exercises 13–16, find the value of each of the expressions in Exercises 9–12, given that $\cos x = \dfrac{1}{3}$ and $0 < x < \dfrac{\pi}{2}$.

17. Use the fact that tangent is an odd function to prove that
$$\tan (a - b) = \frac{\tan a - \tan b}{1 + \tan a \tan b}.$$

18. Derive identity (22) from identity (20).

Prove each identity.

B 19. $\tan \left(x + \dfrac{\pi}{4} \right) = \dfrac{\cos x + \sin x}{\cos x - \sin x}$
20. $\tan \left(x - \dfrac{\pi}{4} \right) = \dfrac{\sin x - \cos x}{\cos x + \sin x}$

21. $\dfrac{1}{1 - \tan x} - \dfrac{1}{1 + \tan x} = \tan 2x$
22. $\tan 3x = \dfrac{3 \tan x - \tan^3 x}{1 - 3 \tan^2 x}$

23. $\tan \dfrac{\theta}{2} + \cot \dfrac{\theta}{2} = 2 \csc \theta$ $\left(\text{Hint: } \cot \dfrac{\theta}{2} = \dfrac{1}{\tan \dfrac{\theta}{2}} \right)$

24. $\cot 2x = \dfrac{\cot^2 x - 1}{2 \cot x}$

C 25. $(\cos \theta - \sin \theta)(\sec \theta + \csc \theta) = 2 \cot 2\theta$

26. $\dfrac{2 \tan \left(\dfrac{x}{2} \right)}{1 + \tan^2 \left(\dfrac{x}{2} \right)} = \sin x$ (Hint: Use identity (8).)

27. $\tan \left(\dfrac{x}{2} + \dfrac{\pi}{4} \right) = \sec x + \tan x$

$\left(\text{Hint: Use } \tan \theta = \dfrac{\sin \theta}{\cos \theta} \text{ and see Exercise 47 of Section 3-4.} \right)$

28. Find a formula for $\tan \dfrac{x}{2}$ *in terms of tan x only* by using the following method:

(a) Rewrite identity (22) in the form $\tan x = \dfrac{2 \tan \dfrac{x}{2}}{1 - \tan^2 \dfrac{x}{2}}$.

(b) Regarding (a) as a quadratic equation in $\tan \dfrac{x}{2}$ solve the equation for $\tan \dfrac{x}{2}$.

29. Derive the same identity requested in Exercise 28 using the following method: Divide the numerator and denominator of identity (23b) by cos x and make use of the identity:

$$\sec \theta = \pm\sqrt{\tan^2 \theta + 1}$$

30. Prove that if γ in the given diagram is the angle between two non-vertical lines L_1 and L_2, then

$$\tan \gamma = \frac{m_1 - m_2}{1 + m_1 m_2}$$

where m_1 is the slope of L_1 and m_2 is the slope of L_2. (Hint: If θ is the angle made by the graph of $y = mx + b$ and the positive x-axis, then $\tan \theta = m$.)

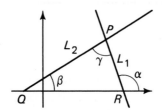

TRIGONOMETRIC EQUATIONS

3-6

Solving Trigonometric Equations

Objective: To solve trigonometric equations.

A *trigonometric equation* is simply an equation involving one or more circular or trigonometric functions. This term usually refers to a *conditional equation*, an equation true for only some values of the variable. The following examples illustrate the distinction between a conditional equation and an identity.

Conditional Equation	*Identity*
$2x + 7 = 13$	$2x + 3x = 5x$
(true only for $x = 3$)	(true for all x)
$\sin^2 x = 1$	$\sin^2 x + \cos^2 x = 1$
$\left(\text{true only for odd multiples of } \dfrac{\pi}{2}\right)$	(true for all x)

We solve an equation involving a single trigonometric function by first solving for the function, and then finding the values of the variable that satisfy this simplified equation.

Example 1   Find the general solution of $2 \cos \theta = 1$.

Solution   Since no domain of θ is specified, we must consider all values of θ. First, solve the equation for $\cos \theta$.

$$2 \cos \theta = 1$$

$$\cos \theta = \frac{1}{2}$$

On a unit-circle diagram, draw the line $x = \frac{1}{2}$ to help locate the values of θ for which $\cos \theta = \frac{1}{2}$. We see that two particular values of θ that satisfy $\cos \theta = \frac{1}{2}$ are $\theta = \frac{\pi}{3}, \frac{5\pi}{3}$.

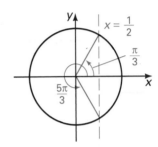

Clearly, adding any integral multiple of 2π to either of these values will produce another solution. Therefore, the general solution of the equation is

$$\theta = \frac{\pi}{3} + 2k\pi, \frac{5\pi}{3} + 2k\pi, \text{ where } k \text{ is any integer.} \quad \blacksquare$$

Example 2   Solve $2 \sin^2 t - \sin t = 1$ for $0 \le t < 2\pi$.

Solution   First solve for $\sin t$.

$$2 \sin^2 t - \sin t = 1$$
$$2 \sin^2 t - \sin t - 1 = 0$$

This is a quadratic equation in $\sin t$. We can factor the left side as we would $2x^2 - x - 1$. We get

$$(2 \sin t + 1)(\sin t - 1) = 0.$$

Therefore, $\sin t = -\frac{1}{2}$ or $\sin t = 1$.

By drawing two unit-circle diagrams similar to the figure in Example 1, we find that $t = \frac{7\pi}{6}, \frac{11\pi}{6}, \frac{\pi}{2}$. $\quad \blacksquare$

Example 3   Solve $4 \sin^2 \theta - 13 \sin \theta + 3 = 0$ for $0° \le \theta < 360°$.

Solution   As in Example 2, we factor the given equation.

$$4 \sin^2 \theta - 13 \sin \theta + 3 = 0$$
$$(4 \sin \theta - 1)(\sin \theta - 3) = 0$$

$$\sin \theta = \frac{1}{4} = 0.25 \text{ or } \sin \theta = 3$$

(Solution continued next page.)

From a sketch, there are two solutions of $\sin \theta = 0.25$ in the given interval. Estimate the first to be one-half of 30°, or 15°. Estimate the other solution to be $180° - 15°$, or 165°. Using a calculator, $\theta = 14.5°$ or $165.5°$ to the nearest tenth of a degree. There are no θ such that $\sin \theta = 3$. ∎

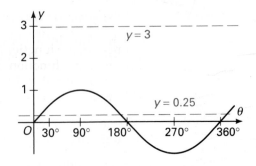

When solving an equation by division, we must check whether the divisor set equal to 0 yields any solutions of the original equation. If it does, these solutions should be recorded before dividing. If it does not, then no solutions are lost in the division process.

Example 4   Solve $\sin x = \sqrt{3} \cos x$ for $0 \le x < 2\pi$.

Solution   Before dividing by $\cos x$, check whether $\cos x = 0$ yields a solution. If $\cos x = 0$, then $x = \dfrac{\pi}{2}, \dfrac{3\pi}{2}$. Now $\sin \dfrac{\pi}{2} = 1$ and $1 \ne \sqrt{3} \cdot 0$, so $\dfrac{\pi}{2}$ is not a solution. Similarly, $\dfrac{3\pi}{2}$ is not a solution. Dividing by $\cos x$, we obtain:

$$\frac{\sin x}{\cos x} = \frac{\sqrt{3} \cos x}{\cos x}$$

Hence, $\tan x = \sqrt{3}$ and $x = \dfrac{\pi}{3}, \dfrac{4\pi}{3}$. ∎

If an equation is solved by squaring both sides, we must check to see whether all resulting solutions satisfy the original equation.

Example 5   Solve $\sec x = 1 - \tan x$ for $0 \le x < 2\pi$.

Solution
$$\sec^2 x = (1 - \tan x)^2$$
$$= 1 - 2 \tan x + \tan^2 x$$
$$1 + \tan^2 x = 1 - 2 \tan x + \tan^2 x$$
$$\tan x = 0$$
$$x = 0 \text{ and } \pi$$

Checking in the original equation we have:
$$\tan (0) + \sec (0) = 0 + 1 = 1$$
$$\tan \pi + \sec \pi = 0 - 1 \ne 1$$

Therefore we must reject the value $x = \pi$, which was actually introduced in the squaring process. The solution is $x = 0$. ∎

The expression or variable upon which a function operates is called the **argument** of the function. (For example, in the function $\sin 2x$, the argument of the function is $2x$.) When a trigonometric equation contains more than one function or argument, a good procedure for solving the equation is to transform it, if possible, into an equation containing only one function of one argument.

Example 6

Solve $\sin \theta + \cos 2\theta = 0$ for $0° \le \theta < 360°$.

Solution

Begin by transforming the equation so that it contains only one function of one argument.

$$\sin \theta + \cos 2\theta = \sin \theta + (1 - 2 \sin^2\theta) \qquad \text{by identity (17a)}$$
$$= -2 \sin^2 \theta + \sin \theta + 1$$

Hence, $2 \sin^2 \theta - \sin \theta - 1 = 0$.
From Example 2, the solution is $\theta = 90°, 210°, 330°$. ■

If only one argument appears in the equation, we do not need to transform the equation.

Example 7

Solve $6 \sin 2\theta = 3$ for $0° \le \theta < 360°$.

Solution

Dividing both sides by 6, we have $\sin 2\theta = \dfrac{1}{2}$. Since we want all values of θ such that $0° \le \theta < 360°$, we must consider values of 2θ such that $0° \le 2\theta < 720°$. Therefore,

$$2\theta = 30°, 150°, 390°, 510° \text{ and } \theta = 15°, 75°, 195°, 255°. \quad ■$$

Calculators and computers can be used to solve trigonometric equations and inequalities. This is illustrated in Example 8.

Example 8

(a) Solve $\sin 2x = \sin x$ graphically over the interval $0 \le x < 2\pi$.
(b) Solve $\sin 2x \le \sin x$ over $0 \le x < 2\pi$.

Solution

(a) By calculator or computer, sketch $y = \sin 2x$ and $y = \sin x$.

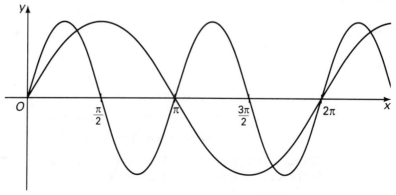

(Solution continued next page.)

The *x*-coordinates of the points of intersection are the solutions of the given equation. From the graphs, the approximate solutions over the given interval are 0, 1.05, 3.14, and 5.24. (The exact solutions are 0, $\frac{\pi}{3}$, π, and $\frac{5\pi}{3}$.)

(b) The intervals over which $\sin 2x \leq \sin x$ are easy to find once the roots of $\sin 2x = \sin x$ have been located.

$$1.05 \leq x \leq 3.14 \text{ and } 5.24 \leq x < 6.28 \quad \blacksquare$$

EXERCISES 3-6

Solve each equation for $0° \leq \theta < 360°$.

A 1. $\sin \theta = -\cos \theta$

2. $2\sqrt{3} \cos \theta - 6 \sin \theta = 0$

3. $\sin \theta + 2 \cos \theta = 0$

4. $4 \sec \theta - \csc \theta = 0$

5. $4 \sin^2 \theta - 3 = 0$

6. $2 \sin \theta = \csc \theta$

7. $1 - 3 \cos \theta = \sin^2 \theta$

8. $\tan^2 \theta = 2 \sec \theta - 1$

9. $\cot^2 \theta = 3(\csc \theta - 1)$

10. $2 \cos^2 \theta + \sin \theta = 1$

11. $\tan \theta = 2 \sin \theta$

12. $\sqrt{2} \sin \theta = \cot \theta$

Solve each equation for $0 \leq x < 2\pi$.

13. $\cos 2x = \sin x$

14. $\cos 2x = -\cos x$

15. $\sin 2x = -\sin x$

16. $\sin 2x = \cos x$

17. $\sin 2x = -\cos 2x$

18. $2 \sin^2 2x = 1$

Give the general solution for each equation.

19. $\sin 2x = \cos 4x$

20. $\tan \left(x - \frac{\pi}{4} \right) = 2 \sin \left(x - \frac{\pi}{4} \right)$

21. $4(\sin x + 1) = 3 \csc x$

22. $\tan x + \cot x = -2$

23. $1 + \cos x = 4 \sin^2 x$

24. $1 + 2 \cot^2 x + \csc x = 0$

25. $\tan^2 x - \sec x = 1$

26. $\cos x + \sec x = 2$

27. $\sec^2 x = 3 - \tan^2 x$

28. $\sqrt{3} \tan x = 2 \sin x$

Solve each trigonometric inequality over the specified interval.

B 29. $\sin x \geq \frac{1}{2}$ over $0 \leq x \leq 2\pi$

30. $\cos x - \sin x \geq 0$ over $0 \leq x \leq 2\pi$

31. $2 \cos x \le \sec x$ over $0 \le x < \dfrac{\pi}{2}$

32. $\csc x > 2 \sin x$ over $0 < x < \dfrac{\pi}{2}$

Solve each equation for $0 \le x < 2\pi$.

33. $3 \sin x + 2 = \cos 2x$

34. $3 \cos 2x + 2 \sin^2 x = 0$

35. $4 \sin^2 2x + 4 \cos 2x = 1$

36. $2 \cos^2 2x = 3 \sin 2x$

37. $2 \sin 2x \sin x = 3 \cos x$

38. $\sin 2x \sin x = \cos x$

Solve each inequality over the specified interval.

39. $\cos 2x \ge 0$ over $0 \le x \le \dfrac{\pi}{2}$

40. $\sin^2 x - \cos^2 x > 0$ over $0 \le x \le \pi$

41. $4 \sin^2 2x \le 1$ over $0 \le x \le \pi$

42. $\cos^2 x \ge \sin 2x$ over $0 \le x < \dfrac{\pi}{2}$

Exercises 43–48: Use the trigonometric addition formulas or the double-angle formulas to solve each equation over $0 \le x < 2\pi$.

43. $4 \sin x \cos x = \sqrt{3}$

44. $4 \sin x \cos x = -\sqrt{2}$

45. $\cos 2x \cos x + \sin 2x \sin x = -\dfrac{1}{2}$

46. $2 \cos 3x \cos x - 2 \sin 3x \sin x = \sqrt{3}$

47. $\sqrt{2}\,(\sin x + \cos x) = \sqrt{3}$ (Hint: Square both sides.)

48. $2(\sin x - \cos x) = \sqrt{2}$

Exercises 49 and 50 use the following information. The approximate distance s in meters that an object will travel if given an initial linear speed v_0 at an angle of elevation θ is given by the formula

$$s = \frac{v_0^2 \sin \theta \cos \theta}{5}$$

where v_0 is in meters per second.

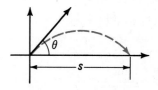

Exercises 49 and 50

49. At what angle must a football be thrown at 20 m/s in order to travel 20 m? (Disregard the height of the person throwing the ball.)

50. For what value of θ will the football in Exercise 49 travel the farthest? How far can the football travel?

51. The area of a right triangle is $\frac{1}{2}$ and the hypotenuse has length 2. Find the angles of the triangle.

52. Solve the equation $\sin \theta + \cos \theta = \sqrt{\dfrac{2 + \sqrt{3}}{2}}$ by squaring both sides. Be sure to check your solutions.

Solve each equation for $0° \le \theta < 360°$.

C 53. $2(\cos^4 \theta - \sin^4 \theta) = 1$

54. $4 \cos^4 \theta - 4 \cos^2 \theta = -\dfrac{1}{2}$ (Hint: Add 1 to both sides.)

55. $\sqrt{1 - \cos 2\theta} = 2 \sin^2 \theta$

56. $\sqrt{\cos 2\theta + 1} = 2 \cos^2 \theta$

Self Quiz

3-4 | 3-5 | 3-6

Prove each identity.

1. $\dfrac{1 - \cos 2\theta}{1 - \cos^2 \theta} = 2$

2. $\dfrac{\cos 2x + \sin^2 x}{\sin 2x} = \dfrac{1}{2} \cot x$

3. Given that $\sin x = \dfrac{5}{13}$ and $\dfrac{\pi}{2} < x < \pi$, find $\sin \dfrac{x}{2}$.

4. Find $\tan 105°$ in simplest radical form.

Given that $\sin x = \dfrac{3}{5}$ and $\dfrac{\pi}{2} < x < \pi$, find:

5. $\tan \dfrac{x}{2}$

6. $\tan 2x$

7. Find the general solution of $2 \sin^2 x - \cos x = 1$.

Solve each equation for $0 \le x < 2\pi$.

8. $\sin x = -\sin 2x$

9. $2 \sin 2x = \tan x$

Basic Trigonometric Identities

$$\tan x = \frac{\sin x}{\cos x} \qquad (1)$$

$$\cot x = \frac{\cos x}{\sin x} \qquad (2)$$

$$\sec x = \frac{1}{\cos x} \qquad (3)$$

$$\csc x = \frac{1}{\sin x} \qquad (4)$$

$$\tan x = \frac{1}{\cot x} \qquad (5)$$

$$\cot x = \frac{1}{\tan x} \qquad (6)$$

The Pythagorean Identities

$$\sin^2 x + \cos^2 x = 1 \qquad (7)$$

$$1 + \tan^2 x = \sec^2 x \qquad (8)$$

$$1 + \cot^2 x = \csc^2 x \qquad (9)$$

Sum and Difference Formulas

$$\cos (a - b) = \cos a \cos b + \sin a \sin b \qquad (10)$$

$$\cos (a + b) = \cos a \cos b - \sin a \sin b \qquad (11)$$

$$\cos \left(\frac{\pi}{2} - x \right) = \sin x \qquad (12)$$

$$\sin \left(\frac{\pi}{2} - x \right) = \cos x \qquad (13)$$

$$\sin (a \pm b) = \sin a \cos b + \cos a \sin b \qquad (14)$$

$$\sin (a - b) = \sin a \cos b - \cos a \sin b \qquad (15)$$

Double-Angle Formulas

$$\sin 2x = 2 \sin x \cos x \qquad (16)$$

$$\cos 2x = \cos^2 x - \sin^2 x \qquad (17)$$

$$\cos 2x = 1 - 2 \sin^2 x \qquad (17a)$$

$$\cos 2x = 2 \cos^2 x - 1 \qquad (17b)$$

Half-Angle Formulas

$$\cos \frac{x}{2} = \pm \sqrt{\frac{1 + \cos x}{2}} \qquad (18)$$

$$\sin \frac{x}{2} = \pm \sqrt{\frac{1 - \cos x}{2}} \qquad (19)$$

Tangent Formulas

$$\tan (a + b) = \frac{\tan a + \tan b}{1 - \tan a \tan b} \qquad (20)$$

$$\tan (a - b) = \frac{\tan a - \tan b}{1 + \tan a \tan b} \qquad (21)$$

$$\tan 2x = \frac{2 \tan x}{1 - \tan^2 x} \qquad (22)$$

$$\tan \frac{x}{2} = \frac{\sin x}{1 + \cos x} \qquad (23a)$$

$$\tan \frac{x}{2} = \frac{1 - \cos x}{\sin x} \qquad (23b)$$

$$\tan \frac{x}{2} = \pm \sqrt{\frac{1 - \cos x}{1 + \cos x}} \qquad (24)$$

ADDITIONAL PROBLEMS

1. (a) Find $\sin (a + b)$ if $\sin a = -\frac{4}{5}$, $\cos b = \frac{15}{17}$, $\pi < a < \frac{3\pi}{2}$, and

   $0 < b < \frac{\pi}{2}$.

   (b) Find $\cos 2\theta$ and $\cos \frac{\theta}{2}$ if $\sin \theta = \frac{3}{5}$ and $0 < \theta < \frac{\pi}{2}$.

2. Prove the identity: $\sin (r + s) \sin (r - s) = \sin^2 r - \sin^2 s$

3. Prove the identity: $\dfrac{\cos x}{1 - \tan x} + \dfrac{\sin x}{1 - \cot x} = \sin x + \cos x$

4. Using calculus, one can show that the area of the isosceles trapezoid shown is maximized when an acute base angle θ satisfies the equation

   $$r^2 (\cos \theta + \cos^2 \theta - \sin^2 \theta) = 0.$$

   What angle θ makes the trapezoid have
   maximum area?

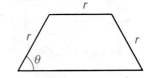

5. Derive a formula for $\tan \left(x - \dfrac{\pi}{4} \right)$ in terms of $\tan x$ only.

6. Simplify $\dfrac{(\sin x + \cos x)^2 - 1}{\sin x \cos x}$.

7. Simplify $(\sin x + \sec x)^2 - (\sin x - \sec x)^2$.

8. Find all possible values of $x - y$ if $\sin x \cos y - \cos x \sin y = \dfrac{1}{2}$.

9. Prove the identity: $\sin \left(\dfrac{\pi}{4} + x \right) \cos \left(\dfrac{\pi}{4} + x \right) = \dfrac{1}{2} \cos 2x$

10. Solve $\cos^2 x - \sin^2 x = \dfrac{1}{2}$ for $0 \le x < 2\pi$.

11. Simplify $\sin 2x \cos x - \cos 2x \sin x$.

12. Find all real numbers x for which $\sin 2x = 2 \sin x$.

13. Find a formula for $\cos 3x$ in terms of $\cos x$ only.

14. (a) Justify: If $\cos x$ is a rational number, then $\cos 2x$ is also.
    (b) Is this statement still true if "cos" is replaced by "sin"? Justify the latter or give a counterexample to show that it is false.

15. Prove the identity: $\dfrac{\cos 2\theta + \sin^4 \theta}{\sin 2\theta} = \dfrac{1}{2} \cot \theta \cos^2 \theta.$

16. Find the value of $\tan \left(x - \dfrac{\pi}{3} \right)$ given that $\sin x = \dfrac{4}{5}$ and $0 < x < \dfrac{\pi}{2}$.

CHAPTER SUMMARY

1. An identity is an equation that is true for all values of the variables for which both sides are defined. Known algebraic and trigonometric identities can be used to simplify expressions involving trigonometric functions.

2. The reciprocal identities (3)–(6) express the trigonometric functions other than sine and cosine in terms of these two functions or as reciprocals of each other. The Pythagorean identities (7)–(9), whose proofs are based on the Pythagorean theorem, state relationships between the squares of trigonometric functions.

3. To prove that an equation is an identity, one must show, using known identities and working with only one side at a time, that the two sides of the given equation are equivalent.

4. For any real numbers a and b, the sine and the cosine of $a + b$ and $a - b$ can be expressed in terms of sin a, sin b, cos a, and cos b.

5. For any real number x the sine and cosine of $2x$ and $\dfrac{x}{2}$ can be expressed in terms of sin x and cos x. Alternative formulas express cos $2x$ in terms of sin x alone or cos x alone. The use of one or the other of these is dictated by convenience. The sign of a function of $\dfrac{x}{2}$ is determined by deciding in which quadrant $\dfrac{x}{2}$ lies.

6. Sum and difference formulas and half-angle and double-angle formulas analogous to those for the sin and cosine functions can be derived for the tangent function.

7. A trigonometric equation usually refers to an equation that is not an identity. To find the values of the variable for which such an equation is true, you may need to use identities to transform the equation so that all the functions involved have the same argument. Often it is also wise to transform the equation so that only a single trigonometric function (for example, sine or cosine) is involved.

8. The final steps in solving a trigonometric equation consist of reducing the equation to one or more equations of the form $f(x) =$ a constant, where $f(x)$ is a trigonometric function, and then finding the values of x that satisfy this equation, by means of a unit-circle diagram.

CHAPTER TEST

3-1 1. Express in terms of cos x: $\dfrac{\tan^2 x + 1}{\sec x}$.

2. Express in terms of a single trigonometric function:
$$\sin^2 x(\tan x + \cot x)^2$$

3-2 3. Prove the identity: $(\tan x - \sin x)(\sec x + 1) = \sin x \tan^2 x$.

4. Prove the identity: $\dfrac{\tan x}{1 - \cos x} - \dfrac{\tan x}{1 + \cos x} = 2 \csc x$.

3-3 5. Find the exact value of $\cos(-15°)$.

6. Find $\sin(s + t)$ if $\sin s = -\dfrac{3}{5}$ and $\sin t = \dfrac{5}{13}$, with $\dfrac{3\pi}{2} \le s \le 2\pi$ and $\dfrac{\pi}{2} \le t \le \pi$.

3-4 7. Find (a) $\sin 2\theta$ and (b) $\sin \dfrac{\theta}{2}$, if $\cos \theta = -\dfrac{7}{25}$ and θ is in Quadrant II.

8. Prove the identity: $\sec^2 x \cos 2x = 1 - \tan^2 x$.

3-5 9. Find $\tan\left(x + \dfrac{x}{2}\right)$ if $\tan x = \dfrac{5}{12}$ and $0 < x < \dfrac{\pi}{2}$.

10. Prove the identity: $\tan \dfrac{x}{2} = \csc x - \cot x$

3-6 11. Find the general solution of the equation:
$$2 \sin x = 3 + 2 \csc x.$$

12. Solve for $0° \le \theta < 360°$: $4 \sin \theta \cos \theta = \tan \theta$.

Using Graphs to Solve Equations

The program below can be used to solve equations like $\sin x = \cos x$. The left and right sides of the equation are entered as functions: $y = \sin x$ and $z = \cos x$ in lines **50** and **60**, respectively. A vertical scale factor of 30 and a horizontal stretch factor of 0.025 are used to magnify the graphs on the screen display. Include these factors in the functions you enter.

In line **90** the distance between the graphs of Y and Z is computed. If it is sufficiently small, then $\sin x$ and $\cos x$ are close and x is an approximate solution of $\sin x = \cos x$.

```
10 HGR : HCOLOR = 3
20 HPLOT 0, 95 TO 279, 95
30 HPLOT 10, 0 TO 10, 190
40 FOR X = -10 TO 269 STEP .25
50 Y = 30 * SIN (.025 * X)
60 Z = 30 * COS (.025 * X)
70 HPLOT 10 + X, 95 - Y
80 HPLOT 10 + X, 95 - Z
90 D = ABS (Y - Z) /30
100 D1 = INT (1000 * D + .5)/1000
110 VTAB 22: HTAB 1 : PRINT "X = "; .025 * X, "|X - Y| = "; D1
120 NEXT X
```

EXERCISES

Use the program to find approximate solutions of each equation in the interval $0 \le x < 2\pi$.

1. $\cos x = 0.5$

2. $\sin x = 1$

3. $\sin^2 x = \cos^2 x$

4. $\cos 2x = \cos x$

5. $\sin 2x = \sin x \cos x$

6. $2 \sin^2 x = \sin x + 1$

Trigonometric Equations

Many trigonometric equations in applications cannot be solved by simple methods. In the study of light bent, or refracted, by a prism, trigonometric equations play an important role. A computer or graphing calculator can often be used to approximate solutions of such equations. Consider the following problems.

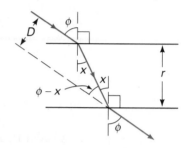

A ray of light passing through a transparent medium (such as a pane of glass) with parallel faces is refracted as it enters and again as it leaves the medium. Its incoming and outgoing paths are parallel but displaced. The displacement, D, is given by

$$D = r\frac{\sin (\phi - x)}{\cos x}.$$

Given $D = 1$ mm, $r = 3$ mm, and $\phi = \dfrac{\pi}{4}$, find x.

To find x, we must solve

$$\cos x = 3 \sin \left(\frac{\pi}{4} - x\right).(*)$$

By graphing both $y = \cos x$ and $y = 3 \sin \left(\dfrac{\pi}{4} - x\right)$ on the same set of axes,

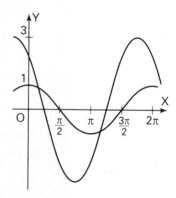

we can find the number of solutions in the interval $0 \le x < 2\pi$, estimate those solutions, and approximate them. From the figure at the right, there are two roots of $(*)$ between 0 and 2π. The first seems to be a little less than $\dfrac{\pi}{4}$.

A heated rod with uniform density and small cross-section is insulated so that it can lose heat only at its ends. To determine a function that gives temperature at a given position and time, we must (among other tasks) solve an equation such as $- \tan x = x. (**)$

The graphs of $y = x$ and $y = -\tan x$ are shown in the figure at the right. The figure indicates that there is a root of (∗∗) between 0 and π. The first positive root, x_1, is a little greater than $\dfrac{\pi}{2} \approx 1.57$.

Using 1.8 as a first approximation of x_1, we can get better and better approximations.

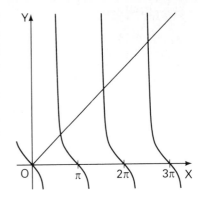

EXERCISES

1. Refer to (∗). Using $\dfrac{0.9\pi}{4}$ as a first estimate of the root x_1, between 0 and $\dfrac{\pi}{2}$ and a calculator, approximate that root by comparing each pair of values.

   (a) $\cos \dfrac{0.9\pi}{4}$; $3 \sin\left(\dfrac{\pi}{4} - \dfrac{0.9\pi}{4}\right)$ (b) $\cos \dfrac{0.7\pi}{4}$; $3 \sin\left(\dfrac{\pi}{4} - \dfrac{0.7\pi}{4}\right)$

2. Of the values $\dfrac{0.9\pi}{4}$ and $\dfrac{0.7\pi}{4}$, which is the better approximation of x_1?

3. Refer to (∗∗). Using 1.8 as a first estimate of x_1 and a calculator, approximate x_1 by finding $-\tan 1.8$, $-\tan 2.0$, $-\tan 2.02$, and $-\tan 2.0288$.

MIXED REVIEW: Chapters 1–3

1. Find the exact values of the six trigonometric functions of 60°.

2. Draw a detailed graph of $y = 2 |\sin x|, 0 \le x \le 2\pi$.

3. Use Table 3, pages 403–406 to find cos 6.03 to two significant digits.

4. If $\csc \theta = -\dfrac{13}{12}$ and $90° < \theta < 270°$, find $\tan \theta$.

5. Prove: $\csc 2x + \cot 2x = \cot x$

6. Find the degree measure of an angle between 0° and 360° that is coterminal with $-412°39'$.

7. (a) Find the radius of a circle in which a central angle of 144° intercepts an arc of length 8π cm. (b) Find the area of the sector.

8. Find the exact values of sin 285° and tan 285°.

9. Find all solutions of $2 \cos 2x = \sin 4x$.

10. A spectator stands 70 m from the base of a 100 m tall building under construction. At what angle does the spectator spot a hoist sitting on the top of the building?

11. The base of an isosceles triangle is 20.8 cm long and each base angle measures 41°. Find the length of the altitude to the base.

12. Is $\text{Tan}^{-1}(-x) = -\text{Tan}^{-1} x$ for all x in the domain of Tan^{-1}? Explain.

13. If $\sin \theta = \dfrac{2}{3}$, find the value of θ between 0° and 90°. Round your answer to the nearest tenth of a degree.

14. Express in terms of a single trigonometric function:
$$\frac{\sec \theta + 1}{\tan \theta} + \frac{\tan \theta}{\sec \theta + 1}$$

15. Evaluate: (a) $\text{Sin}^{-1}\left(\tan \dfrac{7\pi}{4}\right)$

   (b) $\cos\left(\text{Tan}^{-1} \dfrac{3}{4}\right)$

16. Solve $\triangle ABC$ in which $\angle C = 90°$, $a = 12.0$, and $c = 37.0$.

17. Find the exact values of the six trigonometric functions for an angle $\theta, 0° \le \theta \le 360°$, in standard position whose terminal side passes through $(-6, -9)$. Give the measure of θ to the nearest tenth of a degree.

18. Prove that $\dfrac{\cos^3 x}{(\cot x - \cos x)(1 + \sin x)} = \sin x$ is an identity.

19. Express 5°20′42″ in decimal degrees.

20. Express $\sin \theta (\sin \theta - \csc \theta)$ in terms of $\cos \theta$.

21. A point P travels uniformly at an angular speed of $\omega = \dfrac{5\pi}{3}$ radians/s counterclockwise around a circular path of radius 2 with center at the origin. Find the coordinates of P and the distance that P has traveled 6 s after P moves from $(\sqrt{3}, -1)$.

22. Use the formula for $\sin (a - b)$ to find the exact value of $\sin 75°$. (Use $a = 135°$ and $b = 60°$.)

23. Solve for $0 \le x < 2\pi$: $\cos 3x \cos x + \sin 3x \sin x = -\dfrac{\sqrt{2}}{2}$.

24. (a) Convert $\dfrac{5\pi}{9}$ to degree measure.   (b) Convert $420°$ to radian measure.

25. Prove: $\dfrac{\cot \alpha - \tan \alpha}{\sin \alpha \cos \alpha} = \csc^2 \alpha - \sec^2 \alpha$

If $\cos \theta = -\dfrac{3}{4}$ and $180° \le \theta < 360°$, evaluate each expression.

26. $\cos 2\theta$

27. $\tan 2\theta$

28. $\sin \dfrac{\theta}{2}$

29. $\tan \dfrac{\theta}{2}$

30. (a) Find x to three significant digits.
    (b) Find θ to the nearest 10′.

31. Part of the graph of a function f is shown below. Graph f in the interval $-4 \le x \le 4$ given that f is even.

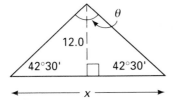

32. Draw the graph of $y = -|\tan x|$, $0 \le x \le 2\pi$, where $x \ne \dfrac{\pi}{2}$ or $\dfrac{3\pi}{2}$.

33. Is $y = \cos x + \sin x$ even, odd, or neither?

34. Prove: $\left(\sin \dfrac{\alpha}{2} + \cos \dfrac{\alpha}{2} \right)^2 = \sin \alpha + 1$

35. Prove: $\cot \left(\dfrac{\pi}{2} - x \right) = \tan x$

4 Solving Oblique Triangles

Many construction projects, such as the bridge shown in the photograph, require trigonometry applied to oblique triangles. In this chapter, methods for measuring and solving such triangles are presented.

THE LAWS OF COSINES AND SINES

4-1 The Law of Cosines

Objective: To use the law of cosines to solve certain oblique triangles.

Each problem stated below involves an **oblique triangle**, that is, a triangle having no right angle.

Problem 1 Ships *A* and *B* leave port *C* at noon and travel along straight paths making an angle of 120° with each other. How far apart are they at 2 P.M. if the speed of ship *A* is 25 km/h and that of ship *B* is 15 km/h?

Problem 2 A triangular lot has frontages of 40 m and 70 m on two streets, and its third side is 60 m long. At what angle do the streets intersect?

Both of these problems can be solved by using the **law of cosines** stated and proved at the top of the next page.

The law of cosines states that in any triangle *ABC* (Figure 4-1):

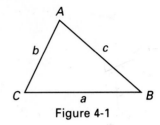

$$c^2 = a^2 + b^2 - 2ab \cos C$$
$$b^2 = c^2 + a^2 - 2ca \cos B$$
$$a^2 = b^2 + c^2 - 2bc \cos A$$

Figure 4-1

It is sufficient to prove the first of these formulas. Introduce a coordinate system with the origin at the vertex *C*, the *x*-axis along the side $\overline{CB}$ (Figure 4-2). Coordinates (x, y) of *A* are $(b \cos C, b \sin C)$ (Recall Section 2-2). Using the distance formula, we have:

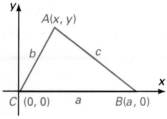

Figure 4-2

$$c^2 = (x - a)^2 + y^2$$
$$= (b \cos C - a)^2 + (b \sin C)^2$$
$$= (b^2 \cos^2 C - 2ab \cos C + a^2) + (b^2 \sin^2 C)$$
$$= b^2 (\cos^2 C + \sin^2 C) - 2ab \cos C + a^2$$
$$= a^2 + b^2 - 2ab \cos C$$

Example 1   Solve Problem 1 stated on the preceding page.

Solution   At the end of 2 hours, ships *A* and *B* have traveled 50 km and 30 km, respectively, as shown at the right. We use the law of cosines to find the distance $c = AB$.

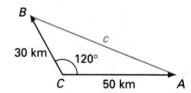

$$c^2 = a^2 + b^2 - 2ab \cos C$$
$$= 30^2 + 50^2 - 2 \cdot 30 \cdot 50 \cos 120°$$
$$= 900 + 2500 - 2 \cdot 30 \cdot 50\left(-\frac{1}{2}\right)$$
$$= 4900$$

Therefore, $c = \sqrt{4900} = 70$ km.   ■

Example 2   Solve Problem 2, stated on the preceding page.

Solution   Draw and label a figure. Use the law of cosines:

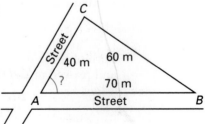

$$a^2 = b^2 + c^2 - 2bc \cos A$$
$$60^2 = 40^2 + 70^2 - 2 \cdot 40 \cdot 70 \cos A$$

$$3600 = 1600 + 4900 - 5600 \cos A$$

$$36 = 65 - 56 \cos A$$

$$\cos A = \frac{65 - 36}{56} = \frac{29}{56} = 0.5178$$

$$\angle A = 58.8° \quad \blacksquare$$

Example 3   Find the lengths of the sides of a parallelogram if its diagonals are 30 cm and 18 cm long and intersect at a 60° angle.

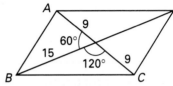

Solution   Use the fact that the diagonals of a parallelogram bisect each other and apply the law of cosines twice.

$AB^2 = 9^2 + 15^2 - 2(9)(15) \cos 60°$	$BC^2 = 15^2 + 9^2 - 2(15)(9) \cos 120°$
$\quad = 81 + 225 - 270 \times \dfrac{1}{2}$	$\quad = 225 + 81 - 270\left(-\dfrac{1}{2}\right)$
$\quad = 171$	$\quad = 441$
$AB = \sqrt{171} = 13.1$ cm	$BC = 21$ cm

Solving some problems requires several major steps.

Example 4   In the truss for a bridge shown at the right, $\triangle ABD$ is congruent to $\triangle CBE$, $AB = 8$ m, $AD = 7$ m and $\angle A = 45°$. Find the length DE.

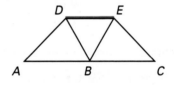

Solution   We plan our strategy by reasoning as follows:

"I can find DE if I can find $DQ = \frac{1}{2}DE$."

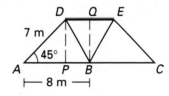

"I can find DQ (by using the Pythagorean theorem in right $\triangle DQB$) if I can find DB and BQ."

"I can find DB by using the law of cosines in $\triangle ABD$."

"I can find PD equal to BQ from right $\triangle APD$."

We now carry out our strategy by executing its four steps in reverse order.

Step 1 $BQ = PD = AD \sin A = 7 \sin 45° = 4.950$

Step 2 $BD^2 = 8^2 + 7^2 - 2 \cdot 8 \cdot 7 \cos 45° = 33.80$;
$\quad\quad\quad BD = \sqrt{33.80} = 5.814$

Step 3 $DQ^2 = BD^2 - BQ^2 = 33.80 - 4.950^2 = 9.298$;
$\quad\quad\quad DQ = \sqrt{9.298} = 3.049$

Step 4 $DE = 2DQ = 6.10$ m $\blacksquare$

In Examples 2, 3, and 4, we assumed for simplicity that the given data were exact but gave our answers to three significant digits and the nearest 0.1°. Continue this practice in the exercises unless you can give an exact answer, as in Example 1.

EXERCISES 4-1

In Exercises 1-8 find the asked-for part of $\triangle ABC$.

A 1. $a = 5; b = 8; \angle C = 60°; c = ?$
2. $b = 15; c = 7; \angle A = 60°; a = ?$

3. $b = 7; c = 8; \angle A = 120°; a = ?$
4. $a = 9; c = 15; \angle B = 120°; b = ?$

5. $a = 3; b = 8; c = 7; \angle C = ?$
6. $a = 5; b = 19; c = 16; \angle B = ?$

7. $a = 7; b = 14; c = 10; \angle B = ?$
8. $a = 5; b = 7; c = 9; \angle A = ?$

Find the lengths of the diagonals of a parallelogram having sides of the given lengths and an angle of the given measure.

9. 6 cm, 10 cm, 60°
10. 30 m, 40 m, 110°

Find the lengths of the sides of a parallelogram if its diagonals have the given lengths and intersect at the given angle.

11. 30 cm, 14 cm, 72°
12. 10 cm, 16 cm, 120°

In Exercises 13 and 14, find the angles of a parallelogram if its sides have lengths a and b and a diagonal has length d.

13. $a = 7$ m, $b = 8$ m, $d = 13$ m

14. $a = 5$ km, $b = 8$ km, $d = 7$ km

15. Solve Problem 1, page 125, assuming that the slower ship leaves port at 11 A.M.

16. The angle between the straight highways joining town A to towns B and C measures 60°. How far apart are B and C if their distances from A are 16 km and 21 km, respectively?

17. A triangular course for a 30 km yacht race has sides 7 km, 9 km, and 14 km long. Find the largest angle of the course.

18. A cruise ship and a freighter leave port at the same time and travel straight-line courses at 30 km/h and 10 km/h, respectively. Two hours later they are 50 km apart. Find the angle between their courses.

19. Points A and B on opposite sides of a pond are sighted from point C as shown at the right. If $\angle C = 30°$, $CA = 25$ m, and $CB = 30$ m, how far apart are A and B?

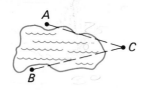

20. A tunnel is to be dug from point A to point B. The distances from A and B to a point C are 400 m and 700 m and $\angle C = 52°$. How long will the tunnel be? (Suggestion: To work with smaller numbers, use 100 m as the unit of length.)

21. A baseball diamond is a 90-ft square. The mound is 60.5 ft from home plate. How far is it from the mound to first base?

22. The supports for a basketball backboard are parallel and meet the backboard at a 38° angle, as shown at the right. If $BD = 140$ cm and the supports are 150 cm long, find the length AD of the diagonal brace.

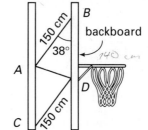

23. A diagonal brace $\overline{BC}$ is to be added to the structure in Exercise 22. How long will it be?

24. A water molecule consists of two hydrogen atoms and one oxygen atom joined as shown. The distance between the nucleus of the oxygen atom and the nucleus of each hydrogen atom is 9.58×10^{-9} cm. The bond angle θ is 104.8°. How far apart are the hydrogen nuclei?

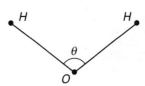

25. What well-known theorem does the law of cosines generalize? (Hint: Apply the law of cosines to a right triangle.)

26. Show that $\angle A$ of $\triangle ABC$ is obtuse if $a^2 > b^2 + c^2$.

27. Show that in $\triangle ABC$, $\angle A = \text{Cos}^{-1}\left(\dfrac{b^2 + c^2 - a^2}{2bc}\right)$.

B 28. Use Exercise 27 to show that if $a > b + c$, then no triangle can be formed with sides of lengths a, b, and c. (Hint: If $a > b + c$, then $b^2 + c^2 - a^2 < b^2 + c^2 - (b + c)^2$.)

29. The sides of a parallelogram have lengths a and b, and its diagonals have lengths p and q. Show that $p^2 + q^2 = 2(a^2 + b^2)$.

30. A flagpole 4 m tall stands on a sloping roof. A guy wire 5 m long joins the top of the pole to the roof at a point 6 m up from the base of the pole. What angle does the roof make with the horizontal?

31. A vertical pole 20 m tall on a 15° slope is to be braced by two cables extending from the top of the pole to points on the ground 30 m up and 30 m down the slope. How long will the cables be?

In Exercises 32–35, the lengths of the sides of $\triangle ABC$ are given. Find the length of the median to side $\overline{AB}$.

32. $a = 4; b = 8; c = 8$

33. $a = 20; b = 18; c = 18$

34. $a = 17; b = 19; c = 16$

35. $a = 10; b = 13; c = 16$

Exercises 36–39 refer to $\triangle ABC$ for which b, c, and $\angle B$ or $\cos B$ are given. Find a. There may be two answers.

36. $b = 7; c = 3; \angle B = 120°$

37. $b = 7; c = 8; \angle B = 60°$

38. $b = 2; c = 4; \cos B = 0.875$

39. $b = 7; c = 5; \cos B = 0.2$

Exercises 40–43 refer to the figure at the right. Lengths a, b, c, and d are given. Find x in simplest radical form.

40. $a = 2; b = 6; c = 5; d = 9$

41. $a = 4; b = 21; c = 15; d = 20$

42. $a = 3; b = 2; c = 6; d = 4$

43. $a = 3; b = 6; c = 7; d = 11$

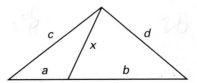

44. Shown below is a folding chair in which $DC = 20$ cm, $EC = 25$ cm, $CA = 16$ cm, and $CB = 20$ cm. If $DE = 30$ cm, find AB.

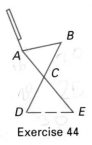 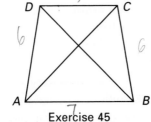 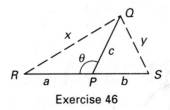

Exercise 44   Exercise 45   Exercise 46

45. A trapezoidal section of a steel tower is shown above. In it, $AB = 7$ m, $DC = 5$ m, $AD = BC = 6$ m. Find AC.

46. In the figure at the right above, $\overline{RS}$ and $\overline{PQ}$ are rods, arranged so that they can pivot about P. Show that $bx^2 + ay^2$ is a constant, independent of what the angle θ is.

C 47. Towns A, B, C, and D are joined by straight roads. A mileage chart is shown at the right. (For example the distance between B and D is 7 miles.) Find the distance between C and D. (Hint: Draw a figure, using the fact that $AC + CB = AB$.)

	A	B	C	D
A	0	10	4	9
B	10	0	6	7
C	4	6	0	?
D	9	7	?	0

48. The lengths of two sides and a diagonal of a parallelogram are 16 cm, 21 cm, and 19 cm, respectively. How long is the other diagonal?

49. In the isosceles $\triangle ABC$, $CA = CB = 14$, and the median from A has length 9. Find AB.

50. In the figure at the left below, $AB = DC$, $AP = PD = 3$, $BP = 1$, and $PC = 4$. Find AB.

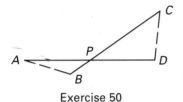

Exercise 50

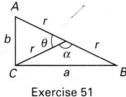

Exercise 51

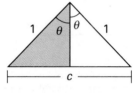

Exercise 52

51. Use the law of cosines and the middle figure above to prove that if the median to the longest side of a triangle has half the length of that side, then the triangle is a right triangle. (Hint: Use the converse of the Pythagorean theorem.)

52. Prove that $\cos 2\theta = 1 - 2 \sin^2 \theta$ for $0° < \theta < 90°$. (Hint: Using the triangle shown above, express c^2 by using the law of cosines, and find $\dfrac{c}{2}$ from the shaded right triangle. Then equate the expressions for c^2.)

4-2

The Law of Sines

Objective: To use the law of sines to solve certain oblique triangles.

In the problem below only one length is given, so that the law of cosines cannot be applied.

Problem   A weather balloon is directly over a 2000 m long airstrip. The angles of elevation of the balloon from the ends of the strip are 42° and 63°. How far is the balloon from the nearer end of the strip?

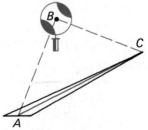

This problem can be solved by using the **law of sines** stated and proved at the top of the next page.

The law of sines states that in any triangle ABC:

$$\frac{\sin A}{a} = \frac{\sin B}{b} = \frac{\sin C}{c}$$

To prove the first equality, draw the altitude $\overline{CD}$ to $\overline{AB}$ (Figure 4-3). From right $\triangle ADC$ we have $CD = b \sin A$, and from right $\triangle BDC$ we have $CD = a \sin B$. Thus, $b \sin A = a \sin B$. Dividing by ab, we get $\dfrac{\sin A}{a} = \dfrac{\sin B}{b}$.

The other equalities follow similarly.

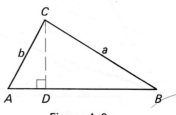

Figure 4–3

Example 1 Solve the weather balloon problem stated on page 131.

Solution Draw and label a sketch. Note that $\angle B = 180° - (42° + 63°) = 75°$. (Since $\angle A = 42° < 75° = \angle B$, you can expect a to be less than 2000.)

To find a, we use $\dfrac{\sin A}{a} = \dfrac{\sin B}{b}$, or

$$a = \frac{b \sin A}{\sin B}$$

$$= \frac{2000 \sin 42°}{\sin 75°} = 1385$$

The balloon is 1385 m from C. ∎

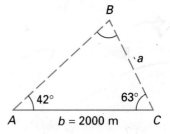

Example 2 A guy wire bracing a transmission tower is 20 m long and makes an angle of 50° with the ground. It is to be replaced by a 30 m wire starting from the same point on the ground. How much farther up the tower will the new wire reach?

Solution [Strategy: Draw and label a sketch. Find, in order, α, β, γ (using the law of sines), δ, and CD (using the law of sines again).]

Step 1 $\alpha = 90° - 50° = 40°$

Step 2 $\beta = 180° - \alpha = 140°$

Step 3 $\dfrac{\sin \gamma}{20} = \dfrac{\sin \beta}{30}$

$$\sin \gamma = \frac{20 \sin 140°}{30} = 0.4285$$

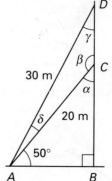

Thus, $\gamma = 25.37°$ or $154.63°$. However γ cannot equal $154.63°$, since $\triangle ACD$ already has one obtuse angle.

Step 4 $\quad \delta = 180° - (\beta + \gamma) = 180° - (140° + 25.37°) = 14.63°$

Step 5 $\quad \dfrac{\sin \delta}{CD} = \dfrac{\sin \beta}{AD}$

$$CD = \frac{AD \sin \delta}{\sin \beta} = \frac{30 \sin 14.63°}{\sin 140°} = 11.8$$

The new wire will reach 11.8 m farther up the tower than the old wire. ■

Surveyors use **triangulation** to find the distance between two inaccessible points P and Q. They lay off a base line $\overline{AB}$ whose length can be measured accurately (Figure 4-4). They then measure each of the angles at A and B. These angles can be used to find $\angle APB$ and $\angle AQB$. The distance AP can be found from $\triangle APB$ by using the law of sines ($\angle APB$, $\angle ABP$, and AB are known). Similarly, AQ can be found from $\triangle AQB$. Since AP, AQ, and $\angle PAQ$ are now known, the required distance PQ can be found by applying the law of cosines to $\triangle PAQ$.

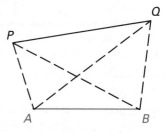

Figure 4-4

EXERCISES 4-2

Find the indicated part of $\triangle ABC$.

A 1. $a = 16; \angle A = 35°; \angle B = 65°; b = ?$ 2. $c = 10; \angle A = 48°; \angle C = 63°; a = ?$

3. $b = 2.1; \angle A = 110°; \angle C = 40°; a = ?$ 4. $a = 2.5; \angle B = 50°; \angle C = 100°; c = ?$

5. $c = 30; \angle A = 42°; \angle C = 98°; b = ?$ 6. $b = 120; \angle B = 105°; \angle C = 25°; a = ?$

7. $a = 18; b = 15; \angle A = 110°; \angle B = ?$ 8. $b = 8.5; c = 6.4; \angle B = 115°; \angle C = ?$

In Exercises 9 and 10, find the exact values of the indicated parts of $\triangle ABC$.

9. $b = 2\sqrt{6}; \angle A = 60°; \angle B = 45°; a = ?$ 10. $a = 6\sqrt{6}; \angle B = 45°; \angle C = 15°; b = ?$

In Exercises 11–14, find the exact value of $\dfrac{a}{b}$ in $\triangle ABC$. (Recall that $\sin^2 \theta + \cos^2 \theta = 1$.)

11. $\sin A = \dfrac{2}{3}; \cos B = \dfrac{3}{5}$ 12. $\cos A = \dfrac{4}{5}; \sin B = \dfrac{4}{5}$

13. $\cos A = \dfrac{12}{13}; \cos B = -\dfrac{4}{5}$ 14. $\cos A = \dfrac{15}{17}; \cos B = -\dfrac{8}{17}$

15. Two angles of a triangle measure 72° and 61°. If the longest side is 30 m long, how long is the shortest side?

16. Two angles of a triangle measure 37° and 48°. If the shortest side is 10.5 m long, how long is the longest side?

17. To find the distance between two points, A and B, on opposite sides of a swamp, a surveyor laid off a base line $\overline{AC}$ 25 m long and found that $\angle BAC = 82°$ and $\angle BCA = 69°$. Find AB.

18. A six-meter long loading ramp that makes a 25° angle with the horizontal is to be replaced by a ramp whose angle of inclination is only 10°. How long will the new ramp be?

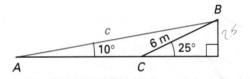

19. A pilot approaching a 3000-meter runway finds that the angles of depression of the ends of the runway are 14° and 20°. How far is the plane from the nearer end of the runway?

20. From the top of a building 25 m high the angle of elevation of a weather balloon is 54° and from the bottom of the building it is 61°. How high is the balloon above the ground?

Let △ABC be any triangle and let h be the altitude to side $\overline{AB}$. Show that the following are true.

B 21. $\dfrac{\sin A + \sin B}{\sin B} = \dfrac{a + b}{b}$

22. $\dfrac{\sin A - \sin B}{\sin B} = \dfrac{a - b}{b}$

23. $a = \dfrac{c \sin A}{\sin (A + B)}$

24. $h = \dfrac{c \sin A \sin B}{\sin (A + B)}$

25. A tree stands on an 18° slope. From a point 20 m down the slope, the angle of elevation of the top of the tree is 32°. How tall is the tree?

26. The skipper of a sailboat 6 km from the nearer of two towers on shore 10 km apart finds that the angle between the lines of sight to the towers is 35°. How far is the boat from the farther tower?

27. To find the height of a building across a canal, Lee laid off a base line, $\overline{AB}$, 50 m long and measured the angles shown. Lee then found the angle of elevation of the top of the building from A to be 20°. How tall is the building?

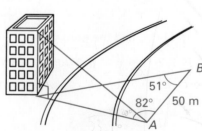

28. Points *P* and *Q* on the far bank of a river are sighted from points *A* and *B* on the near bank as shown at the right. The banks of the river are parallel and *AB* = 25 m. Find *PQ*.

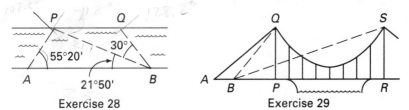

Exercise 28   Exercise 29

29. Find the distance *PR* between the equally-tall support towers of the suspension bridge shown at the right. Use these measurements: *AB* = 50 m, ∠ *PAQ* = 40°, ∠ *PBQ* = 52°, and ∠ *RBS* = 18°

Exercises 30 and 31 refer to triangulation (Figure 4-4, page 133). In each case *AB* = 120 m. Find *PQ* given the following angles.

C 30. ∠ *PAQ* = 80°, ∠ *QAB* = 35°, ∠ *ABP* = 28°, and ∠ *PBQ* = 85°

31. ∠ *PAQ* = 75°, ∠ *PAB* = 102°, ∠ *ABQ* = 127°, and ∠ *PBQ* = 81°

32. A tree stands on a 15° slope. Diana, standing directly down the slope from the tree, finds that the angle of elevation of its top is 35°. When she moves 50 m closer to the tree, the angle of elevation of its top is 45°. How tall is the tree?

33. A pilot of a transoceanic jet flying at 11,500 m finds that a stationary ship is in the same vertical plane as the jet's course. He measures the ship's angle of depression to be 16° and two minutes later finds it to be 46°. Find the speed of the jet.

34. Prove that sin 2θ = 2 sin θ cos θ (0° < θ < 90°) by following the pattern of Exercise 52, Section 4-1, but using the law of sines.

The circumscribed circle of △*ABC* is shown at the right.

35. Let *O* be the center of the circumscribed circle and let *R* be its radius. Show that

$$R = \frac{1}{2}\left(\frac{a}{\sin A}\right); R = \frac{1}{2}\left(\frac{b}{\sin B}\right); R = \frac{1}{2}\left(\frac{c}{\sin C}\right).$$

$$\left(\text{Hint: In the figure } \angle A = \frac{1}{2} \angle COB.\right)$$

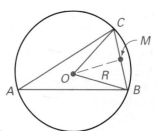

36. Explain how to use the formulas for *R* in Exercise 35 to give another proof of the law of sines.

37. Describe a geometric construction to locate the center of the circumscribed circle.

Self Quiz

In Exercises 1–4, find the asked-for part of $\triangle ABC$.

1. $a = 8.5; \angle B = 28°; \angle C = 42°; b = ?$

2. $a = 18.0; b = 30.5; \angle C = 115°; c = ?$

3. $a = 12.0; b = 28.0; c = 32.0; \angle B = ?$

4. $a = 120; b = 175; \angle B = 65°; \angle A = ?$

5. A vertical pole stands on a 20° slope. From a point 50 m down the slope from the pole, the angle of elevation of the top of the pole is 35°. How tall is the pole?

6. The sides of a triangular city lot have lengths 64 m, 76 m, and 20 m. Find the largest angle the sides make with each other.

GENERAL OBLIQUE TRIANGLES

| **4-3** | **Solving General Triangles** |

Objective: To analyze and handle the four cases in triangle solving.

The problem of solving triangles can be divided into four cases according to which measurements are given:

    SSS: Given three sides.
    SAS: Given two sides and the included angle.
    SSA: Given two sides and the angle opposite one of them.
    SAA: Given one side and two angles.

You may recall that SSS, SAS, and SAA are used in geometry to indicate ways of proving triangles congruent. For example, two triangles that have corresponding sides of the same length must be congruent, and we say that three sides *determine* a triangle. But SSA does not necessarily determine a unique triangle. As you will see, there may be no triangle with given SSA measurements, or there may be one, or there may be two.

    Recall that to *solve* a triangle means to find all three missing measurements. The SSS and SAS cases can be solved using only the law of cosines. It usually is simpler, however, to start the solution using the law of cosines but continue it with the law of sines.

Example 1   (The SSS case) Solve $\triangle ABC$ given $a = 53$, $b = 70$, and $c = 64$.

Solution   First find the largest angle. Since b is greater than a and c, $\angle B$ is the largest angle.

Step 1   Use the law of cosines to find $\angle B$.

$$\cos B = \frac{a^2 + c^2 - b^2}{2ac} = \frac{53^2 + 64^2 - 70^2}{2 \cdot 53 \cdot 64} = 0.2955$$

$\angle B = 72.8°$

Step 2   Use the law of sines to find $\angle A$.

$$\sin A = \frac{a \sin B}{b} = \frac{53 \sin 72.8°}{70} = 0.7233$$

There are two angles between $0°$ and $180°$ having 0.7233 as sine, namely $46.3°$ and $133.7°$, but since $\angle B$ is the largest angle of $\triangle ABC$, we have $\angle A = 46.3°$.

Step 3   Use the fact that $\angle A + \angle B + \angle C = 180°$ to find $\angle C$.
$\angle C = 180° - (\angle A + \angle B) = 180° - (46.3° + 72.8°) = 60.9°$

Therefore, $\angle A = 46.3°$, $\angle B = 72.8°$, and $\angle C = 60.9°$.   ■

A solution of a triangle can be checked by using a formula that involves all six parts of the triangle. For example, we could use the formula:

$$\frac{\sin A + \sin B}{\sin C} = \frac{a + b}{c}$$

Checking Example 1 we find that:

$$\frac{\sin A + \sin B}{\sin C} = \frac{\sin 46.3° + \sin 72.8°}{\sin 60.9°} \qquad \left| \quad \frac{a + b}{c} = \frac{53 + 70}{64} \right.$$

$$= \frac{0.7230 + 0.9553}{0.8738} = 1.921 \qquad \left| \qquad\qquad = 1.922 \right.$$

This is a satisfactory check.

Example 2   (The SAS case) Solve $\triangle ABC$ given $b = 3.5$, $c = 2.5$, and $\angle A = 37°$.

Solution   **Step 1**   Use the law of cosines to find a.

$$a^2 = b^2 + c^2 - 2bc \cos A$$
$$= 3.5^2 + 2.5^2 - 2 \cdot 3.5 \cdot 2.5 \cos 37° = 4.52$$

Hence, $a = \sqrt{4.52} = 2.13$.

Step 2   Use the law of sines to find the smaller of $\angle B$ and $\angle C$.

$$\sin C = \frac{c \cos A}{a} = \frac{2.5 \sin 37°}{2.13} = 0.7064$$

Hence, $\angle C = 44.9°$.

Step 3 $\angle B = 180° - (\angle A + \angle C) = 180° - (37° + 44.9°) = 98.1°$
Hence, $a = 2.1$, $\angle B = 98.1°$, $\angle C = 44.9°$.   ■

Example 3   (The SAA case) Solve $\triangle ABC$ given $\angle A = 42°$, $\angle B = 34°$, $c = 42.5$.

Solution   **Step 1** $\angle C = 180° - (\angle A + \angle B) = 180° - (42° + 34°) = 104°$

Step 2 $a = \dfrac{c \sin A}{\sin C} = \dfrac{42.5 \sin 42°}{\sin 104°} = 29.3$

Step 3 $b = \dfrac{c \sin B}{\sin C} = \dfrac{42.5 \sin 34°}{\sin 104°} = 24.5$

Thus, $a = 29.3$, $b = 24.5$, $\angle C = 104°$.   ■

SSA is called the **ambiguous case** because there may be two, one, or no solutions. Figure 4-5 illustrates this fact for the case when a, b, and $\angle A$ are given and $\angle A$ is acute.

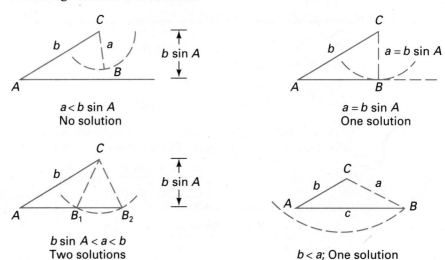

Figure 4-5

If $\angle A$ is obtuse, there is one solution if $a > b$ and no solution if $a \leq b$.

Example 4   In $\triangle ABC$, $\angle A = 42°$ and $b = 35$. Find $\angle B$ given that:
(a) $a = 20$ (b) $a = 30$ (c) $a = 40$

Solution   Use $\dfrac{\sin A}{a} = \dfrac{\sin B}{b}$ in the form $\sin B = \dfrac{b \sin A}{a}$.

(a) $\sin B = \dfrac{35 \sin 42°}{20} = 1.1710$

There is no solution because $\sin B$ cannot be greater than 1.

(b) $\sin B = \dfrac{35 \sin 42°}{30} = 0.7807$. Two solutions: $\angle B = 51.3°$ or $128.7°$.

(c) $\sin B = \dfrac{35 \sin 42°}{40} = 0.5855$; $\angle B = 35.8°$ or $144.2°$

But if $\angle B = 144.2°$, then $\angle A + \angle B = 42° + 144.2° = 186.2°$. Since $186.2° > 180°$, $\angle B$ cannot equal $144.2°$. Therefore, there is only one solution: $\angle B = 35.8°$.   ■

Complete solutions of $\triangle ABC$ in parts (b) and (c) can be obtained by using $\angle C = 180° - (\angle A + \angle B)$ and $c = \dfrac{a \sin C}{\sin A}$. There will, of course, be two complete solutions in part (b).

In most applications involving triangle solving it is not necessary to find a complete solution. Example 5 is an exception because all parts of a triangle must be found in order to find the one part needed.

Example 5

A derrick at the end of a dock has an arm 25 m long that makes an angle of 122° with the floor of the dock. The arm is to be braced with a cable 40 m long from the end of the arm back to the dock. How far from the edge of the dock will the cable be fastened?

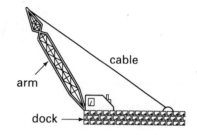

Solution

Draw and label a sketch for this SSA problem. The goal is to find c.

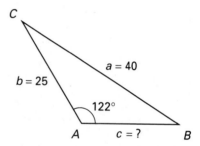

Step 1 Find $\angle B$.

$$\sin B = \frac{b \sin A}{a}$$

$$= \frac{25 \sin 122°}{40} = 0.5300$$

$$\angle B = 32.0°$$

Step 2 Find $\angle C$.
$$\angle C = 180° - (\angle A + \angle B) = 180° - (122° + 32°) = 26°$$

Step 3 Find c: $c = \dfrac{a \sin C}{\sin A} = \dfrac{40 \sin 26°}{\sin 122°} = 20.7$

Therefore, the cable is fastened 20.7 m from the edge of the dock. ■

EXERCISES 4-3

Solve $\triangle ABC$. If no solution exists, so state. If there are two solutions, find both.

A 1. $a = 21$; $c = 30$; $\angle B = 42°$

2. $a = 16$; $\angle B = 32°$; $\angle C = 50°$

3. $b = 14$; $\angle B = 25°$; $\angle C = 110°$

4. $a = 5$; $b = 8$; $c = 10$

5. $a = 2.3$; $b = 3.7$; $c = 5.0$

6. $b = 120$; $c = 145$; $\angle A = 100°$

7. $b = 20$; $c = 15$; $\angle B = 115°$

8. $a = 30$; $b = 20$; $\angle A = 130°$

9. $a = 12; b = 15; \angle A = 55°$ 10. $a = 12; b = 7; \angle B = 35°$

11. $a = 5.2; b = 3.9; c = 6.5$ 12. $b = 13.4; c = 6.7; \angle C = 30°$

13. $b = 15; c = 13; \angle C = 50°$ 14. $b = 1.1; c = 1.8; \angle B = 40°$

15. If $\angle B$ is acute, what condition must b, c, and $\angle B$ satisfy in order that there be at least one triangle having these parts?

16. Draw diagrams similar to those in Figure 4-5 to illustrate the two SSA cases where $\angle A$ is obtuse.

17. A monument consists of a 20 m flagpole standing on a mound in the shape of a cone with vertex angle 140°. How long a shadow does the pole cast on the cone when the angle of elevation of the sun is 58°?

18. Ann is flying a plane on a triangular course at 400 km/h. She flies due east for two hours and then turns left through a 15° angle measured clockwise from north. How long after turning will she be exactly northeast of where she started?

19. John is flying a plane from Upton to Vista, a distance of 500 km. Because of a storm between the two cities he has flown 17.5° off course for 300 km. How far is he now from Vista and through what angle should he turn to fly directly there?

20. Maria hears the 4:00 P.M. whistle of Wilson Industries at 10 seconds after the hour and she hears the 4:00 P.M. whistle of Ramos Manufacturing 8 seconds later. If the angle between Maria's lines of sight to the two plants is 56°, how far apart are they? (The speed of sound is 340 m/s.)

B 21. A communication satellite is in orbit 35,800 km above the equator. It completes one orbit every 24 hours, so that from Earth it appears to be stationary above a point on the equator. If this point has the same longitude as Houston, find the measure of θ, the satellite's angle of elevation from Houston. The latitude of Houston is 29.7° N; take the radius of Earth to be 6400 km.

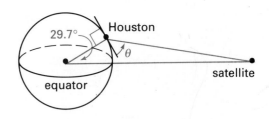

22. What is the greatest latitude from which a signal can travel to the satellite of Exercise 21 in a straight line?

23. A kite 2.5 m long is a quadrilateral having two sides each 1 m long and two sides each 2 m long. How wide is the kite? (That is, what is the length of the shorter diagonal?)

24. From the top of a tower 80 m above sea level, an observer sights a sailboat at an angle of depression of 9°. Turning in a different direction he sights another sailboat at an angle of depression of 12°. The angle between these lines of sight is 36°. How far apart are the boats?

25. To cross a river, an explorer swings on a 100-foot vine attached to a tree leaning over the river at a 45° angle, as shown at the right. The vine is attached to the tree 120 feet from its base. How wide is the river?

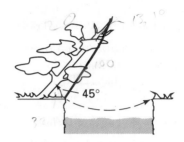

26. Show that in any triangle ABC, $c = a \cos B + b \cos A$. (Hint: Consider separately the cases where both $\angle A$ and $\angle B$ are acute and where one of them is obtuse. Draw figures.)

Exercises 27–30 refer to the figure at the right where $CB' = CB$.

27. Given that $a = 8$, $b = 13$, and $\angle A = 30°$, find $\angle BCB'$.

28. Given the measures in Exercise 27, find $\angle ACB'$.

C 29. Show that $\dfrac{\text{area } \triangle ABC}{\text{area } \triangle AB'C} = \dfrac{\sin \angle ACB}{\sin \angle ACB'}$.

Exercises 27-30

30. Show that the ratio in Exercise 29 equals $\dfrac{\sin (B + A)}{\sin (B - A)}$.

31. In $\triangle ABC$, $a = 3$, $b = 5$, and $\angle C = 120°$. Find the length of the median to the longest side.

32. In $\triangle ABC$, $c = 10$, and $\angle A = \angle B = 50°$. Find the length of the median to $\overline{AC}$.

33. Find the lengths of the diagonals of the trapezoid shown below.

34. Find the lengths of the diagonals of the quadrilateral shown below.

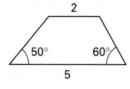

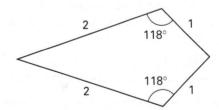

35. Given $\triangle ABC$ with $\angle C \neq 90°$, show that $\dfrac{\cos A}{\cos B} = \dfrac{b}{a}$ implies that $\triangle ABC$ is isosceles. (Hints: Explain why the given equation implies that $\angle A$ and $\angle B$ are acute. Then combine the equation with the law of sines and use a double-angle formula.)

4-4   Areas of Triangles

Objective: To apply formulas for the areas of triangles.

In geometry you learned that the area K of a triangle is given by the formula $K = \frac{1}{2}$(base)(height). This formula can easily be transformed to find the area of any SAS-case triangle: Suppose that b, c, and $\angle A$ are known. Let b be the base. Then the height $h = c \sin A$ (Figure 4-6), and therefore the area $K = \frac{1}{2} b \cdot c \sin A$, or $K = \frac{1}{2} bc \sin A$. This is the first of the following.

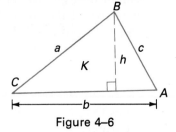

Figure 4–6

The area K of $\triangle ABC$ is given by each of the following:

$$K = \frac{1}{2} bc \sin A \qquad K = \frac{1}{2} ac \sin B \qquad K = \frac{1}{2} ab \sin C$$

$$K = \frac{1}{2} a^2 \frac{\sin B \sin C}{\sin A} \qquad K = \frac{1}{2} b^2 \frac{\sin A \sin C}{\sin B} \qquad K = \frac{1}{2} c^2 \frac{\sin A \sin B}{\sin C}$$

$$K = \sqrt{s(s - a)(s - b)(s - c)}, \text{ where } s = \frac{1}{2}(a + b + c)$$

From the law of sines we find that $b = \dfrac{a \sin B}{\sin A}$, and when this is substituted into $K = \frac{1}{2} ab \sin C$, we obtain $K = \frac{1}{2} a^2 \dfrac{\sin B \sin C}{\sin A}$, the first formula on the second line of the box above. These second-line formulas are well suited to the ASA case.

The last formula in the box above is **Hero's** (or **Heron's**) **formula** and is proved in Exercises 36–38, page 145. It clearly fits the SSS case. An SSS problem can, of course, be solved by first using the law of cosines to find an angle and then using an SAS area formula.

Example 1   Find the area of $\triangle ABC$ in which $a = 14.8$, $c = 23.5$, and $\angle B = 148.5°$.

Solution   Use $K = \frac{1}{2} ac \sin B$. Thus, $K = \frac{1}{2}(14.8)(23.5) \sin 148.5° = 90.9$.

The area of $\triangle ABC$ is 90.9 square units. ∎

Example 2   Find the area of $\triangle ABC$ given $b = 3.16$ cm, $\angle B = 62.7°$, $\angle C = 65.8°$.

Solution   First find $\angle A$. Then use the formula $K = \frac{1}{2} b^2 \dfrac{\sin A \sin C}{\sin B}$.

$$\angle A = 180° - (\angle B + \angle C) = 180° - (62.7° + 65.8°) = 51.5°$$

$$K = \frac{1}{2}(3.16)^2 \frac{\sin 51.5° \sin 65.8°}{\sin 62.7°} = 4.01$$

The area is 4.01 cm². ∎

Example 3 Find the area of a lot having sides 41.4 m, 27.3 m, and 38.7 m.

Solution Use $K = \sqrt{s(s - a)(s - b)(s - c)}$, where $s = \frac{1}{2}(a + b + c)$.

$$s = \frac{1}{2}(41.4 + 27.3 + 38.7) = 53.7$$

$$s - a = 12.3, \quad s - b = 26.4, \quad s - c = 15.0$$
$$K = \sqrt{53.7 \cdot 12.3 \cdot 26.4 \cdot 15.0} = 511$$

The area is 511 m². ∎

Examples 1, 2, and 3 illustrate how to find areas of triangles in cases SAS, ASA, and SSS, respectively. In the ambiguous case, SSA, the triangle must be partially solved before its area, if any, can be found.

Example 4 Find the area of $\triangle ABC$ given $a = 16.0$, $b = 26.2$, and $\angle A = 32.5°$.

Solution Start to solve $\triangle ABC$. Then use $K = \frac{1}{2}ab \sin C$.

$$\sin B = \frac{b \sin A}{a} = \frac{26.2 \sin 32.5°}{16.0} = 0.8798$$

$\angle B = 61.6°$	or $\angle B = 118.4°$
$\angle C = 180° - (32.5° + 61.6°)$	$\angle C = 180° - (32.5° + 118.4°)$
$= 85.9°$	$= 29.1°$
$K = \frac{1}{2}(16.0)(26.2) \sin 85.9°$	$K = \frac{1}{2}(16.0)(26.2) \sin 29.1°$
$= 209$ square units or	$= 102$ square units

Example 5 Find the area of a regular pentagon inscribed in a circle of radius 7.

Solution Pentagon $ABCDE$ can be partitioned into five congruent triangles.

$$\text{Area} = 5\left[\frac{1}{2}(OA)(OB)(\sin \angle AOB)\right]$$

$$= 5\left[\frac{1}{2}(7)(7)(\sin 72°)\right]$$

$$= 5[23.30] = 116.5$$

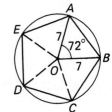

The area is 116.5 square units. ∎

EXERCISES 4-4

A 1–14. Find the areas of the triangles described in Exercises 1–14 on pages 139–140. If no triangle exists, so state. If there are two triangles, find the area of each.

15. Find the area of a parallelogram that has a 62° angle and sides of lengths 16 and 30.

16. Find the area of a rhombus having perimeter 72 cm and a 56° angle.

17. Find the area of a parallelogram that has sides of lengths 20 m and 30 m and a diagonal 40 m long.

18. Find the area of a parallelogram whose diagonals have lengths 24 and 36 and cross at a 48° angle.

Find the area of a regular polygon of *n* sides inscribed in a circle of radius *r*.

19. $n = 12; r = 4$ 20. $n = 9; r = 9$ 21. $n = 8; r = 10$ 22. $n = 15; r = 8$

B 23. A triangle has area 24 m², and two of its angles measure 62° and 78°. How long is its longest side?

24. How long is the shortest side of the triangle in Exercise 23?

25. Find the area of the quadrilateral shown at the left below.

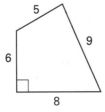

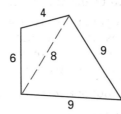

Exercise 25   Exercise 26

26. Find the area of the quadrilateral shown at the right above.

27. Explain how to find the area of a convex quadrilateral knowing the lengths of its sides and one of its angles.

28. Explain how to find the area of a convex quadrilateral knowing the lengths of its sides and one of its diagonals.

29. Let P be any point on side $\overline{AB}$ of $\triangle ABC$. Let $l = CP$ and $\phi = \angle APC$. Express the area K of $\triangle ABC$ in terms of c, l, and ϕ.

C 30. The diagonals of a convex quadrilateral have lengths p and q and meet at an angle ϕ. Express the area K of the quadrilateral in terms of p, q, and ϕ.

31. A pyramid has a square base 314 cm on a side as shown. Each lateral edge makes an angle of 22°20′ with a diagonal of the base. Find the lateral area.

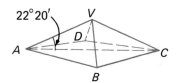

The **inscribed circle** of △ABC is the circle inside △ABC that is tangent to the three sides of the triangle as shown at the right.

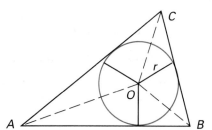

32. Let O be the center of the inscribed circle and let its radius be r. Show that

$$r = \frac{\sqrt{s(s-a)(s-b)(s-c)}}{s}$$

where $s = \frac{1}{2}(a + b + c)$. (Hint: Use area △ABC = area △BOC + area △COA + area △AOB and apply Hero's formula.)

33. Find the area of the inscribed circle of △ABC in which $a = 9$, $b = 13$, and $\angle C = 38°$.

34. Find the area of the region inside △ABC but outside the inscribed circle of △ABC in which $a = 12$, $b = 15$, and $c = 20$.

35. (a) Give a formula for the area of a regular n-sided polygon inscribed in a circle of radius r, in terms of r and n.

   (b) Change the formula in part (a) so that any angles are given in radian measure. Then use the approximation $\sin x \approx x$ for small angles in radian measure to show that as n gets large, the area of the polygon approaches πr^2. Explain this geometrically.

Exercises 36–38 give a sequential proof of Hero's formula.

36. Prove that the area of △ABC is given by the formula:

$$\text{Area} = \frac{1}{2}ab \sqrt{1 - \left(\frac{a^2 + b^2 - c^2}{2ab}\right)^2} = \frac{1}{4}\sqrt{4a^2b^2 - (a^2 + b^2 - c^2)^2}$$

(Hint: Use the law of cosines to solve for $\cos C$.)

37. Show that in △ABC with $s = \dfrac{a + b + c}{2}$,

$$s(s - c) = \frac{(a + b)^2 - c^2}{4} \quad \text{and} \quad (s - a)(s - b) = \frac{c^2 - (a - b)^2}{4}.$$

38. (a) Show that $s(s - a)(s - b)(s - c) = \dfrac{4a^2b^2 - (a^2 + b^2 - c^2)^2}{16}$.

   (b) Use this and the result of Exercise 37 to prove Hero's formula.

39. Prove that the area of △ABC is given by $\text{Area} = \dfrac{1}{2}c^2 \dfrac{\sin A \sin B}{\sin C}$.

Self Quiz

1–4. Find the areas of the triangles described in Exercises 1–4 of the Self Quiz on page 136.

5. If $\angle B = 30°$ and $c = 16$, tell the number of triangles ABC with (a) $b = 20$ (b) $b = 5$ (c) $b = 10$ (d) $b = 8$.

6. A triangle has area 72 cm², and two of its angles measure 44° and 62°. How long is its longest side?

7. A ship was heading for port 600 km away when to avoid a storm it sailed for 400 km 18° off course. How far is the ship now from port and through what angle should it turn to head directly there?

ADDITIONAL PROBLEMS

In Exercises 1–6, solve △ABC. If no triangle exists, so state. If there are two triangles, find the parts of both.

1. $a = 18; c = 15; \angle B = 116°$

2. $a = 42; c = 38; \angle A = 58°$

3. $c = 5.8; \angle B = 48°; \angle A = 100°$

4. $b = 28; c = 25; \angle C = 118°$

5. $b = 54; c = 64; \angle B = 51°$

6. $a = 4.8; b = 3.7; c = 7.3$

7–12. Find the areas of the triangles in Exercises 1–6 in the cases where triangles exist.

13. A pilot approaching a 5000 m runway finds that the angles of depression of the ends of the runway are 19° and 11°. How far is the plane from the farther end of the runway?

14. A jib sail is in the form of a triangle having sides 2.5 m, 3.8 m, and 4.9 m long. Find its largest angle.

15. Find the area of the sail in Exercise 14.

16. A kite is in the form of a quadrilateral $ABCD$ with $AB = AD = 60$ cm, $CB = CD = 100$ cm, and the long diagonal $AC = 120$ cm. How wide is the kite?

17. Two planes leave the same airport at 1:00 P.M., one flying due east at 500 km/h, and the other flying due northwest at 600 km/h. How far apart are the planes at 3:00 P.M.?

18. Seaport A is 60 km due north of seaport B. At noon a ship leaves A and travels due southeast at 20 km/h. At what time(s) is the ship 50 km from B?

19. Two fire lookout stations, A and B, are 8.6 km apart. They spot a plume of smoke at point F. How far is F from the nearer station if $\angle FAB = 32°$ and $\angle FBA = 75°$?

20. A tetrahedron is a solid having four triangular faces. Find its total area if each of its edges is 10 cm long.

21. In an SSA case $\angle A$, a, and b are given. There are two solutions: $\triangle ABC$ and $\triangle AB'C$. Express BB' in terms of the given parts.

22. In a trapezoid $ABCD$, $\overline{AB}$ and $\overline{CD}$ are parallel, $AB = 5$, $CD = 2$, $\angle A = 60°$, and $\angle B = 40°$. Find the lengths of the nonparallel sides.

CHAPTER SUMMARY

1. The law of cosines states that in any $\triangle ABC$,

$$a^2 = b^2 + c^2 - 2bc \cos A,$$

with similar equations involving $\angle B$ and $\angle C$. It often is useful to put the law into the form

$$\cos A = \frac{b^2 + c^2 - a^2}{2bc}.$$

2. The law of sines states that in any $\triangle ABC$,

$$\frac{\sin A}{a} = \frac{\sin B}{b} = \frac{\sin C}{c}.$$

This law is most frequently used in one of the forms

$$a = \frac{b \sin A}{\sin B} \text{ or } \sin A = \frac{a \sin B}{b}.$$

3. Depending on the parts given, triangle-solving problems fall into four cases:

    SSS   Three sides given

    SAS   Two sides and the included angle given

    SSA   Two sides and the angle opposite one of them given

    SAA   One side and two angles given

The ambiguous case, SSA, may determine no triangle, one triangle, or two triangles.

4. The area K of $\triangle ABC$ may be found by using a formula of one of the following types:

$$K = \frac{1}{2}\,ab\,\sin C$$

$$K = \frac{1}{2}\,a^2\,\frac{\sin B \sin C}{\sin A}$$

$$K = \sqrt{s(s-a)(s-b)(s-c)},\ \text{where } s = \frac{1}{2}(a+b+c)$$

CHAPTER TEST

In Exercises 1–5, solve each triangle.

4-1
4-2
4-3

1. $b = 20;\ c = 15;\ \angle B = 115°$
2. $b = 12;\ \angle B = 25°;\ \angle C = 110°$
3. $a = 20;\ b = 30;\ c = 40$

4. $b = 15;\ c = 13;\ \angle C = 50°$

5. $a = 20;\ c = 30;\ \angle B = 40°$

4-1

6. Towns A and B are 8 km apart, and Lyn is 5 km from A and 7 km from B. How far is she from the midpoint of $\overline{AB}$?

4-2

7. Guy wires bracing a flagpole start at the same point on the ground and make angles of 28° and 40° with the pole. The shorter wire is 18 m long. How long is the longer wire?

4-4

8–12. Find the areas of the triangles in Exercises 1–5 above.

Changing Angles and Changing Areas

Is the following statement true? If the measure of the angle C between two sides $\overline{AC}$ and $\overline{BC}$ of a triangle increases, then the area K of the triangle increases.

We might be tempted to say "yes." However, the following computer-generated tables help us see that the statement is sometimes true and sometimes false. In the tables, C is the measure of the angle between $\overline{AC}$ with length 3 and $\overline{BC}$ with length 2, and K is the area of $\triangle ABC$.

C	K
51°	2.3314
52°	2.3640
53°	2.3959
54°	2.4270
55°	2.4575
56°	2.4871
57°	2.5160
58°	2.5441
59°	2.5715

C	K
86°	2.9927
87°	2.9959
88°	2.9982
89°	2.9995
90°	3.0000
91°	2.9995
92°	2.9982
93°	2.9959
94°	2.9927

C	K
121°	2.5715
122°	2.5441
123°	2.5160
124°	2.4871
125°	2.4575
126°	2.4271
127°	2.3959
128°	2.3640
129°	2.3314

The tables suggest that K increases when C increases from 0° to 90° and that K decreases when C increases from 90° to 180°.

EXERCISES

1. Graph $K = \dfrac{1}{2}(3)(2) \sin C = 3 \sin C$ for $0° \le C \le 180°$.

2. Show that if two sides of one triangle are congruent to two sides of another triangle and the included angles are supplementary, then the triangles have the same area. Give an example from the tables to confirm this.

3. Show that if $\overline{AC}$ and $\overline{BC}$ have fixed lengths, then the area of $\triangle ABC$ is maximum when $\angle C = 90°$.

149

Measurement and Error

Surveyors, carpenters, engineers, machinists, and others must be concerned with error in measurement, since every measurement is an approximation of true measure.

The following figure shows a ruler marked off in whole-number units and a line segment about 12 units long.

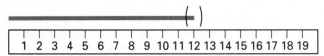

We cannot reliably say that the segment is 12.0, 12.1, or 11.9 units long, because the ruler is not marked in tenths of a unit. Rather, we can say that it is 12 ± 0.5 units long. That is, the length of the segment is somewhere between 11.5 and 12.5 units.

What difference does error in measurement make? To find out, suppose that two sides of a triangle measure 12 ± 0.5 and 15 ± 0.5 and that the angle between the sides measures $32° \pm 0.5°$. The figure at the right shows the minimum measures in black and the maximum measures in red. Then the third side has length between

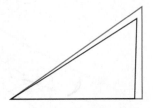

$$\sqrt{11.5^2 + 14.5^2 - 2(11.5)(14.5)\cos 31.5°} = 7.63 \text{ units}$$

and

$$\sqrt{12.5^2 + 15.5^2 - 2(12.5)(15.5)\cos 32.5°} = 8.35 \text{ units}.$$

Using 12 units, 15 units, and 32°, we obtain 7.98 units as the length of the third side. Notice that each of these three calculations is within 8 ± 0.5 units. Thus, we can write the length of the third side as 8 units and make no confident claim about further precision.

If, however, we compute the area of the triangle, we find that it is between

$$\frac{1}{2}(11.5)(14.5)\sin 31.5° = 43.56 \text{ square units}$$

and

$$\frac{1}{2} (12.5) (15.5) \sin 32.5° = 52.05 \text{ square units.}$$

Using 12 units, 15 units, and 32°, we obtain 47.81 square units. These three calculations are within 5 square units of 48 square units. We can safely say that the area is 48 square units while acknowledging that it could be as much as 10% more or 10% less than 48 square units.

EXERCISES

1. A vertical pole 19 ± 0.5 feet tall is supported by a guy wire that is 10 ± 0.5 feet from the base of the pole. What are the minimum and maximum angles that the guy wire makes with the ground?

2. Two sides of a triangle measure $15{,}000 \pm 0.5$ and 4 ± 0.5. The included angle measures $2° \pm 0.5°$.   (a) Find the minimum and maximum lengths of the third side.   (b) Find the minimum and the maximum area.

5 *Sinusoidal Variation*

Trigonometric functions are useful in the study of sounds produced by musical instruments, since physicists view sounds as waves having definite frequencies, periods, and cycles.

GRAPHING SINUSOIDS

| 5-1 | **Amplitude and Period** |

Objective: To graph sinusoids having various amplitudes and periods.

Many applications of trigonometry lead to functions defined by equations of the form

$$y = a \sin b(x - c) + d \text{ or } y = a \cos b(x - c) + d.$$

We call such functions, as well as their graphs, **sinusoids**, and we say that y **varies sinusoidally** with x. In this section we shall study the effects of constants a and b, leaving c and d to the next section.

Example 1　In the same coordinate plane, graph: $y = \sin x$, $y = 2 \sin x$, $y = \frac{1}{2} \sin x$.

Solution　We have already graphed $y = \sin x$ in Section 2-4. To obtain the sinusoid $y = 2 \sin x$, we simply double each y-coordinate of the graph of $y = \sin x$. Similarly, to graph $y = \frac{1}{2}\sin x$, we halve each y-coordinate of $y = \sin x$.

(Solution continued next page.)

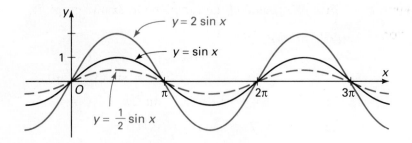

The three functions in Example 1 have the same period, namely 2π. However, they differ in how far the highest and lowest points of their graphs deviate from their central axis (Figure 5-1). We call this deviation the *amplitude* of the sinusoid.

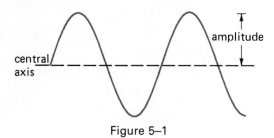

Figure 5–1

In general,

> The amplitude of any periodic function f is
> $$\frac{\text{maximum value of } f - \text{minimum value of } f}{2}.$$

For example, in the case of the sinusoid $y = 2 \sin x$, the amplitude is $\frac{2 - (-2)}{2} = \frac{4}{2} = 2.$

Since the maximum and minimum values of both $\sin x$ and $\cos x$ are 1 and -1 respectively, we have:

> Let $g(x)$ be any linear function. Then any function of the form
> $$a \sin (g(x)) \qquad \text{or} \qquad a \cos (g(x))$$
> has amplitude $|a|$.

Recall from Section 2-3 that the (fundamental) period of a periodic function f is the least positive number p such that $f(x + p) = f(x)$ for all x in the domain of f.

Example 2   Find the period of the function sin 3x and graph it.

Solution   Let $f(x) = \sin 3x$. Then,

$$f(x + p) = \sin 3(x + p)$$
$$= \sin (3x + 3p).$$

Therefore, $f(x + p) = f(x)$ if $\sin (3x + 3p) = \sin 3x$. This will hold if $3p = 2\pi$. Therefore, the period is $p = \dfrac{2\pi}{3}$.

To graph $y = \sin 3x$, note that:
(1) The amplitude of sin 3x is 1.
(2) The period of sin 3x is one third the period of sin x. That is, sin 3x completes three cycles while sin x completes one.

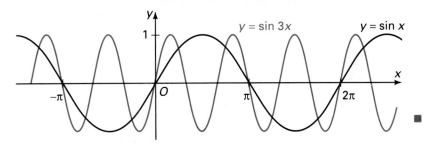

Using the method of Example 2, you can show that:

> Any function of the form $a \sin bx$ or $a \cos bx$, where $b > 0$, has period $\dfrac{2\pi}{b}$.

Example 3   Graph $y = 2 \cos \dfrac{\pi}{2}x$.

Solution   The amplitude is 2, and the period is $\dfrac{2\pi}{\dfrac{\pi}{2}}$, or 4. The graph is shown below.

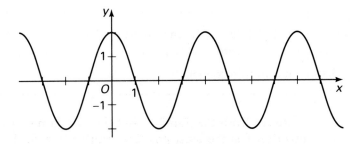

Example 4   Graph $y = -2 \operatorname{Sin}^{-1} \dfrac{x}{3}$.

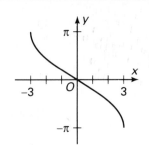

Solution

$$-\frac{y}{2} = \operatorname{Sin}^{-1} \frac{x}{3}$$

By definition of the inverse sine (page 69),

$$\sin\left(\frac{-y}{2}\right) = \frac{x}{3} \text{ and } -\frac{\pi}{2} \le -\frac{y}{2} \le \frac{\pi}{2}.$$

Since $\sin(-\theta) = -\sin\theta$, this becomes:

$$-\sin\frac{y}{2} = \frac{x}{3} \quad \text{and} \quad \pi \ge y \ge -\pi$$

Therefore, graph $x = -3\sin\dfrac{y}{2}$ in the interval $-\pi \le y \le \pi$. ∎

EXERCISES 5-1

Find the amplitude and period of each function. Then match each function with its graph.

A 1. $y = 2\cos x$

2. $y = -3\sin x$

3. $y = 3\cos 2x$

4. $y = 2\sin \pi x$

(a)

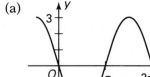

(b)

(c)

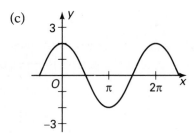

(d)

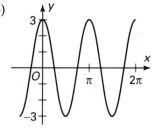

Sketch two cycles of each graph.

5. $y = 3\sin x$ 6. $y = 4\cos x$ 7. $y = -2\cos x$

8. $y = -\dfrac{3}{2}\sin x$ 9. $y = \sin 3x$ 10. $y = \cos 2x$

11. $y = 3\sin 2x$ 12. $y = 2\cos\dfrac{x}{2}$ 13. $y = \dfrac{3}{2}\cos\dfrac{2x}{3}$

14. $y = \dfrac{2}{3} \sin \dfrac{3x}{2}$ **15.** $y = -\sin \dfrac{5x}{2}$ **16.** $y = -3 \cos 3x$

17. $y = 2 \sin \pi x$ **18.** $y = \cos 2\pi x$ **19.** $y = -\cos \dfrac{\pi x}{2}$

20. $y = -2 \sin \dfrac{2\pi x}{3}$ **21.** $y = |2 \sin x|$ **22.** $y = |-\cos x|$

B 23. $y = |\cos 2x|$ **24.** $y = -\dfrac{1}{2} |\sin x|$ **25.** $y = 2 \left| \cos \dfrac{x}{2} \right|$

Graph each function in the interval $-2\pi \le x \le 2\pi$.

26. $y = \cos 2 |x|$ **27.** $y = \sin 2 |x|$ **28.** $y = \sin \pi |x|$

Find an equation of the form $y = a \sin bx$ or $y = a \cos bx$ for each graph given.

29.

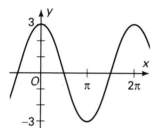

30.

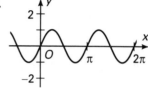

31.

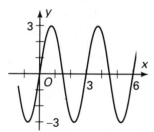

32.

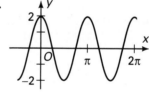

Give the domain and range of the function defined by each equation. Graph the equation.

33. $y = 2 \text{Cos}^{-1} x$ **34.** $y = 2 \text{Sin}^{-1} x$ **35.** $y = \text{Sin}^{-1} \dfrac{x}{2}$

36. $y = \text{Cos}^{-1} \dfrac{x}{2}$ **37.** $y = 2 \text{Tan}^{-1} x$ **38.** $y = \dfrac{1}{2} \text{Cot}^{-1} x$

Graph each of the following.

39. $y = 3 \text{Sin}^{-1} \dfrac{x}{2}$ **40.** $y = 2 \text{Cos}^{-1} \dfrac{x}{3}$ **41.** $y = \dfrac{2}{\pi} \text{Cos}^{-1} \dfrac{x}{4}$

42. $y = \dfrac{1}{\pi} \text{Sin}^{-1} 2x$ **43.** $y = -2 \text{Sin}^{-1} \dfrac{x}{2}$ **44.** $y = -\text{Cos}^{-1} \dfrac{x}{2}$

5-2   Phase Shift and Vertical Shift

Objective: To graph the general sinusoid $y = a \sin b(x - c) + d$.

Recall from algebra that you can obtain the graph of $y = (x - 3)^2$ by shifting the graph of $y = x^2$ to the right 3 units (Figure 5-2). The graph of $y = x^2 + 2$ can be obtained by shifting the graph of $y = x^2$ up 2 units (Figure 5-3).

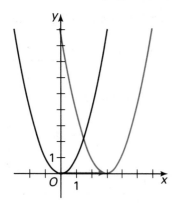   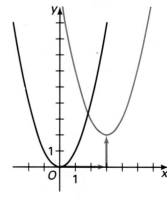

Figure 5-2   Figure 5-3   Figure 5-4

A shift of a curve in which its final position is parallel to or coincident with its original position is called a **translation**. The effect of a horizontal translation followed by a vertical one is illustrated by the graph of $y = (x - 3)^2 + 2$ in Figure 5-4. In general, we have the following rule.

The graph of $y = f(x - c) + d$ can be obtained by translating the graph of $y = f(x)$

to the right c units if $c > 0$ or to the left $|c|$ units if $c < 0$,

and

d units upward if $d > 0$ or $|d|$ units downward if $d < 0$.

Example 1   Graph $y = \cos\left(x - \dfrac{\pi}{2}\right) + 2$.

Solution   The given equation is of the form $y = f(x - c) + d$, where $f(x) = \cos x$, $c = \dfrac{\pi}{2}$ and $d = 2$. Therefore, translate the curve $y = \cos x$ (shown in black) to the right $\dfrac{\pi}{2}$ units and upward 2 units to obtain the desired graph (shown in red).

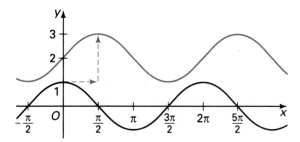

In the case of sinusoids, the rule regarding translations, stated on the preceding page, becomes:

The graph of $y = a \sin b(x - c) + d$ (or $y = a \cos b(x - c) + d$) can be obtained by translating the graph of $y = a \sin bx$ (or $y = a \cos bx$)

> c units to the right if $c > 0$ or $|c|$ units to the left if $c < 0$,

and

> d units upward if $d > 0$ or $|d|$ units downward if $d < 0$.

In the case of $y = a \sin b(x - c) + d$, the number c is called the **phase shift** and d the **vertical shift**. These shifts are understood to be relative to $y = a \sin bx$. The same is true if *sin* is replaced by *cos*.

Example 2   Graph $y = 3 \sin \left(\dfrac{\pi}{4}x + \dfrac{3\pi}{4} \right) + 1$.

Solution   Put the equation in the form $y = a \sin b(x - c) + d$.

$$y = 3 \sin \left(\frac{\pi}{4}x + \frac{3\pi}{4} \right) + 1$$

$$= 3 \sin \frac{\pi}{4}(x + 3) + 1$$

$$= 3 \sin \frac{\pi}{4}(x - (-3)) + 1$$

The amplitude is 3, the period is $\dfrac{2\pi}{\dfrac{\pi}{4}}$, or 8, the phase shift relative to

$y = 3 \sin \dfrac{\pi}{4}x$ is -3, and the vertical shift is 1. Therefore graph

$y = 3 \sin \dfrac{\pi}{4}x$. Then translate that graph to the left 3 units and up 1

unit to obtain the desired graph (shown in red).

(Solution continued next page.)

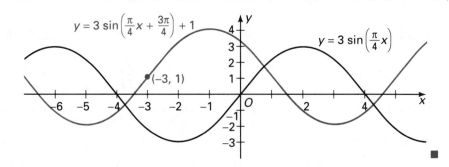

Example 3   Find the *x*-intercepts of the graph found in Example 2.

Solution   Recall that the *x*-intercepts of the graph of *f* are all the values of *x* for which $f(x) = 0$.

$$3 \sin \frac{\pi}{4}(x + 3) + 1 = 0$$

$$\sin \frac{\pi}{4}(x + 3) = -\frac{1}{3}$$

Thus,

$$\frac{\pi}{4}(x + 3) = -0.3398 \text{ or } \frac{\pi}{4}(x + 3) = \pi - (-0.3398) = 3.4814$$

$$x + 3 = \frac{4}{\pi}(-0.3398) \qquad\qquad x + 3 = \frac{4}{\pi}(3.4814)$$

$$= -0.4326 \qquad\qquad x + 3 = 4.4326$$

$$x = -3.4326 \qquad\qquad x = 1.4326$$

Since the function has period $\dfrac{2\pi}{\dfrac{\pi}{4}}$, or 8, the intercepts are

$-3.4326 + 8n$ and $1.4326 + 8n$, where n is an integer.   ■

EXERCISES 5-2

Graph all three equations in the same coordinate system. Label each graph.

A 1. (a) $y = x^2$ (b) $y = (x + 2)^2$ (c) $y = x^2 - 1$

2. (a) $y = \dfrac{x^2}{2}$ (b) $y = \dfrac{(x - 2)^2}{2}$ (c) $y = \dfrac{x^2 + 4}{2}$

3. (a) $y = \cos x$ (b) $y = \cos x - 1$ (c) $y = \cos (x - \pi)$

4. (a) $y = \sin x$ $\qquad$ (b) $y = \sin x + 1$ $\qquad$ (c) $y = \sin\left(x + \dfrac{\pi}{2}\right)$

5. (a) $y = 2 \sin x$ $\qquad$ (b) $y = 2 \sin\left(x - \dfrac{\pi}{2}\right)$ $\qquad$ (c) $y = 2 \sin x - 1$

6. (a) $y = 3 \cos x$ $\qquad$ (b) $y = 3 \cos(x + \pi)$ $\qquad$ (c) $y = 3 \cos x + 2$

7. (a) $y = \cos 2x$ $\qquad$ (b) $y = \cos 2x + 1$ $\qquad$ (c) $y = \cos 2\left(x - \dfrac{\pi}{2}\right)$

8. (a) $y = \sin 2x$ $\qquad$ (b) $y = \sin 2x - 1$ $\qquad$ (c) $y = \sin 2\left(x - \dfrac{\pi}{4}\right)$

Graph each of the following.

9. $y = \sin\left(2x - \dfrac{\pi}{2}\right)$ $\qquad$ 10. $y = \cos\left(2x + \dfrac{\pi}{3}\right)$

11. $y = \cos\left(\pi x + \dfrac{\pi}{2}\right)$ $\qquad$ 12. $y = \sin\left(\dfrac{\pi}{2}x - \dfrac{\pi}{2}\right)$

13. $y = 2 \cos\left(x - \dfrac{\pi}{3}\right) + 1$ $\qquad$ 14. $y = 3 \sin\left(x + \dfrac{\pi}{4}\right) + 3$

15. $y = -2 \sin\left(\dfrac{\pi}{2}x + \dfrac{3\pi}{4}\right) - 1$ $\quad$ 16. $y = 2 \cos\left(\dfrac{\pi}{4}x - \dfrac{\pi}{4}\right) + 1$

17–24. Find the x-intercepts of the graphs in Exercises 9–16.

In Exercises 25–28, express:

B 25. $\sin\left(2x + \dfrac{\pi}{3}\right)$ using the cosine $\quad$ 26. $\sin\left(\dfrac{x}{2} - \dfrac{\pi}{4}\right)$ using the cosine

27. $\cos\left(\dfrac{x}{3} - \dfrac{\pi}{2}\right)$ using the sine $\qquad$ 28. $\cos(3x + \pi)$ using the sine

29. Show that $y = \cos^2 x$ is a sinusoid. Find its amplitude and period and graph it. (Hint: Use a double-angle formula.)

30. Repeat Exercise 29 for $y = \sin^2 x$.

Find the domain and range of the function defined by the given equation and graph it.

C 31. $y = \dfrac{2}{\pi} \text{Cos}^{-1} x$ $\qquad$ 32. $y = \dfrac{4}{\pi} \text{Sin}^{-1} x$

33. $y = \text{Sin}^{-1}(x - 1)$ $\qquad$ 34. $y = \text{Cos}^{-1}(x + 1)$

35. $y = \text{Cos}^{-1}(x - 1) - \dfrac{\pi}{2}$ $\qquad$ 36. $y = \text{Sin}^{-1}(x + 1) + \dfrac{\pi}{2}$

5-3   The Expression $A \cos \theta + B \sin \theta$ (Optional)

Objective: To write $A \cos \theta + B \sin \theta$ in the form $C \cos (\theta - \gamma)$ and to use the result in graphing and solving equations.

In certain areas of applied mathematics, expressions of the form

$$A \cos \theta + B \sin \theta$$

occur frequently. For example, the expression for the force necessary to slide a crate up a ramp has this form if friction is taken into account (See Exercise 37, page 165).

The expression above can be put into a form that is more useful for most purposes, namely:

$$A \cos \theta + B \sin \theta = C \cos (\theta - \gamma)$$

where $C = \sqrt{A^2 + B^2}$ and γ is such that

$$\cos \gamma = \frac{A}{C} \qquad \text{and} \qquad \sin \gamma = \frac{B}{C}$$

To express $A \cos \theta + B \sin \theta$ in terms of cosine, let $C = \sqrt{A^2 + B^2}$. Then,

$$\left(\frac{A}{C}\right)^2 + \left(\frac{B}{C}\right)^2 = \frac{A^2 + B^2}{C^2}$$

$$= \frac{C^2}{C^2}$$

$$= 1$$

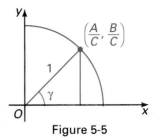

Figure 5-5

Hence, as shown in Figure 5-5, the point $\left(\dfrac{A}{C}, \dfrac{B}{C}\right)$ is on the unit circle and there is a γ such that

$$\cos \gamma = \frac{A}{C} \qquad \text{and} \qquad \sin \gamma = \frac{B}{C}.$$

Therefore,

$$A \cos \theta + B \sin \theta = C \left(\frac{A}{C} \cos \theta + \frac{B}{C} \sin \theta\right)$$

$$= C(\cos \gamma \cos \theta + \sin \gamma \sin \theta)$$

$$= C \cos (\theta - \gamma)$$

We usually choose γ in the interval $-\pi < \gamma < \pi$ or $-180° < \gamma < 180°$.

Example 1   Express $\cos 25° + \sqrt{3} \sin 25°$ as a cosine.

Solution   With $A = 1$ and $B = \sqrt{3}$,

$$C = \sqrt{1^2 + \sqrt{3}^2} = \sqrt{4} = 2, \cos \gamma = \frac{1}{2}, \text{ and } \sin \gamma = \frac{\sqrt{3}}{2}.$$

Since $\cos 60° = \frac{1}{2}$ and $\sin 60° = \frac{\sqrt{3}}{2}$, we may choose $\gamma = 60°$. Hence $\cos 25° + \sqrt{3} \sin 25° = 2 \cos (-35°)$, or $2 \cos 35°$. ■

Example 2   Express $y = 4 \cos x + 3 \sin x$ in the form $y = C \cos (x - \gamma)$ and draw its graph.

Solution   Since $C = \sqrt{4^2 + 3^2} = 5$, $\cos \gamma = \frac{4}{5} = 0.8$ and $\sin \gamma = \frac{3}{5} = 0.6$.

Using a calculator or Table 3, we have $\gamma = 0.64$ (to two decimal places). Therefore $y = 5 \cos (x - 0.64)$.

To graph this sinusoid, note that the amplitude is 5, the period is 2π, and the phase shift relative to $y = 5 \cos x$ is 0.64. The graph is shown in red below along with the graph of $y = 5 \cos x$ in black. ■

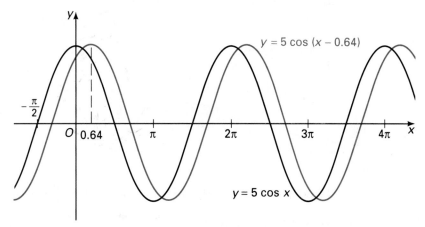

Example 2 suggests that the sum of two sinusoids having the same period T is itself a sinusoid having period T. This fact is proved in Exercise 40.

Example 3   Solve $2 \cos 2\phi - 3 \sin 2\phi = 1.7$, where $0° \le \phi < 360°$.

Solution   **Step 1**   Put $2 \cos 2\phi - 3 \sin 2\phi$ into the form $C \cos (\theta - \gamma)$:

$$C = \sqrt{2^2 + (-3)^2} = \sqrt{13}$$

$$\cos \gamma = \frac{2}{\sqrt{13}} \quad \text{and} \quad \sin \gamma = \frac{-3}{\sqrt{13}}$$

(Solution continued next page.)

Since $\cos \gamma > 0$ and $\sin \gamma < 0$, γ is in the fourth quadrant. We may take

$$\gamma = \text{Sin}^{-1}\left(\frac{-3}{\sqrt{13}}\right) = \text{Sin}^{-1}(-0.8321) = -56.3°.$$

Therefore, $2 \cos 2\phi - 3 \sin 2\phi = \sqrt{13} \cos (2\phi + 56.3°)$.

Step 2 Solve $\sqrt{13} \cos (2\phi + 56.3°) = 1.7$, where $0° \le \phi < 360°$.

$$\cos (2\phi + 56.3°) = \frac{1.7}{\sqrt{13}} = 0.4715$$

Since $\text{Cos}^{-1} \, 0.4715 = 61.9°$, solve the equations:
$2\phi + 56.3° = 61.9°, 360° - 61.9°, 360° + 61.9°, 720° - 61.9°$
$$= 61.9°, 298.1°, 421.9°, 658.1°$$

Thus,
$$2\phi = 5.6°, 241.8°, 365.6°, 601.8°$$
$$\phi = 2.8°, 120.9°, 182.8°, 300.9° \quad \blacksquare$$

Example 4 Find the x-intercepts of the graph of Example 2.

Solution Since the equation of the graph is $y = 5 \cos (x - 0.64)$, we need to solve the equation $\cos (x - 0.64) = 0$. We obtain $x - 0.64 = \frac{\pi}{2} + n\pi$, where n is an integer. Thus, $x = 0.64 + 1.57 + n\pi = 2.21 + n\pi$. $\quad \blacksquare$

EXERCISES 5-3

Rewrite each expression in the form $C \cos \phi$, where $0° < \phi < 360°$.

A 1. $\cos 100° + \sin 100°$ 2. $\cos 100° - \sin 100°$

3. $\sqrt{3} \cos 40° - \sin 40°$ 4. $\cos 70° + \sqrt{3} \sin 70°$

5. $\cos 10° + \sqrt{3} \sin 10°$ 6. $\sin 25° + \cos 25°$

7. $\sin 75° - \sqrt{3} \cos 75°$ 8. $\sqrt{3} \cos 20° + \sin 20°$

Express each of the following in the form $C \cos (x - \gamma)$ and draw its graph. Find γ to the nearest hundredth, where necessary.

9. $y = 2 \cos x + 2 \sin x$ 10. $y = -\cos x + \sin x$

11. $y = \sqrt{3} \cos x - \sin x$ 12. $y = \cos x + \sqrt{3} \sin x$

13. $y = 3 \cos x + 4 \sin x$ 14. $y = 3 \sin x - 4 \cos x$

15. $y = 5 \cos x - 12 \sin x$ 16. $y = 12 \cos x + 5 \sin x$

17–24. Find the *x*-intercepts of the graphs in Exercises 9–16. Give answers to the nearest hundredth.

Find all solutions of each equation in the interval $0° \le \phi < 360°$. Give answers to the nearest 0.1°.

25. $2 \sin \phi + 2 \cos \phi = \sqrt{2}$ 26. $\cos \phi + \sqrt{3} \sin \phi = \sqrt{2}$

B 27. $\sqrt{3} \cos \phi - \sin \phi = 1.5$

28. $\cos \phi - \sin \phi = 1.2$

29. $3 \cos \phi + 4 \sin \phi = 2$

30. $4 \cos \phi - 3 \sin \phi = 4.5$

31. $2 \cos 2\phi - 3 \sin 2\phi = 1$

32. $3 \cos 2\phi + \sin 2\phi = 2$

In Exercises 33 and 34, use the figure below, which shows a rectangle inscribed in a circle of radius 1.

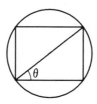

33. Find θ so that the perimeter of the rectangle will be $2\sqrt{6}$.

34. Find θ so that the perimeter of the rectangle will be 5.

35. If $A \cos \theta + B \sin \theta$ is written in the form $C \sin (\theta + \gamma)$, give equations to find C and γ. (Hint: Use $\sin (\alpha + \beta) = \sin \alpha \cos \beta + \cos \alpha \sin \beta$.)

36. If $A \cos \theta + B \sin \theta$ is written in the form $C \sin (\theta - \gamma)$, give formulas to find C and γ.

C 37. The force necessary to accelerate a 1 kg mass 1 m/s² is called a **newton** (N). The force in newtons necessary to keep an object moving at constant speed up a ramp inclined at an angle θ with the horizontal is given (approximately) by

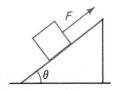

$$F = 9.8 \, m(\sin \theta + k \cos \theta),$$

where *m* is the mass of the object in kilograms and *k* is the coefficient of friction between the object and the ramp. Find θ if a force of 130 N moves a 20 kg crate up the ramp at constant speed, the coefficient of friction being $\frac{5}{12}$.

38. The force F in newtons necessary to keep an object with mass m kg moving at constant speed on a horizontal surface is given by

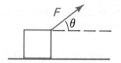

$$F = \frac{9.8\ mk}{\cos\theta + k\sin\theta}$$

where θ is the angle the force makes with the horizontal and k is the coefficient of friction. Find θ if a force of 60 N moves a 12 kg sled at constant speed, the coefficient of friction being $\dfrac{8}{15}$.

39. In Exercise 38, at what angle is the force required to move the sled at constant speed a minimum? What is this minimum force?

40. Show that for any constants a, b, α and β,

$$a \sin(\theta + \alpha) + b \sin(\theta + \beta) = c \cos(\theta - \gamma),$$

where $\qquad c^2 = a^2 + 2ab\cos(\alpha - \beta) + b^2$

$$\cos\gamma = \frac{a\sin\alpha + b\sin\beta}{c} \quad \text{and} \quad \sin\gamma = \frac{a\cos\alpha + b\cos\beta}{c}$$

(Hint: Expand $\sin(\theta + \alpha)$ and $\sin(\theta + \beta)$ using an addition formula. Collect coefficients of $\cos\theta$ and $\sin\theta$ and use them as A and B, respectively, in the boxed statement on page 162. Then simplify.)

Find expressions in terms of a, b, α, and β that determine c and γ if:

41. $a\cos(\theta + \alpha) + b\sin(\theta + \beta) = c\cos(\theta - \gamma)$

42. $a\cos(\theta + \alpha) + b\cos(\theta + \beta) = c\cos(\theta - \gamma)$

Self Quiz

5-1	5-2	5-3

Graph each of the following.

1. $y = 1 - \sin 2x$ 　　　　2. $y = \sin\dfrac{\pi}{4}(x - 1) + 1$

3. Express $\sqrt{3}\cos 20° + \sin 20°$ as a cosine.

4. Solve $\cos\phi - \sqrt{3}\sin\phi = \sqrt{2}$ in the interval $0° \le \phi < 360°$.

5. Express the function $\sin\left(2x - \dfrac{\pi}{3}\right)$ in terms of the cosine.

6. Graph $y = \cos 2x$ and $y = 2\cos x$ in the same coordinate plane.

APPLYING SINUSOIDS

5-4

Simple Harmonic Motion

Objective: To find and apply equations for simple harmonic motion.

Simple harmonic motion (SHM) is the motion of a particle when it is displaced from a central (or equilibrium) position, and the force tending to restore it to that position is proportional to the displacement.

For example, suppose that an object is attached to the bottom of a coil spring whose top is fastened to the ceiling. If the object is pulled down from its rest position and then released, it will describe SHM.

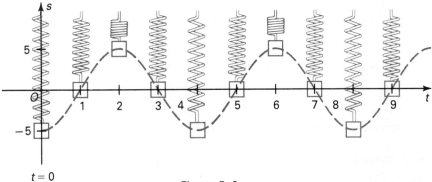

Figure 5–6

Figure 5-6 shows its position s (in centimeters) at various times t (in seconds).

It can be shown (using calculus) that in every SHM, position, s, varies sinusoidally with time, t, and if $s = 0$ is assigned to the central position, then the equation of motion is of the form

$$s = a \sin (\omega t + \beta), \text{ where } \omega > 0.$$

Useful special cases correspond to $\beta = 0$ and $\beta = \dfrac{\pi}{2}$:

$$s = a \sin \omega t \text{ and } s = a \cos \omega t.$$

Since the period T of $a \sin (\omega t + \beta)$ is $\dfrac{2\pi}{\omega}$, this is the time it takes for the particle to make one "back-and-forth" cycle. The **frequency**, f, of the motion is the number of cycles described in unit time, that is, $f = \dfrac{1}{T}$. If time is measured in seconds, we have the reciprocal relationships stated on the next page for the SHM $s = a \sin (\omega t + \beta)$.

Period: $T = \dfrac{2\pi}{\omega}$ seconds per cycle

Frequency: $f = \dfrac{\omega}{2\pi}$ cycles per second

(One cycle per second = one hertz (Hz).)

Example 1 An object suspended from a spring is pulled down 3 cm and released. It oscillates with frequency 2 cycles per second.
(a) Sketch the displacement of the object over the first 1 second after it is pulled down.
(b) Find an equation of the SHM.

Solution (a) When $t = 0$, $s = -3$, since initially the object is pulled down 3 units. Since the object oscillates with frequency 2 cycles per second, it completes 1 cycle every half second. The following graph shows s over the interval $0 \le t \le 1$.

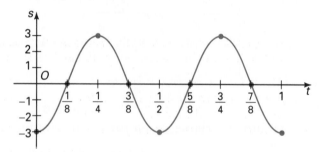

(b) The graph suggests an equation of the form $s = a \cos \omega t$. Since the period is $\dfrac{1}{2}$, $\omega = \dfrac{2\pi}{\dfrac{1}{2}} = 4\pi$. Thus, $s = -3 \cos 4\pi t$. ∎

It can be shown (using calculus) that if a particle is describing the SHM $s = a \sin (\omega t + \beta)$, then its velocity v can be found from

$$v^2 = \omega^2 (a^2 - s^2).$$

Notice that the velocity depends on amplitude, period, and position.
Since the speed of the particle is $|v|$, we see that the greatest speed $|\omega a|$ occurs at the central position ($s = 0$) and that the speed is zero at the extreme positions ($s = \pm a$).

Example 2 A particle moving in SHM has speed 5 cm/s when it is 6 cm from its central position and has speed 3 cm/s when it is 10 cm from its central position. Find (a) the period and (b) the amplitude of the SHM. (c) Find the greatest speed of the particle.

Solution   Substitute the given pairs of $|v|$, s values into $v^2 = \omega^2(a^2 - s^2)$.

$$|v| = 5 \text{ when } s = 6 \qquad\qquad |v| = 3 \text{ when } s = 10$$
$$25 = \omega^2(a^2 - 6^2) \qquad\qquad 9 = \omega^2(a^2 - 10^2)$$
$$= \omega^2 a^2 - 36\omega^2 \qquad\qquad = \omega^2 a^2 - 100\omega^2$$

(a) Subtract to obtain:

$$25 - 9 = (\omega^2 a^2 - 36\omega^2) - (\omega^2 a^2 - 100\omega^2)$$
$$16 = 64\omega^2$$

Hence, $\omega = \dfrac{1}{2}$ and the period is $\dfrac{2\pi}{\omega} = 4\pi$ seconds.

(b) Substitute $\omega^2 = \dfrac{1}{4}$ into $25 = \omega^2(a^2 - 36)$ to obtain $25 =$

$\dfrac{1}{4}(a^2 - 36)$, or $100 = a^2 - 36$.

Thus, the amplitude a is $\sqrt{136}$ cm, or $2\sqrt{34}$ cm.

(c) The greatest speed, $|\omega a|$, is $\dfrac{1}{2} \cdot 2\sqrt{34} = \sqrt{34}$ cm/s. ■

The motion of a pendulum bob is very nearly a SHM if the angle through which it swings is small, say less than 10°.

Example 3   The bob of a pendulum hanging
at rest is given an initial velocity
of 3π cm/s to the right. Thereafter it oscillates with amplitude
3 cm.
(a) Find an equation of the SHM.
(b) Find its period and frequency.

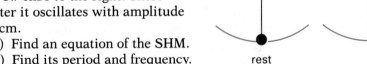

rest   ←s→

Solution   (a) The SHM has an equation of the form $s = a \sin(\omega t + \beta)$ where s is
the distance between the bob and the rest position at time t. Find
a, β, and ω. Since the amplitude is 3, $s = 3 \sin(\omega t + \beta)$.
When $t = 0$, $s = 0$. Therefore,

$$0 = 3 \sin \beta.$$

To satisfy this equation, we may take $\beta = 0$. Then the equation for
s becomes

$$s = 3 \sin \omega t.$$

Since $v = 3\pi$ when $s = 0$, $(3\pi)^2 = \omega^2(3^2 - 0^2)$. Therefore, $\omega = \pi$.
Hence, $s = 3 \sin \pi t$.

(b) The period T is $\dfrac{2\pi}{\omega} = \dfrac{2\pi}{\pi} = 2$ seconds.

The frequency $f = \dfrac{1}{T} = \dfrac{1}{2}$ cycle per second. ■

EXERCISES 5-4

Exercises 1–5: For each of the following cases of simple harmonic motion, (a) sketch the position *s* as a function of time *t*. (b) Find an equation for *s* in the form *s* = *a* cos (*ωt* + *β*). (c) Find the frequency of the motion. (Neglect the effects of friction and air resistance.)

A 1. An object on a coiled spring is pulled down 5 cm and released. It makes a complete oscillation every 3 seconds.

2. An object on a coiled spring is pushed up 4 cm and released. It makes a complete oscillation every $\frac{2}{3}$ second.

3. A pendulum bob is pulled 3 cm to the right of its rest position and released. One second later it is 3 cm to the left of its rest position.

4. The spar buoy shown at the right is lifted until its marked water line is 40 cm above the surface of the water and then released. Four seconds later the water line is 40 cm below the surface.

5. A spar buoy is pushed down 50 cm and released. Its water line is at the surface 1.5 seconds later.

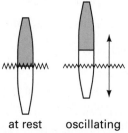

at rest   oscillating

Exercises 6–10: (a) Find an equation of the form *s* = *a* cos (*ωt* + *β*) for each SHM. (b) Sketch *s* as a function of time. (c) Find the frequency of the motion.

6. An object on a coiled spring, hanging at rest, is given an initial upward velocity of 4π cm/s. Thereafter, it oscillates with amplitude 2 cm.

7. An object, hanging at rest, is given an initial upward velocity of 4π cm/s. Thereafter, it oscillates with amplitude 4 cm.

8. A pendulum hanging at rest is given an initial velocity of 12 cm/s to the right. Thereafter it oscillates with amplitude 3 cm.

9. If a hole could be drilled through the center of Earth, a golf ball dropped into the hole would reach the other end, 8000 miles away, in about 42.5 minutes. It then describes SHM.

10. Suppose that at *t* = 0 the ball in Exercise 9 is 2000 miles below the hole and moving toward the center of Earth. (Hint: Use the result of Exercise 9.)

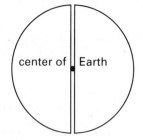

center of   Earth

11–20. Express the equations describing the simple harmonic motions in Exercises 1–10 in the form *y* = *a* sin (*ωt* + *β*).

B 21. The line voltage v of a standard 120 V (volt) alternating current varies sinusoidally with time t from a maximum value of $120 \sqrt{2}$ V (volts) to a minimum of $-120 \sqrt{2}$ V (the current changes direction) and has frequency 60 Hz. (a) Find an equation of the form $v = a \sin (\omega t + \beta)$ giving the voltage v at time t. (b) Find β if $v = 120$ V when $t = 0$.

22. (a) Repeat Exercise 21(a) using the cosine function.

(b) Find β if $v = 120$ V when $t = \dfrac{1}{192}$ s.

23. Find the greatest speed of a particle moving in SHM with amplitude 3 cm and frequency 4 Hz.

24. A particle describing SHM has speed 5 ft/min when it is 5 ft from its central position and 13 ft/min when it is 4 ft from the central position. Find (a) its frequency, (b) its amplitude, and (c) its greatest speed.

25. The maximum speed of a particle is 10 cm/s, and when it is 20 cm from its central position, its speed is 6 cm/s. Find (a) its frequency and (b) its amplitude.

26. The figure at the right shows a point P moving counterclockwise around the circle $x^2 + y^2 = a^2$ with constant angular speed ω radians/s. (Recall the discussion of uniform circular motion in Section 2-2.) P starts at $P_0(a \cos \beta, a \sin \beta)$. Let Q be the projection of P on the x-axis. Show that Q describes SHM by noting that the x-coordinate of Q is the x-coordinate of P.

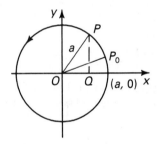

27. The period T of a pendulum of length l cm is $2\pi \sqrt{\dfrac{l}{g}}$ seconds, where $g = 980$ cm/s^2. Find the length of the pendulum in Example 3. (It is called a **seconds pendulum** because it describes half a period in one second.)

C 28. That a pendulum describes SHM requires the assumption that $\sin \theta \approx \theta$ (radians) for small angles. Find the percentage error in this assumption when $\theta = 10°$.

29. If the amplitude of a particle in SHM is a, how far from its central position is it when it is traveling with half its maximum speed?

30. It can be shown (using calculus) that if a particle is describing the SHM $s = a \sin (\omega t + \beta)$, then its velocity v is given by $v = a\omega \cos (\omega t + \beta)$. Use this fact to prove that $v^2 = \omega^2(a^2 - s^2)$. (Hint: For brevity, let $\theta = \omega t + \beta$. Use $\cos^2 \theta = 1 - \sin^2 \theta$.)

5-5   Sums of Sinusoids

Objective: To graph sums of sinusoids of different frequencies.

There are many important periodic functions that are not sinusoids but are sums of sinusoids of different frequencies.

Consider, for example, the sound produced by a violin string. This sound is produced by a combination of sinusoidal vibrations of the string. The note produced by the vibration of lowest frequency, called the **fundamental**, determines the *pitch* of the sound. But the string simultaneously produces other weaker notes, called **overtones**, or **harmonics**, and the combination of these determines the *quality* of the sound. It is quality that enables us to distinguish between middle C played on a violin from middle C played on a flute.

The first and usually strongest overtone is the note an octave higher than the fundamental. Since the frequency of $a \sin \omega t$ is $\dfrac{\omega}{2\pi}$, an equation describing the combined sounds of the fundamental and first overtone would have the form

$$s = a \sin \omega t + b \sin 2\omega t.$$

It can be shown that if f has period $\dfrac{2\pi}{p}$ and g has period $\dfrac{2\pi}{q}$, then $f + g$ has period $\dfrac{2\pi}{d}$, where d is the greatest common divisor of p and q (Exercise 32 on page 174).

A method of graphing such a function, called **addition of ordinates**, is illustrated in Example 1.

Example 1   Graph $y = 2 \sin x + \cos 2x$ showing at least one cycle.

Solution

Step 1   Since $\sin x$ has period 2π and $\cos 2x$ has period π, then $2 \sin x + \cos 2x$ has period 2π.

Step 2   Graph $y_1 = 2 \sin x$ and $y_2 = \cos 2x$ over $0 \le x \le 2\pi$.

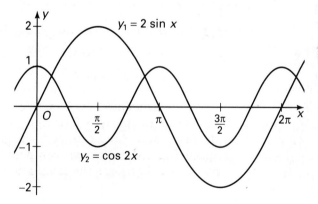

Step 3 For a chosen x, start from either graph, say (x, y_1), and measure off the ordinate y_2 of the other graph, up if $y_2 > 0$ and down if $y_2 < 0$. The point obtained has ordinate $y_1 + y_2$ and therefore is on the desired graph. Repeat the procedure as many times as you need to draw the final graph.

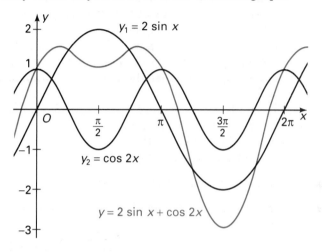

Example 2

For the graph obtained in Example 1, find the x-intercepts that occur in some one cycle.

Solution

Over the interval $0 \leq x < 2\pi$, there are two x-intercepts. The smaller is a little more than $\pi \approx 3.14$. The larger is a little less than $2\pi \approx 6.28$. Solve $2 \sin x + \cos 2x = 0$ over the interval $0 \leq x < 2\pi$.

$$2 \sin x + (1 - 2 \sin^2 x) = 0$$
$$2 \sin^2 x - 2 \sin x - 1 = 0$$

Now use the quadratic formula:

$$\sin x = \frac{2 \pm \sqrt{4 + 8}}{4} = \frac{1 \pm \sqrt{3}}{2}$$

Since $\dfrac{1 + \sqrt{3}}{2} > 1$, we have $\sin x = \dfrac{1 - \sqrt{3}}{2} = -0.3660$.

Since $\sin x < 0$ when x is in the third or fourth quadrant,

$$x = \pi - (-0.3747) \text{ or } 2\pi - (0.3747).$$
$$= 3.5163 \text{ or } 5.9085. \quad \blacksquare$$

You can check your solution to Example 1 by using a graphing calculator or computer. Graph $y_1 = 2 \sin x$, $y_2 = \cos 2x$, and $y = 2 \sin x + \cos 2x$ on the same axes. Check to see if the graph of y resembles your sketch. A graphing calculator or computer can be used to solve $2 \sin x + \cos 2x = 0$. Compare the calculator results with your answer.

EXERCISES 5-5

Graph each equation showing at least one complete cycle.

A 1. $y = \cos x + \sin 2x$ 2. $y = \sin x + \cos 2x$ 3. $y = \cos x - \cos 2x$

4. $y = \sin x - \sin 2x$ 5. $y = 2 \sin x + \sin 2x$ 6. $y = 2 \cos x - \cos 2x$

7. $y = 2 \sin x - \cos 2x$ 8. $y = 2 \cos x + \sin 2x$

9. $y = 2 \sin \frac{\pi}{2}x + \cos \pi x$ 10. $y = 2 \cos \frac{\pi}{2}x - \sin \pi x$

11–20. For the graphs in Exercises 1–10, find the x-intercepts in some cycle.

Graph each equation over the interval $0 \le x \le 2\pi$.

B 21. $y = x + \cos x$ 22. $y = x - \sin x$

23. $y = \cos x + |\cos x|$ 24. $y = \sin x - |\sin x|$

In Exercises 25–30, use one of the following identities to express each function as a sum instead of a product and graph the equation over one cycle.
$$2 \sin \alpha \sin \beta = -\cos (\alpha + \beta) + \cos (\alpha - \beta)$$
$$2 \sin \alpha \cos \beta = \sin (\alpha + \beta) + \sin (\alpha - \beta)$$
$$2 \cos \alpha \cos \beta = \cos (\alpha + \beta) + \cos (\alpha - \beta)$$

25. $y = 2 \cos x \cos 2x$ 26. $y = 2 \sin x \sin 2x$ 27. $y = 2 \sin x \cos 2x$

28. $y = \sin 3x \cos x$ 29. $y = \sin \frac{x}{2} \cos \frac{3x}{2}$ 30. $y = \sin \frac{x}{2} \sin \frac{3x}{2}$

Let f and g be functions. Let m be the least common multiple of the positive integers p and q, and let d be their greatest common divisor. Show that:

C 31. If f has period p and g has period q, then $f + g$ has period m.

32. If f has period $\frac{2\pi}{p}$ and g has period $\frac{2\pi}{q}$, then $f + g$ has period $\frac{2\pi}{d}$.

Challenge

Graph (a) $y = \sin \frac{1}{x}$ and (b) $y = x \sin \frac{1}{x}$.

5-6   Graph Interpretation and Modeling

Objective: To obtain equations of sinusoids from their graphs and from real-life situations.

Example 1 illustrates how we can find an equation of a sinusoid given its graph.

Example 1   (a) Find an equation of the sinusoid shown at the right.
(b) Find y when $x = 40$.

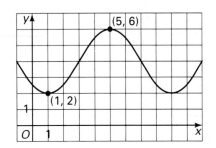

Solution   (a) The given sinusoid has amplitude $\dfrac{6 - 2}{2}$, or 2. Its period is $2(5 - 1)$, or 8. Using the formula *period* $= \dfrac{2\pi}{\omega}$, we see that $8 = \dfrac{2\pi}{\omega}$, or $\omega = \dfrac{\pi}{4}$.

The sinusoid crosses its central axis going upward midway between (1, 2) and (5, 6), namely, at (3, 4). It can therefore be obtained by translating the curve $y = 2 \sin \dfrac{\pi}{4}x$ to the right 3 units and up 4 units.

The given sinusoid therefore has the equation $y = 2 \sin \dfrac{\pi}{4}(x - 3) + 4$.

(b) Setting $x = 40$, we have

$$y = 2 \sin \frac{\pi}{4}(37) + 4$$

$$= 2 \sin \left(8\pi + \frac{5\pi}{4} \right) + 4$$

$$= 2 \sin \frac{5\pi}{4} + 4$$

$$= 2\left(-\frac{\sqrt{2}}{2} \right) + 4$$

$$= 4 - \sqrt{2}. \quad \blacksquare$$

(Note that the answer given for Example 1(a) is not unique. Another answer is $y = 2 \cos \dfrac{\pi}{4}(x - 5) + 4$.)

In Examples 2 and 3 we set up mathematical models. Note that we use for variables letters that suggest the quantities involved.

Example 2

A Ferris wheel rotating at 4 rpm is 6 m in radius and has its axle 8 m above the ground. At time 0, a chair on the rim of the wheel is 5 m above the ground and going up. Find an equation giving the height h of the seat above the ground t seconds later.

Solution

Step 1 Sketch the situation at time 0.

Step 2 Since the wheel and thus the seat travel at 4 rpm, the seat (initially at C) is moving counterclockwise around a circle of radius 6 m centered at the wheel's axle.

Step 3 Introduce an xh-coordinate system with the x-axis coincident with the ground and the h-axis containing the axle. Then, from Section 2-2,

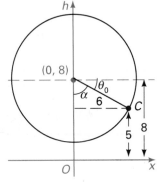

$$h = r \sin(\omega t + \theta_0) + 8.$$

Since $\omega = 4 \text{ rpm} = \dfrac{2\pi}{15}$ radians per second and $r = 6$,

$$h = 6 \sin\left(\frac{2\pi}{15}t + \theta_0\right) + 8.$$

From the figure, $\alpha = 60°$, or $\dfrac{\pi}{3}$. Thus $\theta_0 = -\dfrac{\pi}{6}$ and we have

$$h = 6 \sin\left(\frac{2\pi}{15}t - \frac{\pi}{6}\right) + 8 = 6 \sin\frac{2\pi}{15}\left(t - \frac{5}{4}\right) + 8. \qquad \blacksquare$$

Example 2 could be solved by the following analysis. Since the wheel revolves at 4 rpm, one revolution every 15 seconds, $h = 5$ when $t = 0$ and 15. From the figure, $\alpha = 60°$. Thus the seat reaches its highest and lowest points after turning through 120° and 300°, respectively. Thus, $h = 14$ when $t = 5$ and $h = 2$ when $t = 12.5$.

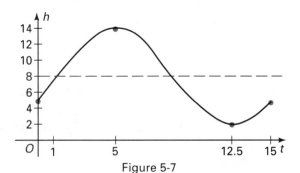

Figure 5-7

The graph in Figure 5-7 shows a sinusoid with amplitude 6 and period 15 that is a translation of $y = 6 \cos \dfrac{2\pi}{15}t$ to the right 5 units and up 8 units. Therefore an equation for h is

$$h = 6 \cos \frac{2\pi}{15}(t - 5) + 8.$$

This equation is equivalent to the one found in Example 2.

Example 3

The graph shows how the price p of PQR stock varied during the ten years after it was issued. (a) Find a function of the form

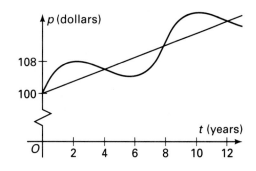

$$p(t) = k + mt + a \sin \omega t$$

that approximates the price t years after the issue date given that $p\,(0) = 100$, $p(2) = 108$, and $p(4) = 106$. Assuming that the stock continues to perform in this way, approximate the price (b) 22 years and (c) 25 years after its issue date.

Solution

(a) The sinusoidal part of $p(t) = k + mt + a \sin \omega t$ has period 8 yrs, so that $8 = \dfrac{2\pi}{\omega}$, or $\omega = \dfrac{\pi}{4}$. Therefore,

$$p(t) = k + mt + a \sin \frac{\pi}{4}t.$$

Substitute in turn $t = 0$ and $t = 4$ to get $100 = k$ and $106 = k + m \cdot 4$. Hence $k = 100$ and $m = 1.5$, and

$$p(t) = 100 + 1.5t + a \sin \frac{\pi}{4t}.$$

Finally, substitute $t = 2$: $108 = 100 + 1.5 \cdot 2 + a \sin \dfrac{\pi}{2} = 103 + a$

Therefore, $a = 5$, and

$$p(t) = 100 + 1.5t + 5 \sin \frac{\pi}{4}t.$$

(b) $p(22) = 100 + 1.5 \cdot 22 + 5 \sin \left(\dfrac{\pi}{4} \cdot 22 \right)$

$$= 133 + 5 \sin \left(4\pi + \frac{3\pi}{2} \right) = 133 + 5(-1) = 128$$

(Solution continued next page.)

(c) $p(25) = 100 + 1.5 \cdot 25 + 5 \sin\left(\dfrac{\pi}{4} \cdot 25\right)$

$= 137.5 + 5 \sin\left(6\pi + \dfrac{\pi}{4}\right) = 137.5 + 5\left(\dfrac{\sqrt{2}}{2}\right) \approx 141$ ■

EXERCISES 5-6

Find an equation of each sinusoid shown.

A 1.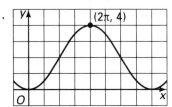

2.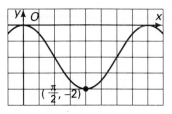

3.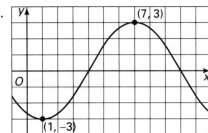

4.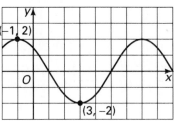

5.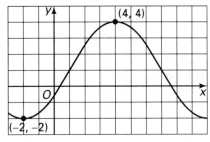

6.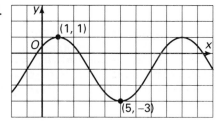

7–12. Rework Exercises 1–6. If originally you used a sine (cosine) function, this time use a cosine (sine) function.

Find an equation of the sinusoid having the given adjacent minimum and maximum points. (Hint: Draw a sketch.)

13. Min(0, 0), max$\left(\dfrac{\pi}{2}, 3\right)$

14. Max(0, 5), min(π, 1)

15. Max(1, 0), min(5, −1)

16. Min(−1, 1), max(5, 3)

Repeat Example 2 for a seat that was, at time 0,

17. at the lowest possible point.   18. at the highest possible point.

Repeat Example 3 with the given data.

19. $p(0) = 50, p(2) = 53, p(4) = 52$ 20. $p(4) = 88, p(6) = 88, p(8) = 96$

B 21. The depth of the water at a certain anchorage varies sinusoidally with time because of ocean swells. The least depth is 6 m: the greatest is 10 m; and the least time between these is 30 s. Find an equation expressing the depth d in terms of time t.

22. The swells in Exercise 21 move at 1.5 m/s. Find an equation of the profile of the water. (Hint: Use rate × time = distance to relate time t to a variable x measuring distance in the direction the swell is moving.)

The figures below show three ways in which a guitar string can vibrate (the *y*-displacement is exaggerated). In (a) the string is vibrating in its **fundamental**; in (b) and (c) it is vibrating in its **first** and **second harmonics**, respectively.

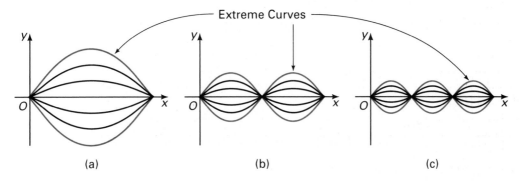

Extreme Curves

(a)   (b)   (c)

The curves have equations of the form $y = a \sin \dfrac{n\pi x}{l}, n = 1, 2, 3$. Usually a string is not vibrating in just one of these modes, but in a linear combination of them. Find and graph an equation of the extreme curve of a string of length 60 cm that has its fundamental and harmonics present with the given amplitudes (in millimeters).

Example   Fundamental 1.0, second harmonic 0.5.

Solution   The equation is

$$y = 1.0 \sin \frac{\pi x}{60} + 0.5 \sin \frac{\pi x}{20}.$$

Use the method of Section 5-5 to obtain the graph.

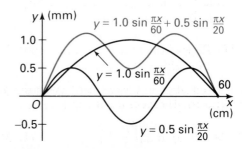

$y = 1.0 \sin \frac{\pi x}{60} + 0.5 \sin \frac{\pi x}{20}$

$y = 1.0 \sin \frac{\pi x}{60}$

$y = 0.5 \sin \frac{\pi x}{20}$

23. Fundamental 1.0, first harmonic 0.5.

24. Fundamental 1.0, first harmonic 1.0.

25. First harmonic 1.0, second harmonic 1.0.

26. Fundamental 1.0, first harmonic 0.5, second harmonic 0.5.

C 27. The displacement y of a point on a string vibrating in its fundamental mode depends on its distance x from one end of the string and on time t according to an equation of the form $y = a \sin hx \sin kt$. Find the constants h and k in terms of the length l of the string and frequency f.

28. The sketches at the right show a carnival ride consisting of a beam with a Ferris wheel at each end. The beam rotates at 6 rpm and the wheels are driven by electric motors fixed to the ends of the beam and rotating at 12 rpm. Find and graph an equation giving the height h of a point on the rim of a wheel t seconds after it was at its lowest point.

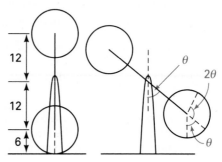

Self Quiz

5-4 5-5 5-6

1. An object hanging from a coil spring is pulled down 6 cm and released. One second later it is passing through its rest position. Find an equation of this simple harmonic motion.

2. Graph at least one cycle of $y = \cos x - \sin 2x$.

3. Find the x-intercepts of the curve in Exercise 2.

4. Find an equation of the sinusoid shown at the right
   (a) using the cosine function
   (b) using the sine function.

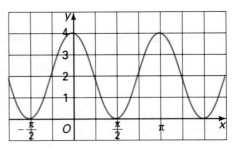

5. A particle describing simple harmonic motion has speed 6 cm/s when it is 1 cm from its central position and speed 2 cm/s when it is 3 cm from its central position. Find (a) its period, (b) its amplitude, and (c) its maximum speed.

6. Adjacent minimum and maximum points of a sinusoid are $(1, -1)$ and $(3, 3)$. Find its equation.

ADDITIONAL PROBLEMS

1. Express $\sqrt{3} \cos 10° + \sin 10°$ as a cosine.

2. Express $\sqrt{3} \cos 10° + \sin 10°$ as a sine.

3. A pendulum is pulled from its rest position 3 cm to the right and released. Thereafter it swings back and forth with frequency $\frac{1}{2}$ cycle per second. Find an equation describing this SHM.

4. Graph $y = \cos \pi x$ and $y = \cos 2\pi x$ in the same coordinate plane.

5. Find the domain and range of $y = -\mathrm{Sin}^{-1} \dfrac{x}{2}$. Draw its graph.

6. Graph the equation $y = 2 \sin \left(x - \dfrac{\pi}{4} \right) + 2$.

7. Find an equation of the sinusoid having (0, 1) and (2, 3) as adjacent minimum and maximum points.

8. Express $\cos 2\left(x - \dfrac{\pi}{4} \right)$ using the sine function.

9. Graph $y = |1 + 2 \sin x|$ in the interval $0 \le x \le 2\pi$.

10. Put $y = 3 \cos x + 4 \sin x$ into the form $y = C \cos (x - \gamma)$. Then graph it.

11. Find the x-intercepts of the graph of Exercise 10 in the interval $0 \le x < 2\pi$. Give answers to the nearest hundredth.

12. Graph $y = \cos x + \cos 2x$.

13. Find the x-intercepts of the graph in Exercise 12 in some one cycle.

14. The sinusoid below indicates the temperature T on a winter day in Minneapolis over a 24-hour period beginning at midnight. Find its equation and use it to find the times when the temperature is 0° C.

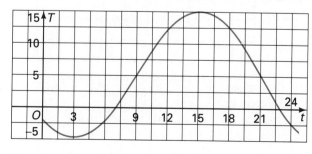

15. Graph $y = \cos^2 \dfrac{x}{2}$.   16. Graph $y = \dfrac{1}{2}x + \sin x$ in $0 \le x \le 2\pi$.

17. An object hanging at rest from the end of a coil spring is given an upward initial velocity of 12 cm/s. Thereafter it oscillates with amplitude 3 cm. Find its equation and period.

18. Graph at least one cycle of $y = \sin x + \cos x$.

19. Find the x-intercepts of the graph in Exercise 18 in some one cycle.

20. Solve $\sqrt{3} \cos 2x - \sin 2x = \sqrt{2}$ over the interval $0° \le x < 360°$.

21. The maximum speed of a particle in simple harmonic motion is 10 cm/s, and its speed is 6 cm/s when it is 4 cm from its central position. Find (a) its period and (b) its amplitude.

CHAPTER SUMMARY

1. Functions of the form
$$a \sin b(x - c) + d \text{ or } a \cos b(x - c) + d, b > 0,$$
as well as their graphs, are called sinusoids. Since the amplitude of any periodic function f is given by
$$\frac{\text{maximum value of } f - \text{minimum value of } f}{2},$$
the amplitude of each sinusoid above is $|a|$. The period of each is $\frac{2\pi}{b}$.

2. The graph of $y = a \sin b(x - c) + d$ can be obtained by translating the graph of $y = a \sin bx$. The translation is
    to the right c units if $c > 0$ or to the left $|c|$ units if $c < 0$,
and
    up d units if $d > 0$ or down $|d|$ units if $d < 0$.
The numbers c and d are called the phase shift and vertical shift, respectively, both relative to $y = a \sin bx$. The same is true if sine is replaced by cosine throughout.

3. The expression $A\cos \theta + B\sin \theta$ can be put into the form $C \cos (\theta - \gamma)$, where $C = \sqrt{A^2 + B^2}$, $\cos \gamma = \dfrac{A}{C}$, $\sin \gamma = \dfrac{B}{C}$.

4. Simple harmonic motion (SHM) can be described by an equation of the form $s = a \sin (\omega t + \beta)$. The frequency of the motion is the number of cycles described in unit time. Period T and frequency f are related by $f = \dfrac{1}{T}$. Velocity and position are related by $v^2 = \omega^2(a^2 - s^2)$.

5. Sums of sinusoids of different periods can be graphed by the method of addition of ordinates.

CHAPTER TEST

5-1   1. Graph two cycles of $y = 2 \cos \dfrac{\pi}{2} x$.

2. Give the domain and range of the function defined by $y = 2 \operatorname{Sin}^{-1} \dfrac{x}{2}$ and draw its graph.

5-2   3. Graph $y = \sin \left(\pi x - \dfrac{\pi}{4} \right) + 1$.

4. Express $\cos \left(2x - \dfrac{\pi}{3} \right)$ using the sine function.

5-3   5. Express $\cos 35° + \sin 35°$ in the form $C \cos \phi$, where $0° < \phi < 360°$.

6. Find all solutions of $\cos \theta - \sqrt{3} \sin \theta = \sqrt{2}$ in the interval $0° \le \theta < 360°$.

5-4   7. A pendulum hanging at rest is pulled 3 cm to the right and released. Thereafter it describes simple harmonic motion with period 4 seconds. Find an equation of the motion.

8. A particle describing simple harmonic motion has speed 6 cm/s when it is 1 cm from its central position and has speed 3 cm/s when it is 2 cm from the central position. Find its period and amplitude.

5-5   9. Graph at least one cycle of $y = \sin x - \cos 2x$.

10. Find the intercepts of the graph in Exercise 9.

5-6   11. The cross section of an ocean swell is sinusoidal. The distance from crest to crest is 80 m, and the troughs are 6 m lower than the crests. Find an equation of the cross section.

12. Find an equation of the sinusoid shown below.

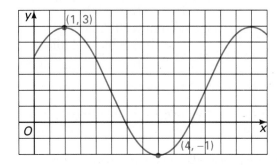

A Simple Graphing Utility

The following program and its modifications provide a handy graphing utility for many applications.

```
10 HGR : HCOLOR = 3
20 HPLOT 0, 95 TO 279, 95          Draw x-axis.
30 HPLOT 140, 0 TO 140, 190        Draw y-axis.
40 FOR X = − 140 TO 139
50 Y(1) = 20 * SIN (.05 * X)       Evaluate the first function.
60 Y(2) = 30 * SIN (.05 * X)       Evaluate the second function.
70 Y(3) = 40 * SIN (.05 * X)       Evaluate the third function.
200 HPLOT 140 + X, 95 − Y(1)       Graph the first function.
210 HPLOT 140 + X, 95 − Y(2)       Graph the second function.
220 HPLOT 140 + X, 95 − Y(3)       Graph the third function.
230 NEXT X
```

To explore amplitude change, run the program as it is given here. Notice that the graph of Y(2) = 30 * SIN (.05 * X) is taller than the graph of Y(1) = 20 * SIN (.05 * X). The graph of Y(3) = 40 * SIN(.05 * X) is taller still.

To explore period change, modify the given program as follows.

```
50 Y(1) = 30 * SIN (1 * .05 * X)
60 Y(2) = 30 * SIN (2 * .05 * X)
70 Y(3) = 30 * SIN (4 * .05 * X)
```

Notice that as the multiplier of .05 * X increases, the period decreases.

To graph a function such as $y = \sin 2x + \cos x$ by addition of ordinates, modify the original program as follows.

```
50 Y(1) = 30 * SIN (2 * .05 * X)
60 Y(2) = 30 * COS (.05 * X)
70 Y(3) = Y(1) + Y(2)
```

The program can also be used to verify that a graph you draw is correct. For example, the function $y = |\sin x|$ has the following properties.

1. For all x, $y \geq 0$ since the absolute value of any real number is nonnegative.

2. The function $y = |\sin x|$ is an even function, since
$$|\sin (-x)| = |-\sin x| = y = |\sin x|.$$
Thus, the graph should be symmetric about the y-axis.

3. The period of $y = |\sin x|$ is 2π, just as the period of $y = \sin x$ is 2π, since
$$|\sin (x + 2\pi)| = |\sin x|.$$

4. The maximum value of $y = |\sin x|$ is the same as the maximum value of $y = \sin x$.

Modify the original program as follows.

```
50 Y(1) = 30 * SIN (.05 * X)
60 Y(2) = ABS (30 * SIN (.05 * X))
```

Delete line 70. Then run the program. From the screen display, you can tell whether your graph has the features it should have.

EXERCISES

Use the graphing utility to graph each collection of functions.

1. Y(1) = 20 * COS (.05 * X)
   Y(2) = 30 * COS (.05 * X)
   Y(3) = 40 * COS (.05 * X)

2. Y(1) = 15 * COS (.05 * X)
   Y(2) = 25 * SIN (2 * .05 * X)
   Y(3) = Y(1) + Y(2)

Graph each of the following functions. Then use the graphing utility to check your answer.

3. $y = |\cos x|$

4. $y = \sin |x|$

5. $y = -\sin x$

6. (a) Use the graphing utility to graph
$$y = \sin 2x \text{ and } y = 2 \sin x \cos x.$$
   Enter the functions in lines 50 and 60 as
   ```
   Y1 = 20 * SIN (2 * .05 * X)
   ```
   and
   ```
   Y2 = 20 * 2 * SIN (.05 * X) * COS (.05 * X)
   ```
   Delete line 70.

   (b) What relationship exists between $\sin 2x$ and $2 \sin x \cos x$?

Radio Waves

Each radio station is assigned a fixed frequency, called the frequency of the *carrier wave*. This frequency corresponds to the number on the dial you select. For example, a disk jockey might announce that you are tuned to "85 on your AM dial" or "94.6 on your FM dial." Fire departments, police departments, and other city, state and federal agencies that use radio transmission as a means of communication are also assigned specific radio frequencies. The Federal Communications Commission (FCC) regulates the assignment of frequencies.

The figure below shows a carrier wave, having the form $y = A \sin qt$.

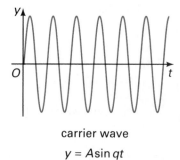

carrier wave

$y = A \sin qt$

The radio signal must also carry the information that your radio can decode as sounds. This information is also a wave, but one whose frequency must vary in the audio range. Such frequencies are all much lower than that of the carrier wave.

There are two ways that a single wave can have both a fixed carrier frequency and a variable audio frequency. The first of these is *amplitude modulation* (AM). By this means, the amplitude of the carrier wave varies periodically to produce an envelope. The frequency of this envelope is the audio frequency of the sound being transmitted. See the figure at the right on the next page in which the dashed line represents the envelope. In the second method of radio transmission, called frequency modulation (FM), the amplitude of the carrier wave is held constant while the frequency changes (by a relatively small amount) in accordance with the sound being transmitted. See the figure on the following page.

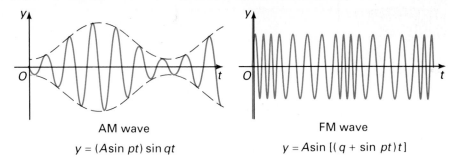

AM wave
$y = (A\sin pt) \sin qt$

FM wave
$y = A\sin [(q + \sin pt)t]$

In both types of modulation, $\dfrac{q}{2\pi}$, the average frequency of the carrier wave, is large relative to $\dfrac{p}{2\pi}$, the frequency of the sound transmitted.

EXERCISES

1. Write an equation for an AM wave of amplitude 1 transmitting a musical tone with a frequency of 256 Hz on a radio station whose carrier frequency is 800 kHz.
2. Use a computer graphing program such as the one on pages 184-185 to graph each of the following.
   (a) (AM wave): $y = (50 \sin 0.03t)(\sin 0.3t)$
   (b) (FM wave): $y = 50 \sin ((0.4 + 0.02 \sin 0.08t)t)$

CUMULATIVE REVIEW   Chapters 3–5

Chapter 3

1. Express $\tan x(\sin x - \sec x \cot x)$ in terms of $\cos x$.

2. Express $\dfrac{\sin \theta}{1 + \cos \theta} + \cot \theta$ in terms of a single trigonometric function.

Prove each identity.

3. $\dfrac{(\sin x + \cos x)^2}{\cos x} = \sec x + 2 \sin x$

4. $\dfrac{\cot x}{\csc x - 1} = \dfrac{1 + \sin x}{\cos x}$

5. If $\sin x = -\dfrac{3}{5}$ and $\pi < x < \dfrac{3\pi}{2}$, find each of the following.

   (a) $\sin\left(x + \dfrac{\pi}{6}\right)$ (b) $\cos\left(x - \dfrac{\pi}{6}\right)$ (c) $\tan\left(x + \dfrac{\pi}{6}\right)$

6. Prove $\cos 3x = \cos x(\cos^2 x - 3 \sin^2 x)$.

7. Evaluate $\tan 22.5°$ using identities.

Solve each of the following for $0 \le x < 2\pi$.

8. $1 + 2 \tan^2 x + \sec x = 0$

9. $1 - \sin 2x = \cos x - 2 \sin x$

Chapter 4

1. In $\triangle ART$, $\angle R = 135°$ and $\angle T = 15°$. If $AT = 6\sqrt{3}$, find RT in simplest radical form.

2. Find all possible values for $\angle B$ in $\triangle ABC$ if $\angle A = 27°$, $BC = 15$, and $AB = 24$. Round answers to the nearest tenth of a degree.

3. In order to determine the angle formed by two walls, a meter stick is placed as shown with one end touching one wall 20 cm from the intersection of the walls. If the stick touches the other wall at a point 105 cm from the intersection, find the angle formed by the walls to the nearest degree.

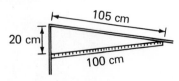

4. Find the area of $\triangle ABC$ if $\angle B = 30°$, $a = 10$, and $c = 15$.

5. Find the area of a triangle with sides of 25 cm, 39 cm, and 40 cm.

6. In the figure at the right,
   $AD = 20, BC = 60$, and
   $\angle BDC = 50°$.
   (a) Find the measure of $\angle ABD$.
   (b) Find the area of $\triangle BCD$.

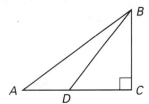

Chapter 5

In Exercises 1–4, (a) give the period and amplitude of each function and (b) graph the function over the interval $-2\pi \le x \le 2\pi$.

1. $y = 2 \sin 3x$

2. $y = 4 \cos \dfrac{x}{2}$

3. $y = 5 - 2 \sin 3x$

4. $y = \cos\left(2x - \dfrac{\pi}{3}\right)$

5. (Optional) Rewrite $8 \sin \theta - 6 \cos \theta$ in the form $C \cos (\theta - \phi)$, where $-180° \le \phi \le 180°$.

6. An object is pushed upward 5 cm and released. It makes one complete oscillation every 3 seconds. (a) Sketch the displacement as a function of time. (b) Find an equation of the form $s = a \cos \omega t$ for the SHM. (c) Find the frequency of the motion.

7. Graph $y = \sin\left(x + \dfrac{\pi}{2}\right) + \sin 2x$ for $0 \le x \le 2\pi$.

8. Find an equation of the form $y = a \cos b (x - c) + d$ for the sinusoid shown below.

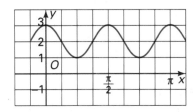

6 Vectors in the Plane

The photograph illustrates power being generated by the force of wind. Vectors are important tools in the study of problems involving wind speed and direction, force, and work and energy.

GEOMETRIC VECTORS AND APPLICATIONS

6-1 Geometric Vectors

Objective: To define vectors and basic vector operations
geometrically.

What will happen if a force **F**, like that shown in Figure 6-1, acts on a
heavy box resting on the floor? If the magnitude of the force is small,
the box will not move at all. If the magnitude is increased enough, the
box will slide along the floor. If the mag-
nitude of the force is still greater, the box
will be lifted from the floor. Similarly,
varied effects can be achieved by changing the *direction* of the force **F**. Knowing
the magnitude and direction of **F**, the
weight of the box, and the frictional force
produced by the box and the floor, how
can we predict which of these possible
outcomes will occur?

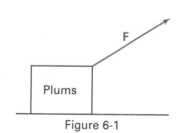

Figure 6-1

To answer the question posed on the preceding page and others like it, we introduce the mathematical notion of a *vector*. A **vector** is a quantity that has both magnitude and direction. Force and velocity are examples of vector quantities. On the other hand, a physical quantity, such as speed, mass, or length, that can be described by magnitude alone, is called a *scalar*. (In this book, scalars will always be real numbers.)

A natural way to represent a vector quantity geometrically is by a *directed line segment*, or *arrow*, whose length is the magnitude of the quantity and whose direction is the direction of the quantity. We call these arrows vectors and use boldface letters such as **u** or **v** to denote them. (In handwritten work, we often write **v** as $\vec{v}$.) The tail of the arrow is called the *initial point* of the vector and the tip is called the *terminal point*. We regard two vectors as *equal* if they have the same length and direction. In Figure 6-2, **u** = **v** and **F** = **H**.

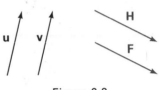

Figure 6-2

One reason for the usefulness of vectors is that they can be added mathematically in a way that describes how pairs of physical quantities combine. For example, the force **F** pictured in Figure 6-1 is equivalent to the sum of forces, $\mathbf{F}_x$ and $\mathbf{F}_y$, (pictured in Figure 6-3) acting simultaneously. Together, these two forces cause the same motion as **F**. Therefore, in order to answer the question that began this section, we have only to check whether the magnitude of vector $\mathbf{F}_x$ is greater than the frictional force tending to prevent the box's horizontal motion and whether $\mathbf{F}_y$ is greater than the gravitational force on the box.

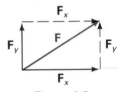

Figure 6-3

Note that, in Figure 6-3, $\mathbf{F}_x$ and $\mathbf{F}_y$ are represented both by solid and dotted arrows. Thus, the following definition of *vector addition* should seem reasonable.

Definition of Vector Addition

To find the sum of the vectors **u** and **v**, draw **u** and draw **v** with the initial point of **v** at the terminal point of **u**. The sum, **u** + **v**, is the vector whose tail is the initial point of **u** and whose tip is the terminal point of **v**. (See Figure 6-4.)

Figure 6-4

This definition is equivalent to the **parallelogram rule** illustrated in Figure 6-5. That is, **u** + **v** is the diagonal of the parallelogram having **u** and **v** as adjacent sides.

Figure 6-5

If **w** = **u** + **v**, we sometimes refer to **u** and **v** as **components** of **w**, and we refer to **w** as the **resultant** of **u** and **v**.

We can also define multiplication of a vector by a scalar.

Definition of Scalar Multiplication

To find the product $r\mathbf{v}$ of a scalar r and a vector **v**, multiply the length of **v** by $|r|$ and reverse the direction if $r < 0$.

Figure 6-6 illustrates the definition of scalar multiplication for $r = 2$, $\frac{1}{2}$, -1.5, and -1. Note the direction of the arrowhead in each case. If **u** = $r\mathbf{v}$ for some scalar r, we say that **u** and **v** are **parallel**. Thus, each two of the vectors in Figure 6-6 are parallel.

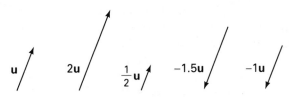

Figure 6-6

The **zero vector**, denoted **0**, has magnitude 0 and no direction. Note that **0** = $r\mathbf{v}$ when $r = 0$ or **v** = **0**. The zero vector is parallel to every vector.

We write $(-1)\mathbf{v}$ as $-\mathbf{v}$ and define **u** − **v** as **u** + $(-\mathbf{v})$. Thus, **u** − **v** is the vector that must be added to **v** to produce **u**. Vector subtraction is illustrated in Figure 6-7.

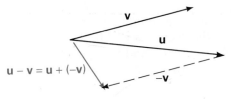

Figure 6-7

It can be shown that vector addition and scalar multiplication obey many algebraic laws. In particular, vector addition is commutative and associative, and scalar multiplication distributes over addition:

$$\mathbf{u} + \mathbf{v} = \mathbf{v} + \mathbf{u}$$
$$(\mathbf{u} + \mathbf{v}) + \mathbf{w} = \mathbf{u} + (\mathbf{v} + \mathbf{w})$$
$$r(\mathbf{u} + \mathbf{v}) = r\mathbf{u} + r\mathbf{v}$$

Figure 6-8 illustrates associativity. Figure 6-9 illustrates the fact that $2(\mathbf{u} + \mathbf{v}) = 2\mathbf{u} + 2\mathbf{v}$.

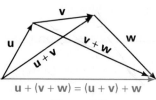

$\mathbf{u} + (\mathbf{v} + \mathbf{w}) = (\mathbf{u} + \mathbf{v}) + \mathbf{w}$

Figure 6-8

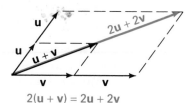

$2(\mathbf{u} + \mathbf{v}) = 2\mathbf{u} + 2\mathbf{v}$

Figure 6-9

The expression $2\mathbf{u} + 2\mathbf{v}$ is an example of a *linear combination* of vectors. In general, an expression of the form

$$r_1\mathbf{v}_1 + r_2\mathbf{v}_2 + \cdots + r_n\mathbf{v}_n$$

where $\mathbf{v}_1, \mathbf{v}_2, \ldots, \mathbf{v}_n$ are vectors and $r_1, r_2, \ldots, r_n$ are scalars is called a **linear combination** of $\mathbf{v}_1, \mathbf{v}_2, \ldots, \mathbf{v}_n$.

Example 1   Using the vectors $\mathbf{u}$ and $\mathbf{v}$ shown at the right, sketch the linear combinations (a) $-\mathbf{u} + 2\mathbf{v}$ and (b) $\mathbf{u} - 3\mathbf{v}$.

Solution   You can use the definition of addition or the parallelogram rule.

(a)   (b)

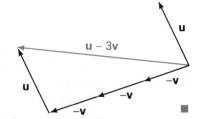

Example 2   (a) Express the vectors $\mathbf{u}$ and $\mathbf{v}$ as linear combinations of the vectors $\mathbf{a}$ and $\mathbf{b}$.
(b) Express $\mathbf{a}$ and $\mathbf{b}$ as linear combinations of $\mathbf{u}$ and $\mathbf{v}$.

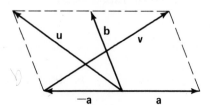

Solution   (a) From the figure, $\mathbf{u} = -\mathbf{a} + \mathbf{b}$ and $\mathbf{v} = 2\mathbf{a} + \mathbf{b}$.

(b) Set up a system of simultaneous equations from the linear combinations in part (a). Solve this system first for $\mathbf{a}$ and then for $\mathbf{b}$.

$$
\begin{aligned}
-\mathbf{a} + \mathbf{b} &= \mathbf{u} \\
2\mathbf{a} + \mathbf{b} &= \mathbf{v} \\
\hline
-3\mathbf{a} \phantom{+\mathbf{b}} &= \mathbf{u} - \mathbf{v} \\
\mathbf{a} &= -\frac{1}{3}\mathbf{u} + \frac{1}{3}\mathbf{v}
\end{aligned}
\qquad
\begin{aligned}
-2\mathbf{a} + 2\mathbf{b} &= 2\mathbf{u} \\
2\mathbf{a} + \mathbf{b} &= \mathbf{v} \\
\hline
3\mathbf{b} &= 2\mathbf{u} + \mathbf{v} \\
\mathbf{b} &= \frac{2}{3}\mathbf{u} + \frac{1}{3}\mathbf{v} \quad \blacksquare
\end{aligned}
$$

The magnitude of a vector $\mathbf{v}$ is also called its **norm**, denoted $\|\mathbf{v}\|$. Three important properties of the norm are:

$$
\begin{aligned}
&\|\mathbf{v}\| = 0 \text{ if and only if } \mathbf{v} = \mathbf{0}. &(1)\\
&\|r\mathbf{v}\| = |r|\,\|\mathbf{v}\| &(2)\\
&\|\mathbf{u} + \mathbf{v}\| \le \|\mathbf{u}\| + \|\mathbf{v}\| &(3)
\end{aligned}
$$

Property (1) is obviously true. Properties (2) and (3) will be proved algebraically in Exercise 31 of Section 6-4 and Exercise 46 of Section 6-5, respectively. Figure 6-10 shows that property (3) is a formulation of the familiar triangle inequality of geometry.

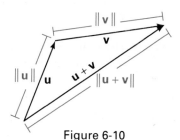

Figure 6-10

EXERCISES 6-1

Copy vectors **u** and **v** shown at the right. Sketch each linear combination.

A 1. $2\mathbf{u} + \mathbf{v}$ 2. $\mathbf{u} - \mathbf{v}$

3. $-3\mathbf{u} + \mathbf{v}$ 4. $2\mathbf{u} - 3\mathbf{v}$

5. $-\frac{1}{2}\mathbf{u} - 2\mathbf{v}$ 6. $2\mathbf{u} + 1.5\mathbf{v}$ 7. $-\mathbf{u} + 2.5\mathbf{v}$ 8. $3\mathbf{u} - \frac{1}{2}\mathbf{v}$

Express **u** as a linear combination of **v** and **w**.

9.

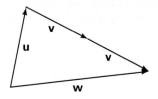

10.

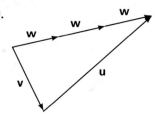

11.

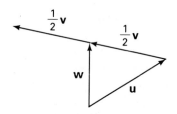

12.

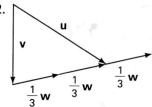

13. Draw any two nonparallel vectors **u** and **v** having the same initial point. On the same diagram, draw 2**u** − **v** and **v** − 2**u**. Are these vectors parallel?

14. Draw any two nonparallel vectors **u** and **v** having the same initial point. On the same diagram, draw **u** − 2**v** and **v** − $\frac{1}{2}$**u**. Are these vectors parallel?

Draw diagrams with vectors **u**, **v**, and **w** satisfying the given conditions.

15. $\|\mathbf{u}\| = 3\|\mathbf{v}\|$ and **w** = **u** + **v**

16. $\|\mathbf{u}\| = \frac{1}{2}\|\mathbf{v}\|$ and **w** = **v** − **u**

17. $\|\mathbf{w}\| = \|\mathbf{u}\| + \|\mathbf{v}\|$ and **w** = **u** + **v**

18. $\|\mathbf{w}\| = \|\mathbf{u}\| + \|\mathbf{v}\|$ and **w** = **u** − **v**

Copy the vectors shown below. Sketch each linear combination.

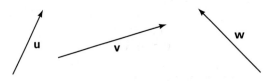

B 19. **u** + **v** − **w**   20. −2**u** − **v** + **w**   21. **v** − **u** + 2**w**   22. −$\frac{1}{2}$**u** + **v** + **w**

23. Draw two nonparallel vectors **u** and **v** with a common initial point and with equal magnitudes. On the same diagram, sketch **u** + **v** and **u** − **v**. What geometric relationship seems to hold between **u** + **v** and **u** − **v**? Referring to the parallelogram determined by **u** and **v**, what theorem from geometry have you illustrated?

24. Draw two vectors **u** and **v** that have the same initial point and are perpendicular. Sketch **u** + **v**. State an algebraic relationship among $\|\mathbf{u}\|$, $\|\mathbf{v}\|$, and $\|\mathbf{u} + \mathbf{v}\|$.

25. Give a necessary and sufficient condition on the vectors **u** and **v** for equality to hold in norm property (3), the triangle inequality.

26. Given that $\|\mathbf{u}\| = \|\mathbf{v}\|$ and $\mathbf{u} = r\mathbf{v}$ for some scalar r, what can you conclude about r?

In Exercises 27–30, (a) express **u** and **v** in terms of **a** and **b**. Then (b) express **a** and **b** in terms of **u** and **v**.

27.

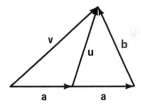

28.

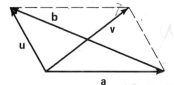

29.

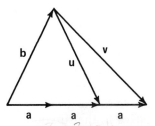

30.

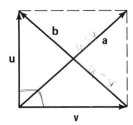

C 31. Show that for any vectors **u** and **v**, $\|\mathbf{u} - \mathbf{v}\| \geq \|\mathbf{u}\| - \|\mathbf{v}\|$. (Hint: Use **u** − **v** for **u** in norm property (3).)

32. (a) Referring to the diagram at the right, express the vector **b** in terms of **u** and **v**.
    (b) Express **a** in terms of **u** and **v**.
    (c) By solving simultaneously the equations you derived in (a) and (b), eliminate **u** and **v**, and thus express **b** in terms of **a** alone.
    (d) What theorem from geometry have you demonstrated?

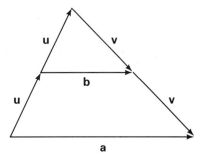

33. A theorem of geometry states that the sum of the squares of the lengths of the sides of a parallelogram equals the sum of the squares of the lengths of the diagonals. Draw a diagram illustrating the parallelogram determined by two vectors **u** and **v**. State the theorem as a vector equation.

6-2

Vectors and Navigation

Objective: To solve navigation problems by using vectors.

In certain applications it is convenient to denote the directed line segment with initial point A and terminal point B by $\overrightarrow{AB}$. For example, if an object moves from A to B, its **displacement** is given by the vector $\overrightarrow{AB}$. In navigation, the direction of a vector is specified by giving its *bearing*. The *bearing* of a vector is the angle, measured clockwise, between due north and the vector. Figure 6-11 illustrates bearings of 20°, 100°, 260°, and 300°. Bearing and some other useful navigation terms are defined in the following list.

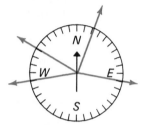

Figure 6–11

bearing	the angle θ, $0° \leq \theta < 360°$, measured clockwise, that a vector makes with due north
heading	the bearing of the vector that points in the direction in which a craft, such as a ship or a plane, is aimed
true course	the bearing of the vector that points in the direction in which a craft is actually traveling
ground speed	the speed of a plane relative to the ground
air speed	the speed of a plane relative to the surrounding air, that is, the speed the plane would actually have if there were no wind

Figure 6-12 illustrates a plane with an air speed of 300 km/h and a heading of 40°. This velocity is represented by $\overrightarrow{AB}$. Note that $\|\overrightarrow{AB}\|$ is the air speed. The figure also illustrates a wind with a bearing of 90°: $\overrightarrow{BC}$. The plane has a true course with a bearing of 50°: $\overrightarrow{AC}$.

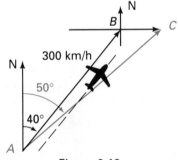

Figure 6-12

Example 1

A ship leaves port A and sails 150 km with heading 115° to point B. It then changes its heading to 75° and sails 120 km to C. Find its distance and bearing from A.

Solution

The situation is illustrated in the figure.

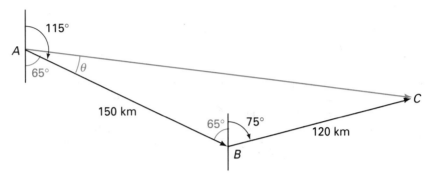

To find AC, use the law of cosines. The angle B in triangle ABC is 140°. The lengths of the adjacent sides are 150 km and 120 km. By the law of cosines,

$$(AC)^2 = 150^2 + 120^2 - 2 \cdot 150 \cdot 120 \cos 140°$$
$$= 22{,}500 + 14{,}400 - 36{,}000(-0.7660)$$
$$= 64{,}476$$

Therefore, $AC = 254$ km.

To find the bearing of $\overrightarrow{AC}$, use the law of sines to find θ. Applying the law of sines to $\triangle ABC$, we have:

$$\frac{\sin \theta}{120} = \frac{\sin 140°}{254}$$

$$\sin \theta = \frac{120(0.6428)}{254}$$

$$= 0.3037$$

Therefore, $\theta = 17.7°$, or $17°40'$. Thus, the bearing of $\overrightarrow{AC} = 115° - \theta = 115° - 17.7° = 97.3°$, or $97°20'$. ∎

Example 2

The air speed of a light plane is 200 km/h and its heading is 90°. A 40 km/h wind is blowing with a bearing of 160°.
(a) Find the ground speed and the true course of the plane.
(b) What heading should the pilot use so that the true course will be 90°? What will the ground speed be then?

Solution

(a) In the figure, $\mathbf{u}$ and $\mathbf{v}$ are the velocities of the plane relative to the air and the ground, respectively, and $\mathbf{w}$ is the wind velocity. The ground speed is $\|\mathbf{v}\|$ and the true course is θ.

(Solution continued next page.)

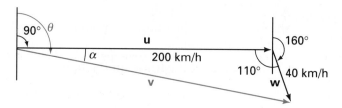

From the figure, $\|\mathbf{v}\|$ should be a little more than 200.
To find $\|\mathbf{v}\|$, apply the law of cosines.

$$\|\mathbf{v}\|^2 = 200^2 + 40^2 - 2 \cdot 200 \cdot 40 \cos 110°$$
$$= 40{,}000 + 1600 - (16{,}000)(-0.3420) = 47{,}072$$
$$\|\mathbf{v}\| = 217$$

Therefore, the ground speed is 217 km/h.

To find θ, apply the law of sines: $\dfrac{\sin \alpha}{40} = \dfrac{\sin 110°}{217}$

$$\sin \alpha = \frac{40(0.9397)}{217} = 0.1732$$

$$\alpha = 10°$$

Therefore, $\theta = 90° + 10° = 100°$ and the true course is 100°.

(b) Denote the plane's air and ground velocities by $\mathbf{u}'$ and $\mathbf{v}'$ respectively.
The required heading is θ and the new ground speed is $\|\mathbf{v}'\|$.
From the figure, θ should be a little less than 90°.

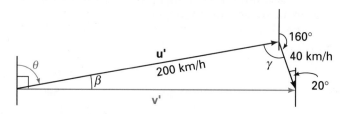

To find θ, apply the law of sines: $\dfrac{\sin \beta}{40} = \dfrac{\sin 70°}{200}$

$$\sin \beta = \frac{40(0.9397)}{200} = 0.1879$$

$$\beta = 10.8° \text{ or } 10°50'$$

Thus, $\theta = 90° - 10.8° = 79.2°$. The heading should be 79.2°, or
79°10'. To find $\|\mathbf{v}'\|$, note $\gamma + 10.8° + 70° = 180°$. Hence, $\gamma = 99.2°$.

We apply the law of sines again: $\dfrac{\|\mathbf{v}'\|}{\sin 99.2°} = \dfrac{200}{\sin 70°}$

$$\|\mathbf{v}'\| = \frac{200(0.9871)}{0.9397} = 210$$

Therefore, the new ground speed would be 210 km/h. ∎

EXERCISES 6-2

Give distance to the nearest km, mi, or m. Give speeds to the nearest km/h or mi/h. Give angle measures to the nearest tenth of a degree.

A 1. A ship leaves port and sails west for 120 km, then south for 40 km. What are the distance and the bearing of the port from the ship?

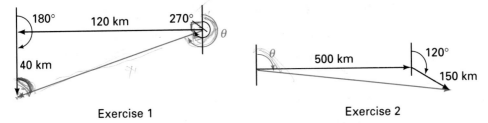

Exercise 1   Exercise 2

2. A plane flies due east for 500 km and then on a heading of 120° for 150 km. What are its distance and bearing from its starting point?

3. A plane heads due east with an air speed of 300 km/h. A 45 km/h wind is blowing with a bearing 150°. Find the plane's true course and ground speed.

4. If the plane in Exercise 3 headed north while everything else remained the same, what would be its ground speed and true course?

5. A lake ferry leaves port *A* bound for port *B*, which is 80 km away and bears 330° from *A*. Because of shallow water the ferry first travels 20 km with heading 40°. What heading should it then use to proceed directly to *B* and how far must it travel?

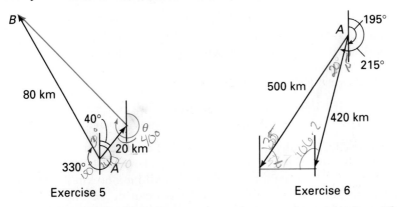

Exercise 5   Exercise 6

6. Two planes take off in still air from an airport at the same time. The speed of one plane is 420 km/h and its heading is 195°. The speed of the other plane is 500 km/h and its heading is 215°. How far apart are they after 1 h? What is the bearing of each from the other?

7. A current is flowing so that if an outboard motor boat is pointed 21° upstream (measured from a line perpendicular to the river bank) and sails with a velocity of 12 km/h it will land directly opposite its starting point. What is the speed of the current?

8. The speed in still water of a power boat is 20 km/h. The boat heads directly west across an 80 m wide river that is flowing due south at 6 km/h. (a) How far downstream will the boat land? (b) What heading should the operator have used in order to land directly opposite the starting point?

9. A plane with a heading of 30° has an air speed of 120 mi/h. A 15 mi/h wind is blowing with a bearing of 315°. Find the plane's true course and ground speed.

10. An airplane pilot whose craft has an air speed of 150 km/h wishes to fly on a true course of 65°. A 15 km/h wind is blowing with a bearing of 135°. (a) In what direction should the plane be pointed? (b) What will be the plane's ground speed?

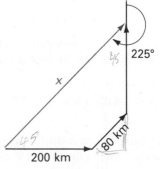

B 11. A ship leaves Port Cod and sails due east for 200 km, then on a heading of 45° for 80 km, and then due north until the bearing of Port Cod from the ship is 225°. How far away is the ship from Port Cod?

Exercise 11

12. A search plane leaves its base and flies due north for 400 km. It then flies due east for 100 km and then on a heading of 210° for 200 km. At this point what are the plane's distance and bearing from its base?

13. A plane flies with an air speed of 140 mi/h and a heading of 75° for an hour, while a 10 mi/h wind is blowing from the north. (A north wind blows southward.) With the same wind speed and air speed, what should the heading be so that the true course will take the plane back to its starting point on the return trip?

14. Answer Exercise 13 if the wind has shifted and blows at 25 mi/h from the east on the return trip. (An east wind blows westward.)

C 15. Airport *SFO* bears 320° from airport *LAX* and is 540 km away. A pilot is planning a straight-line flight from *LAX* to *SFO* to leave at 1 P.M. The plane's air speed will be 650 km/h and there will be a 60 km/h wind blowing from 280°. What will the compass heading be and what is the plane's ETA (estimated time of arrival) to the nearest minute?

16. When the pilot in Exercise 15 plans the return trip to leave *SFO* at 6 P.M. conditions are the same as before. What will be the heading and ETA?

6-3 Vectors and Force

Objective: To apply vector concepts to solve problems involving force.

Let **v** be a nonzero vector placed so that its tail is at the origin of a two-dimensional coordinate system. We can write **v** as a sum

$$\mathbf{v} = \mathbf{v}_x + \mathbf{v}_y,$$

where $\mathbf{v}_x$ and $\mathbf{v}_y$ are parallel to the *x*- and *y*-axes, respectively. (See Figure 6-13.) We call $\mathbf{v}_x$ and $\mathbf{v}_y$ the *x*- and *y-components* of **v**, and we say that **v** has been *resolved into horizontal and vertical components*. Trigonometry shows that

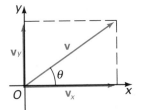

$$\|\mathbf{v}_x\| = \|\mathbf{v}\| \cos \theta$$

$$\|\mathbf{v}_y\| = \|\mathbf{v}\| \sin \theta$$

Figure 6-13

where θ is the angle between **v** and the positive *x*-axis.

The foregoing analysis is useful in dealing with physical situations that involve forces. So also is the following law of mechanics.

> When a body is at rest or is moving at constant velocity, the vector sum of the forces acting on it is zero.

This means that for a stationary object (or one moving at a constant velocity), the sum of all horizontal components is 0 and the sum of all vertical components is 0. This discussion applies equally well to any resolution into perpendicular components.

The basic unit of force in the kilogram-meter-second system is the *newton*, defined as the force necessary to give a 1 kg mass an acceleration of 1 m/s². The symbol for the newton is N. Near the surface of Earth the gravitational force exerted on an object of mass *m* kilograms is about 9.80*m* newtons. Thus, for example, the gravitational force on an object of 12.4 kg is about 9.80 × 12.4 newtons, about 122 newtons.

Example 1   A loading ramp makes an angle of 24° with the horizontal. What is the frictional force that will keep a 150 kg crate from sliding down?

Solution   The figure on the following page shows the gravitational force that is exerted on the crate resolved into components **F** and **H** parallel and perpendicular to the ramp, respectively. Notice that the angle between vectors **F** and **G** is 90° − 24° = 66°. Therefore, the angle between **G** and **H** is 24°. If the crate is not moving, the frictional force on it, **F′**,

(Solution continued next page.)

and the component, **F**, of the gravitational force parallel to the ramp must satisfy

$$\mathbf{F} + \mathbf{F}' = \mathbf{0}.$$

That is, $\mathbf{F}' = -\mathbf{F}$.

Since $\sin 24° = \dfrac{\|\mathbf{F}\|}{\|\mathbf{G}\|}$ and $\|\mathbf{G}\| = (9.80)(150)$,

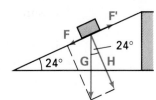

$$\|\mathbf{F}\| = (9.80)(150)(0.4067) = 598.$$

The frictional force **F**' is therefore 598 N. ∎

The *tension* in a rope or cable is the magnitude of the force it transmits.

Example 2   A 250 kg piano is suspended from a cable that makes a 25° angle with a building and from a second cable that makes a 40° angle with a crane as shown. Find the tension in each cable.

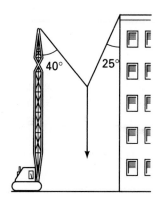

Solution   The second figure shows **F**, **G**, and **H** in red and copies of **F** and **G** in black. The angle between **F** and **G** is 25° and therefore, the angle between **H** and **F** is $180° - (25° + 40°) = 115°$. Also $\|\mathbf{G}\| = (9.80)(250) = 2450$. To find $\|\mathbf{F}\|$ and $\|\mathbf{H}\|$, apply the law of sines to the triangle formed by **F**, **G**, and **H**.

$$\frac{\|\mathbf{F}\|}{\sin 40°} = \frac{\|\mathbf{G}\|}{\sin 115°} = \frac{2450}{\sin 115°}$$

Hence, $\|\mathbf{F}\| = \dfrac{2450 \sin 40°}{\sin 115°} = 1738.$

$$\frac{\|\mathbf{H}\|}{\sin 25°} = \frac{\|\mathbf{G}\|}{\sin 115°} = \frac{2450}{\sin 115°}$$

Hence, $\|\mathbf{H}\| = \dfrac{2450 \sin 25°}{\sin 115°} = 1142.$

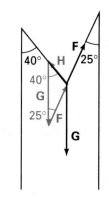

The tensions in **F** and **H** are 1738 N and 1142 N, respectively. ∎

Example 3   A force **F** is exerted on a sled rope that makes a 30° angle with the horizontal. If the sled has mass 20 kg and the frictional force tending to prevent horizontal motion is 0.3 of the net downward force under the sled, what is the minimum value of $\|\mathbf{F}\|$ that will move the sled?

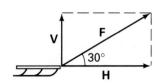

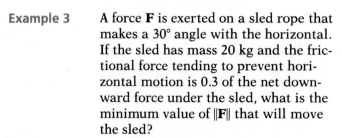

Solution   The magnitude of the horizontal component of **F**, $\|\mathbf{F}\|\cos 30°$, must be great enough to offset the frictional force. The vertical component of **F** has magnitude

$$\|\mathbf{F}\| \sin 30°$$

and the force of gravity acting on the sled is

$$(9.80)(20).$$

Hence, the net downward force under the sled is

$$(9.80)(20) - \|\mathbf{F}\| \sin 30°.$$

Therefore, the frictional force that must be overcome in order to move the sled horizontally is

$$0.3[(9.80)(20) - \|\mathbf{F}\| \sin 30°].$$

The magnitude of the force **F** that will move the sled is therefore the solution of the equation:

$$\|\mathbf{F}\| \cos 30° = 0.3[(9.80)(20) - \|\mathbf{F}\| \sin 30°]$$
$$0.8660\|\mathbf{F}\| = 58.8 - 0.15\|\mathbf{F}\|$$
$$1.016\|\mathbf{F}\| = 58.8$$
$$\|\mathbf{F}\| = 57.9$$

At least 57.9 N of force must be applied to move the sled. ■

EXERCISES 6-3

In Exercises 1–19, solve each problem.

A 1. A 30 kg object is on a ramp inclined at 22° with the horizontal. Find the magnitudes of the components of the gravitational force on the object parallel and perpendicular to the ramp.

2. A 100 kg box slides with constant velocity down a ramp inclined at 20° with the horizontal. Find the frictional force acting on the box.

3. In Example 2, what force must the cable attached to the crane exert in order to have the cable make an angle of 35° with the building?

4. A 300 kg object is suspended by two cables each of which makes an angle of 45° with the horizontal. Find the tension in each cable.

5. A 500 kg motor is suspended by two cables which make angles of 40° and 60° with the horizontal. Find the tension in each cable.

6. Work Example 2 assuming that the cable attached to the crane makes an angle of 30° with the horizontal.

7. A 250 kg piano is suspended from a cable attached to a winch on top of a building and from a second cable attached to a vertical crane. The second cable makes a 25° angle with the crane. That cable exerts a force of 900 N. Find the tension in the first cable and the angle it makes with the building.

8. Find the magnitude of the force that must be exerted on the rope in Example 3 if the rope makes a 45° angle with the horizontal.

9. A space vehicle's main engine produces a thrust of 1000 N in the direction opposite to the vehicle's course. If a directional rocket is fired with a thrust of 300 N in a direction perpendicular to the course, by how many degrees will the vehicle's course change? What will be the effective thrust in the direction opposite to the new course?

B 10. Three forces **F**, **G**, and **H** act on an object P as shown in the figure at the right. If the object does not move under the effect of the three forces, find
   (a) the magnitude of the component of **H** in the horizontal direction.
   (b) ‖**H**‖.
   (c) the magnitude of the component of **H** in the vertical direction.
   (d) ‖**F**‖.

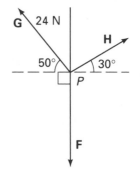

In Exercises 11–14, the given forces act on a particle P. Find the magnitude and direction of a force **F** that will keep P stationary. (Each direction, θ, is measured counterclockwise as shown at the right.) Use the fact that if P is stationary, then the sum of all horizontal components and the sum of all vertical components of forces acting on P are 0.

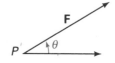

11. **F**$_1$: 12 N; 50°
    F$_2$: 10 N; 155°
    F$_3$: 20 N; 250°

12. **F**$_1$: 10 N; 80°
    F$_2$: 15 N; 200°
    F$_3$: 5 N; 340°

13. **F**$_1$: 14 N; 95°
    F$_2$: 20 N; 200°
    F$_3$: 16 N; 300°

14. **F**$_1$: 70 N; 60°
    F$_2$: 50 N; 90°
    F$_3$: 70 N; 240°
    F$_4$: 80 N; 330°

15. A crate rests on an incline that makes an angle θ with the horizontal. Find θ if the magnitude of the frictional force preventing the crate from sliding down the incline equals 0.2 the magnitude of the component of the gravitational force perpendicular to the incline.

16. A weight is suspended from two ropes, as shown at the left below. If **G** is a force of 20 N and **H** is a force of 10 N, find ‖**F**‖. (Hint: Sketch a triangle with **F**, **G**, and **H** as sides.)

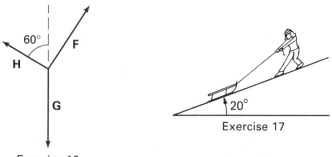

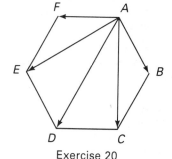

Exercise 17

Exercise 16

17. The figure at the right above represents a 50 kg sled on a 20° slope. What force making a 45° angle with the horizontal is necessary to keep the sled from sliding down? Assume that friction is negligible.

C 18. What is the change of course of the rocket in Exercise 9 if the directional rocket is fired at an angle of 40° to the main engine thrust?

19. A box is on a ramp making a 15° angle with the horizontal. A force of 480 N *up* the ramp moves the box with constant velocity and a force of 20 N *down* the ramp also moves it with constant velocity. Find the mass of the box.

20. Five forces act at one vertex A of a regular hexagon, one force directed toward each of the other vertices as shown. Show that
$$\vec{AB} + \vec{AC} + \vec{AD} + \vec{AE} + \vec{AF} = 3\vec{AD}.$$

Exercise 20

Self Quiz

6-1 6-2 6-3

Copy the vectors **u** and **v** shown below. Then sketch each linear combination.

1. $3\mathbf{u} - \mathbf{v}$
2. $-2\mathbf{u} + \mathbf{v}$

3. Refer to the diagram at the right. Express **w** and **z** as linear combinations of **u** and **v**.

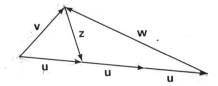

4. A plane heads due west at 200 km/h while a wind is blowing from the north at 30 km/h. Find the plane's true course and ground speed.

5. A ship sails on a heading of 120° for 50 km and then changes to a heading of 40° for the next 20 km. Find its distance and bearing from its starting point.

6. What frictional force will keep a 50 kg object from sliding down a ramp that makes an angle of 28° with the horizontal?

7. A 300 kg beam is hoisted by two cables that make angles of 35° and 130°, respectively, with the horizontal. Find the tension in each cable.

ALGEBRAIC VECTORS AND APPLICATIONS

6-4 Algebraic Vectors

Objective: To study vectors and vector operations algebraically.

In this section, we consider vectors that lie in a fixed plane into which we have introduced a coordinate system. We denote by **i** and **j** the vectors having initial point the origin and terminal points $(1, 0)$ and $(0, 1)$, respectively. Any vector **v** can now be expressed as a linear combination of **i** and **j**:

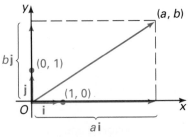

Figure 6-14

$$\mathbf{v} = a\mathbf{i} + b\mathbf{j}$$

where the terminal point of **v** is (a, b) as illustrated in Figure 6-14. For example, the vector $\mathbf{v} = 3\mathbf{i} + 2\mathbf{j}$ represents the vector whose initial point is the origin and whose terminal point is $(3, 2)$.

When the initial point of **v** is the origin, we say that **v** is in **standard position**. In this context, we call a and b the **scalar components** of **v**.

In terms of scalar components, we have the following:

> If $\mathbf{u} = a\mathbf{i} + b\mathbf{j}$ and $\mathbf{v} = c\mathbf{i} + d\mathbf{j}$, then:
> $\mathbf{u} = \mathbf{v}$ if and only if $a = c$ and $b = d$ (1)
> $\mathbf{u} + \mathbf{v} = (a + c)\mathbf{i} + (b + d)\mathbf{j}$ (2)
> $r\mathbf{u} = (ra)\mathbf{i} + (rb)\mathbf{j}$ for every scalar r (3)

Example 1 Find r and s such that $(r\mathbf{i} + 4\mathbf{j}) + (s\mathbf{i} + r\mathbf{j}) = 7\mathbf{i} + 3\mathbf{j}$.

Solution Using (2), we have:

$$(r\mathbf{i} + 4\mathbf{j}) + (s\mathbf{i} + r\mathbf{j}) = 7\mathbf{i} + 3\mathbf{j}$$
$$(r + s)\mathbf{i} + (4 + r)\mathbf{j} = 7\mathbf{i} + 3\mathbf{j}$$

Using (1), we have:

$$r + s = 7 \text{ and } 4 + r = 3$$

Solving this system, we find that $r = -1$ and $s = 8$. ∎

Example 2 Let $\mathbf{u} = -2\mathbf{i} + 3\mathbf{j}$ and $\mathbf{v} = 2\mathbf{i} + \mathbf{j}$. Find $\mathbf{u} + 2\mathbf{v}$ and $\mathbf{u} - 2\mathbf{v}$ and draw all the vectors involved with their initial points at the origin.

Solution

$$\begin{aligned}
\mathbf{u} + 2\mathbf{v} &= (-2\mathbf{i} + 3\mathbf{j}) + 2(2\mathbf{i} + \mathbf{j}) \\
&= -2\mathbf{i} + 3\mathbf{j} + 4\mathbf{i} + 2\mathbf{j} \\
&= 2\mathbf{i} + 5\mathbf{j} \\
\mathbf{u} - 2\mathbf{v} &= (-2\mathbf{i} + 3\mathbf{j}) - 2(2\mathbf{i} + \mathbf{j}) \\
&= -2\mathbf{i} + 3\mathbf{j} - 4\mathbf{i} - 2\mathbf{j} \\
&= -6\mathbf{i} + \mathbf{j} \quad ∎
\end{aligned}$$

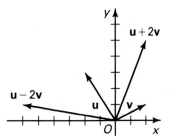

The following formula gives the norm of a vector $\mathbf{v}$ in terms of its scalar components.

> If $\mathbf{v} = a\mathbf{i} + b\mathbf{j}$, then $\|\mathbf{v}\| = \sqrt{a^2 + b^2}$.

Notice that $\|\mathbf{v}\|^2 = a^2 + b^2$.

Example 3 Let $\mathbf{u} = -9\mathbf{i} + 12\mathbf{j}$ and $\mathbf{v} = 4\mathbf{i} - 12\mathbf{j}$. Find $\|\mathbf{u}\|$, $\|\mathbf{v}\|$, and $\|\mathbf{u} + \mathbf{v}\|$.

Solution

$$\|\mathbf{u}\| = \sqrt{(-9)^2 + 12^2} = \sqrt{81 + 144} = \sqrt{225} = 15$$
$$\|\mathbf{v}\| = \sqrt{4^2 + (-12)^2} = \sqrt{16 + 144} = \sqrt{160} = 4\sqrt{10}$$
$$\|\mathbf{u} + \mathbf{v}\| = \|(-9\mathbf{i} + 12\mathbf{j}) + (4\mathbf{i} - 12\mathbf{j})\| = \|-5\mathbf{i}\| = 5 \quad ∎$$

A vector having norm 1 is called a **unit vector**. The unit vector in the direction of a nonzero vector **v** is

$$\frac{1}{\|\mathbf{v}\|}\mathbf{v}, \text{ or, more simply, } \frac{\mathbf{v}}{\|\mathbf{v}\|}.$$

Example 4   Find a unit vector parallel to $2\mathbf{i} - \mathbf{j}$.

Solution   Letting $\mathbf{v} = 2\mathbf{i} - \mathbf{j}$, a vector **u** having the desired properties is given by

$$\mathbf{u} = \frac{\mathbf{v}}{\|\mathbf{v}\|} = \frac{2\mathbf{i} - \mathbf{j}}{\sqrt{2^2 + (-1)^2}} = \frac{2}{\sqrt{5}}\mathbf{i} - \frac{1}{\sqrt{5}}\mathbf{j} = \frac{2\sqrt{5}}{5}\mathbf{i} - \frac{\sqrt{5}}{5}\mathbf{j}.$$

Of course, $-\mathbf{u}$ also has the desired properties.   ∎

EXERCISES 6-4

For each pair of vectors **u** and **v**, find (a) $\mathbf{u} + \mathbf{v}$, (b) $\mathbf{u} - \mathbf{v}$, and (c) a unit vector parallel to $\mathbf{u} + \mathbf{v}$.

A 1. $\mathbf{u} = 2\mathbf{i} + \mathbf{j}; \mathbf{v} = \mathbf{i} - 5\mathbf{j}$

2. $\mathbf{u} = -3\mathbf{i} + 2\mathbf{j}; \mathbf{v} = 8\mathbf{i} + 10\mathbf{j}$

3. $\mathbf{u} = \mathbf{i} - 4\mathbf{j}; \mathbf{v} = -3\mathbf{i} + 2\mathbf{j}$

4. $\mathbf{u} = \mathbf{i} + 3\mathbf{j}; \mathbf{v} = \mathbf{i} - \mathbf{j}$

5. $\mathbf{u} = 7\mathbf{i} - 5\mathbf{j}; \mathbf{v} = \mathbf{i} - 10\mathbf{j}$

6. $\mathbf{u} = \sqrt{2}\mathbf{i} + \sqrt{7}\mathbf{j}; \mathbf{v} = \sqrt{2}\mathbf{i} + \sqrt{7}\mathbf{j}$

7. $\mathbf{u} = \mathbf{i} + 3\sqrt{3}\mathbf{j}; \mathbf{v} = \mathbf{i} - \sqrt{3}\mathbf{j}$

8. $\mathbf{u} = -5\mathbf{i} + 2\sqrt{7}\mathbf{j}; \mathbf{v} = 3\mathbf{i} - 3\sqrt{7}\mathbf{j}$

In Exercises 9–12, find the specified linear combinations of $\mathbf{u} = 3\mathbf{i} + \mathbf{j}$ and $\mathbf{v} = 2\mathbf{i} - 3\mathbf{j}$.

9. $3\mathbf{u} + \mathbf{v}$ 10. $2\mathbf{u} - 3\mathbf{v}$ 11. $\frac{1}{2}\mathbf{u} + \frac{1}{2}\mathbf{v}$ 12. $\frac{1}{3}\mathbf{u} - \frac{2}{3}\mathbf{v}$

In Exercises 13–18, solve for the scalars s and t.

13. $(s\mathbf{i} + t\mathbf{j}) + (2\mathbf{i} - \mathbf{j}) = \mathbf{i}$

14. $(s\mathbf{i} + t\mathbf{j}) + (2\mathbf{i} - \mathbf{j}) = \mathbf{j}$

15. $(s\mathbf{i} - 3\mathbf{j}) + [(s - t)\mathbf{i} + t\mathbf{j}] = 5\mathbf{i} + 4\mathbf{j}$

16. $[(s + t)\mathbf{i} + 4\mathbf{j}] + [2\mathbf{i} + (s - t)\mathbf{j}] = 0$

17. $(3s - t)\mathbf{i} + (2t - s)\mathbf{j} = 7\mathbf{i} - 4\mathbf{j}$

18. $s\mathbf{i} + (3t - 1)\mathbf{j} = (t + 1)\mathbf{i} + 2s\mathbf{j}$

A system of objects satisfying each of the properties listed in Exercises 19–26 is called a *vector space*. Verify each of the following for vectors $\mathbf{u} = a\mathbf{i} + b\mathbf{j}$, $\mathbf{v} = c\mathbf{i} + d\mathbf{j}$, and $\mathbf{w} = e\mathbf{i} + f\mathbf{j}$ and scalars s and t.

B 19. $\mathbf{u} + \mathbf{v} = \mathbf{v} + \mathbf{u}$

20. $(\mathbf{u} + \mathbf{v}) + \mathbf{w} = \mathbf{u} + (\mathbf{v} + \mathbf{w})$

21. $\mathbf{u} + \mathbf{0} = \mathbf{u}$

22. $\mathbf{u} + (-\mathbf{u}) = \mathbf{0}$

23. $s(\mathbf{u} + \mathbf{v}) = s\mathbf{u} + s\mathbf{v}$

24. $(s + t)\mathbf{u} = s\mathbf{u} + t\mathbf{u}$

25. $s(t\mathbf{u}) = (st)\mathbf{u}$

26. $1\mathbf{u} = \mathbf{u}$

Vectors $\mathbf{v}_1$, $\mathbf{v}_2$, . . ., $\mathbf{v}_n$ are said to be *linearly independent* if $r_1\mathbf{v}_1 + r_2\mathbf{v}_2 + \cdots + r_n\mathbf{v}_n = \mathbf{0}$ implies that r_1, r_2, . . . , r_n are zero.

27. Show that if a set of vectors contains **0**, the set cannot be linearly independent.

28. Show that **i**, **j**, and **v** are *not* linearly independent for any vector **v**.

29. Show that if **u** and **v** are independent, $\mathbf{w} = r\mathbf{u} + s\mathbf{v}$, and $\mathbf{w} = r'\mathbf{u} + s'\mathbf{v}$, then $r = r'$ and $s = s'$.

30. Show that **u** and $r\mathbf{u}$ are not linearly independent.

C 31. Prove algebraically that $\|r\mathbf{v}\| = |r|\,\|\mathbf{v}\|$.

32. Show that $\mathbf{i} + \mathbf{j}$ and $\mathbf{i} - \mathbf{j}$ are linearly independent. (Hint: Solve $a(\mathbf{i} + \mathbf{j}) + b(\mathbf{i} - \mathbf{j}) = \mathbf{0}$ for a and b.)

33. Show that if **u** and **v** are nonparallel, that is $\mathbf{v} \neq k\mathbf{u}$ for any k, then **u** and **v** are linearly independent. (Hint: Suppose that $r\mathbf{u} + s\mathbf{v} = \mathbf{0}$. Consider Case 1: $r = 0$ and $s \neq 0$; Case 2: $r \neq 0$ and $s \neq 0$.)

6-5 The Dot Product

Objective: To define and apply the dot product.

In working with vectors, it is often useful to isolate the component of a given vector in a certain direction. A mathematical notion that facilitates this task, as well as many others involving vectors, is the *dot product*. Let **u** and **v** be two nonzero vectors. The **angle between u** and **v** is defined to be the angle θ, $0° \leq \theta \leq 180°$, that **u** and **v** determine when their initial points are placed together. We define the **dot product** of **u** and **v**, denoted **u · v**, by

$$\mathbf{u} \cdot \mathbf{v} = \|\mathbf{u}\|\,\|\mathbf{v}\|\cos\theta.$$

If **u** or **v** is the zero vector, we define **u · v** to be 0. The dot product is also called the *inner*, or *scalar*, product. It is important to note that the dot product of two vectors is a *scalar*, not another vector.

A very useful formula expresses the dot product of two vectors in terms of their scalar components.

If $\mathbf{u} = a\mathbf{i} + b\mathbf{j}$ and $\mathbf{v} = c\mathbf{i} + d\mathbf{j}$, then

$$\mathbf{u} \cdot \mathbf{v} = ac + bd.$$

To illustrate the formula on the preceding page, let $\mathbf{u} = 2\mathbf{i} + 5\mathbf{j}$ and $\mathbf{v} = 3\mathbf{i} - 2\mathbf{j}$, then $\mathbf{u} \cdot \mathbf{v} = (2)(3) + 5(-2) = -4$.

To prove the formula, let θ be the angle between $\mathbf{u}$ and $\mathbf{v}$. Applying the law of cosines to the triangle shown in Figure 6-15, whose sides have lengths $\|\mathbf{u}\|$, $\|\mathbf{v}\|$, and $\|\mathbf{u} - \mathbf{v}\|$, we have

$$\|\mathbf{u} - \mathbf{v}\|^2 = \|\mathbf{u}\|^2 + \|\mathbf{v}\|^2 - 2\|\mathbf{u}\| \|\mathbf{v}\| \cos \theta,$$

or,

$$\|\mathbf{u}\| \|\mathbf{v}\| \cos \theta = \frac{1}{2}[\|\mathbf{u}\|^2 + \|\mathbf{v}\|^2 - \|\mathbf{u} - \mathbf{v}\|^2].$$

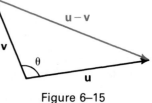

Figure 6–15

Recalling that $\|\mathbf{u}\| \|\mathbf{v}\| \cos \theta = \mathbf{u} \cdot \mathbf{v}$ and that $\|p\mathbf{i} + q\mathbf{j}\|^2 = p^2 + q^2$, we can write the equation above as

$$\mathbf{u} \cdot \mathbf{v} = \frac{1}{2}[(a^2 + b^2) + (c^2 + d^2) - ((a - c)^2 + (b - d)^2)]$$

$$= \frac{1}{2}[a^2 + b^2 + c^2 + d^2 - (a^2 - 2ac + c^2) - (b^2 - 2bd + d^2)]$$

$$= \frac{1}{2}[2ac + 2bd]$$

$$= ac + bd.$$

It is easy to see that

$$\mathbf{i} \cdot \mathbf{i} = \mathbf{j} \cdot \mathbf{j} = 1 \text{ and } \mathbf{i} \cdot \mathbf{j} = \mathbf{j} \cdot \mathbf{i} = 0$$

and that

$$\mathbf{v} \cdot \mathbf{v} = \|\mathbf{v}\|^2.$$

The dot product has three other important properties:

$\mathbf{u} \cdot \mathbf{v} = \mathbf{v} \cdot \mathbf{u}$	commutative property
$\mathbf{u} \cdot (\mathbf{v} + \mathbf{w}) = \mathbf{u} \cdot \mathbf{v} + \mathbf{u} \cdot \mathbf{w}$	distributive property
$r(\mathbf{u} \cdot \mathbf{v}) = (r\mathbf{u}) \cdot \mathbf{v} = \mathbf{u} \cdot (r\mathbf{v})$	bilinearity property

Example 1   Prove the distributive property: $\mathbf{u} \cdot (\mathbf{v} + \mathbf{w}) = \mathbf{u} \cdot \mathbf{v} + \mathbf{u} \cdot \mathbf{w}$

Solution   Let $\mathbf{u} = a\mathbf{i} + b\mathbf{j}$, $\mathbf{v} = c\mathbf{i} + d\mathbf{j}$, and $\mathbf{w} = e\mathbf{i} + f\mathbf{j}$.

$$\mathbf{u} \cdot (\mathbf{v} + \mathbf{w}) = (a\mathbf{i} + b\mathbf{j}) \cdot [(c\mathbf{i} + d\mathbf{j}) + (e\mathbf{i} + f\mathbf{j})]$$

$$= (a\mathbf{i} + b\mathbf{j}) \cdot [(c + e)\mathbf{i} + (d + f)\mathbf{j}]$$

$$= a(c + e) + b(d + f)$$

$$= (ac + bd) + (ae + bf) = \mathbf{u} \cdot \mathbf{v} + \mathbf{u} \cdot \mathbf{w}. \quad \blacksquare$$

The commutative property and the property of bilinearity will be proved as Exercises 35 and 36, respectively.

The definition of the dot product provides us with a formula for finding the angle θ between two nonzero vectors $\mathbf{u}$ and $\mathbf{v}$. We have

$$\cos \theta = \frac{\mathbf{u} \cdot \mathbf{v}}{\|\mathbf{u}\| \, \|\mathbf{v}\|}.$$

Therefore,

$$\theta = \text{Cos}^{-1}\left(\frac{(\mathbf{u} \cdot \mathbf{v})}{\|\mathbf{u}\| \, \|\mathbf{v}\|}\right).$$

Example 2   Find the angle between $\mathbf{u} = 2\mathbf{i} - 3\mathbf{j}$ and $\mathbf{v} = 5\mathbf{i} + 7\mathbf{j}$ to the nearest tenth of a degree.

Solution

$$\cos \theta = \frac{2 \cdot 5 + (-3)7}{\sqrt{13} \, \sqrt{74}} = \frac{-11}{\sqrt{13 \cdot 74}}$$

$$\theta = \text{Cos}^{-1}\left(\frac{-11}{\sqrt{13 \cdot 74}}\right) = 110.8° \quad \blacksquare$$

Two nonzero vectors are **orthogonal** if they are perpendicular, that is, if the angle between them is 90°. Two vectors are parallel if the angle between them is 0° or 180°. Using the definition of dot product:

> $\mathbf{u}$ and $\mathbf{v}$ are orthogonal if and only if $\mathbf{u} \cdot \mathbf{v} = 0$.
>
> $\mathbf{u}$ and $\mathbf{v}$ are parallel if and only if $\mathbf{u} \cdot \mathbf{v} = \pm \|\mathbf{u}\| \, \|\mathbf{v}\|$.

The zero vector is both parallel and orthogonal to every vector.

Example 3   Find a unit vector orthogonal to $\mathbf{u} = 3\mathbf{i} + 5\mathbf{j}$.

Solution   First find *some* vector orthogonal to $\mathbf{u}$. The vector $\mathbf{v} = x\mathbf{i} + y\mathbf{j}$ will be orthogonal to $\mathbf{u}$ if $\mathbf{u} \cdot \mathbf{v} = 0$, that is, if

$$(3\mathbf{i} + 5\mathbf{j}) \cdot (x\mathbf{i} + y\mathbf{j}) = 0$$
$$3x + 5y = 0.$$

One solution of this equation is $x = 5$ and $y = -3$. Thus, $\mathbf{v} = 5\mathbf{i} - 3\mathbf{j}$ is orthogonal to $\mathbf{u}$ and

$$\frac{\mathbf{v}}{\|\mathbf{v}\|} = \frac{5\mathbf{i} - 3\mathbf{j}}{\sqrt{34}} = \frac{5\sqrt{34}}{34}\mathbf{i} - \frac{3\sqrt{34}}{34}\mathbf{j}$$

is a unit vector orthogonal to $\mathbf{u}$. Of course, $-\mathbf{v}$ is also a unit vector orthogonal to $\mathbf{u}$. $\blacksquare$

As stated at the beginning of this section, the dot product is useful in finding the component of a vector $\mathbf{v}$ in a given direction. If $\mathbf{v}$ is any vector, the **vector projection** of $\mathbf{v}$ onto a vector $\mathbf{a}$, denoted $\mathbf{v_a}$, is the vector whose terminal point is the foot of the perpendicular from $\mathbf{v}$ to the line containing $\mathbf{a}$. The norm of $\mathbf{v_a}$ is called the **scalar projection** of $\mathbf{v}$ onto $\mathbf{a}$.

Figure 6-16 illustrates projections where θ, the angle between **v** and **a**, is acute and where θ is obtuse.

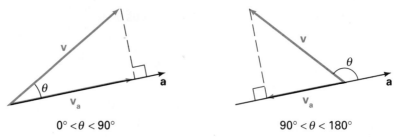

$0° < \theta < 90°$ $\qquad\qquad$ $90° < \theta < 180°$

Figure 6-16

To find a formula for $\mathbf{v_a}$ involving the dot product, refer to Figure 6-16. Note that if θ is acute, then $\dfrac{\mathbf{a}}{\|\mathbf{a}\|}$ is a unit vector in the direction of **a** and therefore in the direction of $\mathbf{v_a}$. Since $\|\mathbf{v_a}\| = \|\mathbf{v}\| \cos \theta$, we have that:

$$\mathbf{v_a} = (\|\mathbf{v}\| \cos \theta)\left(\frac{\mathbf{a}}{\|\mathbf{a}\|}\right) = \left(\frac{\|\mathbf{a}\|\,\|\mathbf{v}\| \cos \theta}{\|\mathbf{a}\|\,\|\mathbf{a}\|}\right)\mathbf{a} = \left(\frac{\mathbf{a} \cdot \mathbf{v}}{\mathbf{a} \cdot \mathbf{a}}\right)\mathbf{a}$$

If θ is obtuse, we use $\cos (180° - \theta) = -\cos \theta$ and the unit vector $-\dfrac{\mathbf{a}}{\|\mathbf{a}\|}$ in Figure 6-16 to obtain the same conclusion.

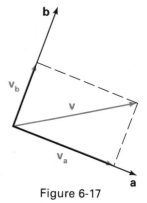

Let **a** and **b** be mutually orthogonal vectors and let **v** be an arbitrary vector (Figure 6-17). By the definition of vector addition we have

$$\mathbf{v} = \mathbf{v_a} + \mathbf{v_b}.$$

In terms of the dot product, we have

$$\mathbf{v} = \frac{\mathbf{a} \cdot \mathbf{v}}{\mathbf{a} \cdot \mathbf{a}}\mathbf{a} + \frac{\mathbf{b} \cdot \mathbf{v}}{\mathbf{b} \cdot \mathbf{b}}\mathbf{b}.$$

We call the vectors $\mathbf{v_a}$ and $\mathbf{v_b}$ the **components** of **v** parallel to **a** and **b**, respectively, and say that **v** has been **resolved** into these components.

Figure 6-17

Example 4   Resolve $\mathbf{v} = -2\mathbf{i} + 5\mathbf{j}$ into components parallel to $\mathbf{a} = 6\mathbf{i} - 2\mathbf{j}$ and to $\mathbf{b} = \mathbf{i} + 3\mathbf{j}$.

Solution   Since $\mathbf{a} \cdot \mathbf{b} = 6 \cdot 1 + (-2)3 = 0$, the vectors **a** and **b** are orthogonal, and we may apply the formula derived above.

$$\mathbf{a} \cdot \mathbf{v} = -12 - 10 = -22 \qquad \mathbf{b} \cdot \mathbf{v} = -2 + 15 = 13$$
$$\mathbf{a} \cdot \mathbf{a} = 36 + 4 = 40 \qquad \mathbf{b} \cdot \mathbf{b} = 1 + 9 = 10$$

Therefore, $\mathbf{v} = -\dfrac{11}{20}\mathbf{a} + \dfrac{13}{10}\mathbf{b}.$ ■

EXERCISES 6-5

In Exercises 1–8, find the dot product of the given vectors.

A 1. $u = i + j$; $v = i - j$ 2. $u = 2i - j$; $v = 2i + j$

3. $u = 4i - 5j$; $v = 5i + 4j$ 4. $u = -3i + 6j$; $v = -9i - 3j$

5. $u = i$; $v = 3i - 7j$ 6. $u = -4i + j$; $v = 4i - j$

7. $u = pi - qj$; $v = pi + qj$ 8. $u = pi + pj$; $v = qi - qj$

9. Which pairs of the given vectors are (a) parallel? (b) orthogonal?
   $u = 4i + 6j$ $v = -4i + 6j$ $w = 3i + 2j$ $t = 2i - 3j$

10. Which pairs of the given vectors are (a) parallel? (b) orthogonal?
   $u = 2i - 4j$ $v = i + 2j$ $w = 8i + 4j$ $t = 2i + j$

Use the distributive property to find $u \cdot (v + w)$ for the following vectors.

11. $u = 2i + 3j$ 12. $u = 5i + 2j$ 13. $u = 3i - 4j$ 14. $u = -3i + 2j$
      $v = 3i + 8j$ $v = 2i - 6j$ $v = 8i - 4j$ $v = -7i - 3j$
      $w = i + 2j$ $w = 4i + j$ $w = -3i - 5j$ $w = 2i + 5j$

In Exercises 15–20, find the angle between the given vectors.

15. $i + \sqrt{3}j$; $-i + \sqrt{3}j$ 16. $\sqrt{3}i - j$; $-i + \sqrt{3}j$ 17. $i + 2j$; $4i + 3j$

18. $4i + 3j$; $4i - 3j$ 19. $2i + 5j$; $3i - 4j$ 20. $5i + j$; $i - 2j$

In Exercises 21–26, find unit vectors (a) parallel, and (b) perpendicular to v.

21. $v = -i + j$ 22. $v = 2i + j$ 23. $v = 4i - 3j$

24. $v = 5i + 12j$ 25. $v = xi + yj$ 26. $v = xi - yj$

In Exercises 27–30, find the scalar projection of v onto a.

27. $v = j$ 28. $v = i$ 29. $v = i + j$ 30. $v = -i + j$
     $a = i + j$ $a = -i + j$ $a = 4i - 3j$ $a = 3i + 4j$

In Exercises 31–34, resolve v into components parallel to a and to b.

31. $v = j$ 32. $v = i + 2j$ 33. $v = 2i - 3j$ 34. $v = i + 5j$
     $a = i + j$ $a = i + j$ $a = 4i + j$ $a = 3i - 2j$
     $b = i - j$ $b = -i + j$ $b = i - 4j$ $b = 2i + 3j$

In Exercises 35–38, let $u = ai + bj$ and $v = ci + dj$. Let r and s be real numbers. Give an algebraic proof for each of the following.

B 35. $u \cdot v = v \cdot u$ 36. $r(u \cdot v) = (ru) \cdot v = u \cdot (rv)$

37. $v \cdot v = \|v\|^2$ 38. $(ru) \cdot (sv) = (rs)(u \cdot v)$

In Exercises 39–43, use properties of the dot product to prove the given statement.

39. $(\mathbf{u} + \mathbf{v}) \cdot (\mathbf{u} + \mathbf{v}) = \mathbf{u} \cdot \mathbf{u} + 2\mathbf{u} \cdot \mathbf{v} + \mathbf{v} \cdot \mathbf{v}$

40. $(\mathbf{u} + \mathbf{v}) \cdot (\mathbf{u} - \mathbf{v}) = \mathbf{u} \cdot \mathbf{u} - \mathbf{v} \cdot \mathbf{v}$

41. $\|\mathbf{u} + \mathbf{v}\|^2 = (\mathbf{u} + \mathbf{v}) \cdot (\mathbf{u} + \mathbf{v})$

42. $\|\mathbf{u} + \mathbf{v}\|^2 = \|\mathbf{u}\|^2 + 2\mathbf{u} \cdot \mathbf{v} + \|\mathbf{v}\|^2$

C 43. $(\mathbf{u} \cdot \mathbf{v})^2 \leq \|\mathbf{u}\|^2 \|\mathbf{v}\|^2$ (Hint: Use the definition of dot product.)

44. Show that $\mathbf{u} + \mathbf{v}$ and $\mathbf{u} - \mathbf{v}$ are orthogonal if and only if $\|\mathbf{u}\| = \|\mathbf{v}\|$.

45. Show that $\|\mathbf{u} + \mathbf{v}\|^2 + \|\mathbf{u} - \mathbf{v}\|^2 = 2\|\mathbf{u}\|^2 + 2\|\mathbf{v}\|^2$.
    (Hint: $\|\mathbf{u} + \mathbf{v}\|^2 = (\mathbf{u} + \mathbf{v}) \cdot (\mathbf{u} + \mathbf{v})$.)

46. Use the results of Exercises 42 and 43 to prove that $\|\mathbf{u} + \mathbf{v}\| \leq \|\mathbf{u}\| + \|\mathbf{v}\|$.

6-6   Work and Energy

Objective: To apply vectors to problems involving work and energy.

When a force acts to move an object, it does *work* and *energy* is expended. In physics the terms "work" and "energy" are interchangeable. Note that if an object is not moved, no work is done, regardless of how much force is exerted.

If the force **F** applied to an object is constant (that is, it does not change with time or position) and the direction of **F** is the same as the direction of the object's motion (Figure 6-18), then the work, W, done by **F** in moving the object from point A to point B is defined to be

$$W = \|\mathbf{F}\| \, \|\mathbf{d}\|,$$

where $\mathbf{d} = \vec{AB}$ is the displacement vector of the object.

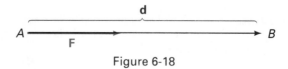

Figure 6-18

The basic unit of work (or energy) is the *joule* (J), defined as the work done by a force of one newton in moving an object a distance of one meter. Another unit is the *kilowatt-hour* (kW·h). One kilowatt-hour equals 3.6×10^6 J.

Example 1   How much energy (in kW·h) is needed to lift a 2000 kg elevator 60 m?

Solution   The motor lifting the elevator must overcome the force of gravity and therefore must exert an upward force, **F**, of magnitude

$$\|\mathbf{F}\| = 2000 \times 9.80 = 1.96 \times 10^4.$$

The displacement **d** is also upward and has magnitude

$$\|\mathbf{d}\| = 60.$$

Thus, the energy expended (in joules) is

$$W = \|\mathbf{F}\|\,\|\mathbf{d}\|$$
$$= (1.96 \times 10^4) \times 60$$
$$= 1.176 \times 10^6.$$

Expressed in (kW·h), this is

$$\frac{1.176 \times 10^6}{3.6 \times 10^6} = 0.3267 \text{ (kW·h)}. \quad \blacksquare$$

Let us now consider a constant force **F** that moves an object from A to B but does not have the same direction as $\mathbf{d} = \vec{AB}$. The situation is shown in Figure 6-19. In this case only the projection of **F** onto **d**, $\mathbf{F_d}$, is effective in doing work. Thus, the work done by **F** is

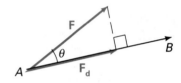

Figure 6-19

$$W = \|\mathbf{F}\| \cos\theta\,\|\mathbf{d}\|.$$

The formula can be written as a dot product.

$$W = \mathbf{F} \cdot \mathbf{d} = \|\mathbf{F}\|\,\|\mathbf{d}\| \cos\theta.$$

Notice that W may be 0 if $\theta = 90°$ or may be negative if $90° < \theta \le 180°$.

Example 2   Find the work done by the force $\mathbf{F} = \mathbf{i} + 3\mathbf{j}$ in moving an object from $A(-1, 0)$ to $B(2, 4)$. Assume that the force is given in newtons and the distance in meters.

Solution   The displacement of the object is given by

$$\mathbf{d} = AB = (2 - (-1))\mathbf{i} + (4 - 0)\mathbf{j} = 3\mathbf{i} + 4\mathbf{j}.$$

Thus, the work done by **F** is:

$$W = \mathbf{F} \cdot \mathbf{d} = (\mathbf{i} + 3\mathbf{j}) \cdot (3\mathbf{i} + 4\mathbf{j}) = 1 \cdot 3 + 3 \cdot 4 = 15$$

The work done is 15 J. $\blacksquare$

Example 3   How much energy does the engine of a 1200 kg car expend in moving the car up a 5000 m grade that makes an angle of 8.5° with the horizontal? (Ignore friction and consider only the force of gravity.)

Solution   The work W done by the engine equals $\mathbf{G'} \cdot \mathbf{d}$ where $\mathbf{G'}$ is the force needed to overcome $\mathbf{G}$, the force of gravity.

Notice that

$$\|\mathbf{G'}\| = \|\mathbf{G}\|.$$

Since

$$\|\mathbf{G}\| = 1200 \cdot 9.80$$
$$= 11{,}760,$$
$$\|\mathbf{d}\| = 5000$$

and

$$\theta = 81.5°,$$
$$W = \mathbf{G'} \cdot \mathbf{d}$$
$$= 11{,}760 \cdot 5000 \cdot \cos 81.5°$$
$$= 8{,}700{,}000$$
$$= 8.7 \times 10^6$$

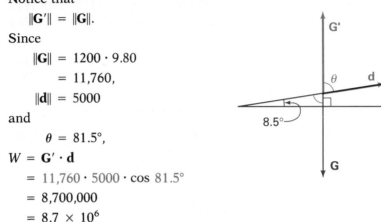

The engine must exert 8.7×10^6 J, about 2.42 kW·h.   ■

EXERCISES 6-6

In Exercises 1–5, forces are measured in newtons and distances in meters. Give answers in joules.

A 1. Find the work done by the force $\mathbf{F} = 2\mathbf{i} + 3\mathbf{j}$ in moving a particle from $A(1, 0)$ to $B(0, 4)$.

2. Find the work done by the force $\mathbf{F} = 5\mathbf{j}$ in moving an object from $A(-1, 4)$ to $B(5, 2)$.

3. How much work is done by force $\mathbf{F} = 10\mathbf{i} - 5\mathbf{j}$ in moving an object (a) from $A(-2, 0)$ to $B(3, 7)$ to $C(5, 0)$? (b) from $A(-2, 0)$ to $C(5, 0)$?

4. What is the combined work done by the forces $\mathbf{F} = 6\mathbf{i} + 5\mathbf{j}$ and $\mathbf{G} = 2\mathbf{i} - 7\mathbf{j}$ in moving an object from $A(-4, 1)$ to $B(1, -4)$? What is the work done by $\mathbf{F} + \mathbf{G}$ in moving an object from A to B?

5. Answer the questions in Exercise 4 for the forces $\mathbf{F} = 3\mathbf{i} - 5\mathbf{j}$ and $\mathbf{G} = 2\mathbf{i} + \mathbf{j}$ that move an object from $A(4, 7)$ to $B(6, -1)$.

6. The mass of a loaded helicopter is 1250 kg. How much energy does its engine expend in ascending vertically for 50 m?

7. How much energy does a crane expend by lifting 25 buckets of concrete from street level to a point 20 m above street level? The mass of a bucket of concrete is 1800 kg.

B 8. How much energy does a 90 kg man expend in climbing a 10 m ladder that makes an angle of 60° with the horizontal?

9. A rock is dragged for 60 m along a horizontal sidewalk by a force of 220 N exerted on a rope that is tied to the rock. If the rope makes an angle of 32° with the sidewalk, how much work is done?

10. A 200 m long conveyer belt carries coal up an 11° slope. How many kilowatt-hours of energy are used in transporting 100 metric tons of coal? (1 metric ton = 1000 kilograms.)

11. A rope tow transports skiers up a 25° slope for a horizontal distance of 600 m. How much energy does the tow use in one day if it transports 235 skiers with an average mass of 62 kg?

12. A student weighing 132 pounds climbs a flight of 11 stairs each being 1 ft deep and 1 ft high. How many Calories does that student burn up?

(Use the following conversion factors: 1 lb = 0.454 kg; 1 ft = 0.305 m, and 1 Calorie = 4184 joules.)

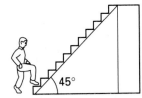

13. A sled is dragged up a 30° hill by a rope that makes an angle of 45° with the plane of the hill. If the force exerted on the rope is 40 N, how much work is done in raising the sled's vertical position 10 m?

14. Answer the question in Exercise 13 if the grade of the hill is 45° and the angle of the sled-rope with the plane of the hill is 30°.

C 15. Work Example 3 assuming that there is a frictional force of magnitude 500 N. Remember that friction always acts in direct opposition to motion.

In Exercises 16 and 17 prove the stated facts.

16. The work done by a constant force **F** in moving an object along $\overrightarrow{AB}$ and then along $\overrightarrow{BC}$ is the same as the work **F** does in moving the object along $\overrightarrow{AC}$.

17. Let **F** and **G** be constant forces. The work done by **F** + **G** in moving an object along $\overrightarrow{AB}$ equals the combined work done by **F** and by **G** in moving the object along $\overrightarrow{AB}$.

The *coefficient of friction, k,* of two surfaces is

$$k = \frac{\|\mathbf{M}\|}{\|\mathbf{H}\|},$$

where **H** is a force holding the surfaces together and **M** is the force just sufficient to make one surface move relative to the other. (In Example 1, page 203, **M** = **F**.) The *angle of repose* ϕ of the two surfaces is

$$\phi = \mathrm{Tan}^{-1}\, k.$$

Give a physical interpretation of this angle. Describe a way to find k without measuring forces.

Self Quiz

6-4	6-5	6-6

For the vectors **u** = 3**i** + 4**j** and **v** = 2**i** − **j**, find:

1. 2**u** − **v**

2. **u** + 3**v**

3. a unit vector parallel to **u**

4. Solve for r and s if $(r + s)\mathbf{i} + (2r - 3s)\mathbf{j} = -2\mathbf{i} + 11\mathbf{j}$.

5. Find the dot product of 7**i** + 3**j** and −2**i** + 4**j**.

6. Find the angle between the vectors **i** + **j** and −**i** + 4**j** to the nearest degree.

7. Find the scalar projection of **v** = 2**i** − **j** onto **a** = 3**i** + 4**j**.

8. How much work (in kW·h) is done by a crane in lifting 20 stone slabs each weighing 1500 kg a vertical distance of 25 m?

9. Neglecting friction, how much work (in joules) is done in moving a 300 kg piano up a 10 m ramp that makes an angle of 9° with the horizontal?

ADDITIONAL PROBLEMS

Express each vector in terms of **u** and **v**.

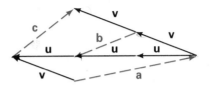

1. **a**

2. **b**

3. **c**

Let $u = -i + 3j$ and $v = 2i + 4j$. Find r and s such that $ru + sv = w$.

4. $w = -7i + 7j$ 5. $w = 9i + j$

6. Two planes take off from an airfield at P with headings of 40° and 150° from P, respectively, and speeds of 200 km/h and 300 km/h, respectively. How far apart are the planes after 30 min? What is the bearing of the second plane from the first?

7. Draw two arrows with a common initial point and representing nonparallel vectors **u** and **v**. Draw $v - u$. Referring to your diagram, state the law of cosines in terms of $\|u\|$, $\|v\|$, $\|u - v\|$, and $u \cdot v$.

8. Find the angle between the vectors $u = 3i + j$ and $v = -5i + 2j$.

9. Find a unit vector in the direction of $17i - 8j$.

10. Find a unit vector orthogonal to $u = 2i + j$.

11. In the diagram, forces **F** and **H** act along lines that make angles of 25° and 40° with the horizontal, respectively. If $\|F\| = 50$, what should the norm of **H** be in order that the resultant force be directed vertically upward?

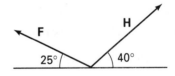

12. Let $u = 15i - 8j$. Find a vector **v** satisfying the given conditions.
    (a) **u** and **v** are orthogonal; $\|v\| = \|u\|$.

    (b) **u** and **v** have opposite direction; $\|u\| = \|v\|$.

13. Find the vector projection of $2i + j$ onto $5i - 12j$.

14. Draw two nonparallel vectors **u** and **v** with a common initial point such that $\|u - v\| = \|u + v\|$. How do **u** and **v** appear to be related? Justify your answer by a theorem from geometry.

15. Two ropes that make constant angles of 155° and 75° with the horizontal are being used to lift heavy beams. If a beam produces a tension of 3000 N in the first rope and 5000 N in the second rope, how much work in kilowatt-hours is done in lifting 10 beams 40 m?

16. Let a and b be real numbers. Show that **i** and $ai + bj$ are linearly independent if $b \neq 0$.

17. Resolve $3\mathbf{i} + 4\mathbf{j}$ into components parallel to $\mathbf{a} = \mathbf{i} + \mathbf{j}$ and to $\mathbf{b} = -\mathbf{i} + \mathbf{j}$.

18. Show, using the dot product, that $\mathbf{u}$ and $\mathbf{v}$ are orthogonal if and only if $\|\mathbf{u}\|^2 + \|\mathbf{v}\|^2 = \|\mathbf{u} + \mathbf{v}\|^2$. Interpret this equation geometrically.

19. Explain why a bullet fired horizontally from a gun will hit the ground at the same time as another bullet that is dropped at the same moment from the same height and falls freely. Assume that the ground is perfectly flat and there is no air resistance.

20. Show that $\|\mathbf{u} - \mathbf{v}\| = \|\mathbf{u} + \mathbf{v}\|$ if and only if $\mathbf{u} \cdot \mathbf{v} = 0$.

21. Find the work done by $\mathbf{F} = 4\mathbf{i} - 2\mathbf{j}$ in moving an object from $A(0, 4)$ to $B(6, 2)$.

CHAPTER SUMMARY

1. A vector is a quantity that has both magnitude and direction. A quantity that has only magnitude (a real number) is called a scalar.

2. If arrows represent vectors, vector sums can be found by the parallelogram rule. If $\mathbf{w} = \mathbf{u} + \mathbf{v}$, $\mathbf{u}$ and $\mathbf{v}$ are called components of $\mathbf{w}$, and $\mathbf{w}$ is the resultant of $\mathbf{u}$ and $\mathbf{v}$. If r is a scalar, $r\mathbf{v}$ has magnitude $|r|$ times the magnitude of $\mathbf{v}$, and its direction is opposite to that of $\mathbf{v}$ if $r < 0$. The magnitude of $\mathbf{v}$ is its norm, $\|\mathbf{v}\|$.

3. The displacement of an object is the vector from the object's starting point to its termination point. The bearing of a vector is the angle of the vector, measured clockwise from due north. Other terms used in navigation are heading, true course, ground speed, and air speed.

4. Velocities can also be represented in a vector diagram. The speed of an object is the norm of its velocity vector.

5. A vector representing a force is often resolved into vertical and horizontal components. When an object is at rest or is moving at a constant velocity, the vector sum of the forces acting on it is 0.

6. Any vector $\mathbf{v}$ in a fixed plane can be written as $\mathbf{v} = a\mathbf{i} + b\mathbf{j}$, where $\mathbf{i}$ and $\mathbf{j}$ are the unit vectors in the directions of the positive axes of a coordinate system for the plane. Vector addition and scalar multiplication can then be carried out in terms of the scalar components a and b. Also, $\|\mathbf{v}\| = \sqrt{a^2 + b^2}$.

7. The dot product of $\mathbf{u}$ and $\mathbf{v}$ is defined by $\mathbf{u} \cdot \mathbf{v} = \|\mathbf{u}\| \|\mathbf{v}\| \cos \theta$; they are perpendicular if and only if $\mathbf{u} \cdot \mathbf{v} = 0$; they are parallel if and only if $\mathbf{u} \cdot \mathbf{v} = \pm \|\mathbf{u}\| \|\mathbf{v}\|$.

8. For any orthogonal vectors **a** and **b**, a vector **v** can be resolved into components parallel, respectively, to **a** and **b**. These components are called the vector projections of **v** onto **a** and **b**; their norms are called the scalar projections of **v** onto **a** and **b**.

9. For a constant force moving in a straight line, work is defined as the dot product of the force vector and the displacement vector.

CHAPTER TEST

6-1

1. Copy vectors **u** and **v** shown at the right. Sketch the vector $2\mathbf{u} - \mathbf{v}$.

2. Draw two nonparallel vectors **u** and **v** with the same initial point. In the same diagram draw the vector $\frac{1}{2}(\mathbf{u} + \mathbf{v})$ and the vector $\frac{1}{2}(\mathbf{u} - \mathbf{v})$. How are $\frac{1}{2}(\mathbf{u} + \mathbf{v})$ and $\frac{1}{2}(\mathbf{u} - \mathbf{v})$ related?

6-2

3. A plane with a heading of 70° has an air speed of 400 km/h while a 55 km/h wind is blowing from the west. Find the plane's ground speed and true course.

6-3

4. A 50-kg box sits on a ramp inclined at 14° with the horizontal. Find the frictional force that keeps the box from sliding down the ramp and the force perpendicular to the ramp that the box exerts.

5. A 450-kg object is suspended from two cables each of which makes an angle of 60° with the horizontal. Find the tension in each cable.

6-4

6. Solve for the scalars s and t if $s\mathbf{i} + (t + 1)\mathbf{j} = (t - 7)\mathbf{i} - 3s\mathbf{j}$.

7. Show that the vectors $\mathbf{i}$, $\mathbf{i} + \mathbf{j}$, and $\mathbf{i} - \mathbf{j}$ are not linearly independent.

6-5

8. Resolve $\mathbf{v} = \mathbf{i} - 2\mathbf{j}$ into components parallel, respectively, to $\mathbf{i} + \mathbf{j}$ and $\mathbf{i} - \mathbf{j}$.

9. Find the angle between the vectors $3\mathbf{i} + \mathbf{j}$ and $-\mathbf{i} + 2\mathbf{j}$.

10. Show that $\|\mathbf{a} - \mathbf{b}\|^2 = \|\mathbf{a}\|^2 - 2\mathbf{a} \cdot \mathbf{b} + \|\mathbf{b}\|^2$.

6-6

11. How much work is done by a force $\mathbf{F} = 5\mathbf{i} - 3\mathbf{j}$ in moving an object from $(-1, 2)$ to $(5, 10)$?

12. How much energy in kW·h does it take to pull a 6-metric-ton cable car up a 20° incline 750 m long? (1 metric ton = 1000 kg.)

Vector Resultants

A computer can be used to solve the following problem and others like it.

> Forces $\mathbf{F}_1$ with magnitude 120 N and direction 68° and $\mathbf{F}_2$ with magnitude 90 N and direction 135° act on a particle O. What are the magnitude and direction of the resultant $\mathbf{F}_1 + \mathbf{F}_2$? (Angles are measured counterclockwise from the horizontal.)

To answer the question, we can write $\mathbf{F}_1$ as (x_1, y_1) and $\mathbf{F}_2$ as (x_2, y_2) and add their coordinates to obtain $(x_1 + x_2, y_1 + y_2)$ the coordinates of the resultant. (See the two figures that follow.) From these coordinates, we can get $\|\mathbf{F}_1 + \mathbf{F}_2\|$, the magnitude of $\mathbf{F}_1 + \mathbf{F}_2$, and θ, the direction of $\mathbf{F}_1 + \mathbf{F}_2$.

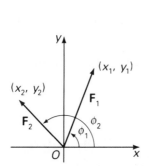
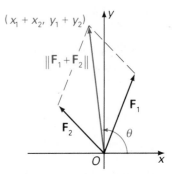

The program that follows will find the magnitude and direction of the resultant of a set of vectors whose magnitude and directions you supply.

The diagram below helps to explain lines **120** through **170**. Notice from the diagram that it is the quadrant in which (x, y) lies and therefore, the algebraic signs of x and y that tell us how to get θ from $\text{Tan}^{-1} \left| \dfrac{y}{x} \right|$.

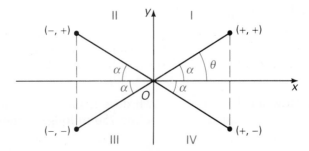

```
10 SX(0) = 0 : SY(0) = 0
20 INPUT N                          Enter the number of vectors.
30 FOR I = 1 TO N
40 INPUT N(I) : INPUT D(I)          Enter vector magnitude and
                                    direction.
50 M(I) = D(I) * 3.14159 / 180      Convert degrees to radians.
60 X(I) = N(I) * COS (M(I))         Find each x-coordinate.
70 Y(I) = N(I) * SIN (M(I))         Find each y-coordinate.
80 SX(I) = SX(I − 1) + X(I)         Add x-coordinates.
90 SY(I) = SY(I − 1) + Y(I)         Add y-coordinates.
100 NEXT I
110 R = SQR (SX(N) ↑ 2 + SY(N) ↑ 2)    Find the magnitude of the sum.
120 IF R < .01 THEN TH = 0             Use Tan⁻¹ to find θ.
125 IF R < .01 THEN GOTO 200
130 A1 = ABS (SY(N) / SX(N))
140 A = ATN (A1)
150 IF SX(N) > 0 AND SY(N) > 0 THEN TH = A          Quadrant I
160 IF SX(N) < 0 AND SY(N) > 0 THEN TH = 3.14159 − A   Quadrant II
170 IF SX(N) < 0 AND SY(N) < 0 THEN TH = 3.14159 + A   Quadrant III
180 IF SX(N) > 0 AND SY(N) < 0 THEN TH = 6.28319 − A   Quadrant IV
190 TH = TH * 180 / 3.14159
200 PRINT "MAGNITUDE = " ; R ;"        " ; "DIRECTION =" ; TH
```

The computer program is particularly helpful when three or more vectors are given and we wish to find the net effect of their action together.

EXERCISES

Use the computer program to find the resultant of each set of vectors.

1. (a) Vector F_1 has magnitude 72 N and direction 45°.
      Vector F_2 has magnitude 72 N and direction 225°.
   (b) Compare the printout for Exercise 1 (a) with what intuition tells you $F_1 + F_2$ should be.

2. F_1: magnitude 130 N and direction 43°
   F_2: magnitude 100 N and direction 140°
   F_3: magnitude 90 N and direction 230°
   F_4: magnitude 160 N and direction 180°

3. Use the result of Exercise 2 to find the magnitude and direction of an additional force F_5 that will balance the resultant $F_1 + F_2 + F_3 + F_4$ and therefore keep O stationary.

Testing Conjectures About Forces

Intuition tells us that if two forces $\mathbf{F}_1$ and $\mathbf{F}_2$ are applied at a point, then the magnitude of the resultant will be maximum when the forces act in the same direction and minimum when they act in opposite directions. Intuition says that if $\mathbf{F}_1$ and $\mathbf{F}_2$ form an angle α ($0° < \alpha < 180°$), then the resultant has magnitude somewhere between the minimum and the maximum.

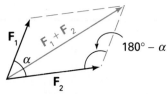

With the law of cosines, we can express $\|\mathbf{F}_1 + \mathbf{F}_2\|$ as a function of α.

$$\|\mathbf{F}_1 + \mathbf{F}_2\| = \sqrt{\|\mathbf{F}_1\|^2 + \|\mathbf{F}_2\|^2 - 2\|\mathbf{F}_1\|\,\|\mathbf{F}_2\| \cos(180° - \alpha)}$$

For discussion, suppose that $\|\mathbf{F}_1\| = 3$ and that $\|\mathbf{F}_2\| = 7$. The following table illustrates that as α decreases, $\|\mathbf{F}_1 + \mathbf{F}_2\|$ increases, and that as α increases, $\|\mathbf{F}_1 + \mathbf{F}_2\|$ decreases. The table was made by computing $\|\mathbf{F}_1 + \mathbf{F}_2\|$ for different values of α.

α	$\|\mathbf{F}_1 + \mathbf{F}_2\|$	α	$\|\mathbf{F}_1 + \mathbf{F}_2\|$
20°	9.8711	160°	4.3181
15°	9.9281	165°	4.1752
10°	9.9680	170°	4.0790
5°	9.9920	175°	4.0199
0°	10.0000	180°	4.0000

Using the table, we can estimate $\|\mathbf{F}_1 + \mathbf{F}_2\|$ for certain values of α not in the table. For example, if $\alpha = 12.5°$, the average of 10° and 15°, we can estimate $\|\mathbf{F}_1 + \mathbf{F}_2\|$ to be about 9.94, the average of 9.9680 and 9.9281. On the other hand, if $\|\mathbf{F}_1 + \mathbf{F}_2\|$ is known to be 4.25, then α is between 160° and 165°, and by linear interpolation (page 39), $\alpha = 162°$ to the nearest degree.

Another problem is to see how $\|\mathbf{F}_1 + \mathbf{F}_2\|$ changes as $\mathbf{F}_1$ and $\mathbf{F}_2$ are applied at a constant angle such as 60°, $\|\mathbf{F}_1\|$ is fixed at 3, and $\|\mathbf{F}_2\|$ varies. With the law of cosines, $\|\mathbf{F}_1 + \mathbf{F}_2\|$ as a function of $\|\mathbf{F}_2\|$ is given by

$$\|\mathbf{F}_1 + \mathbf{F}_2\| = \sqrt{9 + \|\mathbf{F}_2\|^2 - 6\|\mathbf{F}_2\| \cos 120°}$$
$$= \sqrt{\|\mathbf{F}_2\|^2 + 3\|\mathbf{F}_2\| + 9}.$$

A calculator or computer can be used to make a table of values of $\|\mathbf{F}_1 + \mathbf{F}_2\|$ for different values of $\|\mathbf{F}_2\|$.

$\|\mathbf{F}_2\|$	$\|\mathbf{F}_1 + \mathbf{F}_2\|$
4.0	6.0828
4.1	6.1733
4.2	6.2642
4.3	6.3553
4.4	6.4467
4.5	6.5383

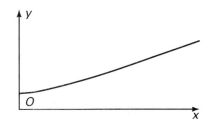

It is not surprising that when $\|\mathbf{F}_1\|$ and α are held constant, $\|\mathbf{F}_1 + \mathbf{F}_2\|$ increases as $\|\mathbf{F}_2\|$ increases. This fact is reflected in the graph at the right above.

Many problems involve several forces acting simultaneously and in different directions. Computer simulations, models, and numerical data in tabular form enable engineers to experiment with various magnitudes and angles of application to maximize or minimize forces. Simulation makes it possible to test conjectures without actually building the structure that is to exert or sustain those forces.

Guidance and maneuvering engines must produce the right thrust in the right direction for the lunar lander to position itself for a safe landing on the moon.

7 *Complex Numbers*

Mathematicians and engineers use methods involving complex numbers to solve problems involving the transmission of electrical energy. Functions involving complex numbers are also found in the study of elasticity, aerodynamics, and hydrodynamics.

POLAR COORDINATES

The Polar Coordinate System

Objective: To understand and use polar coordinates.

We can locate a point in the plane as the intersection of two lines parallel to two perpendicular axes (rectangular coordinates as in Figure 7-1), or as the intersection of a circle centered at the origin and a ray whose endpoint is the origin (*polar coordinates* as in Figure 7-2).

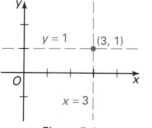

Figure 7-1

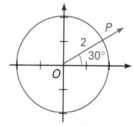

Figure 7-2

The point *P* in Figure 7-2 can be identified with the *polar coordinates* (2, 30°), with reference to the *pole*, or *origin O*, and the *polar axis* $\overrightarrow{OA}$. In general, a point with polar coordinates (*r*, *θ*), like *Q* in Figure 7-3, is at a distance *r* from the pole and on a ray whose rotation from the polar axis is *θ*.

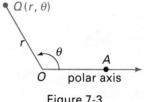

Figure 7-3

Each point has infinitely many pairs of polar coordinates. For example, (4, 210°), (4, 570°), and (4, −150°) are all polar coordinates of the same point *P* since the angles are coterminal.

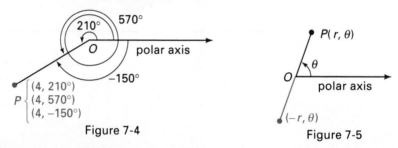

Figure 7-4

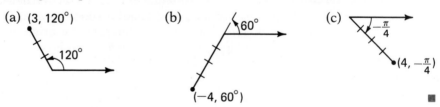

Figure 7-5

Moreover, it will be convenient to let a negative value of *r* denote the *negative* of the distance *OP*. In this case, *P* is on the extension of the terminal side of *θ* through *O*, and is |*r*| units from *O* (Figure 7-5). Thus, (−4, 30°) and (−4, −330°) are also polar coordinates of *P*.

Example 1 Graph: (a) (3, 120°)   (b) (−4, 60°)   (c) $\left(4, -\dfrac{\pi}{4}\right)$

Solution Find the ray from the pole making angle *θ* with the polar axis. Then measure *r* units either along this ray or along the extension of that ray through the pole. In (c) the absence of ° indicates *radians*.

(a) (3, 120°)   (b)   (c)

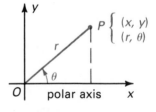

When polar coordinates are used with rectangular coordinates, the polar axis is taken to coincide with the nonnegative *x*-axis as shown in Figure 7-6. The equations in the following table can be derived from Figure 7-6. They enable us to change from one coordinate system to the other.

Figure 7-6

Coordinate Changes	
From polar to rectangular	**From rectangular to polar**
$x = r \cos \theta$ $y = r \sin \theta$	$r = \pm\sqrt{x^2 + y^2}$ $\cos \theta = \dfrac{x}{r}, \ \sin \theta = \dfrac{y}{r}$

Example 2

(a) Convert $(3, -3)$ to polar coordinates and (b) convert $\left(-2, \dfrac{2\pi}{3}\right)$ to rectangular coordinates.

Solution

(a) Since $r = \pm\sqrt{3^2 + (-3)^2}$, we may take $r = 3\sqrt{2}$. Then, since

$$\cos \theta = \frac{3}{3\sqrt{2}} = \frac{\sqrt{2}}{2} \text{ and } \sin \theta = \frac{-3}{3\sqrt{2}} = -\frac{\sqrt{2}}{2}, \text{ we may take } \theta = -\frac{\pi}{4}$$

to obtain $\left(3\sqrt{2}, -\dfrac{\pi}{4}\right)$ as a pair of polar coordinates.

(b) Since $x = -2 \cos \dfrac{2\pi}{3} = -2\left(-\dfrac{1}{2}\right) = 1$

and $y = -2 \sin \dfrac{2\pi}{3} = -2\left(\dfrac{\sqrt{3}}{2}\right) = -\sqrt{3}$, we obtain $\left(1, -\sqrt{3}\right)$. ∎

Some curves have very simple polar-coordinate equations. Consider, for example, the circle C of radius a with center at the origin. A point is on C if and only if it has polar coordinates of the form (r, θ), where $r = a$. Thus, a polar equation of C is $r = a$. Similarly, the line through the origin making an angle α with the polar axis has polar equation $\theta = \alpha$.

The coordinate-change equations are useful in transforming an equation written in one coordinate system into an equation involving another coordinate system.

Example 3

(a) Transform $r = 2 \cos \theta$ to rectangular form.
(b) Transform $x^2 - y^2 = 1$ to polar form.

Solution

(a) Transforming $r = 2 \cos \theta$ to rectangular coordinates is somewhat easier if we multiply both sides by r.

$$r^2 = 2r \cos \theta$$

Then, $x^2 + y^2 = 2x$.
By completing the square in x, we obtain

$$x^2 - 2x + 1 + y^2 = 1$$
$$(x - 1)^2 + y^2 = 1,$$

an equation of the circle centered at $(1, 0)$ and having radius 1.

(Solution continued next page.)

(b) To transform $x^2 - y^2 = 1$ to polar form, use the coordinate-change equations and a double-angle formula.

$$x^2 - y^2 = 1$$
$$(r \cos \theta)^2 - (r \sin \theta)^2 = 1$$
$$r^2(\cos^2 \theta - \sin^2 \theta) = 1$$
$$r^2 \cos 2\theta = 1$$
$$r^2 = \frac{1}{\cos 2\theta} = \sec 2\theta \quad \blacksquare$$

Transforming a polar equation to rectangular form often makes it easier to graph the equation. Figure 7-7 shows the graph of $r = 2 \cos \theta$ obtained by plotting points and by transforming coordinates.

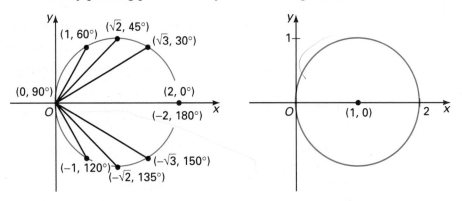

Figure 7-7

EXERCISES 7-1

Plot each point. Find its rectangular coordinates.

A 1. $(2, -30°)$ 2. $(3, 150°)$ 3. $(-2, 30°)$ 4. $(-3, 150°)$

5. $\left(2, -\frac{\pi}{6}\right)$ 6. $\left(3, -\frac{2\pi}{3}\right)$ 7. $(-2, -30°)$ 8. $(-3, -150°)$

9. $(2, 135°)$ 10. $(-3, 135°)$ 11. $\left(4, -\frac{\pi}{3}\right)$ 12. $\left(-3, -\frac{\pi}{4}\right)$

Graph each pair of rectangular coordinates. Find a pair of polar coordinates.

13. $\left(\sqrt{3}, 1\right)$ 14. $\left(-1, -\sqrt{3}\right)$ 15. $\left(-2, 2\sqrt{3}\right)$ 16. $\left(-\sqrt{2}, -\sqrt{2}\right)$

17. $\left(-\sqrt{2}, \sqrt{2}\right)$ 18. $\left(\sqrt{3}, -\sqrt{3}\right)$ 19. $\left(\sqrt{6}, \sqrt{2}\right)$ 20. $\left(\sqrt{2}, -\sqrt{6}\right)$

Find a polar equation for each of the following.

21. $y = -2$ 22. $x = 4$ 23. $x + 2 = 0$ 24. $3 - y = 0$

25. $y = x$ 26. $x + y = 0$ 27. $x^2 + y^2 = 4$ 28. $x^2 + y^2 = 3$

29. $x + \sqrt{3}y = 0$ 30. $y = x\sqrt{3}$ 31. $x + y = 4$ 32. $x - y = 2$

Graph the following polar equations.

33. $r = 1$ 34. $\theta = 90°$ 35. $r \cos \theta = 2$

36. $r \sin \theta = 3$ 37. $\theta = \dfrac{3\pi}{4}$ 38. $\theta + \dfrac{\pi}{6} = 0$

Transform the equation to rectangular form. Identify and sketch the graph.

B 39. $r = 2 \sin \theta$ 40. $r = -4 \cos \theta$ 41. $r = \cos \theta + \sin \theta$

42. $r = \tan \theta \sec \theta$ 43. $r^2 = \sec \theta \csc \theta$ 44. $r = \cot \theta \csc \theta$

45. Prove the following **distance formula:** The distance between the points
$P_1(r_1, \theta_1)$ and $P_2(r_2, \theta_2)$ is $P_1P_2 = \sqrt{r_1^2 + r_2^2 - 2r_1r_2 \cos (\theta_2 - \theta_1)}$.

46. Use Exercise 45 to show that
(a) if $\theta_1 = \theta_2$, then $P_1P_2 = |r_1 - r_2|$ and

(b) if $r_1 = r_2 = r_0$, then $P_1P_2 = 2\left|r_0 \sin \dfrac{1}{2}(\theta_2 - \theta_1)\right|$.

47. Find a polar equation of the circle with radius a and center (r_0, θ_0).

C 48. Let (r, θ) be polar coordinates of a point
on l, a line not passing through the ori-
gin O. Let (p, ω) be the foot of the perpendicular from O to l. (a) Show that
a polar equation of l is

$$r \cos (\omega - \theta) = p.$$

(b) Show that a rectangular-coordi-
nate equation of l is

$$(\cos \omega)x + (\sin \omega)y = p.$$

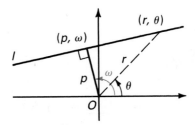

Exercise 48

This is called the **normal form** of the equation of the line.
(c) Show that when the equation $Ax + By = C$ is put into the form
given in (b), the result is

$$\frac{A}{\pm\sqrt{A^2 + B^2}}\, x + \frac{B}{\pm\sqrt{A^2 + B^2}}\, y = \frac{C}{\pm\sqrt{A^2 + B^2}},$$

where the sign of the radical is chosen so that the right-hand
member is positive. This shows that the distance p between the
origin and line l is given by $\dfrac{C}{\pm\sqrt{A^2 + B^2}}$.

7-2

Graphs of Polar Equations

Objective: To graph polar equations.

We often graph polar equations by observing how the trigonometric functions vary within quarters of a period.

Example 1   Graph $r = 2 \cos 3\theta$.

Solution   We set up a table that shows how r varies as θ increases. Notice that we have chosen intervals of θ for which 3θ varies by $90°$, that is, on which $\cos 3\theta$ runs through a quarter of its period. Since $|r| \le 2$, the graph is inside the circle of radius 2 centered at the origin.

θ	3θ	$\cos 3\theta$	$r = 2 \cos 3\theta$
$0° \rightarrow 30°$	$0° \rightarrow 90°$	$1 \rightarrow 0$	$2 \rightarrow 0$
$30° \rightarrow 60°$	$90° \rightarrow 180°$	$0 \rightarrow -1$	$0 \rightarrow -2$
$60° \rightarrow 90°$	$180° \rightarrow 270°$	$-1 \rightarrow 0$	$-2 \rightarrow 0$
$90° \rightarrow 120°$	$270° \rightarrow 360°$	$0 \rightarrow 1$	$0 \rightarrow 2$
$120° \rightarrow 150°$	$360° \rightarrow 450°$	$1 \rightarrow 0$	$2 \rightarrow 0$
$150° \rightarrow 180°$	$450° \rightarrow 540°$	$0 \rightarrow -1$	$0 \rightarrow -2$

In the first entry of the table, we see that as θ increases from $0°$ to $30°$, r decreases from 2 to 0. Then as θ continues to increase from $30°$ to $60°$, r decreases from 0 to -2. The corresponding portions of the graph are shown at the left below. The complete graph, called a **three-leaved rose**, appears at the right below. (As θ increases from $180°$, the curve is repeated.)

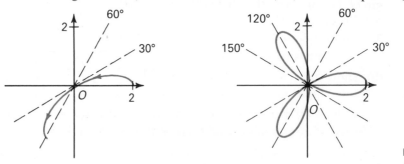

Several tests for symmetry of polar curves may be deduced from Figure 7-8.

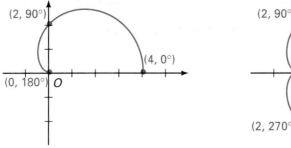

Figure 7-8

Symmetric about	if the equation is unchanged by
polar axis	replacing θ by $-\theta$
origin	replacing r by $-r$
vertical line $\theta = \dfrac{\pi}{2}$	replacing θ by $\pi - \theta$

In Example 1, for instance, the identity $2 \cos (3(-\theta)) = 2 \cos (-3\theta) = 2 \cos 3\theta$ shows that the graph is symmetric with respect to the polar axis. Testing for symmetry reduces the work needed to graph a polar equation.

Example 2 Graph $r = 2(1 + \cos \theta)$, a heart-shaped curve, called a **cardioid**.

Solution Since $\cos (-\theta) = \cos \theta$, $2(1 + \cos (-\theta)) = 2(1 + \cos \theta) = r$. Hence the graph is symmetric with respect to the polar axis. We first draw that part of the graph that is above the polar axis.

θ	$\cos \theta$	$r = 2(1 + \cos \theta)$
$0° \rightarrow 90°$	$1 \rightarrow 0$	$4 \rightarrow 2$
$90° \rightarrow 180°$	$0 \rightarrow -1$	$2 \rightarrow 0$

The portion of the graph above the polar axis is shown at the left below. The completed graph is shown at the right below.

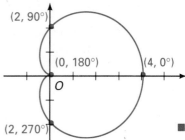

Example 3   Graph $r^2 = 4 \sin 2\theta$.

Solution   Replacing r by $-r$ leaves the equation unchanged. The graph, therefore, is symmetric with respect to the origin. We first draw the part of
the curve above the polar axis. In doing this, it is sufficient to graph
$r = \sqrt{4 \sin 2\theta} = 2\sqrt{\sin 2\theta}$.

θ	2θ	$\sin 2\theta$	$4 \sin 2\theta$	$r = 2\sqrt{\sin 2\theta}$
$0° \rightarrow 45°$	$0° \rightarrow 90°$	$0 \rightarrow 1$	$0 \rightarrow 4$	$0 \rightarrow 2$
$45° \rightarrow 90°$	$90° \rightarrow 180°$	$1 \rightarrow 0$	$4 \rightarrow 0$	$2 \rightarrow 0$
$90° \rightarrow 135°$	$180° \rightarrow 270°$	negative	negative	—
$135° \rightarrow 180°$	$270° \rightarrow 360°$	negative	negative	—

From the table, we have the portion of the curve shown at the left
below. The complete graph, called a **lemniscate**, is obtained by using
symmetry with respect to the origin and is shown at the right below.

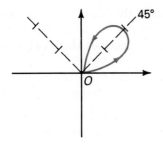
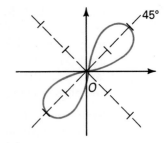

Example 4   Graph $r = 1 - 2 \sin \theta$.

Solution   Replacing θ by $\pi - \theta$ leaves the equation unchanged since $\sin (\pi - \theta) = \sin \theta$. The graph, therefore, is symmetric with respect to the vertical
line $\theta = 90°$. We shall first draw the portion of the graph correspond-
ing to $-90° \le \theta \le 90°$ and then obtain the rest of the graph by using
symmetry.

θ	$\sin \theta$	$-2 \sin \theta$	$r = 1 - 2 \sin \theta$
$-90° \rightarrow 0°$	$-1 \rightarrow 0$	$2 \rightarrow 0$	$3 \rightarrow 1$
$0° \rightarrow 90°$	$0 \rightarrow 1$	$0 \rightarrow -2$	$1 \rightarrow -1$

Notice that when $0° \le \theta \le 90°$, r changes from positive to negative. The change of sign occurs when $1 - 2 \sin \theta = 0$, that is, when $\theta = 30°$. From the table, we obtain the curve at the left below. The complete graph, called a **limaçon**, is shown at the right below.

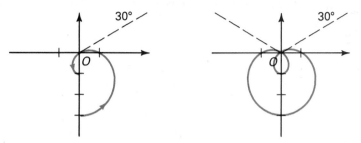

Many software packages have the capacity to graph a polar equation. They do this by finding r for various values of θ, and then converting (r, θ) to (x, y). This conversion to rectangular coordinates is necessary, since screen coordinates are rectangular.

For example, a computer would graph $r = \cos \dfrac{\theta}{3}$ by computing

$$x = r \cos \theta \qquad\qquad y = r \sin \theta$$
$$= \cos \frac{\theta}{3} \cos \theta \qquad\qquad = \cos \frac{\theta}{3} \sin \theta$$

and then plotting (x, y). Figures 7-9 and 7-10 are computer-generated graphs of $r = \cos \dfrac{\theta}{3}$ and $r = \cos \dfrac{\theta}{4}$.

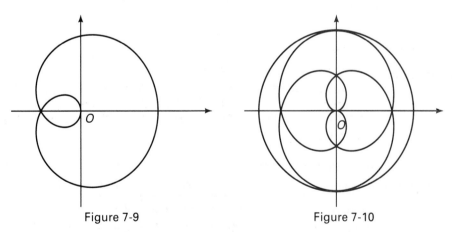

Figure 7-9 Figure 7-10

Sometimes, as in the next example, the variable θ appears independently of a trigonometric function. In such a case, θ denotes *radian* measure.

Example 5   Graph $r = \dfrac{1}{\theta}$ where $\theta > 0$.

Solution   We see that the larger θ is, the smaller r is. The following graph, a *spiral*, was drawn by computer. Note that

$$x = r \cos \theta = \frac{\cos \theta}{\theta} \text{ and } y = r \sin \theta = \frac{\sin \theta}{\theta}.$$

The spiral winds around the origin infinitely many times, although a drawing cannot show this.

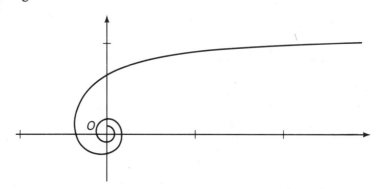

EXERCISES 7-2

Graph the following polar equations.

A **1.** $r = 1 - \sin \theta$ (cardioid)

  2. $r = 1 - \cos \theta$ (cardioid)

  3. $r = 1 + \cos \theta$ (cardioid)

  4. $r = 1 + \sin \theta$ (cardioid)

  5. $r = 2 \sin 3\theta$ (three-leaved rose)

  6. $r = 2 \cos 2\theta$ (four-leaved rose)

  7. $r = 6 \sin 2\theta$ (four-leaved rose)

  8. $r = 2 \cos 5\theta$ (five-leaved rose)

  9. $r^2 = 4 \cos 2\theta$ (lemniscate)

 10. $r = 1 + 2 \cos \theta$ (limaçon with small loop)

 11. $r = 1 + 2 \sin \theta$ (limaçon with small loop)

 12. $r = 2 - 4 \sin \theta$ (limaçon with small loop)

13. $r = \theta, \theta \geq 0$ (spiral)

14. $r = \dfrac{2}{\theta}, \theta > 0$ (spiral)

B 15. $r = |\cos \theta|$ (two circles)

16. $r = |\sin \theta|$ (two circles)

17. $r^2 - 3r + 2 = 0$ (two circles)

18. $r^2 - r - 2 = 0$ (two circles)

19. $r^2 - 3r = 0$ (circle and a point)

20. $r = \cos \theta + \sin \theta$ (circle)

21. $r = 2 \cos \theta + 2 \sin \theta$ (circle)

22. $r = 2 \cos \dfrac{1}{2}\theta$

23. $r = 2 \sin \dfrac{1}{2}\theta$

C 24. The diagram at the right illustrates a fixed circle A and another circle B of equal radius, that rolls around circle A without slipping, so that a fixed point P on circle B describes a closed curve as θ runs from $0°$ to $360°$. When $\theta = 0$, P is at the origin. Both circles have radius 1. Let $r = OP$.

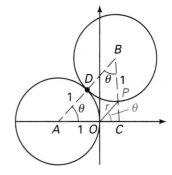

    (a) Show that $\angle POC = \theta$. (Hint: First show that $\angle ABP = \theta$. Then show that triangles ABC and OPC are similar isosceles triangles.)

    (b) Show that $r = 2(1 - \cos \theta), 0 \leq \theta < 90°$.

    (c) Judging from the result of (b), what kind of curve do you think is traced by P as circle B rolls all the way around circle A?

25. A line segment l of length 2 moves so that one endpoint is on the x-axis and the other is on the y-axis. Let P be the foot of the perpendicular dropped from the origin to l. Show algebraically that P traces out a four-leaved rose. (Hint: Let a and b be the x- and y-intercepts of the line containing l. Express these in terms of r and θ, and derive a relationship between r and θ from a relationship between a and b.)

26. Let F_1 and F_2 be the points $(1, 0°)$ and $(1, 180°)$, respectively. A point P moves so that $PF_1 \cdot PF_2 = 1$. Show that P moves on a "horizontal" lemniscate (See Exercise 9.).

7-3

Conic Sections (Optional)

Objective: To find polar equations of conic sections and sketch their graphs.

You may have already studied parabolas, ellipses, and hyperbolas. These curves, formed when a plane intersects a cone, are called *conic sections*. An ellipse is often defined as the set of all points in the plane such that the sum of the distances from each point in the set to two fixed points is a constant. A hyperbola can be defined similarly (with "absolute value of the difference" replacing "sum"). A parabola is usually defined as a set of all points equidistant from a fixed point and a fixed line. All three types of curves, however, share a property that will be more convenient to adopt as a definition for the purpose of studying their polar equations.

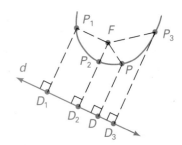

In a plane, let *d* be a fixed line and let *F* be a fixed point not on *d*. Let *e* be a positive number. Then the set of all points *P* such that

$$\frac{PF}{PD} = e$$

where *PD* is the perpendicular distance from the point *P* to the line *d* (Figure 7-11), is called a **conic section** with **eccentricity** *e*, **focus** *F*, and **directrix** *d*. We often abbreviate *conic section* to *conic*.

$$\frac{P_1F}{P_1D_1} = \frac{P_2F}{P_2D_2} = \frac{P_3F}{P_3D_3}$$

Figure 7-11

If *e* = 1, the conic section is a parabola. In this case *P* moves so that *PF* = *PD*. A conic is an ellipse if 0 < *e* < 1 and a hyperbola if *e* > 1.

Example 1   Find a polar equation of the parabola having the pole as focus and the line *x* = −4 as directrix.

Solution   For any point *P*(*r*, *θ*), we see from the diagram that *PF* = *r* and *PD* = 4 + *r* cos *θ*. Thus, *P* is on the parabola if and only if

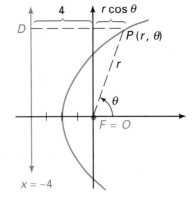

$$\frac{PF}{PD} = \frac{r}{4 + r\cos\theta} = 1$$

Hence,

$$r = 4 + r\cos\theta$$
$$r - r\cos\theta = 4$$
$$r = \frac{4}{1 - \cos\theta}.$$ ■

We can use the method of Example 1 to derive equations of conics having the pole as a focus and either a horizontal or a vertical directrix. For example, if the directrix is the line $y = p$ with $p > 0$, we have (Figure 7-12)

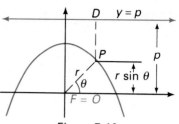

$$PF = r \text{ and } PD = p - r \sin \theta.$$

If the eccentricity of the conic is e, then

$$\frac{PF}{PD} = \frac{r}{p - r \sin \theta} = e.$$

Figure 7-12

When we solve this equation for r, we obtain $r = \dfrac{ep}{1 + e \sin \theta}$.

The various cases are illustrated in Figure 7-13. In each case, p is the distance from the focus to the directrix, and e is the eccentricity. Each of these conics has one of the lines $\theta = 0°$ or $\theta = 90°$ as an axis of symmetry.

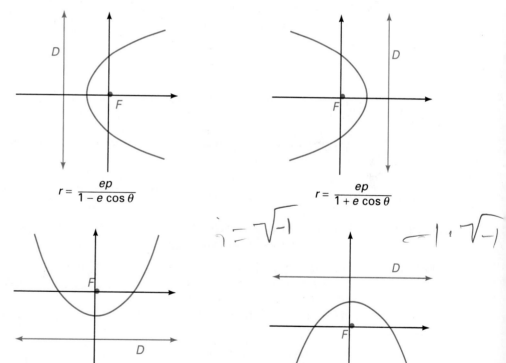

$$r = \frac{ep}{1 - e \cos \theta}$$

$$r = \frac{ep}{1 + e \cos \theta}$$

$$r = \frac{ep}{1 - e \sin \theta}$$

$$r = \frac{ep}{1 + e \sin \theta}$$

Figure 7-13

Example 2   Determine the type of conic and sketch its graph.

(a) $r = \dfrac{6}{2 + \cos \theta}$ (b) $r = \dfrac{10}{2 - 3 \sin \theta}$

Solution   (a) To compare the given equation with those shown in Figure 7-13, we make the constant term of the denominator equal to 1 by dividing numerator and denominator by 2.

$$r = \frac{6}{2 + \cos \theta} = \frac{3}{1 + \dfrac{1}{2} \cos \theta} = \frac{\dfrac{1}{2} \cdot 6}{1 + \dfrac{1}{2} \cos \theta}.$$

Comparing this equation with $r = \dfrac{ep}{1 + e \cos \theta}$, we see that

$e = \dfrac{1}{2} < 1$, so that the conic is an *ellipse*.

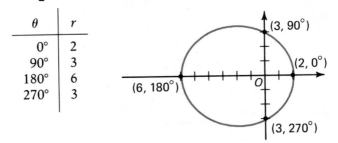

θ	r
0°	2
90°	3
180°	6
270°	3

(b) Put the equation into one of the forms in Figure 7-13.

$$r = \frac{10}{2 - 3 \sin \theta} = \frac{5}{1 - \dfrac{3}{2} \sin \theta} = \frac{\dfrac{3}{2} \cdot \dfrac{10}{3}}{1 - \dfrac{3}{2} \sin \theta}.$$

Since $e = \dfrac{3}{2} > 1$, the conic

is a *hyperbola*.

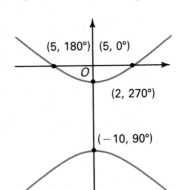

θ	r
0°	5
90°	-10
180°	5
270°	2

We began the section by recalling the following definition of an ellipse.

An *ellipse* is the set of all points such that the sum of the distances from each point in the set to two fixed points is a constant.

We can show that the ellipse

$$r = \frac{6}{2 + \cos \theta}$$

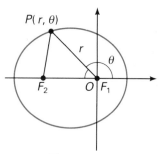

P(r, θ)

Figure 7-14

in Example 2(a) has two foci, F_1 (0, 0°) and F_2 (4, 180°) (Exercise 27) and that the ellipse satisfies this condition. Let $P(r, \theta)$ be any point of the curve above the polar axis. (See Figure 7-14.) Note that $2 \leq r \leq 6$ since $-1 \leq \cos x \leq 1$. Clearly, $PF_1 = r$. Now we apply the law of cosines to $\triangle F_1 F_2 P$.

$$PF_2{}^2 = r^2 + 4^2 - 2 \cdot r \cdot 4 \cos (180° - \theta)$$
$$= r^2 + 16 + 8r \cos \theta$$

From the equation $r = \dfrac{6}{2 + \cos \theta}$, we have $2r + r \cos \theta = 6$, and hence $r \cos \theta = 6 - 2r$. Therefore,

$$PF_2{}^2 = r^2 + 16 + 8(6 - 2r)$$
$$= r^2 - 16r + 64$$
$$= (r - 8)^2.$$

Since $r < 8$, $8 - r > 0$, so that $PF_2 = \sqrt{(r - 8)^2} = 8 - r$.

Therefore,

$$PF_1 + PF_2 = r + (8 - r) = 8, \text{ a constant.}$$

The case where P is below the polar axis is similar.

Note that one advantage of describing a conic section by a polar equation is that, by doing so, we obtain an expression for r as an explicit function of θ. This makes it much easier to write a computer program to draw the conic or to use a commercial graphing program to do so. The graph in Figure 7-15 at the top of the next page was drawn by computer. It shows five members of the family of curves whose general equation is

$$r = \frac{2e}{1 - e \cos \theta}$$

for the values $e = 0.4, 0.6, 0.8, 1$, and 4. The first three conics are ellipses, the fourth is a parabola, and the last one is a hyperbola.

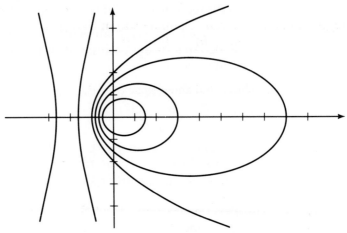

Figure 7-15

EXERCISES 7-3

Determine the type of each conic.

A 1. $r = \dfrac{4}{1 - \sin \theta}$

2. $r = \dfrac{3}{1 - \dfrac{1}{2} \sin \theta}$

3. $r = \dfrac{6}{2 - \cos \theta}$

4. $r = \dfrac{4}{1 - \cos \theta}$

5. $r = \dfrac{8}{1 + 3 \cos \theta}$

6. $r = \dfrac{6}{1 + 2 \sin \theta}$

7. $r = \dfrac{15}{3 + 2 \sin \theta}$

8. $r = \dfrac{3}{2 + 2 \cos \theta}$

9. $r = \dfrac{5}{2 - 2 \sin \theta}$

10. $r = \dfrac{10}{2 + 3 \cos \theta}$

11–20. Sketch the graphs of the conics in Exercises 1–10.

B 21. Identify the curve: $r^2 = \sec 2\theta$.   22. Identify the curve: $r^2 = \csc 2\theta$.

23. Let P be any point on the conic: $r = \dfrac{6}{2 - \cos \theta}$. Let F_1 and F_2 be the points $(0, 0°)$ and $(4, 0°)$, respectively. Show that $PF_1 + PF_2 = 8$.

24. Let P be any point on the conic: $r = \dfrac{12}{3 - \sin \theta}$. Let F_1 and F_2 be the points $(0, 0°)$ and $(3, 90°)$, respectively. Show that $PF_1 + PF_2 = 9$.

25. Write the equation given in Example 2(a) in the form
$$\frac{(x - h)^2}{a^2} + \frac{(y - k)^2}{b^2} = 1.$$

26. Write the equation given in Example 2(b) in the form
$$\frac{(y - k)^2}{b^2} - \frac{(x - h)^2}{a^2} = 1.$$

C 27. Every ellipse (and every hyperbola) has two foci, each with its corresponding directrix. Show that for the ellipse: $r = \dfrac{6}{2 + \cos \theta}$, $(0, 0°)$ is one focus with a corresponding directrix of $x = 6$, while $(4, 180°)$ is the other focus with a corresponding directrix of $x = -10$.

28. Each of the planetary orbits is an ellipse with the sun located at one focus. The distance from the focus to the directrix of the orbit of Mercury is approximately 170 million miles and the eccentricity of the orbit is approximately 0.2. Find a polar equation of the orbit with the sun at the pole. Use this equation to find Mercury's maximum and minimum distances from the sun. (These points of maximum and minimum distance are called the aphelion and the perihelion, respectively.)

Self Quiz   7-1   7-2   7-3

1. Give a pair of polar coordinates of (a) $\left(-5\sqrt{3}, 5\right)$ and (b) $(2, -2)$.

2. Give the rectangular coordinates of (a) $(4, -30°)$ and (b) $(6, 270°)$.

3. Transform $y^2 - x^2 = 1$ to polar form.

4. Transform $r = 6 \sin \theta - 8 \cos \theta$ to rectangular form and sketch the graph.

5. Graph the polar equation $r = 1 - 2 \cos \theta$.

(Optional) Determine the type of each conic and sketch its graph.

6. $r = \dfrac{12}{4 - 2 \sin \theta}$

7. $r = \dfrac{10}{5 + 5 \cos \theta}$

REPRESENTING COMPLEX NUMBERS

7-4

Complex Numbers

Objective: To use complex numbers and algebraic operations involving them.

To solve an equation such as $x^2 = 2$, mathematicians augmented the system of rational numbers to include the number $\sqrt{2}$ (as well as other irrational numbers). Numbers of the form $a + b\sqrt{2}$, where a and b are rational, could then be manipulated according to the usual rules of algebra, with the proviso that $(\sqrt{2})^2 = 2$. Similarly, to solve

$$x^2 = -1,$$

having no solution in the system of real numbers, mathematicians introduced the number i, the **imaginary unit**, with the property that

$$i^2 = -1, \text{ or } i = \sqrt{-1}.$$

This definition and familiar algebraic rules imply that

$$i^3 = i^2 \cdot i = (-1)(i) = -i$$
$$i^4 = i^3 \cdot i = (-i)(i) = -(-1) = 1$$
$$i^5 = i^4 \cdot i = 1 \cdot i = i,$$

and that $i^{4n} = (i^4)^n = (1)^n = 1$, for any integer n. Thus we can easily calculate any integral power of i.

Example 1 Find i^{31}.

Solution $i^{31} = i^{28+3} = i^{28} \cdot i^3 = (i^4)^7 \cdot i^3 = (1)^7 \cdot i^3 = i^3 = -i.$ ■

A **complex number** is any number of the form

$$x + yi, \text{ or } x + iy,$$

where x and y are real numbers. The form $x + yi$ is called the standard form of the complex number. The number x is called the **real part** of $x + iy$, and y (*not iy*) is called the **imaginary part**. Thus,

$$2 - i, 4i, \text{ and } -5$$

have real parts 2, 0, and -5, respectively, and imaginary parts -1, 4, and 0, respectively. An **imaginary number** is one whose imaginary part is nonzero. A number, such as $4i$, whose real part is 0 and whose imaginary part is not zero is called a **pure imaginary** number.

Two complex numbers $x + yi$ and $u + vi$ are **equal** if and only if $x = u$ and $y = v$.

If we treat complex numbers as binomials and replace i^2 by -1 wherever it occurs, we can perform operations like the following:

$$(4 + i) + (3 - 2i) = (4 + 3) + (1 - 2)i$$
$$= 7 - i$$
$$(4 + i)(3 - 2i) = 12 - 8i + 3i - 2i^2$$
$$= 12 - 5i - 2(-1)$$
$$= 14 - 5i$$

More generally:

$$(u + vi)(x + yi) = ux + vxi + uyi + vyi^2$$
$$= ux + (vx + uy)i + vy(-1)$$
$$= (ux - vy) + (vx + uy)i$$

Examples like these suggest the following definitions of the sum and the product of two complex numbers:

$$(u + vi) + (x + yi) = (u + x) + (v + y)i$$
$$(u + vi)(x + yi) = (ux - vy) + (vx + uy)i$$

It can be shown that the set of all complex numbers with these definitions of addition and multiplication satisfies the axioms for a mathematical structure called a *field* (See Exercises 29–32.). The difference and the quotient of two complex numbers are then defined, as usual, in terms of addition and multiplication, respectively. Example 2 illustrates how to find the quotient of two complex numbers.

Example 2   Let $w = 2 - i$ and $z = 3 - 4i$. Find $\dfrac{w}{z}$ as a complex number in the form $x + yi$.

Solution

$$\frac{w}{z} = \frac{2 - i}{3 - 4i} = \frac{2 - i}{3 - 4i} \cdot \frac{3 + 4i}{3 + 4i}$$
$$= \frac{6 + 8i - 3i - 4i^2}{9 - 16i^2}$$
$$= \frac{6 + 5i - 4(-1)}{9 - 16(-1)} = \frac{10 + 5i}{25}$$

Therefore, $\dfrac{w}{z} = \dfrac{2 + i}{5}$, or $\dfrac{2}{5} + \dfrac{1}{5}i$. ■

Notice that in Example 2 we multiplied both numerator and denominator by the *conjugate* of the denominator, that is, the number $3 + 4i$, obtained from $3 - 4i$ by changing the sign of the imaginary part. In general, the **conjugate**, $\bar{z}$, of a complex number $z = x + yi$ is $x - yi$. Furthermore, since

$$z\bar{z} = (x + yi)(x - yi) = x^2 - y^2i^2 = x^2 - y^2(-1) = x^2 + y^2,$$

we see that the product of two conjugate complex numbers is a nonnegative real number.

The **modulus**, or **absolute value**, of $z = x + yi$ is

$$|z| = \sqrt{x^2 + y^2}.$$

From the preceding paragraph we see that $z\bar{z} = |z|^2$. This and three other properties of the modulus of a complex number are listed below. (See Exercises 34 and 35 of this section and Exercises 21–26 of Section 7-5.)

$$z\bar{z} = |z|^2$$
$$|z|\,|w| = |zw|$$
$$\left|\frac{z}{w}\right| = \frac{|z|}{|w|} \qquad (w \neq 0)$$
$$|z + w| \leq |z| + |w| \qquad \text{triangle inequality}$$

In Section 7-5 we shall see that $|z|$ has a geometric interpretation, "the distance from z to the origin," like the absolute value of a real number.

Example 3   Find the reciprocal of $z = 3 + i$.

Solution   Note that $\dfrac{1}{z} = \dfrac{\bar{z}}{z\bar{z}} = \dfrac{\bar{z}}{|z|^2}$. Since $\bar{z} = 3 - i$ and $|z| = \sqrt{3^2 + 1^2} = \sqrt{10}$,

$$\frac{1}{z} = \frac{3 - i}{10}, \text{ or } \frac{3}{10} - \frac{1}{10}i. \quad \blacksquare$$

EXERCISES 7-4

Put all complex-number answers into the form $x + yi$, or $x + iy$. Find (a) $w + z$, (b) wz, and (c) $\dfrac{w}{z}$.

A 1. $w = 3 + i; z = 1 - i$ **2.** $w = 1 - 3i; z = 2 - i$

3. $w = 5i; z = 3 + 4i$ **4.** $w = i; z = 2 + 2i$

5. $w = -1 + i\sqrt{3}; z = -1 - i\sqrt{3}$ 6. $w = 3 + 4i; z = 3 - 4i$

Find (a) the conjugate, (b) the modulus, and (c) the reciprocal of z.

7. $z = 2 + i$ 8. $z = \dfrac{1}{2} + \dfrac{1}{2}i$ 9. $z = \dfrac{\sqrt{3}}{2} - \dfrac{1}{2}i$ 10. $z = \dfrac{3}{5} - \dfrac{4}{5}i$

Find z^2 and z^3.

11. $z = -1 + i$ 12. $z = -1 - i$ 13. $z = -1 + i\sqrt{3}$ 14. $z = 1 + i\sqrt{3}$

Use addition, subtraction, multiplication, and division to solve each equation for the complex number z.

15. $2iz = 3 - 7i$ 16. $(3 - 5i)z = -2 + 4i$

17. $6 - 2i = 8 + i - (1 + 2i)z$ 18. $3 + 2i + (5 - i)z = 7i$

Write each of the following as 1, -1, i, or $-i$.

B 19. (a) i^{25} (b) i^{34} (c) i^{200} 20. (a) i^{55} (b) i^{80} (c) i^{123}

21. $i + i^2 + i^3 + \cdots + i^{21}$ 22. $i + i^2 + i^3 + \cdots + i^{201}$

23. (a) $\dfrac{1}{i}$ (b) $\dfrac{1}{i^2}$ (c) $\dfrac{1}{i^3}$ 24. i^{4n+2}, n a positive integer

25. i^{4n+3}, n a positive integer   26. i^{4n+1}, n a positive integer

27. i^{4n-1}, n a positive integer   28. $i^{36} + i^{37} + i^{38} + i^{39} + i^{40}$

A **field** is an algebraic system that consists of a set F together with two binary operations that satisfy the following axioms. (Note that the system of real numbers is a field.)

The Field Axioms

Let $a, b, c \in F$.

ADDITION AXIOMS		MULTIPLICATION AXIOMS
$a + b \in F$	Closure property	$ab \in F$
$a + b = b + a$	Commutative property	$ab = ba$
$(a + b) + c = a + (b + c)$	Associative property	$(ab)c = a(bc)$
$a + 0 = a$	Existence of identity	$a \cdot 1 = a$
$a + (-a) = 0$	Existence of inverse	$a \cdot \dfrac{1}{a} = 1 \ (a \neq 0)$

DISTRIBUTIVE AXIOM $a(b + c) = ab + ac$, $(b + c)a = ba + ca$

Show that the following axioms hold for complex numbers.

29. Associative property of addition

30. Associative property of multiplication

31. Existence of multiplicative inverse

32. Distributive axiom

33. Show that if z and w are any two nonreal complex numbers such that $z + w$ and zw are both real, then z and w are complex conjugates. (Hint: Write $z = x + yi$ and $w = u + vi$, with $y \neq 0$ and $v \neq 0$.)

C 34. Prove that for any two complex numbers z and w, $|z|\,|w| = |zw|$.

35. Show that if $w \neq 0$, then $\left|\dfrac{z}{w}\right| = \dfrac{|z|}{|w|}$.

36. Find a field other than the systems of real and complex numbers.

| 7-5 | **The Complex Plane and the Polar Form** |

Objective: To represent complex numbers as points in a coordinate plane and in trigonometric form.

We can represent complex numbers as points in the Cartesian plane by letting $x + yi$ correspond to the point (x, y), as indicated in Figure 7-16. When the plane is used for this purpose, it is called the **complex plane**, or the **Argand plane**. The x-axis is called the **real axis**, and the y-axis is called the **imaginary axis**.

Let $z = x + yi$ be any complex number. Graphically, then, $|z|$ is the distance from 0 to the point z. This is indicated in Figure 7-17, which also shows the graphical relationships among z, $\bar{z}$, and $-z$.

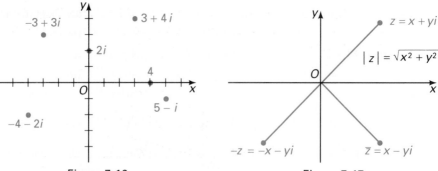

Figure 7-16   Figure 7-17

The sum of any two complex numbers $z = x + yi$ and $w = u + vi$ can be represented graphically by adding the vectors drawn from O to the graphs of z and w (Figure 7-18).

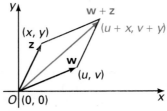

Figure 7-18

A graphical representation of complex multiplication will be presented in Section 7-6. Indeed, if we regard complex numbers as vectors with the set of real numbers as scalars, all the results of Chapter 6 will apply to addition and "scalar multiplication" of complex numbers.

Example 1   Let $w = 4 + i$ and $z = -3 + 2i$. Draw a graphical representation of
(a) $w + z$,   (b) $w - z$, and   (c) $z + \bar{z}$.

Solution   (a)   (b)   (c)

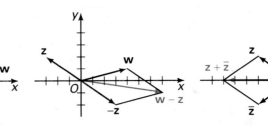

The following property, known as the *triangle inequality*, should look familiar from our work with vectors in Chapter 6. Both its algebraic form and its graphical interpretation are identical to those of the triangle inequality on page 195. It says that for any complex numbers z and w,

$$|z + w| \le |z| + |w|.$$

The triangle inequality is illustrated in Figure 7-19. A proof of the triangle inequality making use of the properties of complex numbers is given in Exercises 21–26 on page 253.

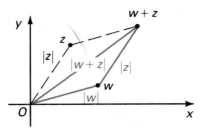

Figure 7–19

Recall that each point (x, y) in the plane has polar coordinates (r, θ) with $r \geq 0$, and

$$x = r \cos \theta, \, y = r \sin \theta.$$

Thus, any complex number $z = x + yi$ can be written as

$$z = (r \cos \theta) + i(r \sin \theta), \text{ or}$$
$$z = r(\cos \theta + i \sin \theta) \text{ for } r \geq 0.$$

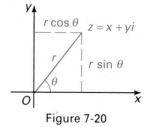

Figure 7-20

This form is called the **trigonometric** or **polar form** of the complex number z (Figure 7-20). The polar form, $r(\cos \theta + i \sin \theta)$, is often abbreviated as r cis θ. It is easy to see that $r = |z|$. The angle θ is called the **argument** of z, written **arg** z, and r is what we have called the **modulus** of z.

Example 2

Write each complex number in polar form. If necessary, express angles to the nearest tenth of a degree.
(a) $z = -3 + 3i$ (b) $w = 4 - 3i$

Solution

(a) From the figure at the right,
$$r = |z| = \sqrt{(-3)^2 + 3^2} = 3\sqrt{2}.$$
Since $\cos \theta = \dfrac{-3}{3\sqrt{2}} = -\dfrac{\sqrt{2}}{2}$ and

$\sin \theta = \dfrac{3}{3\sqrt{2}} = \dfrac{\sqrt{2}}{2}, \theta = 135°.$

Therefore,
$$z = 3\sqrt{2}(\cos 135° + i \sin 135°).$$

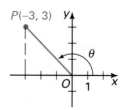

(b) From the figure at the right,
$$r = |w| = \sqrt{(4)^2 + (-3)^2} = \sqrt{25} = 5.$$
Since $\cos \theta = \dfrac{4}{5} = 0.8$ and θ is in the fourth quadrant,
$$\theta = 360° - 36.9° = 323.1°.$$
Therefore,
$$w = 5(\cos 323.1° + i \sin 323.1°). \quad \blacksquare$$

Example 3

Write each complex number in $x + yi$ form.
(a) $\sqrt{2}(\cos 315° + i \sin 315°)$ (b) $10(\cos 90° + i \sin 90°)$

Solution

(a) $\sqrt{2}(\cos 315° + i \sin 315°) = \sqrt{2}\left[\dfrac{\sqrt{2}}{2} + i\left(-\dfrac{\sqrt{2}}{2}\right)\right] = 1 - i$

(b) $10(\cos 90° + i \sin 90°) = 10(0 + i \cdot 1) = 10i \quad \blacksquare$

EXERCISES 7-5

For each pair of complex numbers, draw a graphical representation of the given combinations.

A 1. Let $w = 3 - i$ and $z = 2 + 3i$. Draw $w + z$ and $w - z$ on the same plane.

2. Let $w = -4 + 2i$ and $z = 4i$. Draw $w + z$ and $w - z$ on the same plane.

3. Let $w = 5 + 2i$ and $z = -3 + 3i$. Draw $\overline{w + z}$ and $\overline{w} + \overline{z}$ on the same plane.

4. Let $w = 4 - 3i$ and $z = -2 - i$. Draw $\overline{w + z}$ and $\overline{w} + \overline{z}$ on the same plane.

Graph the given complex number.

5. $2(\cos 30° + i \sin 30°)$

6. $\sqrt{2}(\cos 135° + i \sin 135°)$

7. $3(\cos 270° + i \sin 270°)$

8. $4(\cos 300° + i \sin 300°)$

Write each complex number in polar form. If necessary, express angles to the nearest tenth of a degree.

9. $2 - 2i$

10. $-2 + i\sqrt{3}$

11. $3i$

12. -2

13. $-4 - 2i\sqrt{3}$

14. $1 - i$

Express in $x + yi$ form.

15. $12(\cos 150° + i \sin 150°)$

16. $6(\cos 225° + i \sin 225°)$

17. $10(\cos 320° + i \sin 320°)$

18. $20(\cos 115° + i \sin 115°)$

B 19. For each complex number, graph the given number z and the number iz on one set of axes. What is the graphical relationship between z and iz?
 (a) $z = 3 + 4i$ (b) $z = -2 + i$ (c) $z = 5 - 2i$

20. Using each complex number z as in Exercise 19, graph z and $-iz$ on one set of axes. What is the graphical relationship between z and $-iz$?

Exercises 21–26 lead to a proof of the triangle inequality. We denote the real part of the complex number z by Re z and the imaginary part by Im z.

21. Show that Re $z \le |z|$ and Im $z \le |z|$.

22. Show that $z + \overline{z} = 2$ Re z and that $z - \overline{z} = 2i$ Im z.

23. Show that $|\overline{z}| = |z|$ and $|-z| = |z|$.

24. Show that $\overline{w + z} = \overline{w} + \overline{z}$ and that $\overline{wz} = \overline{w}\,\overline{z}$.

25. Show that $w\overline{z} + \overline{w}z \le 2\,|w|\,|z|$ by justifying the following steps.

$$w\overline{z} + \overline{w}z = w\overline{z} + \overline{w\overline{z}} = 2 \text{ Re } w\overline{z}$$
$$\le 2\,|w|\,|\overline{z}| = 2\,|w|\,|z|$$

26. Prove the triangle inequality by justifying the following steps.

$$|w + z|^2 = (w + z)\overline{(w + z)}$$
$$= (w + z)(\overline{w} + \overline{z})$$
$$= w\overline{w} + w\overline{z} + \overline{w}z + z\overline{z}$$
$$= |w|^2 + (w\overline{z} + \overline{w}z) + |z|^2$$
$$\le |w|^2 + 2|w|\,|z| + |z|^2$$

   Therefore, $|w + z|^2 \le (|w| + |z|)^2$.
   Since both $|w + z|$ and $|w| + |z|$ are nonnegative, the last inequality implies that $|w + z| \le |w| + |z|$.

C 27. Under what conditions is it true that $|w + z| = |w| + |z|$?

28. Prove that $|w| - |z| \le |w - z|$. (Hint: Apply the triangle inequality to $|(w - z) + z|$.)

29. Show that the distance between the graphs of any two complex numbers w and z is $|w - z|$.

30. Show that for any complex number z, $|z|^2 + |iz|^2 = |z + iz|^2$. (Hint: Simplify each side of the equation separately by using Exercise 34 of Section 7-4.) Interpret this equation graphically.

Self Quiz

$\boxed{\text{7-4}}$ $\boxed{\text{7-5}}$

1. Write each complex number in polar form and graph the number.
   (a) $4 - 4i\sqrt{3}$ (b) $-3 - 3i$

2. Simplify: (a) i^7 (b) $\dfrac{1}{i^3}$ (c) $i^3 + i^4 + i^5 + i^6$

3. For $z = 1 - i$ and $w = 3 + i$, find:
   (a) $z + w$ (b) zw (c) $\dfrac{z}{w}$ (d) z^3

4. If $z = r(\cos \theta + i \sin \theta)$ and $\overline{z} = s(\cos \alpha + i \sin \alpha)$, express s and α in terms of r and θ.

7-6

The Polar Form of Products and Quotients

Objective: To find products and quotients of complex numbers in polar form.

In this section we shall investigate how polar form can be used to multiply and divide complex numbers.

Let $z = r(\cos \alpha + i \sin \alpha)$ and $w = s(\cos \beta + i \sin \beta)$ be any two complex numbers. The product zw is then given by:

$$zw = r(\cos \alpha + i \sin \alpha) \cdot s(\cos \beta + i \sin \beta)$$
$$= rs(\cos \alpha \cos \beta - \sin \alpha \sin \beta + i \sin \alpha \cos \beta + i \cos \alpha \sin \beta)$$
$$= rs[(\cos \alpha \cos \beta - \sin \alpha \sin \beta) + i(\sin \alpha \cos \beta + \cos \alpha \sin \beta)]$$

Recalling the formulas for $\sin (\alpha + \beta)$ and $\cos (\alpha + \beta)$, we can state that

---

If $z = r(\cos \alpha + i \sin \alpha)$ and $w = s(\cos \beta + i \sin \beta)$, then
$$zw = rs[\cos (\alpha + \beta) + i \sin (\alpha + \beta)].$$

Example 1 Find zw if $z = 3(\cos 30° + i \sin 30°)$ and $w = 2(\cos 150° + i \sin 150°)$.

Solution $zw = (3 \cdot 2)[\cos (30° + 150°) + i \sin (30° + 150°)]$

$\qquad = 6(\cos 180° + i \sin 180°) = 6(-1 + 0 \cdot i) = -6$ ∎

The polar forms of $z = r(\cos \alpha + i \sin \alpha)$ and $w = s(\cos \beta + i \sin \beta)$ can be used to find the graph of zw. To do so, we can use a geometric construction. Let $A(r, \alpha)$ be the graph of z, $B(s, \beta)$ be the graph of w, and U be the point $(1, 0)$ as illustrated in Figure 7-21. Construct the ray OP' that makes an angle $\alpha + \beta$ with the positive x-axis. At B, construct $\angle OBB'$ congruent to $\angle OUA$. Let P be the point where $\overrightarrow{OP'}$ and $\angle OBB'$ intersect. Then $\angle UOP = \alpha + \beta = \arg (zw)$. By duplicating $\angle OUA$ at B, we create similar triangles OUA and OBP. Then

$$\frac{OP}{r} = \frac{s}{1}.$$

Hence, $OP = rs = |zw|$.

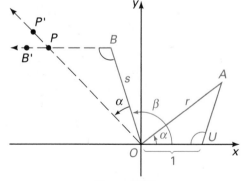

Figure 7-21

The reciprocal of $z = r(\cos \theta + i \sin \theta)$, $z \neq 0$, can be expressed in polar form as follows.

$$\frac{1}{z} = \frac{1}{z} \cdot \frac{\bar{z}}{\bar{z}} = \frac{\bar{z}}{|z|^2} = \frac{r(\cos \theta - i \sin \theta)}{r^2} = \frac{1}{r}(\cos \theta - r \sin \theta)$$

Therefore, since $\cos \theta = \cos (-\theta)$ and $-\sin \theta = \sin (-\theta)$, we have the following.

> If $z = r(\cos \theta + i \sin \theta)$, then $\dfrac{1}{z} = \dfrac{1}{r}[\cos (-\theta) + i \sin (-\theta)]$.

We can now find the polar form of the quotient $\dfrac{z}{w}$ (where $w \neq 0$) by writing it as $z \cdot \dfrac{1}{w}$ and using the reciprocal and product formulas above. The result is the following.

> If $z = r(\cos \alpha + i \sin \alpha)$ and $w = s(\cos \beta + i \sin \beta)$, then
> $$\frac{z}{w} = \frac{r}{s}[\cos (\alpha - \beta) + i \sin (\alpha - \beta)].$$

Example 2 Let $z = 3(\cos 120° + i \sin 120°)$. Find $\dfrac{z}{i}$.

Solution The modulus of i is 1 and its argument is 90°. Hence, in polar form
$$i = 1(\cos 90° + i \sin 90°).$$

Therefore, $\dfrac{z}{i} = \dfrac{3}{1}[\cos (120° - 90°) + i \sin (120° - 90°)]$

$$= 3(\cos 30° + i \sin 30°). \quad \blacksquare$$

Graphically, multiplying or dividing a complex number z by i is the same as rotating the vector representing z through an angle of 90°, counterclockwise in the case of multiplication and clockwise in the case of division (Figure 7-22). (See also Exercises 13–16 on page 257.) Note that

$$|iz| = |i| \, |z| = |z|$$

and

$$\left| \frac{z}{i} \right| = \frac{|z|}{|i|} = \frac{|z|}{1} = |z|.$$

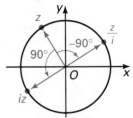

Figure 7-22

EXERCISES 7-6

For each of the following find $z_1 z_2$ in polar form.

A 1. $z_1 = \sqrt{2}(\cos 135° + i \sin 135°)$; $z_2 = 2(\cos 90° + i \sin 90°)$

2. $z_1 = \sqrt{2}(\cos 45° + i \sin 45°)$; $z_2 = \sqrt{2}(\cos 135° + i \sin 135°)$

3. $z_1 = 2(\cos 60° + i \sin 60°)$; $z_2 = 4(\cos 120° + i \sin 120°)$

4. $z_1 = 2\sqrt{3}(\cos 300° + i \sin 300°)$; $z_2 = 2(\cos 30° + i \sin 30°)$

Find the reciprocal of z in (a) polar form and (b) the form x + yi.

5. $z = 2(\cos 120° + i \sin 120°)$ 6. $z = \cos 135° + i \sin 135°$

7. $z = -1 + i$ 8. $z = -\sqrt{3} - i$

9–12. In Exercises 1–4, find $\dfrac{z_1}{z_2}$ in polar form.

Find the polar form of (a) iz and (b) $\dfrac{z}{i}$. Sketch the graphs of these along with the graph of z in the same complex plane.

B 13. $z = 2(\cos 100° + i \sin 100°)$

14. $z = \cos(-40°) + i \sin(-40°)$

15. $z = -2 - 2i$

16. $z = -\sqrt{3} + i$

17. Show that if $z = r(\cos \theta + i \sin \theta)$, then $z^2 = r^2(\cos 2\theta + i \sin 2\theta)$.

18. Show that if $z = r(\cos \theta + i \sin \theta)$, then $z^3 = r^3(\cos 3\theta + i \sin 3\theta)$.

Exercises 19–23: Suppose that $z = r(\cos \theta + i \sin \theta)$ where $r > 1$. Graph each of the following by choosing an arbitrary r and an arbitrary first-quadrant angle θ. Also express each as a complex number in polar form in terms of r and θ.

C 19. $\bar{z}$ 20. z^2 21. $\dfrac{1}{z}$ 22. $\dfrac{1}{\bar{z}}$ 23. $(\bar{z})^2$

24. Let $z = r(\cos \theta + i \sin \theta)$ and let n be an integer. Show that if
$$w = \sqrt[3]{r}\left(\cos \frac{\theta + 2n\pi}{3} + i \sin \frac{\theta + 2n\pi}{3}\right), \text{ then } w^3 = z.$$

25. Let $z = r(\cos \theta + i \sin \theta)$ and $w = s(\cos \phi + i \sin \phi)$, where r and s are nonzero. (a) Under what conditions placed on θ and ϕ will zw be real? (b) Under what conditions placed on θ and ϕ will zw be pure imaginary?

Challenge

We can regard complex numbers as vectors by letting $x\mathbf{i} + y\mathbf{j}$ represent $x + yi$. Let $\mathbf{u} = a\mathbf{i} + b\mathbf{j}$ represent $z = a + bi$ and $\mathbf{v} = c\mathbf{i} + d\mathbf{j}$ represent $w = c + di$. Show that $\mathbf{u}$ and $\mathbf{v}$ are orthogonal if and only if $w = kiz$. (Hint: Begin by solving $w = kiz$ for k.)

7-7

De Moivre's Theorem

Objective: To use De Moivre's theorem to find powers of complex numbers.

In Section 7-6 we derived the formula

$$zw = rs[\cos(\alpha + \beta) + i\sin(\alpha + \beta)].$$

Setting $z = w = r(\cos\theta + i\sin\theta)$, we obtain

$$z \cdot z = r \cdot r[\cos(\theta + \theta) + i\sin(\theta + \theta)]$$
$$z^2 = r^2(\cos 2\theta + i\sin 2\theta)$$

and

$$z^2 \cdot z = r^2(\cos 2\theta + i\sin 2\theta) \cdot r(\cos\theta + i\sin\theta)$$
$$= r^3[\cos(2\theta + \theta) + i\sin(2\theta + \theta)]$$
$$z^3 = r^3(\cos 3\theta + i\sin 3\theta).$$

These results suggest that

$$z^n = r^n(\cos n\theta + i\sin n\theta), \text{ for all positive integers } n.$$

This is indeed true. Furthermore, the formula, called *De Moivre's theorem* holds for *every* integer n.

De Moivre's Theorem

For every integer n, $[r(\cos\theta + i\sin\theta)]^n = r^n(\cos n\theta + i\sin n\theta)$.

When $r = 1$, we can express De Moivre's theorem in the following simplified form:

$$(\cos\theta + i\sin\theta)^n = \cos n\theta + i\sin n\theta$$

Example 1   Express $(-1 + i\sqrt{3})^6$ in the form $x + yi$.

Solution   First express $-1 + i\sqrt{3}$ in polar form.

$$-1 + i\sqrt{3} = 2\left(-\frac{1}{2} + i\frac{\sqrt{3}}{2}\right) = 2(\cos 120° + i \sin 120°)$$

Therefore, $(-1 + i\sqrt{3})^6 = [2(\cos 120° + i \sin 120°)]^6$

$$= 2^6[\cos (6 \cdot 120°) + i \sin (6 \cdot 120°)]$$
$$= 64(\cos 720° + i \sin 720°)$$
$$= 64(\cos 0° + i \sin 0°)$$
$$= 64(1 + i \cdot 0) = 64 \quad \blacksquare$$

De Moivre's theorem can be proved for positive integers n by mathematical induction. To do so, we must first show that the theorem is true for $n = 1$. Then, we must show that if the theorem holds for a given positive integer k, that is,

$$[r(\cos \theta + i \sin \theta)]^k = r^k(\cos k\theta + i \sin k\theta), \qquad (*)$$

then the theorem holds for $k + 1$. That is,

$$[r(\cos \theta + i \sin \theta)]^{k+1} = r^{k+1}[\cos (k + 1)\theta + i \sin (k + 1)\theta].$$

This can be done by rewriting the expression on the left in the preceding equation as

$$[r(\cos \theta + i \sin \theta)]^k \cdot [r(\cos \theta + i \sin \theta)]$$

and using the induction hypothesis (*). You are asked to carry out the details in Exercise 25 on page 261.

In Exercise 22, you are asked to show that De Moivre's theorem holds in the case where $n = 0$. Now suppose n is a *negative* integer. Let $n = -m$, where m is a *positive* integer. Then

$[r(\cos \theta + i \sin \theta)]^n$

$$= [r(\cos \theta + i \sin \theta)]^{-m}$$

$$= \frac{1}{[r(\cos \theta + i \sin \theta)]^m} \qquad \text{by De Moivre's theorem for positive integers}$$

$$= \frac{1}{r^m(\cos m\theta + i \sin m\theta)} \qquad \text{by the reciprocal formula on page 256}$$

$$= r^{-m}[\cos (-m\theta) + i \sin (-m\theta)]$$

$$= r^n(\cos n\theta + i \sin n\theta) \qquad \text{since } n = -m$$

Example 2   Express $(1 + i)^{-5}$ in the form $x + yi$.

Solution   First express $1 + i$ in polar form.

$$1 + i = \sqrt{2}\left(\frac{1}{\sqrt{2}} + i\frac{1}{\sqrt{2}}\right) = \sqrt{2}(\cos 45° + i \sin 45°)$$

(Solution continued next page.)

Then:

$$(1 + i)^{-5} = [\sqrt{2}(\cos 45° + i \sin 45°)]^{-5}$$

$$= (\sqrt{2})^{-5}[\cos (-5 \cdot 45°) + i \sin (-5 \cdot 45°)]$$

$$= \frac{1}{4\sqrt{2}}[\cos (-225°) + i \sin (-225°)]$$

$$= \frac{1}{4\sqrt{2}}\left(-\frac{1}{\sqrt{2}} + i\frac{1}{\sqrt{2}}\right) = \frac{1}{8}(-1 + i) = -\frac{1}{8} + \frac{1}{8}i \quad ∎$$

De Moivre's theorem can be used to derive certain identities; for example, identities which express $\cos n\theta$ and $\sin n\theta$ in terms of functions of θ for any integer n.

Example 3   Use De Moivre's theorem to derive identities for $\cos 3\theta$ and $\sin 3\theta$ in terms of $\cos \theta$ and $\sin \theta$, respectively.

Solution   By De Moivre's theorem, $\cos 3\theta + i \sin 3\theta = (\cos \theta + i \sin \theta)^3$. Expand $(\cos \theta + i \sin \theta)^3$.

$$(\cos \theta + i \sin \theta)^3 = \cos^3 \theta + 3 \cos^2 \theta \cdot i \sin \theta$$
$$+ 3 \cos \theta \cdot i^2 \sin^2 \theta + i^3 \sin^3 \theta$$
$$= (\cos^3 \theta - 3 \cos \theta \sin^2 \theta) + i(3 \cos^2 \theta \sin \theta - \sin^3 \theta)$$

Therefore,

$$\cos 3\theta + i \sin 3\theta = (\cos^3 \theta - 3 \cos \theta \sin^2 \theta) + i(3 \cos^2 \theta \sin \theta - \sin^3 \theta).$$

By the definition of equality of complex numbers (page 247):

$$\cos 3\theta = \cos^3 \theta - 3 \cos \theta \sin^2 \theta \qquad \sin 3\theta = 3 \cos^2 \theta \sin \theta - \sin^3 \theta$$

Using the identity $\sin^2 \theta + \cos^2 \theta = 1$, we can write these formulas as

$$\cos 3\theta = 4 \cos^3 \theta - 3 \cos \theta \qquad \sin 3\theta = 3 \sin \theta - 4 \sin^3 \theta \quad ∎$$

EXERCISES 7-7

Use De Moivre's theorem to express each of the following in the form $x + yi$.

A 1. $[2(\cos 60° + i \sin 60°)]^6$
2. $[2(\cos 30° + i \sin 30°)]^6$

3. $[\cos (-60°) + i \sin (-60°)]^9$
4. $(\cos 36° + i \sin 36°)^{100}$

5. $\left[\frac{1}{2}(\cos 135° + i \sin 135°)\right]^4$
6. $\left[\frac{1}{2}(\cos 45° + i \sin 45°)\right]^7$

7. $(1 + i\sqrt{3})^3$ 8. $(\sqrt{3} - i)^6$ 9. $(1 + i)^{10}$ 10. $(-1 + i)^{12}$

Express each in the form $x + yi$.

11. $[3(\cos 12° + i \sin 12°)]^5 \left[\dfrac{1}{3}(\cos 10° + i \sin 10°)\right]^3$

12. $[5(\cos 25° + i \sin 25°)]^5 \left[\dfrac{2}{3}(\cos 70° + i \sin 70°)\right]^4$

13. $\dfrac{[2(\cos 23° + i \sin 23°)]^6}{[2(\cos 6° + i \sin 6°)]^8}$ 14. $\dfrac{(\cos 25° + i \sin 25°)^5}{(\cos 10° + i \sin 10°)^8}$

15. $\left[\dfrac{1}{2}(1 + i\sqrt{3})\right]^5 \left[\dfrac{1}{2}(\sqrt{3} - i)\right]^4$ 16. $\left[\dfrac{1}{2}(-\sqrt{3} - i)\right]^3 \left[\dfrac{1}{2}(-1 + i\sqrt{3})\right]^4$

17. $\dfrac{(1 + i\sqrt{3})^2}{(-\sqrt{3} + i)^3}$ 18. $\dfrac{(-\sqrt{3} + i)^3}{(1 - i)^6}$

B 19. Use De Moivre's theorem to show that $\cos 2\theta = \cos^2 \theta - \sin^2 \theta$ and $\sin 2\theta = 2 \sin \theta \cos \theta$.

20. Use De Moivre's theorem to derive identities for $\cos 4\theta$ and $\sin 4\theta$ in terms of $\cos \theta$ and $\sin \theta$, respectively.

21. Use De Moivre's theorem to derive identities for $\cos 5\theta$ and $\sin 5\theta$ in terms of $\cos \theta$ and $\sin \theta$, respectively.

22. Define z^0 by $z^0 = 1$. Verify that De Moivre's theorem holds for the case $n = 0$.

C 23. Verify that De Moivre's theorem holds for the case $n = \dfrac{1}{2}$. (Hint: Expand $r\left(\cos \dfrac{1}{2}\theta + i \sin \dfrac{1}{2}\theta\right)^2$ and use double-angle formulas.)

24. Verify that De Moivre's theorem holds for the case $n = \dfrac{1}{3}$. (Hint: Expand $\left(\cos \dfrac{1}{3}\theta + i \sin \dfrac{1}{3}\theta\right)^3$ and use the results of Example 3 on page 260.)

25. Prove De Moivre's theorem for positive integers n by using mathematical induction.

26. Assume that De Moivre's theorem has been proved for all integers m as well as all rational numbers of the form $\dfrac{1}{n}$, where n is a positive integer. Show that De Moivre's theorem holds for any rational number $\dfrac{m}{n}$.

7-8	## Roots of Complex Numbers

Objective: To use De Moivre's theorem to find roots of complex numbers.

The complex number w is called an ***nth root*** of z if $w^n = z$. Example 1 illustrates a method for finding roots.

Example 1   Find the cube roots of -8.

Solution   To find the solutions of $w^3 = -8$, write w and -8 in polar form:

$$w = r(\cos \theta + i \sin \theta) \qquad -8 = 8(\cos 180° + i \sin 180°)$$

Substituting these expressions into $w^3 = -8$, we have

$$[r(\cos \theta + i \sin \theta)]^3 = 8(\cos 180° + i \sin 180°),$$

or, by De Moivre's theorem,

$$r^3(\cos 3\theta + i \sin 3\theta) = 8(\cos 180° + i \sin 180°).$$

Therefore,

$$r^3 = 8 \text{ and } 3\theta = 180° + k \cdot 360°, \text{ where } k \text{ is an integer,}$$

since 3θ and $180°$ must be coterminal if they have equal sines and equal cosines. Solving these equations for r and θ, we have

$$r = 2 \qquad \theta = 60° + k \cdot 120°$$

The three cube roots of -8 are obtained by letting $k = 0, 1,$ and 2.

$$w_1 = 2(\cos 60° + i \sin 60°) = 2\left(\frac{1}{2} + i\frac{\sqrt{3}}{2}\right) = 1 + i\sqrt{3}$$

$$w_2 = 2(\cos 180° + i \sin 180°) = 2(-1 + 0 \cdot i) = -2$$

$$w_3 = 2(\cos 300° + i \sin 300°) = 2\left(\frac{1}{2} - i\frac{\sqrt{3}}{2}\right) = 1 - i\sqrt{3} \quad \blacksquare$$

Note that if, for example, $k = 3$, we get $\theta = 60° + 3 \cdot 120° = 420°$, which is coterminal with $60°$ and would produce w_1 again.

Using the method of Example 1 we can show that every nonzero complex number has exactly n nth roots. Moreover:

The n nth roots of $r(\cos \theta + i \sin \theta)$ are given by

$$r^{\frac{1}{n}}\left(\cos \frac{\theta + k \cdot 360°}{n} + i \sin \frac{\theta + k \cdot 360°}{n}\right),$$

where $k = 0, 1, 2, \ldots, n - 1$.

Example 2   Find the five fifth roots of $-1 + i$ in polar form. Graph the roots along with $-1 + i$.

Solution   In polar form, $-1 + i = \sqrt{2}(\cos 135° + i \sin 135°)$.
Using $r = \sqrt{2}$ and $\theta = 135°$ in the formula on page 262, we have

$$w = (\sqrt{2})^{\frac{1}{5}} \left(\cos \frac{135° + k \cdot 360°}{5} + i \sin \frac{135° + k \cdot 360°}{5} \right)$$

$$= 2^{\frac{1}{10}} [\cos (27° + k \cdot 72°) + i \sin (27° + k \cdot 72°)].$$

Setting k equal, in turn, to 0, 1, 2, 3, and 4, we obtain the five fifth roots of $-1 + i$:

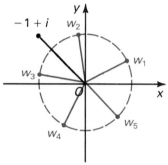

$$w_1 = 2^{\frac{1}{10}} (\cos 27° + i \sin 27°)$$

$$w_2 = 2^{\frac{1}{10}} (\cos 99° + i \sin 99°)$$

$$w_3 = 2^{\frac{1}{10}} (\cos 171° + i \sin 171°)$$

$$w_4 = 2^{\frac{1}{10}} (\cos 243° + i \sin 243°)$$

$$w_5 = 2^{\frac{1}{10}} (\cos 315° + i \sin 315°)$$

The figure shows $-1 + i$ and its five fifth roots.   ∎

Notice that in Example 2 the roots are equally spaced around the circle of radius $2^{\frac{1}{10}} \approx 1.07$, centered at O. In general, the nth roots of $r(\cos \theta + i \sin \theta)$ are equally spaced around the circle of radius $r^{\frac{1}{n}}$ centered at O.

The nth roots of the complex number 1, called the nth roots of unity, are easily found. Since

$$1 = 1(\cos 0° + i \sin 0°),$$

the n nth roots of unity are given by

$$\cos \frac{k \cdot 360°}{n} + i \sin \frac{k \cdot 360°}{n}, \, k = 0, 1, \dots, n - 1,$$

or, if radian measure is used, by

$$\cos \frac{2k\pi}{n} + i \sin \frac{2k\pi}{n}, k = 0, 1, \dots, n - 1.$$

It is easy to prove that if $w = \cos \dfrac{2\pi}{n} + i \sin \dfrac{2\pi}{n}$, then the n nth roots of unity are $w, w^2, \dots, w^{n-1}$, and $w^n = 1$.

Example 3   Find and graph the sixth roots of unity, labeling each as a power of one of them.

Solution   The sixth roots of unity are given by

$$\cos\frac{2k\pi}{6} + i\sin\frac{2k\pi}{6} = \cos\frac{k\pi}{3} + i\sin\frac{k\pi}{3}, \text{ for } k = 0, 1, 2, 3, 4, 5.$$

Hence the roots are:

$$\cos 0 + i\sin 0 = 1$$

$$\cos\frac{\pi}{3} + i\sin\frac{\pi}{3} = \frac{1}{2} + i\frac{\sqrt{3}}{2}$$

$$\cos\frac{2\pi}{3} + i\sin\frac{2\pi}{3} = -\frac{1}{2} + i\frac{\sqrt{3}}{2}$$

$$\cos\pi + i\sin\pi = -1$$

$$\cos\frac{4\pi}{3} + i\sin\frac{4\pi}{3} = -\frac{1}{2} - i\frac{\sqrt{3}}{2}$$

$$\cos\frac{5\pi}{3} + i\sin\frac{5\pi}{3} = \frac{1}{2} - i\frac{\sqrt{3}}{2}$$

In the diagram, $w = \cos\dfrac{\pi}{3} + i\sin\dfrac{\pi}{3} = \dfrac{1}{2} + i\dfrac{\sqrt{3}}{2}.$ ∎

EXERCISES 7-8

In Exercises 1–4, find and graph the indicated roots of unity in the form $x + yi$, labeling each root as a power of one of them as in Example 3.

A   1. The cube roots of unity

2. The fourth roots of unity

3. The fifth roots of unity

4. The eighth roots of unity

Find the indicated roots in the form specified, and sketch their graphs along with that of the given number.

5. The cube roots of i in the form $x + yi$

6. The fourth roots of -1 in the form $x + yi$

7. The fifth roots of $-1 - i$ in polar form

8. The tenth roots of $-\sqrt{3} - i$ in polar form

9. The sixth roots of $-\sqrt{3} + i$ in polar form

10. The fifth roots of $-i$ in polar form

11. The fourth roots of $-2 + 2i\sqrt{3}$ in the form $x + yi$

12. The cube roots of $8i$ in the form $x + yi$

13. The cube roots of $-2 + 2i$ in polar form

14. The fifth roots of $2 - (2\sqrt{3})i$ in polar form

B 15. Without substituting, explain why the cube roots of unity other than 1 satisfy $z^2 + z + 1 = 0$. (Hint: $z^3 - 1 = (z - 1)(z^2 + z + 1)$.)

16. Without substituting, explain why the sixth roots of unity other than 1 satisfy $z^5 + z^4 + z^3 + z^2 + z + 1 = 0$. (Hint: $z^6 - 1 = (z - 1)(?)$.)

17. Generalize Exercises 15 and 16.

18. Show that if k is an integer between 0 and $\dfrac{n}{2}$, then

$$\cos \frac{2(n - k)\pi}{n} = \cos \frac{2k\pi}{n} \text{ and } \sin \frac{2(n - k)\pi}{n} = -\sin \frac{2k\pi}{n}.$$

Explain why this implies that the imaginary nth roots of unity occur in conjugate pairs.

19. Find the roots of the quadratic equation $z^2 + 2z + 1 + 2i = 0$ in the form $x + yi$. (Hint: Use the quadratic formula and find the necessary square roots by the method of this section.)

20. Find the roots of the quadratic equation $z^2 + 2iz - 1 - 8i = 0$ in the form $x + yi$.

An nth root of unity, w, is said to be **primitive** if every nth root of unity is an integral power of w.

C 21. Show that $\cos \dfrac{2\pi}{n} + i \sin \dfrac{2\pi}{n}$ is a primitive nth root of unity.

22. Let $w = \cos \dfrac{2\pi}{8} + i \sin \dfrac{2\pi}{8} = \cos \dfrac{\pi}{4} + i \sin \dfrac{\pi}{4}$. By Exercise 21,

$$w, w^2, w^3, \ldots, w^7, \text{ and } w^8 = 1$$

are all the eighth roots of unity. (a) Show that w^3 is primitive. (Consider $(w^3)^2 = w^6$, $(w^3)^3 = w^9 = w$, etc.) (b) Show that w^2 is *not* primitive. (c) Which of the eighth roots of unity are primitive and which are not?

23. Find the primitive twelfth roots of unity.

24. Explain why, if n is not a prime number, at least one nth root of unity besides 1 is not primitive.

Self Quiz

7-6   7-7   7-8

1. If $z = 2(\cos 30° + i \sin 30°)$, give two values of $n > 0$ such that z^n is a real number.

2. Find the fourth roots of -16 in both polar form and the form $x + yi$.

3. Use De Moivre's theorem to find $(1 - i\sqrt{3})^4$ in the form $x + yi$.

4. If $z = 6(\cos 45° + i \sin 45°)$ and $w = 2(\cos 75° + i \sin 75°)$, express zw in both polar form and the form $x + yi$.

5. For z and w as in Exercise 4, express $\dfrac{z}{w}$ in both polar form and the form $x + yi$.

6. Find the fifth roots of $-16\sqrt{3} + 16i$ in polar form.

ADDITIONAL PROBLEMS

Exercises 1–3: (a) find a rectangular equation for the given figure. (b) Transform your new equation to polar form.

1. The circle with center at the origin and radius 5

2. The line through the origin with slope $\sqrt{3}$

3. The circle with center at $(1, 0)$ and passing through the origin

Graph each polar equation.

4. $r = -\sin \theta$ 5. $r = 4 \sin \theta + 4 \cos \theta$

6. If replacing r by $-r$ *and* θ by $-\theta$ leaves a polar equation unchanged, describe the symmetry of the equation's graph.

7. (*Optional*) Determine the type of the conic given by the polar equation $r = \dfrac{3}{2 - \cos \theta}$ and sketch its graph.

8. Express the reciprocal of $3 - 4i$ in the form $x + yi$.

9. (a) Write $(1 + i)$ in polar form. (b) Show that $(1 + i)^n$ is a real number if and only if n is evenly divisible by 4. (c) For what n will $(1 + i\sqrt{3})^n$ be a real number?

10. Graph $r = 3 \cos \theta$.

11. Find the three cube roots of unity in both polar and $x + yi$ form. Verify in both forms that the product of the three roots is 1.

12. Let $z = 1 + i$. Find z^{-1}, z^0, z^1, z^2, and z^3 in polar form and plot these in the complex plane. (These points lie on a curve called a *logarithmic spiral*.)

13. Repeat Exercise 12 for $z = \sqrt{3} + i$.

14. Let $z = 1 + i\sqrt{3}$ and $w = \sqrt{2} + i\sqrt{2}$. (a) Find zw in $x + yi$ form and polar form. (b) Use part (a) to find $\cos 105°$ and $\sin 105°$.

15. The numbers z and $\dfrac{1}{z}$ are called *reflections* of each other in the unit circle. Verify that a complex number and its reflection in the unit circle have the same argument and that their moduli are reciprocals of each other.

16. Find $i^{-2} + i^{-1} + i^0 + i + i^2$.

17. Let $z = x + yi$. (a) Describe the set of points z in the complex plane such that z^2 has real part 9. (b) Describe the set of points z in the complex plane such that z^2 has imaginary part 6.

18. Show that the product of the nth roots of unity is 1 if n is odd, and -1 if n is even. (Hint: the nth roots of unity have arguments
$$1 \cdot \frac{360°}{n}, 2 \cdot \frac{360°}{n}, 3 \cdot \frac{360°}{n}, \ldots, \text{ and } n \cdot \frac{360°}{n}. \text{ Use the fact that}$$
$$1 + 2 + 3 + \cdots + n = \frac{n(n + 1)}{2}.)$$

19. Let $z = r(\cos \alpha + i \sin \alpha)$ and $w = s(\cos \beta + i \sin \beta)$. Use the conjugate of w to show that $\dfrac{z}{w} = \dfrac{r}{s}[\cos (\alpha - \beta) + i \sin (\alpha - \beta)]$.

20. Find the reciprocal of $\sqrt{3} + i$ in polar form.

CHAPTER SUMMARY

1. Using a polar coordinate system, we can locate a point in a plane by specifying its distance r from the origin and its rotation θ from the polar axis, which coincides with the positive x-axis when a rectangular xy-coordinate system is also introduced into the plane. Conversion formulas that give x and y in terms of r and θ, and vice versa, enable us to change between these coordinate systems.

2. In order to graph a polar equation containing a trigonometric function, it is usually helpful to set up a table which shows intervals for the variable θ that correspond to a quarter of the trigonometric function's period.

3. Any conic section can be defined as the set of points satisfying an equation of the form $PF/PD = e$, where e is a constant and PF and PD are the distances from a variable point P to a fixed point F (the focus) and to a fixed line (the directrix, which contains D), respectively. The conic is an ellipse, parabola, or hyperbola if e is less than, equal to, or greater than 1, respectively.

4. A complex number is a number of the form $a + bi$, where a and b are real numbers and i is the imaginary unit. The set of all complex numbers governed by ordinary addition and multiplication of binomials, with the stipulation that $i^2 = -1$, satisfies the same arithmetic axioms, called the field axioms, as the set of real numbers.

5. The conjugate of a complex number $a + bi$ is the number $a - bi$. The modulus or absolute value of a complex number $a + bi$ is the (nonnegative) quantity $\sqrt{a^2 + b^2}$.

6. Complex numbers can be represented graphically as points in a plane (the Argand or complex plane) by associating the point whose rectangular coordinates are (x, y) with the complex number $x + yi$.

7. Addition of complex numbers is then the same as vector addition of the corresponding vectors, and the modulus of a complex number is the same as the norm of the corresponding vector.

8. Multiplication of complex numbers is most conveniently carried out in polar form. To multiply two complex numbers, we can then multiply their moduli and add their arguments.

9. De Moivre's Theorem says that to find an integral power of a complex number, we raise the modulus of the number by the power.

10. A complex number w is an nth root of z if $w^n = z$. Each complex number has exactly n distinct complex nth roots.
If $z = r(\cos \theta + i \sin \theta)$, the nth roots of z are given by

$$r^{\frac{1}{n}}\left(\cos \frac{\theta + k \cdot 360°}{n} + i \sin \frac{\theta + k \cdot 360°}{n}\right), k = 0, 1, \ldots, n - 1.$$

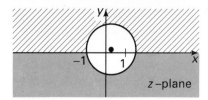
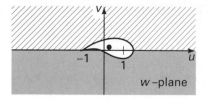

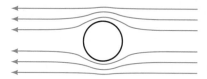

The diagrams below illustrate the flow of air over a circular cylinder and the Joukowski airfoil.

In modern aerodynamic engineering, where testing physical models has become extremely expensive, the first model (produced with the aid of mathematical calculation) must have as near a chance of success as possible.

EXERCISES

Find the image of each complex number under the mapping $w = \frac{1}{2}\left(z + \frac{1}{z}\right)$.

1. 1 2. i 3. $3 - 4i$ 4. $3 + 4i$

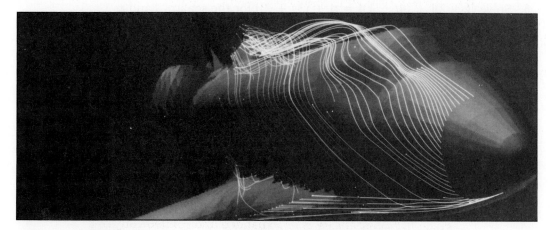

NASA researchers analyze the flow of air around the space shuttle by using supercomputers to translate numerical air-flow data into three-dimensional graphics.

CUMULATIVE REVIEW Chapters 6–7

Chapter 6

1. Using vectors **u** and **v** as shown, sketch:

   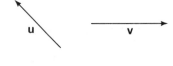

   (a) $\dfrac{3}{2}$**u** – 2**v**

   (b) 3**u** + **v**.

2. A pilot wishes to fly a plane at 500 km/h on a bearing of 280°. However, there is a 40 km/h wind, blowing from 65°. What should be the heading and speed of the plane?

3. A 150 kg crate is resting on a ramp inclined at an angle of 25° with the horizontal. The crate is held in place by a rope parallel to the ramp as shown in the diagram. Find the tension in the rope.

   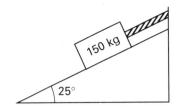

4. Let **a** = 3**i** + 2**j** and **b** = **i** + **j**. Sketch **c** = **a** – 2**b**.

5. Find the norm of **d** = 6**i** – 8**j**.

6. Solve $(s + 2)\mathbf{i} + (4t - 1)\mathbf{j} = t\mathbf{i} + (3s + 2)\mathbf{j}$ for the scalars s and t.

7. Find unit vectors (a) parallel to and (b) perpendicular to 6**i** – 2**j**.

8. Resolve **v** = 3**i** – 2**j** into components parallel to **a** = **i** + 4**j** and to **b** = 4**i** – **j**.

9. Find the angle θ between **c** = **i** + 5**j** and **d** = 2**i** – 3**j**.

10. Use a specific example to test the following for a scalar s and vectors **a** and **b**. State which is true.

    (1) $s(\mathbf{a} \cdot \mathbf{b}) = (s\mathbf{a}) \cdot (s\mathbf{b})$
    (2) $s(\mathbf{a} \cdot \mathbf{b}) = (s\mathbf{a}) \cdot \mathbf{b}$

11. Suppose the crate in Exercise 3 is moved 10 m up the ramp. Ignoring friction, how much energy is expended?

12. A wagon is pulled with a force of 40 N by a handle making an angle of 40° with the horizontal. Neglecting friction, how much work is done if the wagon is pulled 125 m?

Chapter 7

1. Convert $(-7, 300°)$ to rectangular coordinates.

2. Convert $(-3, -3)$ to polar coordinates.

3. Transform
$$r = \frac{\sec \theta}{\tan \theta - 1}$$
   to rectangular form. Identify and sketch the graph.

4. Graph $r = |1 + \theta|$ for $-\pi \le \theta \le \pi$.

5. *(Optional)* Identify and sketch the graph of $r = \dfrac{2}{1 + \cos \theta}$.

6. Let $z = 6 - 6i$ and $w = 4i$.
   (a) Find $z + w$.
   (b) Find $z \cdot w$.

7. Let $z = -\dfrac{5}{13} + \dfrac{12}{13}i$.
   (a) Find the conjugate of z.
   (b) Find the modulus of z.
   (c) Find the reciprocal of z.

Graph the given complex number.

8. $2(\cos 60° + i \sin 60°)$

9. $4(\cos 330° + i \sin 330°)$

10. If $w = 4 - 3i$ and $z = 3 + i$, find $\dfrac{w}{z}$
    (a) in rectangular form.
    (b) in polar form.

Use De Moivre's theorem to express each of the following in $x + yi$ form.

11. $(2 + i)^6$

12. $\left(-\sqrt{3} + i\right)^6$

13. Find the fourth roots of i.

14. Find the roots of the quadratic equation $z^2 - 6z + 9 - 4i = 0$.

MIXED REVIEW: Chapters 1–7

1. Evaluate: (a) $\mathrm{Cos}^{-1}\left(\dfrac{\sqrt{3}}{2}\right)$ (b) $\mathrm{Tan}^{-1}\left(\sin\left(-\dfrac{3\pi}{4}\right)\right)$

2. Give the domain and range of $y = \mathrm{Cos}^{-1} 2x$ and graph the function.

3. An object suspended from a spring is pulled 4 cm below its equilibrium point at time $t = 0$ and released. If the object completes an oscillation in 3 s, find an equation using cosine for the position of the object. Give the frequency of the oscillation.

4. How much energy in kilowatt-hours is needed to lift a 1500 kg girder 50 m?

5. Express $\left(-\sqrt{2} - \left(\sqrt{2}\right)i\right)^8$ in $x + yi$ form.

6. $\triangle XYZ$ has area 21, $\angle Y = 150°$, and $x = 7$. Find z.

7. Find the degree-minute-second measure of the supplement of an angle of $12°5'14''$.

8. A force of 8 N making a 240° angle measured counterclockwise from the horizontal and a force of 6 N making a 310° angle measured the same way act on P. Find the magnitude (to the nearest tenth) and the angle (to the nearest tenth of a degree) that will keep P stationary.

9. A triangle has sides of lengths 4, 7, and 10. Find the length of the median to the shortest side.

10. Find to the nearest cm/s the linear speed of a point on the rim of a record of radius 15 cm turning at an angular speed of $33\frac{1}{3}$ rpm.

11. Prove a distributive property for dot products: $(\mathbf{b} + \mathbf{c}) \cdot \mathbf{a} = \mathbf{b} \cdot \mathbf{a} + \mathbf{c} \cdot \mathbf{a}$.

12. Show that for all real numbers if $f(x) = \sin x - \cos x$, then $f(x + \pi) = -f(x)$.

13. Convert $184.5°$ to radian measure in terms of π.

14. Solve $\triangle RST$ if $\angle R = 57°40'$, $\angle S = 38°30'$, and $t = 50$. Give lengths to the nearest tenth.

15. Find the value of x in the diagram to the nearest tenth.

16. State the domain of the function $\mathrm{Tan}^{-1}(\tan)$. Give a counterexample to show that $\mathrm{Tan}^{-1}(\tan x) = x$ is not true for all x in the domain.

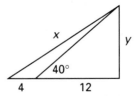

17. Use a half-angle formula to evaluate $\cos 165°$.

18. (Optional) Determine what type of conic has the polar equation $r = \dfrac{2}{1 + \sin \theta}$. Sketch the graph of the conic.

19. Sketch the graph of $r = 6 \sin \theta - 8 \cos \theta$.

20. Find unit vectors (a) parallel to and (b) perpendicular to the vector $-3\mathbf{i} + \mathbf{j}$.

21. Find (a) the conjugate, (b) the modulus, and (c) the reciprocal of $2\sqrt{2} + i$.

22. Each leg of an isosceles triangle is 250 cm long and the vertex angle measures 24°. Find the base and the altitude to the base.

23. Write $-6 - 3i$ in polar form. Express the argument θ to the nearest tenth of a degree.

24. Graph $y = 2 \sin x + \sin 2x$ for $0 \le x \le 2\pi$.

25. Rectangle $ABCD$ is inscribed in a circle. If $DC = 10$ and $\angle DAC = 27°$, find the perimeter of the rectangle to the nearest unit.

26. Prove: $\dfrac{1 - \tan \theta}{1 + \tan \theta} = \dfrac{\cot \theta - 1}{\cot \theta + 1}$

27. Solve for the scalars s and t: $(-2s\mathbf{i} + 3t\mathbf{j}) + (4\mathbf{i} - 2\mathbf{j}) = 3\mathbf{j}$.

28. Find a polar equation for $y = -3x$.

29. If $z = \cos(-100°) + i \sin(-100°)$, find the polar forms of (a) iz and (b) $\dfrac{z}{i}$. Then graph z, iz, and $\dfrac{z}{i}$ in the same complex plane.

30. At what heading and air speed should a plane fly if a wind of 30 km/h is blowing from 300° and the pilot wants to maintain a true course of due south and a ground speed of 400 km/h?

31. Suppose that a point P moves counterclockwise a distance $\dfrac{7\pi}{4}$ on a circle of radius $\dfrac{1}{2}$ with center at origin. If P starts at the point $\left(\dfrac{1}{2}, 0\right)$, what are the coordinates of its final position $P(x, y)$?

32. Rewrite $y = -2 \cos x - 2\sqrt{3} \sin x$ in the form $y = C \cos (x - \phi)$ and graph the function.

33. Find sine, cosine, and tangent of 479°30′ to four significant digits.

34. Use the quadratic formula and De Moivre's theorem to find the roots of $z^2 - 4z + (4 - i) = 0$.

35. Express $\dfrac{\tan^2 \theta + 1}{\tan^2 \theta}$ in terms of a single trigonometric function.

36. Find the general solution: $\cos \theta + \sec \theta = -\dfrac{5}{2}$.

37. If $\csc \theta = \dfrac{29}{20}$ and $90° < \theta < 180°$, find the exact values of the other five trigonometric functions of θ.

38. Solve $\triangle ABC$ if $\angle C = 42°$, $c = 100$, and $a = 120$. If no triangle can be formed, so state. If there is more than one solution, find each one.

39. Prove that $\csc \left(\dfrac{3\pi}{2} - x \right) = -\sec x$.

40. Solve for $0° \le \theta < 360°$ to the nearest degree: $-\sqrt{3} \cos \theta - \sin \theta = \sqrt{2}$.

41. Use the law of sines to show that in $\triangle ABC$,
$$\dfrac{a + b}{b} = \dfrac{\sin A + \sin B}{\sin B}.$$

42. Use the vectors **a** and **b** at the right. Draw a diagram to show that $\mathbf{a} + (\mathbf{b} - \mathbf{a}) = \mathbf{b}$.

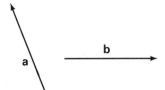

43. Let $\mathbf{a} = a_1\mathbf{i} + a_2\mathbf{j}$ and $\mathbf{b} = b_1\mathbf{i} + b_2\mathbf{j}$. Verify that $\mathbf{a} + (\mathbf{b} - \mathbf{a}) = \mathbf{b}$.

44. If $z_1 = 4(\cos 180° + i \sin 180°)$ and $z_2 = \sqrt{2}(\cos 225° + i \sin 225°)$, find (a) z_1z_2 and (b) $\dfrac{z_1}{z_2}$ in polar form and in the form $x + yi$.

45. Graph $y = -2 \sin \dfrac{\pi}{3}x$ for $0 \le x \le 2\pi$.

46. Express $\dfrac{\tan^2 x}{1 - \cos x}$ in terms of $\sec x$.

47. Find the cube roots of $-i$ in the form $x + yi$.

48. Find the lengths of the diagonals of a parallelogram with sides of lengths 4 and 6 and an angle of measure 128°.

49. Use angle-sum formulas to find the exact values of $\sin 255°$, $\cos 255°$, and $\tan 255°$ in simplest radical form.

50. Find the work done by the force $\mathbf{F} = 7\mathbf{i} - \mathbf{j}$ in moving an object from $A(-7, -3)$ to $B(-4, -5)$.

51. Find all possible values for $\angle R$ in $\triangle RST$ if $\angle S = 120°$, $r = 8$, and $s = 2\sqrt{3}$.

52. Name two different pairs of polar coordinates for point $(-4, 4)$.

53. Describe how to obtain the graph of $y = 1 - \cos(x + \pi)$ from the graph of $y = \cos x$.

54. Solve for $0° \le \theta < 360°$: $3 \sin^2 \theta + \cos \theta = 1$.

55. Graph the polar equation $r = 3 \sin 3\theta$ (three-leaved rose).

56. Find the area to the nearest tenth of $\triangle ABC$ with $a = 20$, $\angle B = 119°20'$, and $\angle C = 9°10'$.

57. Simplify: (a) $\dfrac{1}{i^{203}}$ (b) $\dfrac{1}{i^{4n+2}}$ for a negative integer n.

58. Refer to Exercise 3. Express the equation in terms of the sine function.

59. Express in the form $x + yi$: $\dfrac{[2(\cos 28° + i \sin 28°)]^8}{4(\cos 11° + i \sin 11°)}$

60. A boat sails due east for 80 km and then due northeast for 80 km more. Find the distance and bearing of the boat from its starting point.

61. Prove: $2 \csc 2\theta = \sec \theta \csc \theta$

62. Graph: $y = \cos x - |\cos x|$ for $0 \le x \le 2\pi$.

63. Use a trigonometric identity to find the general solution for $\dfrac{\sin 2x}{1 + \cos 2x} = -1$.

64. If $w = -7 + 2i$ and $z = 3 - 2i$, find (a) $\overline{w + z}$, (b) $\overline{w - z}$, and (c) $\overline{w} - \overline{z}$.

8 Infinite Series

The time-lapse photograph shows the representation of a continuous motion as a sequence of discrete images over time. One of the central notions of this chapter is that of a sequence.

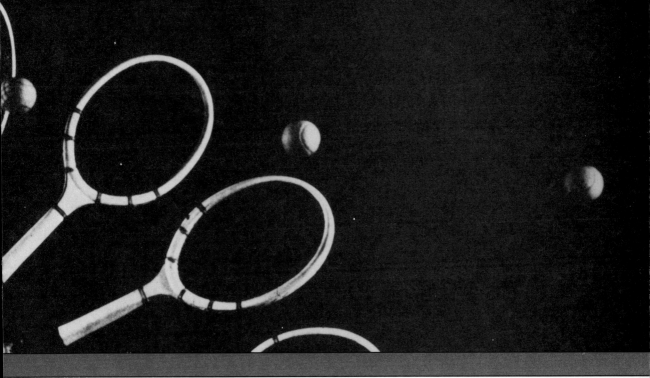

SEQUENCES AND SERIES

Finite Sequences and Series

Objective: To define and use finite sequences and series.

A **finite sequence** is a finite ordered set. We can therefore specify a sequence by listing its **terms** in order. For example,

$$\frac{3}{4}, \frac{3}{2}, 3, 6, 12, 24, 48, 96$$

defines a sequence of 8 terms. To investigate sequences, it is convenient to label the terms with subscripts. For example, in the sequence above, $a_1 = \frac{3}{4}$ and $a_4 = 6$. Note that the subscript of a term specifies its position in the sequence.

In the sequence above, each term after the first is twice as large as the preceding term. We can write this as

$$a_1 = \frac{3}{4} \text{ and } a_n = 2a_{n-1} \text{ for } n = 2, 3, 4, \ldots, 8.$$

A definition in which the first term or terms of a sequence are specified explicitly and the general term, a_n, is given as an expression involving preceding terms is called a **recursive definition**.

Using the recursive definition above, we can also write

$$a_1 = \frac{3}{4}$$

$$a_2 = 2a_1 = 2 \cdot \frac{3}{4}$$

$$a_3 = 2a_2 = 2 \cdot \left(2 \cdot \frac{3}{4}\right) = 2^2 \cdot \frac{3}{4}$$

$$a_4 = 2a_3 = 2 \cdot \left(2^2 \cdot \frac{3}{4}\right) = 2^3 \cdot \frac{3}{4}.$$

In general,

$$a_n = \frac{3}{4} \cdot 2^{n-1} \text{ for } n = 1, 2, 3, \ldots, 8.$$

A definition in which the general term, a_n, is given as a function of n is called an **explicit definition**.

Example 1

Write the first 4 terms of the sequence defined recursively by

$$a_1 = 1 \text{ and } a_n = a_{n-1} + 2(n - 1) + 1$$

and give an explicit definition of this sequence.

Solution

$$a_1 = 1$$
$$a_2 = a_1 + 2 \cdot 1 + 1 = 1 + 2 + 1 = 4$$
$$a_3 = a_2 + 2 \cdot 2 + 1 = 4 + 4 + 1 = 9$$
$$a_4 = a_3 + 2 \cdot 3 + 1 = 9 + 6 + 1 = 16$$

From the list above, $a_2 = 2^2, a_3 = 3^2, a_4 = 4^2$.
Therefore, an explicit definition of this sequence is $a_n = n^2$. ∎

A **finite series** is any expression of the form

$$a_1 + a_2 + a_3 + \cdots + a_n.$$

We call a_n the **nth term** of the series. The **sum** of the series is simply the value of the expression. We can define the **kth partial sum**, S_k, as the sum of the first k terms. A recursive definition of S_k is given by

$$S_1 = a_1 \text{ and } S_k = S_{k-1} + a_k \text{ for } k = 2, 3, \ldots, n.$$

Example 2

Find the first four partial sums of the series

$$\frac{1}{1 \cdot 2} + \frac{1}{2 \cdot 3} + \frac{1}{3 \cdot 4} + \frac{1}{4 \cdot 5} + \cdots + \frac{1}{n(n + 1)}.$$

Solution

$$S_1 = \frac{1}{1 \cdot 2} = \frac{1}{2} \qquad\qquad\qquad S_2 = \frac{1}{2} + \frac{1}{2 \cdot 3} = \frac{2}{3}$$

$$S_3 = \frac{2}{3} + \frac{1}{3 \cdot 4} = \frac{3}{4} \qquad\qquad S_4 = \frac{3}{4} + \frac{1}{4 \cdot 5} = \frac{4}{5} \quad\blacksquare$$

A useful way of writing series is called **sigma notation**. The Greek letter Σ ("sigma") denotes summation. An **index** (k in the example below) is used to indicate the range of summation. The index starts at the number below Σ, increases by increments of 1, and ends at the number above Σ. The values of the index are substituted into the expression following Σ. The following illustrates how this notation is used.

$$\sum_{k=1}^{5} k^2 \text{ means } 1^2 + 2^2 + 3^2 + 4^2 + 5^2$$

A sequence that can be defined recursively by $a_1 = a$ where a is non-zero and $a_{n+1} = ra_n$, for some nonzero constant r is called a **geometric sequence** with **common ratio** r. Notice that $r = \dfrac{a_{n+1}}{a_n}$. The sequence

$$\frac{3}{4}, \frac{3}{2}, 3, 6, 12, 24, 48, 96,$$

already considered, is geometric with $a = \dfrac{3}{4}$ and $r = 2$. We saw that $a_n = \dfrac{3}{4}(2)^{n-1}$. In general:

The nth term, a_n, of the geometric sequence with first term a and common ratio r is given by $a_n = a \cdot r^{n-1}$.

Example 3   Find the 5th term of the geometric sequence whose first term is 6 and whose common ratio is $-\sqrt{3}$.

Solution   We have $a = 6, r = -\sqrt{3}$, and $n = 5$. Therefore, by the formula above,

$$a_5 = 6(\sqrt{3})^4 = 6 \cdot 9 = 54. \quad\blacksquare$$

A series is called a **geometric series** if its terms form a geometric sequence. The partial sums of a geometric series can be found by the following formula, a proof of which is outlined in Exercise 41.

The nth partial sum, S_n, of the geometric series with first term a and common ratio r is given by $S_n = \dfrac{a(1 - r^n)}{1 - r}$.

Example 4   Find the sum of the first 6 terms of the geometric series:

$$10 + 5 + \frac{5}{2} + \frac{5}{4} + \cdots$$

Solution   First note that $a = 10$, $r = \frac{1}{2}$, and $n = 6$. According to the formula above,

$$S_6 = \frac{10\left[1 - \left(\frac{1}{2}\right)^6\right]}{1 - \frac{1}{2}} = \frac{10\left(1 - \frac{1}{64}\right)}{\frac{1}{2}} = 10 \cdot \frac{63}{64} \cdot 2 = \frac{315}{16}. \quad \blacksquare$$

EXERCISES 8-1

Write the first five terms of the sequence defined by each rule.

A 1. $a_1 = 18$; $a_n = \frac{1}{3}a_{n-1}$

2. $a_1 = \frac{1}{40}$; $a_n = -2a_{n-1}$

3. $a_1 = -7$; $a_n = a_{n-1} + 3$

4. $a_1 = 5$; $a_n = -a_{n-1}$

5. $a_1 = 1$; $a_n = 1 - a_{n-1}$

6. $a_1 = 1$; $a_n = na_{n-1}$

7. $a_1 = 1$; $a_2 = 1$; $a_n = a_{n-1} + a_{n-2}$

8. $a_1 = \frac{1}{2}$; $a_n = a_{n-1} + \frac{1}{2^n}$

9. $a_n = 5 \cdot 2^n$

10. $a_n = n^3 + 1$

11. $a_n = 3^{n-2}$

12. $a_n = 3 \cdot \left(-\frac{1}{2}\right)^n$

Find the nth term of each geometric sequence.

13. $a = 5$; $r = -2$; $n = 6$

14. $a = 2$; $r = \sqrt{3}$; $n = 7$

15. $a = 6$; $r = \frac{1}{2}$; $n = 7$

16. $a = 54$; $r = -\frac{1}{3}$; $n = 5$

17. The ninth term of the geometric sequence: $3, -3\sqrt{2}, 6, -6\sqrt{2}, \ldots$

18. The tenth term of the geometric sequence: $2x, -2x^2, 2x^3, -2x^4, \ldots$

Write out explicitly the terms of each series given in sigma notation.

19. $\displaystyle\sum_{k=1}^{5} \frac{1}{(k+1)^2}$

20. $\displaystyle\sum_{k=0}^{4} \sin\left(\frac{k\pi}{6}\right)$

21. $\displaystyle\sum_{k=1}^{5} \frac{k+1}{k^2}$ **22.** $\displaystyle\sum_{k=1}^{6} \frac{(-1)^k}{k}$

Find the sum of the geometric series with the given constants.

23. $a = -4; r = 2; n = 5$

24. $a = 48; r = \dfrac{1}{2}; n = 6$

25. $a = 81; r = 0.1; n = 6$

26. $a = 55; r = -0.1; n = 7$

27. $1 - 2 + 4 - 8 + \cdots$; first 8 terms

28. $162 - 54 + 18 - 6 + \cdots$; first 6 terms

29. $2 + 2 \cdot 10^{-1} + 2 \cdot 10^{-2} + \cdots$; first 5 terms

B 30. Show that $\dfrac{1}{2} + \dfrac{1}{4} + \cdots + \dfrac{1}{2^n} = 1 - \dfrac{1}{2^n} = \dfrac{2^n - 1}{2^n}$.

Find an explicit rule for the general, or *n*th, term of each given sequence.

31. The sequence in Exercise 1

32. The sequence in Exercise 2

33. The sequence in Exercise 3

34. The sequence in Exercise 4 (Hint: $(-1)^n = \pm 1$.)

35. The sequence in Exercise 5 (Hint: See Exercise 34.)

36. The sequence in Exercise 6

37. Find the first term of a geometric sequence whose common ratio is 2 and whose sixth term is -10.

38. Find the first term of a geometric series whose sum of the first four terms is 5 and whose common ratio is $-\dfrac{1}{2}$.

39. Find the common ratio of a geometric sequence whose first term is 12 and whose fifth term is $\dfrac{4}{3}$.

40. Find the sum of the geometric series $250 + 25 + \dfrac{5}{2} + \cdots + \dfrac{1}{400}$.

C 41. (a) Simplify the expression $(1 - r)(r^{n-1} + \cdots + r^2 + r + 1)$.

(b) Use the result of (a) to prove the formula for the sum of a geometric series.

42. (a) Show that
$$\frac{1}{n(n + 1)} = \frac{1}{n} - \frac{1}{n + 1}.$$

(b) Use the result of (a) to show that the kth partial sum of the series in Example 2 is $\dfrac{k}{(k + 1)}.$

(This is an example of what is called a "telescoping series," because its terms "collapse" like the sections of an old-fashioned telescope.)

8-2	**Infinite Sequences and Series**

Objective: To find limits of infinite sequences and sums of infinite series.

The sequence
$$\frac{1}{2}, \frac{2}{3}, \frac{3}{4}, \frac{4}{5}, \ldots, \frac{n}{n + 1}, \ldots$$
is an example of an infinite sequence. In general, an **infinite sequence** is a function whose domain is the set of positive integers and whose range is a subset of real numbers. This function assigns to each integer n the corresponding term a_n.

The terms of the sequence above clearly get closer and closer to 1 as n increases. We say the limit of the sequence as n approaches infinity is 1, or that the sequence **converges** to 1 as n approaches infinity. We write
$$\lim_{n \to \infty} \frac{n}{n + 1} = 1.$$

In general, if the terms a_n of a sequence eventually become and remain arbitrarily close to some real number L as n becomes larger and larger, we say that L is the **limit** of the sequence and write
$$\lim_{n \to \infty} a_n = L.$$

Not every sequence has a limit. For example, the sequence
$$-1, 1, -1, 1, \ldots, (-1)^n, \ldots$$
has terms that clearly do not get closer and closer to any unique real number. We say that such a sequence **diverges**.

Example 1   Find each limit, if it exists. (a) $\lim\limits_{n\to\infty} \dfrac{3n^2 + n + 1000}{n^2}$ (b) $\lim\limits_{n\to\infty} n^2$

Solution

(a) Since $\dfrac{1}{n}$ gets closer and closer to 0 as n increases, $\dfrac{1}{n^2}$ has limit 0 also.

Since $\dfrac{3n^2 + n + 1000}{n^2} = \dfrac{3n^2}{n^2} + \dfrac{n}{n^2} + \dfrac{1000}{n^2} = 3 + \dfrac{1}{n} + \dfrac{1000}{n^2}$,

$\lim\limits_{n\to\infty} \dfrac{3n^2 + n + 1000}{n^2} = \lim\limits_{n\to\infty} \left(3 + \dfrac{1}{n} + \dfrac{1000}{n^2}\right) = 3 + 0 + 0 = 3.$

(b) As n increases, n^2 grows without bound. Therefore, this sequence diverges. ∎

An **infinite series** is simply a series with infinitely many terms. We can use the idea of the limit of a sequence to give a definition of what we mean by the *sum* of an infinite series. If the sequence of partial sums $S_1, S_2, S_3, \ldots, S_n, \ldots$ of an infinite series has a limit, we call that limit the **sum** of the series and say that the series **converges**. Otherwise, we say the series **diverges**.

Using the definition of the sum of an infinite series, we can find the sum of any infinite geometric series whose common ratio r is such that $|r| < 1$. Let a be the first term of the series. Then, according to the formula on page 283,

$$\lim\limits_{n\to\infty} S_n = \lim\limits_{n\to\infty} \dfrac{a(1 - r^n)}{1 - r} = \lim\limits_{n\to\infty} \left(\dfrac{a}{1 - r} - r^n \dfrac{1}{1 - r}\right).$$

Since $|r| < 1$, r^n gets closer and closer to 0 as n increases. Therefore,

$$\lim\limits_{n\to\infty} S_n = \dfrac{a}{1 - r}.$$

We have shown that:

> The sum S of the infinite geometric series with first term a and common ratio r, $|r| < 1$, is given by $S = \dfrac{a}{1 - r}.$

Example 2

(a) Find the sum of the infinite geometric series:
$$0.3 + 0.03 + 0.003 + \cdots$$

(b) Find all x such that the infinite geometric series
$$2 + 2x^2 + 2x^4 + \cdots$$
converges to 4.

(Solution on next page.)

Solution

(a) In this series, $a = 0.3$ and $r = 0.1$. According to the formula above,

$$S = \frac{a}{1 - r} = \frac{0.3}{1 - 0.1} = \frac{0.3}{0.9} = \frac{1}{3}.$$

$$\left(\text{This is a confirmation of the fact that } 0.333 \ldots = \frac{1}{3}. \right)$$

(b) With $a = 2$ and $r = x^2$, the series can be written as $\frac{2}{1 - x^2}$.

Then, $4 = \frac{2}{1 - x^2}$, or $4x^2 - 2 = 0$.

From $4x^2 - 2 = 0$, $x = \pm \frac{\sqrt{2}}{2}$. ■

Example 3   Find the sum of the infinite series $\frac{1}{1 \cdot 2} + \frac{1}{2 \cdot 3} + \frac{1}{3 \cdot 4} + \cdots$.

Solution

The given series is not a geometric one. The general term a_n of this series is given by

$$a_n = \frac{1}{n(n + 1)} = \frac{1}{n} - \frac{1}{n + 1}.$$

(See Exercise 42 on page 286.) Therefore,

$$S_n = \left(1 - \frac{1}{2} \right) + \left(\frac{1}{2} - \frac{1}{3} \right) + \left(\frac{1}{3} - \frac{1}{4} \right) + \cdots + \left(\frac{1}{n} - \frac{1}{n + 1} \right)$$

$$= 1 - \frac{1}{n + 1}$$

$$= \frac{n}{n + 1}$$

since all the other terms cancel in pairs. Therefore, since

$$\lim_{n \to \infty} S_n = \lim_{n \to \infty} \frac{n}{n + 1} = 1,$$

the given series has sum 1. ■

EXERCISES 8-2

Find the limit of each sequence or state that the sequence has no limit.

A 1. $\lim\limits_{n \to \infty} \left(1 - \frac{1}{n} \right)$

2. $\lim\limits_{n \to \infty} \left(\frac{1}{2n + 1} \right)$

3. $\lim\limits_{n \to \infty} \left(\frac{3n - 1}{n} \right)$

4. $\lim\limits_{n \to \infty} \cos \left(\frac{1}{n} \right)$

5. $\lim\limits_{n \to \infty} \sin \left(\frac{n\pi}{2} \right)$

6. $\lim\limits_{n \to \infty} \tan \left(\frac{\pi}{3} - \frac{1}{n} \right)$

7. $\lim\limits_{n\to\infty} \sec\left(\pi + \dfrac{1}{n}\right)$

8. $\lim\limits_{n\to\infty} \operatorname{Sin}^{-1}\left(1 - \dfrac{1}{n}\right)$

9. $\lim\limits_{n\to\infty} \operatorname{Tan}^{-1}\left(1 - \dfrac{1}{n}\right)$

10. $\lim\limits_{n\to\infty}\left(\dfrac{(-1)^n}{\cos\left(\dfrac{1}{n}\right)}\right)$

11. $\lim\limits_{n\to\infty} \dfrac{3n^2}{n^2 + 1}$

12. $\lim\limits_{n\to\infty} \dfrac{n + 1000}{n^2}$

Find the sum of each infinite geometric series.

13. $12 + 6 + 3 + \cdots$

14. $54 - 18 + 6 - 2 + \cdots$

15. $\dfrac{100}{3} + \dfrac{10}{3} + \dfrac{1}{3} + \cdots$

16. $12 + \dfrac{12}{5} + \dfrac{12}{25} + \cdots$

17. $\dfrac{1}{3} - \dfrac{1}{9} + \dfrac{1}{27} - \cdots$

18. $\sqrt{2} + 1 + \dfrac{\sqrt{2}}{2} + \cdots$

19. $\sin^2 x + \sin^4 x + \sin^6 x + \cdots$ for $x \neq \dfrac{\pi}{2} + n\pi$

20. $\tan^2 x - \tan^4 x + \tan^6 x - \cdots$ for $\dfrac{-\pi}{4} < x < \dfrac{\pi}{4}$

Represent each repeating decimal as a fraction, using the method of Example 2(a). (Hint: Rewrite the decimal as an infinite geometric series.)

21. $0.777\ldots$

22. $0.090909\ldots$

23. $0.181818\ldots$

24. $0.135135\ldots$

Find the first four terms of the infinite geometric series with the given first term and given sum.

B 25. $a = 5;\ S = 6$

26. $a = \dfrac{5}{6};\ S = \dfrac{1}{2}$

27. If the infinite geometric series $3x + 3x^3 + 3x^5 + \cdots$ converges to 2, find x.

28. If the infinite geometric series $1 - x + x^2 - x^3 + \cdots$ converges to x, find x.

29. (a) Show that $\dfrac{1}{n^2} - \dfrac{1}{(n + 1)^2} = \dfrac{2n + 1}{n^2(n + 1)^2}$.

(b) Find the sum of the infinite series $\dfrac{3}{4} + \dfrac{5}{36} + \dfrac{7}{144} + \cdots + \dfrac{2n + 1}{n^2(n + 1)^2} + \cdots$.

30. Find the sum of the infinite series
$$\dfrac{1}{1 \cdot 3} + \dfrac{1}{3 \cdot 5} + \dfrac{1}{5 \cdot 7} + \cdots + \dfrac{1}{(2n - 1)(2n + 1)} + \cdots.$$
(Hint: Use the method of Exercise 29.)

C 31. It can be shown that $4 - \dfrac{4}{3} + \dfrac{4}{5} - \dfrac{4}{7} + \cdots$ converges to π. Write a computer program to evaluate the partial sums of this series, and find S_{200} and S_{500}. Compare these with the value of $\pi = 3.1415926$ .... Would you say this series converges "rapidly"?

32. It can be shown that the infinite series whose general term is $\dfrac{1}{n^2}$ converges to $\dfrac{\pi^2}{6}$. Write a computer program to find the partial sums of this series, and find S_{200}, S_{500}, and S_{1000}. Compare these with the value of $\dfrac{\pi^2}{6}$, found from a calculator.

SPECIAL INFINITE SERIES

| 8-3 | **Exponential and Logarithmic Functions** |

Objective: To define and use exponential and logarithmic functions.

A function defined by $f(x) = b^x$, where b is a positive constant other than 1, is called an **exponential function**. The domain is the set of all real numbers and the range is the set of all positive real numbers. As we shall see later in this chapter, there is a close connection between trigonometric functions and exponential functions. This connection, which can be seen through the vehicle of complex numbers, was first discovered by the great Swiss mathematician Leonhard Euler (1707–1783).

Recall the following properties of exponents.

Let a and b be positive real numbers other than 1. Let x and y be real numbers.

1. $b^x b^y = b^{x+y}$

2. $\dfrac{b^x}{b^y} = b^{x-y}$

3. $(b^x)^y = b^{xy}$

4. $a^x b^x = (ab)^x$

5. $b^x = b^y$ if and only if $x = y$.

Property (5) says that b^x is one-to-one and hence, has an inverse, a **logarithmic function** denoted by $\log_b x$. Its domain is all positive real numbers and its range is all real numbers. Since b^x and $\log_b x$ are inverses:

$\log_b x = y$ if and only if $b^y = x$

$b^{\log_b x} = x$ and $\log_b b^y = y$

Figure 8-1 illustrates the graph of $y = 2^x$ together with that of $y = \log_2 x$. The graphs are reflections of one another in the line $y = x$.

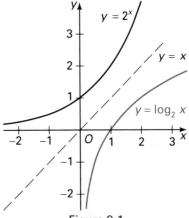

Figure 8-1

We can prove the following laws of logarithms directly from the properties of exponents.

Let b be a positive real number other than 1. Let M and N be positive numbers.

1. $\log_b MN = \log_b M + \log_b N$

2. $\log_b \dfrac{M}{N} = \log_b M - \log_b N$

3. $\log_b M^x = x \log_b M$

4. $\log_b M = \log_b N$ if and only if $M = N$.

We shall prove (1). Proofs of (2) and (3) are requested in Exercises 33 and 34, respectively. Law (4) is just a restatement of property (5) of exponents with $M = b^x$ and $N = b^y$.

Example 1   Prove that $\log_b MN = \log_b M + \log_b N$.

Solution   Let $x = \log_b M$ and $y = \log_b N$.
Then $M = b^x$ and $N = b^y$.

$$\log_b MN = \log_b (b^x b^y)$$
$$= \log_b b^{x+y} \qquad \text{property (1) of exponents}$$
$$= x + y \qquad \log_b x \text{ and } b^x \text{ are inverses}$$
$$= \log_b M + \log_b N \qquad \blacksquare$$

Example 2   Solve each equation.
(a) $\log_3 x = -2$
(b) $\log_{10} x + \log_{10} (x + 2) = \log_{10} 3$.

Solution   (a) By definition, $\log_3 x = -2$ means that $x = 3^{-2} = \dfrac{1}{9}$.

(b) $\log_{10} x + \log_{10} (x + 2) = \log_{10} 3$

$$\log_{10} x(x + 2) = \log_{10} 3 \qquad \text{property (1) of logarithms}$$
$$x(x + 2) = 3 \qquad \text{property (4) of logarithms}$$
$$x^2 + 2x - 3 = 0$$

Hence, $x = -3$ or 1. Reject -3 since $x > 0$. Thus, $x = 1$. ∎

Until the introduction of electronic calculators and computers, logarithms served as an important computational tool and base 10 was convenient. A great deal of scientific and mathematical work, however, makes use of another base, the irrational number e, (named in honor of Euler). The number e is defined as the limit of a sequence.

$$e = \lim_{n \to \infty} \left(1 + \frac{1}{n}\right)^n$$

It is not obvious that the sequence converges, but computation of $\left(1 + \dfrac{1}{n}\right)^n$ for large values of n suggests that it does. The decimal 2.71828 gives a five-place approximation to e.

It can be shown that if x is any real number, then

$$e^x = \lim_{n \to \infty} \left(1 + \frac{x}{n}\right)^n.$$

The function e^x is called the **natural exponential function**. This function can be represented by an infinite series. Using the binomial theorem, we have

$$\left(1 + \frac{x}{n}\right)^n = 1^n + n(1)^{n-1}\left(\frac{x}{n}\right) + \frac{n(n-1)}{2}(1)^{n-2}\left(\frac{x}{n}\right)^2 +$$

$$\frac{n(n-1)(n-2)}{2 \cdot 3}(1)^{n-3}\left(\frac{x}{n}\right)^3 + \cdots + \left(\frac{x}{n}\right)^n$$

$$= 1 + \frac{n}{n}x + \frac{n(n-1)}{n^2}\frac{x^2}{2} + \frac{n(n-1)(n-2)}{n^3}\frac{x^3}{2 \cdot 3} + \cdots + \left(\frac{x}{n}\right)^n.$$

The limit of each rational expression in n as n approaches ∞ is 1. This suggests the important formula:

$$e^x = 1 + x + \frac{x^2}{2!} + \frac{x^3}{3!} + \cdots + \frac{x^n}{n!} + \cdots$$

Recall that $n! = n(n - 1)(n - 2) \cdots 2 \cdot 1$.

Although the foregoing is by no means a proof, it can be shown that the series represents e^x for all real x. Evaluation of a few terms of the series gives a very good approximation to e^x.

The natural exponential function has an inverse called the **natural logarithmic function**, denoted $\ln x$, and read "l, n of x." Many calculators and computers have the natural exponential and logarithmic functions built into them. In BASIC, they are denoted by EXP(X) and LOG(X), respectively. You can find $\log_b x$ by using the so-called change-of-base formula, which we now derive.

Let $y = \log_b x$. Then $x = b^y$. Hence,

$$\log_a x = \log_a b^y = y \log_a b.$$

Solving this equation for y, we obtain $\log_b x = \dfrac{\log_a x}{\log_a b}$.

Example 3

(a) Approximate $\sqrt{e}$ to three decimal places by using the first 5 terms of the series e^x.

(b) Find $\log_2 15$ to three decimal places.

Solution

(a) Since $\sqrt{e} = e^{\frac{1}{2}}$, substitute $\frac{1}{2}$ for x in the series.

$$\sqrt{e} \approx 1 + \frac{1}{2} + \frac{\left(\frac{1}{2}\right)^2}{2!} + \frac{\left(\frac{1}{2}\right)^3}{3!} + \frac{\left(\frac{1}{2}\right)^4}{4!}$$

$$\sqrt{e} \approx 1 + \frac{1}{2} + \frac{\left(\frac{1}{2}\right)^2}{2} + \frac{\left(\frac{1}{2}\right)^3}{6} + \frac{\left(\frac{1}{2}\right)^4}{24}$$

$$\sqrt{e} \approx 1 + 0.5 + 0.125 + 0.02083 + 0.00260$$

$$= 1.648$$

(b) Using ln and a calculator or computer, we have

$$\log_2 15 = \frac{\ln 15}{\ln 2}$$

$$= \frac{2.7081}{0.6931} = 3.907. \quad \blacksquare$$

EXERCISES 8-3

Evaluate each expression.

A 1. $\ln \sqrt{e}$

2. $\ln \left(\dfrac{1}{e^4}\right)$

3. $\ln (e^5 \cdot e^2)$

4. $\ln 1$

5. $e^{\ln 5}$

6. e^0

7. $(e^{\ln 8})^2$

8. $e^{-\ln 3}$

Use the laws of logarithms to write as a single logarithm. Use a calculator or Table 5 on page 408 to approximate result to four decimal places.

9. $\log_{10} 5 + 3 \log_{10} 4$

10. $3(\ln 4 + \ln 5)$

11. $2 \log_3 12 - 3 \log_3 2$

12. $\dfrac{1}{2}(\log_5 25 - \log_5 16)$

13. $\dfrac{\ln 12 + \ln 18}{3}$

14. $2 \ln 10 - \dfrac{1}{2} \ln 25$

Solve for x. Leave answers in terms of e, if appropriate.

15. $e^{x+1} = \sqrt{e}$

16. $e^{2x} \cdot e^3 = e^{x-5}$

17. $e^{2x+3} = 1$

18. $(e^4)^x = e^{6-x}$

19. $\ln (x + 1) = -1$

20. $2 \ln x = \ln 25$

21. $\ln x + \ln (x - 1) = \ln 6$

22. $\ln 10 - \ln x = \ln (x - 3)$

B 23. $\ln e^9 - x^2 = \ln 1$

24. $\ln x + \ln (2x - 1) = 0$

Use a calculator and the first six terms of the series for e^x to approximate each.

25. e

26. $\dfrac{1}{e}$

27. $\sqrt[3]{e}$

28. $\sqrt[5]{e}$

In Exercises 29–32, (a) express each as a power of e, (b) approximate each by using $n = 10$, and (c) approximate each to three decimal places by using the first six terms of the series for e^x.

29. $\displaystyle\lim_{n\to\infty} \left(1 + \dfrac{2}{n}\right)^n$

30. $\displaystyle\lim_{n\to\infty} \left(1 - \dfrac{1}{n}\right)^n$

31. $\displaystyle\lim_{n\to\infty} \left(1 + \dfrac{1}{2n}\right)^n$

32. $\displaystyle\lim_{n\to\infty} \left(1 + \dfrac{1}{n}\right)^{3n}$ (Hint: Let $u = 3n$.)

33. Prove property (2) of logarithms.

34. Prove property (3) of logarithms.

Self Quiz

8-1   8-2   8-3

1. Find the 10th term of the geometric sequence: $40, -20, 10, -5, \ldots$

2. Find S_6 for the geometric series with $a = \dfrac{1}{24}$ and $r = -2$.

3. Evaluate: (a) $e^{\ln 2 + \ln 3}$ (b) $\displaystyle\lim_{n\to\infty} \left(1 + \dfrac{1}{3n}\right)^n$

4. Find $\lim\limits_{n \to \infty} \sin\left(\dfrac{3\pi}{2} + \dfrac{1}{n}\right)$.

5. Find the sum of the infinite geometric series $20 + 4 + 0.8 + \cdots$

6. Convert $0.21212121\ldots$ to a fraction by regarding the decimal as the sum of an infinite geometric series.

7. Write the first five partial sums of the series with
$$a_n = \frac{1}{n} - \frac{1}{n+1}$$
and find an explicit formula for the nth partial sum S_n.

8. Solve for x: $2 \ln (x + 1) = 3 \ln 4$

8-4   Power Series

Objective: To use a power series to represent a function.

Evaluation of a polynomial function is particularly simple. Given a value of x, we need only perform arithmetic operations involving constants and x to evaluate the function. This knowledge prompted mathematicians of the early eighteenth century to seek methods involving polynomials by which functions like $\sin x$, $\cos x$, and e^x could be evaluated. They found that such functions can be represented by polynomials of "infinite degree."

A **power series** is an expression of the form
$$a_0 + a_1x + a_2x^2 + a_3x^3 + \cdots + a_nx^n + \cdots$$
where each a_n is a constant.

Example 1   Find a power series representation of the function $\dfrac{1}{1-x}$.

Solution   The given function fits the formula for the sum of an infinite geometric series with $a = 1$ and $r = x$. Therefore, when $|x| < 1$,
$$\frac{1}{1-x} = 1 + x + x^2 + x^3 + \cdots + x^n + \cdots.$$

For $|x| < 1$, we can evaluate $\dfrac{1}{1-x}$ to any desired accuracy by evaluating sufficiently many terms of the series. ∎

The series in Example 1 converges only when $-1 < x < 1$. When $x = 1$, the series diverges and the rational function is not defined. When, for example, $x = 2$, the power series diverges and $\dfrac{1}{1-x} = -1$. This kind of behavior is typical. In fact, the following can be proved in advanced mathematics courses.

Given a power series, exactly one of the following is true.
1. The series converges for all real x.
2. There is a positive real number r (called the **radius of convergence**) such that the series converges whenever $|x| < r$ and diverges whenever $|x| > r$.
3. The series converges only when $x = 0$.

Note that the theorem makes no mention of either convergence or divergence when $|x| = r$. In fact, the series may converge or diverge when $x = r$ and $x = -r$.

If a power series represents a function on an interval, then the series is called a **power series expansion** of the function on the interval. Such an expansion is also called a **Taylor series**. We also say that the series **converges to the function**. Methods for finding Taylor series involve calculus. In this section, we shall examine some well-known Taylor series without deriving them.

For the function $\sin x$, the Taylor series is given by:

$$\sin x = x - \frac{x^3}{3!} + \frac{x^5}{5!} - \frac{x^7}{7!} + \cdots \qquad \text{all real } x$$

Figure 8-2, produced by a computer, shows the graphs of the first few partial sums of the Taylor series for the function $\sin x$.

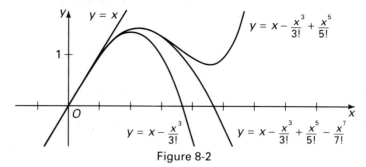

Figure 8-2

The Taylor series for $\cos x$ is given by:

$$\cos x = 1 - \frac{x^2}{2!} + \frac{x^4}{4!} - \frac{x^6}{6!} + \cdots \qquad \text{all real } x$$

Example 2 Approximate $\cos 1.2$ by using the first four terms of the Taylor series.

Solution
$$\cos 1.2 = 1 - \frac{1.2^2}{2!} + \frac{1.2^4}{4!} - \frac{1.2^6}{6!} + \cdots$$

$$\approx 1 - \frac{1.44}{2} + \frac{2.0736}{24} - \frac{2.9860}{720} + \cdots$$

$$\approx 0.362 \quad \blacksquare$$

In Section 8-3, we saw (informally) that the exponential function e^x has a power series expansion given by:

$$e^x = 1 + x + \frac{x^2}{2!} + \frac{x^3}{3!} + \frac{x^4}{4!} + \cdots \qquad \text{all real } x$$

Note that the Taylor series for e^x is made up of the terms of the series for $\sin x$ and $\cos x$, except for sign. This suggests that a relationship exists among these three functions. We shall discover this relationship in Section 8-5.

EXERCISES 8-4

Find a power series expansion for the given function by regarding the function as the sum of an infinite geometric series.

A 1. $\dfrac{1}{1 - x^2}$

2. $\dfrac{3}{1 - 2x}$

3. $\dfrac{2}{1 + x}$ (Hint: $1 + x = 1 - (-x)$.)

4. $\dfrac{5}{1 + 2x^3}$

5. $\dfrac{8}{2 - x}$ (Hint: Divide numerator and denominator by 2.)

6. $\dfrac{6}{3 + x^2}$

7. Use the first four terms of the Taylor series for sin x to approximate each of the following.

   (a) $\sin \dfrac{\pi}{2} \approx \sin 1.57$

   (b) $\sin \dfrac{\pi}{6} \approx \sin 0.52$

   (c) Compare the values you obtained in (a) and (b) with the known values of $\sin \dfrac{\pi}{2}$ and $\sin \dfrac{\pi}{6}$.

8. Use the first four terms of the Taylor series for cos x to approximate each of the following.

   (a) $\cos \dfrac{\pi}{2} \approx \cos 1.57$

   (b) $\cos \dfrac{\pi}{3} \approx \cos 1.05$

   (c) Compare the values you obtained in (a) and (b) with the known values.

9. Use the first four terms of the Taylor series for sin x to approximate each of the following.
   (a) $\sin 2$
   (b) $\sin 1.14$
   (c) Compare the answers to (a) and (b). Give a reason for their relationship.

10. Use the first four terms of the Taylor series for sin x to approximate each of the following.
    (a) $\sin 1.5$
    (b) $\sin 1.64$
    (c) Compare the answers to (a) and (b). Give a reason for their relationship.

B 11. Use the power series expansion of sin x to show that sin x is an odd function: that is, $\sin(-x) = -\sin x$, for all x.

12. Use the power series expansion of cos x to show that cos x is an even function: that is, $\cos(-x) = \cos x$, for all x.

Use the Taylor series given in this section to find a Taylor series for each function.

13. $\sin(x^2)$

14. $\sqrt{e^x}$

15. e^{-x}

16. $\cos \sqrt{x}$

17. e^{x^2}

18. $\dfrac{\sin x}{x}$

8-5 Euler's Formula

Objective: To define and use complex exponents.

In many mathematics books, power series are often used to define functions. For example, e^x could be defined on the set of real numbers by its power series representation, since the power series converges for all real x. The great Swiss mathematician Leonhard Euler was the first to use this idea to define what is meant by a complex exponent.

If we formally substitute ix (where x is a real number and $i = \sqrt{-1}$) for x in the Taylor series for e^x, we get

$$e^{ix} = 1 + ix + \frac{(ix)^2}{2!} + \frac{(ix)^3}{3!} + \frac{(ix)^4}{4!} + \frac{(ix)^5}{5!} + \frac{(ix)^6}{6!} + \cdots$$

$$= 1 + ix - \frac{x^2}{2!} - i\frac{x^3}{3!} + \frac{x^4}{4!} + i\frac{x^5}{5!} - \frac{x^6}{6!} - i\frac{x^7}{7!} + \cdots$$

$$= \left(1 - \frac{x^2}{2!} + \frac{x^4}{4!} - \frac{x^6}{6!} + \cdots\right) + i\left(x - \frac{x^3}{3!} + \frac{x^5}{5!} - \frac{x^7}{7!} + \cdots\right)$$

The two Taylor series in parentheses in the last step are exactly those of $\cos x$ and $\sin x$, respectively. Thus, it seems reasonable to define e^{ix} by

$$e^{ix} = \cos x + i \sin x$$

This equation is known as **Euler's formula**. Note that when $x = \pi$,

$$e^{i\pi} = -1.$$

Example 1 Find: (a) $e^{\frac{i\pi}{2}}$ (b) $e^{\frac{-i\pi}{6}}$

Solution (a) $e^{\frac{i\pi}{2}} = \cos\frac{\pi}{2} + i \sin\frac{\pi}{2} = 0 + i(1) = i$

(b) $e^{\frac{-i\pi}{6}} = \cos\left(-\frac{\pi}{6}\right) + i \sin\left(-\frac{\pi}{6}\right) = \frac{\sqrt{3}}{2} + i\left(-\frac{1}{2}\right) = \frac{\sqrt{3}}{2} - \frac{1}{2}i$ ∎

Now let $z = x + yi$. We define e^z by:

$$e^z = e^{x+yi} = e^x(\cos y + i \sin y)$$

Example 2 Express $e^{2+\frac{i\pi}{3}}$ in $x + yi$ form.

Solution $e^{2+\frac{i\pi}{3}} = e^2\left(\cos\frac{\pi}{3} + i \sin\frac{\pi}{3}\right) = e^2\left(\frac{1}{2} + i\frac{\sqrt{3}}{2}\right) = \frac{e^2}{2} + i\left(\frac{e^2\sqrt{3}}{2}\right)$ ∎

Example 3   Use Euler's formula to show that $e^{ai+bi} = e^{ai} \cdot e^{bi}$.

Solution

$$
\begin{aligned}
e^{ai+bi} &= e^{(a+b)i} \\
&= \cos(a+b) + i\sin(a+b) \\
&= \cos a \cos b - \sin a \sin b + i(\sin a \cos b + \cos a \sin b) \\
&= \cos a(\cos b + i\sin b) + i\sin a(i\sin b + \cos b) \\
&= (\cos a + i\sin a)(\cos b + i\sin b) \\
&= e^{ai} \cdot e^{bi} \quad \blacksquare
\end{aligned}
$$

In Exercise 19, you are asked to show that

$$e^{(a+bi) + (c+di)} = e^{a+bi} \cdot e^{c+di}.$$

This will completely confirm property (1) of exponents. The other properties of exponents (except property (5) on page 290) also apply to complex exponents. It is *not* true that $e^{ai} = e^{bi}$ implies $ai = bi$. For example,

$$e^{2\pi i} = \cos 2\pi + i \sin 2\pi = 1 + 0i = \cos 0 + i \sin 0 = e^{0i}.$$

However, $2\pi i \neq 0i$.

It is worth noting that, since

$$e^{x+yi} = e^x (\cos y + i \sin y),$$

e^{x+yi} is the complex number whose modulus is e^x and whose argument is y:

$$|e^{x+yi}| = e^x \qquad \arg(e^{x+yi}) = y$$

Figure 8-3 illustrates the graph of e^{x+yi} in the complex plane.

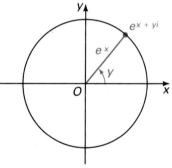

Figure 8-3

EXERCISES 8-5

Express each number in $x + yi$ form.

A 1. $e^{\frac{i\pi}{3}}$

2. $e^{-i\pi}$

3. $e^{\frac{-3i\pi}{4}}$

4. $e^{\frac{7i\pi}{2}}$

5. $e^{2+i\pi}$

6. $e^{-3+\frac{i\pi}{2}}$

7. $e^{-2-i\pi}$

8. $e^{\frac{1}{2}-\frac{i\pi}{3}}$

9. $e^{\ln 3 - \frac{i\pi}{4}}$

10. $e^{-\ln 2 + \frac{2i\pi}{3}}$

11. $e^{-\ln 3 + \frac{\pi i}{2}}$

12. $e^{\ln 4 - i\pi}$

In Exercises 13–16, let $z = e^{a+bi}$. Write what you know about a and/or b under the given conditions.

13. z is real and positive.   14. z is pure imaginary.

15. z is on the circle of radius 5 centered at the origin.

16. z is inside the circle of radius $\sqrt{e}$ centered at the origin.

17. Show that $e^{a+bi} \cdot e^{a-bi}$ is real using
    (a) the properties of exponents on page 290 only.
    (b) the definition of complex exponents only.

18. Show that $e^{a+bi} + e^{a-bi}$ is real using the definition of complex exponents. How are e^{a+bi} and e^{a-bi} related?

B 19. Let $z = a + bi$ and $w = c + di$. Show that $e^{z+w} = e^z \cdot e^w$.

20. Let $z = a + bi$ and $w = c + di$. Show that $e^{z-w} = \dfrac{e^z}{e^w}$.

21. In Section 8-4, we found that

$$\frac{1}{1-x} = 1 + x + x^2 + x^3 + x^4 + \cdots, \qquad |x| < 1.$$

(a) By substituting ix for x in this series, as was done in the text for e^x, find a series for the real part and a series for the imaginary part of $\dfrac{1}{(1-ix)}$.

(b) Use the formula for the sum of an infinite geometric series to simplify the real and imaginary parts you found in (a).

(c) Verify that the function whose real and imaginary parts you found in (b) is equal to $\dfrac{1}{(1-ix)}$.

22. Repeat Exercise 21 using the function

$$\frac{1}{1+x} = 1 - x + x^2 - x^3 + x^4 - \cdots, \qquad |x| < 1,$$

in place of $\dfrac{1}{1-x}$ everywhere.

23. Verify that $\cos x = \dfrac{1}{2}(e^{ix} + e^{-ix})$ and $\sin x = \dfrac{1}{2i}(e^{ix} - e^{-ix})$.

C 24. (a) By subtracting the Taylor series for e^{-x} from the series for e^x, find a Taylor series for the function $\dfrac{1}{2}(e^x - e^{-x})$.

(b) By substituting ix for x in the Taylor series for $\sin x$ given on page 296, show that $\sin (ix) = \dfrac{i}{2}(e^x - e^{-x})$.

| 8-6 | ## Hyperbolic Functions |

Objective: To define and investigate hyperbolic functions.

There are two important functions whose relationship to the unit hyperbola, $x^2 - y^2 = 1$, is analogous to the relationship between the circular functions and the unit circle. These functions are called the *hyperbolic functions.*

The **hyperbolic cosine** function, denoted cosh x (the name rhymes with "Gosh!"), is defined by

$$\cosh x = \frac{1}{2}(e^x + e^{-x}).$$

The **hyperbolic sine** function, denoted sinh x, (pronounced "cinch"), is defined by

$$\sinh x = \frac{1}{2}(e^x - e^{-x}).$$

Hyperbolic functions are used in physics and engineering. For example, a perfectly flexible, inextensible chain or cable suspended from two points hangs naturally in a curve, called a *catenary*, with equation

$$y = k \cosh\left(\frac{x}{k}\right),$$

for some positive real number k.

Using the definitions of cosh x and sinh x, we can show that $\cosh^2 t - \sinh^2 t = 1$.

$$\cosh^2 t - \sinh^2 t = \frac{1}{4}(e^t + e^{-t})^2 - \frac{1}{4}(e^t - e^{-t})^2$$

$$= \frac{1}{4}[(e^{2t} + 2e^t \cdot e^{-t} + e^{-2t}) - (e^{2t} - 2e^t \cdot e^{-t} + e^{-2t})]$$

$$= \frac{1}{4}[(e^{2t} + 2 + e^{-2t}) - (e^{2t} - 2 + e^{-2t})]$$

$$= \frac{1}{4}(4) = 1$$

Therefore,

$$\boxed{\cosh^2 t - \sinh^2 t = 1}$$

Thus, for a given value of t, if $x = \cosh t$ and $y = \sinh t$, the point (x, y) is on the hyperbola $x^2 - y^2 = 1$. Note the similarity between this and the fact that if $x = \cos t$ and $y = \sin t$, the point (x, y) is on the circle $x^2 + y^2 = 1$, and we have the corresponding identity $\cos^2 t + \sin^2 t = 1$. The parameter t can be realized geometrically, in both cases, as twice the area of the shaded regions in Figure 8-4.

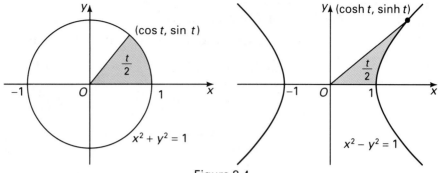

Figure 8-4

Unlike the circular functions, the hyperbolic functions do not have real period. They have period $2\pi i$. The graphs of cosh x and sinh x are shown in Figure 8-5.

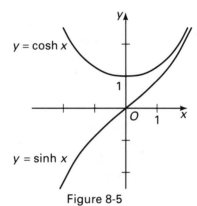

Figure 8-5

Using the definitions of the hyperbolic functions, we can derive their Taylor series directly from the Taylor series for e^x.

Example 1 Find the Taylor series for cosh x.

Solution First find the Taylor series for e^{-x} by substituting $-x$ for x in the Taylor series

$$e^x = 1 + x + \frac{x^2}{2!} + \frac{x^3}{3!} + \frac{x^4}{4!} + \cdots$$

We find that

$$e^{-x} = 1 + (-x) + \frac{(-x)^2}{2!} + \frac{(-x)^3}{3!} + \frac{(-x)^4}{4!} + \cdots$$

$$= 1 - x + \frac{x^2}{2!} - \frac{x^3}{3!} + \frac{x^4}{4!} - \cdots$$

(Solution continued next page.)

Adding these two series and multiplying by $\frac{1}{2}$, we have

$$\cosh x = \frac{1}{2}(e^x + e^{-x})$$

$$= \frac{1}{2}\left(2 + 2\left(\frac{x^2}{2!}\right) + 2\left(\frac{x^4}{4!}\right) + 2\left(\frac{x^6}{6!}\right) + \cdots\right)$$

$$= 1 + \frac{x^2}{2!} + \frac{x^4}{4!} + \frac{x^6}{6!} + \cdots \quad \blacksquare$$

By the same method, we can derive the Taylor series for sinh x. (See Exercise 11.) These results are summarized by the following formulas:

$$\cosh x = 1 + \frac{x^2}{2!} + \frac{x^4}{4!} + \frac{x^6}{6!} + \cdots$$

$$\sinh x = x + \frac{x^3}{3!} + \frac{x^5}{5!} + \frac{x^7}{7!} + \cdots$$

Note that the Taylor series above are exactly the same as those for the sine and cosine functions except for sign, a further motivation for the names of these functions.

By substituting ix for x in the Taylor series for sin x, (much as we did for e^x in Section 8-5), we can derive another relationship between the hyperbolic functions and circular functions.

Example 2   Show that $\sin ix = i \sinh x$.

Solution   The Taylor series for sin x is

$$\sin x = x - \frac{x^3}{3!} + \frac{x^5}{5!} - \frac{x^7}{7!} + \cdots$$

Substituting ix for x in this series, we obtain

$$\sin ix = (ix) - \frac{i^3x^3}{3!} + \frac{i^5x^5}{5!} - \frac{i^7x^7}{7!} + \frac{i^9x^9}{9!} - \cdots$$

$$= ix - \frac{(-i)x^3}{3!} + \frac{ix^5}{5!} - \frac{(-i)x^7}{7!} + \frac{ix^9}{9!} - \cdots$$

$$= i\left(x + \frac{x^3}{3!} + \frac{x^5}{5!} + \frac{x^7}{7!} + \frac{x^9}{9!} + \cdots\right)$$

Since the series in parentheses is exactly the Taylor series of sinh x,

$$\sin ix = i \sinh x. \quad \blacksquare$$

The following table lists some identities that involve hyperbolic functions, along with their counterparts involving circular functions. You are asked to prove some of these in the exercises.

Hyperbolic Functions	Circular Functions
$\cosh x = \dfrac{1}{2}(e^x + e^{-x})$	$\cos x = \dfrac{1}{2}(e^{ix} + e^{-ix})$
$\sinh x = \dfrac{1}{2}(e^x - e^{-x})$	$\sin x = \dfrac{1}{2i}(e^{ix} - e^{-ix})$
$e^x = \cosh x + \sinh x$	$e^{ix} = \cos x + i\sin x$
$\cosh^2 x - \sinh^2 x = 1$	$\cos^2 x + \sin^2 x = 1$
$\cosh 2x = \cosh^2 x + \sinh^2 x$	$\cos 2x = \cos^2 x - \sin^2 x$
$\qquad = 2\cosh^2 x - 1$	$\qquad = 2\cos^2 x - 1$
$\qquad = 1 + 2\sinh^2 x$	$\qquad = 1 - 2\sin^2 x$
$\sinh 2x = 2\sinh x \cosh x$	$\sin 2x = 2\sin x \cos x$

EXERCISES 8-6

A 1. Find the value of (a) $\cosh 0$ and (b) $\sinh 0$.

2. Show that for any x,
   (a) $\cosh x + \sinh x = e^x$ (b) $\cosh x - \sinh x = e^{-x}$.

3. Use the definitions of $\sinh x$ and $\cosh x$ to show that $\cosh x$ is an even function, that is, $\cosh(-x) = \cosh x$; and that $\sinh x$ is an odd function, that is, $\sinh(-x) = -\sinh x$.

4. Repeat Exercise 3 using the Taylor series for $\cosh x$ and $\sinh x$.

Use a calculator to approximate each expression to 3 decimal places by evaluating the first 4 terms of the Taylor series for the function.

5. $\sinh 1.5$ 6. $\cosh 1.5$ 7. $\cosh 2$ 8. $\sinh 2$

9. Find $\sinh x$ if $\cosh x = \dfrac{5}{4}$. (There are two solutions.)

10. Find $\cosh x$ if $\sinh x = \dfrac{5}{12}$. (Note that $\cosh x > 0$.)

11. Use the method of Example 1 to find the Taylor series for $\sinh x$.

12. Use the method of Example 2 to show that $\cos ix = \cosh x$.

B 13. Prove that $\cosh 2x = \cosh^2 x + \sinh^2 x$ from the definitions of the hyperbolic functions. (Hint: Expand $\cosh^2 x + \sinh^2 x$.)

14. Prove that $\sinh 2x = 2\sinh x \cosh x$ from the definitions of $\cosh x$ and $\sinh x$.

15. Express cosh ix in terms of circular functions.

16. Express sinh ix in terms of circular functions.

C 17. Use $\sin x = \dfrac{1}{2i}(e^{ix} - e^{-ix})$ to show that

$$\sin (a + bi) = \sin a \cosh b + i \cos a \sinh b.$$

18. Use $\cos x = \dfrac{1}{2}(e^{ix} + e^{-ix})$ to show that

$$\cos (a + bi) = \cos a \cosh b - i \sin a \sinh b.$$

19. Prove that $\sinh 3x = 3 \sinh x + 4 \sinh^3 x$.

20. Express $(\cosh x + \sinh x)^n$ in terms of $\sinh nx$ and $\cosh nx$.

8-7 Trigonometric Series

Objective: To define and use trigonometric series.

In the late eighteenth century and in the nineteenth century, functions of the form

$$a_0 + a_1 \cos x + b_1 \sin x + a_2 \cos 2x + b_2 \sin 2x +$$

$$\cdots + a_n \cos nx + b_n \sin nx + \cdots$$

arose in the study of vibrating strings, such as guitar strings, and in the study of heat flow.

Leonhard Euler first suggested a method for writing a given periodic function as a **trigonometric series**, a series of the form above. The special type of series he found is called a *Fourier series*, after Jean Baptiste Joseph Fourier (1768–1830) who investigated them.

Finding the a's and the b's in the Fourier series for a given function requires calculus. Therefore, we shall provide a few Fourier series to work with.

Although periodic functions with any positive period can be represented by Fourier series (Exercises 11 and 12), we shall explore functions with period 2π defined on the interval $-\pi \le x \le \pi$.

Example 1   Sketch the graph of the function $f(x)$ having period 2π and defined by:

$$f(x) = \begin{cases} 0 & \text{if } x = -\pi \\ x & \text{if } -\pi < x < \pi \\ 0 & \text{if } x = \pi \end{cases}$$

Solution

For $-\pi < x < \pi$, the graph of $f(x)$ is represented by the line $y = x$. At both π and $-\pi$, $f(x) = 0$. Therefore, by periodicity, the entire graph of $f(x)$ is as indicated below.

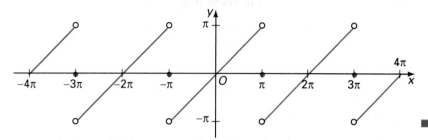

Figure 8-6 illustrates how the 4th, 6th, and 10th partial sums of

$$f(x) = 2\left(\frac{1}{1}\sin x - \frac{1}{2}\sin 2x + \frac{1}{3}\sin 3x - \frac{1}{4}\sin 4x + \cdots\right),$$

the Fourier representation of the function in Example 1, approximate that function. Figure 8-7 shows a computer-generated graph of $f_6(x)$.

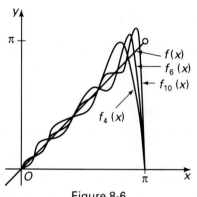

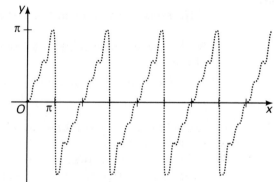

Figure 8-6   Figure 8-7

Our second example is a "square wave," important in electronics.

Example 2

Let $f(x)$ have period 2π and be defined on the interval $0 \le x \le \pi$ by:

$$f(x) = \begin{cases} 1 & \text{if } 0 \le x < \dfrac{\pi}{2} \\ 0 & \text{if } x = \dfrac{\pi}{2} \\ -1 & \text{if } \dfrac{\pi}{2} < x \le \pi \end{cases}$$

If on the interval $-\pi \le x \le \pi$, $f(x)$ is represented by

$$f(x) = \frac{4}{\pi}\left(\cos x - \frac{\cos 3x}{3} + \frac{\cos 5x}{5} - \frac{\cos 7x}{7} + \cdots\right)$$

(a) describe the symmetry of the graph of f and
(b) sketch the graph of the given function for $-\pi \le x \le \pi$.

(Solution on next page.)

Solution

(a) Since the terms of the Fourier series are even functions ($\cos(-\theta) = \cos\theta$), the series representing f is an even function. Thus, the graph of f must be symmetric with respect to the y-axis. That is, the points corresponding to $-\pi \le x < 0$ are reflections in the y-axis of the points corresponding to $0 \le x \le \pi$.

(b) Using the definition of f and the symmetry found in part (a), the graph is as follows.

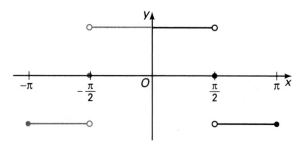

The Fourier series representations of some relatively simple periodic functions are listed below.

(i) If $f(x) = |x|$ for $-\pi \le x < \pi$ and $f(x + 2\pi) = f(x)$, then:

$$f(x) = \frac{\pi}{2} - \frac{4}{\pi}\left(\frac{1}{1^2}\cos x + \frac{1}{3^2}\cos 3x + \frac{1}{5^2}\cos 5x + \cdots\right)$$

(ii) If $f(x) = x^2$ for $-\pi \le x < \pi$ and $f(x + 2\pi) = f(x)$, then:

$$f(x) = \frac{\pi^2}{3} + 4\left(-\frac{1}{1^2}\cos x + \frac{1}{2^2}\cos 2x - \frac{1}{3^2}\cos 3x + \cdots\right)$$

(iii) If $f(x) = 1$ for $0 < x < \pi$, $f(0) = f(\pi) = 0$, then:

$$f(x) = \frac{4}{\pi}\left(\sin x + \frac{\sin 3x}{3} + \frac{\sin 5x}{5} + \frac{\sin 7x}{7} + \cdots\right)$$

Fourier series representations of functions can also be used to represent irrational numbers as infinite sums of rational numbers.

Example 3

Use the Fourier series for $f(x) = x^2$ to represent $\dfrac{\pi^2}{12}$ as a sum of rational numbers.

Solution

From the definition of f and from its Fourier series (ii):

$$x^2 - \frac{\pi^2}{3} = 4\left(-\frac{1}{1^2}\cos x + \frac{1}{2^2}\cos 2x - \frac{1}{3^2}\cos 3x + \cdots\right)$$

$$\frac{x^2}{4} - \frac{\pi^2}{12} = -\frac{1}{1^2}\cos x + \frac{1}{2^2}\cos 2x - \frac{1}{3^2}\cos 3x + \cdots$$

$$\frac{\pi^2}{12} - \frac{x^2}{4} = \frac{1}{1^2} \cos x - \frac{1}{2^2} \cos 2x + \frac{1}{3^2} \cos 3x - \cdots$$

Choose $x = 0$. Then:

$$\frac{\pi^2}{12} = \frac{1}{1^2} \cos 0 - \frac{1}{2^2} \cos 0 + \frac{1}{3^2} \cos 0 - \cdots = 1 - \frac{1}{4} + \frac{1}{9} - \frac{1}{16} + \cdots \quad \blacksquare$$

EXERCISES 8-7

A 1. Sketch the graph of the function f defined by $f(x) = |x|$,
$-\pi \leq x < \pi$ and $f(x + 2\pi) = f(x)$.

2. Use the Fourier series given in (i) to express $\dfrac{\pi^2}{8}$ as an infinite sum of rational numbers.

3. Sketch the graph of the function f defined by $f(x) = x^2$,
$-\pi \leq x < \pi$ and $f(x + 2\pi) = f(x)$.

4. Use the Fourier series given in (ii) to express $\dfrac{\pi^2}{6}$ as an infinite sum of rational numbers.

5. Draw the graph of the function given in (iii) for $0 \leq x \leq \pi$.

6. If the Fourier series given in (iii) represents the function f on the larger interval $-\pi \leq x \leq \pi$, is f even, odd, or neither?

7. Using the results of Exercises 5 and 6, draw the graph of f for $-\pi \leq x \leq \pi$.

B 8. Use the Fourier series given in (iii) to express $\dfrac{\pi}{4}$ as an infinite sum of rational numbers.

9. Use the Fourier series in (ii) to express $\dfrac{11\pi^2}{12}$ as an infinite sum of rational numbers.

10. The function f defined on the interval $0 \leq x \leq \pi$ by $f(x) = x(\pi - x)$ has *two* Fourier series expansions on this interval:

$$f(x) = \frac{\pi^2}{6} - 4\left(\frac{\cos 2x}{2^2} + \frac{\cos 4x}{4^2} + \frac{\cos 6x}{6^2} + \cdots\right) \ and$$

$$f(x) = \frac{8}{\pi}\left(\sin x + \frac{\sin 3x}{3^3} + \frac{\sin 5x}{5^3} + \cdots\right)$$

(a) Sketch the graph of f on the interval $0 \leq x \leq \pi$.

(b) Sketch the graph of the continuation of f to the larger interval $-\pi \le x \le \pi$ defined by the first Fourier series.

(c) On separate axes, sketch f for $-\pi \le x \le \pi$ as defined by the second Fourier series.

(d) Use the first Fourier series to express $\dfrac{\pi^2}{6}$ as an infinite sum of rational numbers.

C 11. Show that if f is a periodic function with fundamental period L, then the function $g(x) = f\left(\dfrac{Lx}{2\pi}\right)$ has period 2π. That is, show that

$g(x + 2\pi) = g(x)$.

12. Suppose that $g(x)$ is as in Exercise 11 and suppose that $g(x)$ has a Fourier series of the form given at the beginning of this section (which is possible, since $g(x)$ has period 2π). Derive the general form of the Fourier series of a periodic function $f(x)$ with fundamental period L (not necessarily equal to 2π). (Hint: Let $u = \dfrac{Lx}{2\pi}$. Use the Fourier series of $g(x)$.)

Challenge

Use a computer or calculator to graph $y = \dfrac{\pi}{2} + 2 \displaystyle\sum_{k=1}^{4} \dfrac{\sin (2k - 1)x}{2k - 1}$.

Self Quiz

 | 8-4 | 8-5 | 8-6 | 8-7 |

1. Prove the identity $\cosh 2x = 1 + 2 \sinh^2 x$ from the definitions of the hyperbolic functions. (Hint: Begin with the right side of the equation.)

2. Approximate $\sin 1$ by evaluating the first four terms of the Taylor series for $\sin x$.

3. Find $\cosh x$ if $\sinh x = -\dfrac{5}{12}$.

4. Use the definition of complex exponents to show that $(e^{ix})^2 = e^{2ix}$.

5. Express in the form $x + yi$: $e^{-1 + \frac{2\pi i}{3}}$.

6. Find a power series expansion for the function

$$\frac{3}{1 - x}$$

by regarding the function as the sum of an infinite geometric series.

7. Use the second Fourier series given in Exercise 10 on page 309 to express $\dfrac{\pi^3}{32}$ as an infinite sum of rational numbers.

8. Let $f(x)$ have period 2π and be defined as the interval $-\pi \le x \le \pi$ by $f(x) = x$ for $-\pi < x < \pi$; $f(\pi) = f(-\pi) = 0$. Use its Fourier series

$$f(x) = 2\left(\sin x - \frac{1}{2} \sin 2x + \frac{1}{3} \sin 3x - \frac{1}{4} \sin 4x + \cdots \right)$$

to write $\dfrac{\pi}{2}$ as an infinite sum of rational numbers.

ADDITIONAL PROBLEMS

1. Find the 9th term of the geometric sequence whose first and second terms are 3 and $-3\sqrt{2}$, respectively.

2. The Taylor series for the function $\ln (1 + x)$ is given by

$$\ln (1 + x) = x - \frac{1}{2}x^2 + \frac{1}{3}x^3 - \frac{1}{4}x^4 - \cdots, \quad -1 < x \le 1.$$

   Use the first four terms of this series to approximate $\ln 2$. The value of $\ln 2$ to four decimal places is 0.6931. How good is your approximation?

3. Write a computer program that uses the Taylor series in Exercise 2 to give an approximate value of $\ln x$ for a given x, where $0 < x \le 2$.

4. Write out the first four partial sums of the series defined by

$$a_1 = \ln \frac{1}{2}$$

$$a_n = \ln \left(\frac{n^2}{(n - 1)(n + 1)} \right), \quad n = 2, 3, \ldots$$

   and simplify your answers.

5. Find a formula for the nth partial sum, S_n, of the series defined in Exercise 4.

6. Complex numbers can help us evaluate the sum of the following (real) infinite series:

$$r \cos \theta + r^2 \cos 2\theta + r^3 \cos 3\theta + \cdots \text{ where } 0 < r < 1$$

   (a) Show that this series is the real part of the series

   $$re^{i\theta} + (re^{i\theta})^2 + (re^{i\theta})^3 + \cdots$$

   (b) Use the formula for the sum of an infinite geometric series (which also holds for complex numbers) to express the sum of the series in (a) in simple form.

   (c) Find the real part of the answer to (b). (Hint: Multiply numerator and denominator by

   $$1 - re^{-i\theta}.$$

   Note that $e^{i\theta} \cdot e^{-i\theta} = e^0 = 1$.)

7. (a) Explain how you know that for any $x > 0$,

   $$\lim_{n \to \infty} \left(1 + \frac{\ln x}{n}\right)^n = x.$$

   (b) You can find a good approximation of $\ln x$ on a calculator that does not have an "$\ln x$" key, as follows: (1) Key in the number x, (2) Press the square root key 11 times, (3) Subtract 1 from your result, and (4) multiply by 2048.
   Explain why the method in (b) works. (Hint: use (a).)

8. Find $\sinh 2x$ if $\sinh x = \dfrac{8}{15}$.

9. Use the Fourier series given in (ii) on page 308 to express $\dfrac{\pi^2}{8}$ as an infinite sum of rational numbers.

10. A unit length is divided into 3 equal parts, and the middle part is discarded. Each of the two remaining parts is similarly divided into thirds, and the two middle thirds of each remaining segment are discarded. If this process is continued indefinitely, what is the total length of the discarded parts?

11. (a) Evaluate $e^{2\pi i}$.
    (b) Use the fact that $e^{x+y} = e^x \cdot e^y$ to evaluate $e^{i\pi}$ and $e^{2\pi i}$. Use Euler's formula to check these results.
    (c) Use the formula in (b) to show that $e^{z + 2\pi i} = e^z$ for any complex number z.

CHAPTER SUMMARY

1. A finite sequence is a finite ordered set. A sequence can be defined recursively by giving the general term as a function of previous terms and the first term or explicitly by giving the general term as a function of its index, that is, its position in the sequence.

2. A finite series is an expression of the form $a_1 + a_2 + a_3 + \cdots + a_n$. The sum of a series is its value. A geometric sequence is a sequence in which each term, after the first, equals a constant (the common ratio) times the preceding term. A geometric series is a series whose terms form a geometric sequence.

3. An infinite sequence is a function whose domain is the set of positive integers. The function assigns to each index n the term a_n. The limit of an infinite sequence is the number to which the terms of the sequence will eventually get arbitrarily close. If such a limit exists, we say the sequence converges. Otherwise, it diverges.

4. An infinite series is a series with infinitely many terms. The limit, or sum, of such a series is defined to be the limit of the sequence of partial sums. A geometric series in which the constant ratio r has absolute value less than 1 converges to $\dfrac{a}{(1 - r)}$, where a is the first term.

5. An exponential function is defined by $f(x) = b^x$, where $b > 0, b \neq 1$. Its inverse, a logarithmic function $\log_b x$, is defined by: $\log_b x = y$ if and only if $b^y = x$. The natural exponential function is e^x. Its inverse is the natural logarithmic function, $\ln x$.

6. A power series is, formally, a polynomial that extends to infinitely many terms. A power series converges (a) for all real x, (b) for x such that $|x| < r$ (and it diverges for $|x| > r$), or (c) only for $x = 0$. The Taylor series for a function on an interval is a power series that converges to the function on the interval. Taylor series for $\sin x$, $\cos x$, and e^x converge to those functions for all x.

7. Euler's formula, $e^{ix} = \cos x + i \sin x$, can be used to define complex exponents.

8. The hyperbolic functions $\sinh x$ and $\cosh x$ are defined in terms of the function e^x. They satisfy many relationships that are analogous to those that hold for the circular functions, but, unlike them, the hyperbolic functions are not periodic.

9. A Fourier series is an infinite series of terms of the form $a_n \cos nx$ or $b_n \sin nx$ ($n = 0, 1, 2, \ldots$) that approximates a given periodic function. A Fourier series contains only cosine terms if the given function is even, or only sine terms if it is odd.

CHAPTER TEST

1. Write the first five terms of the sequence defined by: $a_1 = 1$; $a_n = 2a_{n-1} + 1$, $n = 2, 3, 4, \ldots$. Give an explicit definition of this sequence. (Hint: Note that the terms are close to the powers of 2.)

2. Find the sum of the first six terms of the geometric series:
   $6 - 3 + 1.5 - \cdots$.

3. Find $\displaystyle\lim_{n \to \infty} \frac{\cos\left(\dfrac{1}{n}\right)}{n^2 + 2}$.

4. Find the sum of the *infinite* geometric series $1 - \dfrac{2}{3} + \dfrac{4}{9} - \cdots$.

5. Solve $3 \ln x - 1 = \ln e^5$ for x.

6. Use the laws of logarithms to express $3 \ln 5 - \dfrac{(\ln 16)}{2}$ as the logarithm of a single number.

7. Find a power series expansion for $\dfrac{1}{1 + x^2}$ by regarding the function as the sum of an infinite geometric series.

8. Use the Taylor series for $\cos x$ to write out the first four terms of the Taylor series for $\cos x^2$.

9. If $z = e^{x+yi}$ and z is in the first quadrant of the complex plane and *inside* the unit circle, state the possible values of x and y.

10. The Taylor series of $\text{Tan}^{-1} x$ is given by
$$\text{Tan}^{-1} x = 1 - \frac{x^3}{3} + \frac{x^5}{5} - \frac{x^7}{7} + \cdots.$$

    Use this series to express $\dfrac{\pi}{4}$ as an infinite sum of rational numbers.

11. Find the numerical value of $\sinh (\ln 2)$.

12. Prove that $\cosh 2x = 2 \cosh^2 x - 1$ for all real x.

13. The Fourier series for $f(x) = \dfrac{x}{2}$, $-\pi < x < \pi$, is given by
$$\frac{x}{2} = \sin x - \frac{\sin 2x}{2} + \frac{\sin 3x}{3} - \frac{\sin 4x}{4} + \cdots.$$

    Use this to express $\dfrac{\pi}{4}$ as an infinite sum of rational numbers.

Limits

The notion that sequences are functions enables us to use a computer or a graphing calculator to study them. The following computer program evaluates and graphs the sequence $a_n = \dfrac{35n}{n + 1}$ for $n = 1, 2, 3, \ldots, 279$.

When you run the program, the values of n and a_n are recorded at the bottom of the display screen.

```
10 HOME : HGR : HCOLOR = 3
20 HPLOT 0, 95 TO 279, 95
30 HPLOT 0, 0 TO 0, 190
40 FOR N = 1 TO 279
50 AN = 35 * N / (N + 1)
60 HPLOT N, 95 - AN
70 HTAB1 : VTAB 22 ; PRINT N
80 HTAB 20 : VTAB 22 : PRINT " AN = " ; AN
90 NEXT N
```

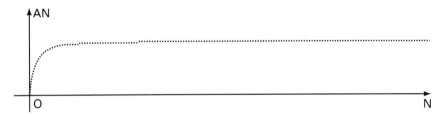

EXERCISES

1. Is it clear from the running computation of n and a_n that as n increases a_n increases? Does the graph clearly confirm this?
2. Use a calculator to find a_n when $n = 1000$, $10,000$, and $100,000$.
3. To see that a_n can never equal 35, try to solve $35 = \dfrac{35n}{n + 1}$. What do you conclude?
4. Change line **50** to read **50 AN = 35 / N**. Run the program to verify that as n increases $\dfrac{35}{n}$ decreases and approaches 0.

3.1415926535897932384626433283279...

Computation of the number π, the ratio of the circumference of any circle to its diameter, has played an important part in the history of mathematics.

About 1700 B.C., Egyptian mathematicians believed that the area of a circle equals that of a square whose side is $\frac{8}{9}$ the diameter of the circle. Using this hypothesis and the formula for the area of a circle, $A = \pi r^2$, they found π to be about $\frac{256}{81}$, or 3.1605. This approximation is correct to one decimal place.

Around 240 B.C., Archimedes attempted to calculate π by inscribing and circumscribing regular polygons about a circle. From his investigations, he was able to write

$$3\frac{10}{71} < \pi < 3\frac{1}{7}.$$

The number $3\frac{1}{7}$, or $\frac{22}{7}$, is an often-used approximation to π.

By the seventeenth and eighteenth centuries, mathematicians had found expressions for π that could be used to approximate that number to any degree of accuracy. Two of these expressions are:

$$\pi = 4\left(1 - \frac{1}{3} + \frac{1}{5} - \frac{1}{7} + \frac{1}{9} - \frac{1}{11} + \cdots\right)$$

$$\pi = 2\left(\frac{2}{1} \cdot \frac{2}{3} \cdot \frac{4}{3} \cdot \frac{4}{5} \cdot \frac{6}{5} \cdot \frac{6}{7} \cdot \frac{8}{7} \cdot \frac{8}{9} \cdots\right)$$

found by Gottlieb Leibniz (one of the founders of calculus) and John Wallis, a British mathematician, respectively. Unfortunately, a three place approximation of π from these formulas requires addition (multiplication) of about 500 terms (factors).

In 1706, the following formula for π was discovered.

$$\frac{\pi}{4} = 4\,\text{Tan}^{-1}\left(\frac{1}{5}\right) - \text{Tan}^{-1}\left(\frac{1}{239}\right)$$

William Shanks, a British mathematician, used that formula to approximate π to 707 decimal places. The task took 15 years and was completed in 1873.

Recently, computers have been used to obtain even finer approximations of π. Using ENIAC, one of the world's first computers, John von Neuman obtained a 2037 place approximation in 70 hours. By 1961, an IBM 7090 computer calculated the first 100,000 places in 9 hours, and by 1973, the million mark was achieved. In 1987, researchers at the University of Tokyo calculated the first 134 million digits of π by using a supercomputer.

Approximation of π is of interest to computer scientists, because it helps them test the efficiency of computers and computer algorithms. Using computer approximations, mathematicians can confirm that the digits of π are patternless, that is, that π is indeed an irrational number.

EXERCISES

1. (a) Use a calculator to find $4 \operatorname{Tan}^{-1}\left(\dfrac{1}{5}\right) - \operatorname{Tan}^{-1}\left(\dfrac{1}{239}\right)$ and $\dfrac{\pi}{4}$.

   (b) Compare the two calculations in part (a).
2. Refer to Wallis's formula. Let P_n be the product of the first n fractions in the continued product.
   (a) Use a calculator to find P_3, P_4, and P_5.
   (b) Let N_n represent the numerator and D_n represent the denominator of the nth fraction in the continued product. Use a calculator to verify that

$$P_n = \frac{N_n}{D_n} P_{n-1} = \frac{D_{n-1} + 1}{N_{n-1} + 1} P_{n-1} \text{ for } n = 3, 4, \text{ and } 5.$$

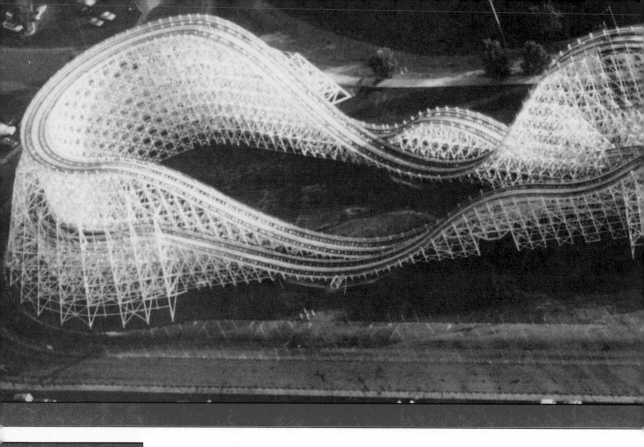

9 *Vectors in Space*

The photograph shows a roller coaster ride, a complex space curve. Design of such rides involves an analysis of forces acting in three dimensions. These forces can be studied and controlled with the aid of vectors. See also the Application Section on page 342.

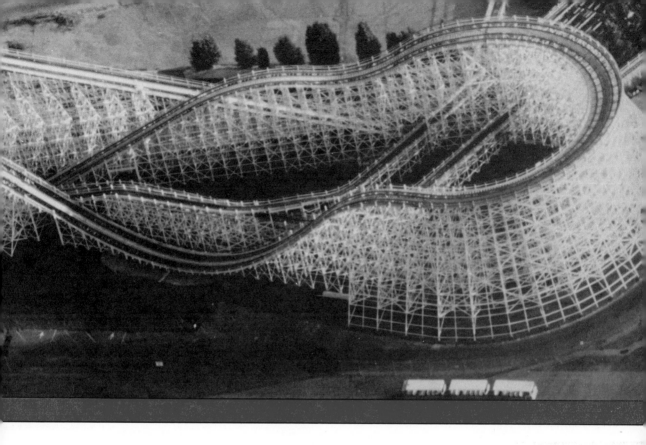

COORDINATES AND VECTORS

Rectangular Coordinates in Space

Objective: To study points in three-dimensional space.

Since we live in three-dimensional space it is not surprising that we need a coordinate system for it. The most commonly used system is obtained by adding a z-axis to an xy-system in the manner shown in Figure 9-1.

The position of each point in space can now be described by an ordered triple (x, y, z) of numbers. Some examples are shown in Figure 9-2.

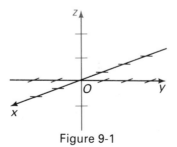

Figure 9-1

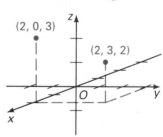

Figure 9-2

The plane determined by the x-axis and the z-axis is called the **xz-plane,** while the planes determined by the x- and y-axes and the y- and z-axes are called the xy- and yz-planes, respectively. These three **coordinate planes** divide space into eight regions, called **octants.** The octant in which points have all positive coordinates is the **first octant.** The other seven octants are not named.

In space the *distance* between the points $P_1(x_1, y_1, z_1)$ and $P_2(x_2, y_2, z_2)$ is given by:

$$P_1P_2 = \sqrt{(x_2 - x_1)^2 + (y_2 - y_1)^2 + (z_2 - z_1)^2}$$

We can prove this distance formula by applying the Pythagorean theorem to the right triangles P_1QR and P_1RP_2 shown in Figure 9-3 to obtain

$$(P_1R)^2 = |x_2 - x_1|^2 + |y_2 - y_1|^2$$

and

$$(P_1P_2)^2 = (P_1R)^2 + |z_2 - z_1|^2 =$$
$$(x_2 - x_1)^2 + (y_2 - y_1)^2 + (z_2 - z_1)^2.$$

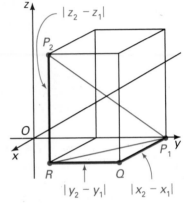

Figure 9-3

Example 1

Show that $A(1, 0, 2)$, $B(3, -1, 4)$, and $C(2, -2, 0)$ are the vertices of an isosceles right triangle.

Solution

Use the distance formula to find AB, BC, and AC.

$$AB = \sqrt{(3 - 1)^2 + (-1 - 0)^2 + (4 - 2)^2} = \sqrt{4 + 1 + 4} = \sqrt{9} = 3$$
$$BC = \sqrt{(2 - 3)^2 + (-2 - (-1))^2 + (0 - 4)^2} = \sqrt{1 + 1 + 16} = 3\sqrt{2}$$
$$AC = \sqrt{(2 - 1)^2 + (-2 - 0)^2 + (0 - 2)^2} = \sqrt{1 + 4 + 4} = \sqrt{9} = 3$$

Since $AB = AC$, $\triangle ABC$ is isosceles, and since $(BC)^2 = (AB)^2 + (AC)^2$, it is also a right triangle. ∎

Spheres are among the most important surfaces in space. The **sphere** of radius r centered at the point $C(h, k, l)$ is the set of all points $P(x, y, z)$ such that

$$CP = r, \text{ or } (CP)^2 = r^2.$$

Using the distance formula, we see that an equation of this sphere is

$$(x - h)^2 + (y - k)^2 + (z - l)^2 = r^2.$$

Spheres with center at the origin have equations of the form

$$x^2 + y^2 + z^2 = r^2.$$

Example 2

Find an equation of the sphere that has center at $(3, -1, 2)$ and is tangent to the xy-plane.

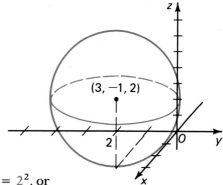

Solution

To fulfill the tangency condition, the radius of the sphere must equal the distance from its center to the xy-plane. Thus $r = 2$. Hence,

$(x - 3)^2 + (y - (-1))^2 + (z - 2)^2 = 2^2$, or

$(x - 3)^2 + (y + 1)^2 + (z - 2)^2 = 2^2$. ∎

The equation $(x - h)^2 + (y - k)^2 + (z - l)^2 = r^2$ may be expanded and simplified, resulting in an equation of the form

$$x^2 + y^2 + z^2 + ax + by + cz + d = 0.$$

Thus, in Example 2, the equation of the sphere may also be given as $x^2 + y^2 + z^2 - 6x + 2y - 4z + 10 = 0$.

When an equation of a sphere is given in this expanded form, we can find its center and radius by completing squares.

Example 3

Find the center and radius of the sphere
$$x^2 + y^2 + z^2 - 2x + 6y + 4z + 5 = 0.$$

Solution

We complete three squares:

$(x^2 - 2x +\ \ ) + (y^2 + 6y +\ \ ) + (z^2 + 4z +\ \ ) = -5$
$(x^2 - 2x + 1) + (y^2 + 6y + 9) + (z^2 + 4z + 4) = -5 + 1 + 9 + 4$
$(x - 1)^2 + (y + 3)^2 + (z + 2)^2 = 9$

Therefore, the sphere has center $(1, -3, -2)$ and radius 3. ∎

EXERCISES 9-1

In Exercises 1–6, (a) plot the given points and draw the triangle that they determine. (b) Use the distance formula to check whether the triangle is isosceles, right, both, or neither.

A 1. $(0, 0, 0), (2, 3, 4), (0, -2, 5)$ **2.** $(2, 1, 3), (4, -1, 4), (0, 0, 5)$

3. $(1, 0, 0), (0, 2, -2), (3, 2, 2)$ **4.** $(0, 1, -1), (1, 1, 2), (3, 4, -2)$

5. $(0, 1, 1), (1, 0, 1), (1, 1, 0)$ **6.** $(1, 0, 0), (1, 1, 1), (2, 2, -2)$

In Exercises 7–10, find an equation in the form $x^2 + y^2 + z^2 + ax + by + cz + d = 0$ of the sphere having the given center and radius.

7. Center $(0, 0, 0)$, radius 5

8. Center $(0, 0, 1)$, radius 1

9. Center $(2, -1, -2)$, radius 3

10. Center $(1, -2, 3)$, radius 4

In Exercises 11–14, find the center and radius of the given sphere.

11. $x^2 + y^2 + z^2 = 2x$

12. $x^2 + y^2 + z^2 + 6x - 16z = 0$

13. $x^2 + y^2 + z^2 - 4x + 6y + 2z + 5 = 0$

14. $x^2 + y^2 + z^2 = x + y + z$

In Exercises 15 and 16 the given points are diagonally opposite vertices of a rectangular solid (box) that has its edges parallel to the coordinate axes. Find the other six vertices.

B 15. $(2, 0, 3), (5, 2, -1)$

16. $(1, 3, 5), (2, 4, 6)$

In Exercises 17 and 18 three vertices of a rectangle are given. Find the fourth vertex.

17. $(0, 2, 4), (3, 2, 1), (3, 0, 1)$

18. $(1, 2, 3), (1, 5, 3), (-1, 2, 5)$

In Exercises 19 and 20 find an equation for the set of points that are equidistant from the given points. (Hint: Let $P(x, y, z)$ be any point equidistant from A and B. Then $AP = BP$.)

19. $A(3, 0, 1), B(0, 3, 1)$

20. $A(0, 0, 0), B(2, 2, 2)$

In Exercises 21–24, find an equation of the sphere having the stated properties.

21. Center at $(3, 0, -4)$, passes through the origin

22. Center at $(2, 3, -3)$, tangent to the xy-plane

23. Radius 2, tangent to the xz-plane at $(2, 0, 3)$ (Two answers)

24. Radius 3, center in the first octant, tangent to all three coordinate planes

C 25. A point P moves so that it is always twice as far from $(5, 0, 0)$ as from $(2, 0, 0)$. Show that P is always the same distance from some point. State the distance and find the coordinates of that point.

26. Show that the midpoint of the segment with endpoints (x_1, y_1, z_1) and (x_2, y_2, z_2) is the point $\left(\dfrac{x_1 + x_2}{2}, \dfrac{y_1 + y_2}{2}, \dfrac{z_1 + z_2}{2} \right)$.

27. Find an equation whose graph is the set of all points the sum of whose distances from $(0, 2, 0)$ and $(0, -2, 0)$ is 5 units.

9-2   Vectors in Space

Objective: To use vectors in three-dimensional space.

In space we represent vector quantities by arrows and denote these vectors by boldface letters, just as we did in the plane. At this point it is a good idea to reread Sections 6-1 and 6-4. As you will see, the definitions given for vectors in the plane can be extended to vectors in space.

Let us denote by **i**, **j**, and **k** the unit vectors having their initial points at the origin and their terminal points at $(1, 0, 0)$, $(0, 1, 0)$, and $(0, 0, 1)$, respectively. From Figure 9-4, any vector **a** is a linear combination of **i**, **j**, and **k**. That is,

$$\mathbf{a} = a_1\mathbf{i} + a_2\mathbf{j} + a_3\mathbf{k}.$$

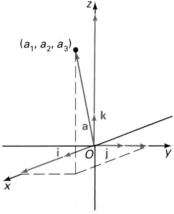

Figure 9-4

Algebraic properties of vectors can be used to show that if $\mathbf{a} = a_1\mathbf{i} + a_2\mathbf{j} + a_3\mathbf{k}$, $\mathbf{b} = b_1\mathbf{i} + b_2\mathbf{j} + b_3\mathbf{k}$, and t is a scalar, then:

$\mathbf{a} = \mathbf{b}$ if and only if $a_1 = b_1$, $a_2 = b_2$, and $a_3 = b_3$	(1)
$\mathbf{a} + \mathbf{b} = (a_1 + b_1)\mathbf{i} + (a_2 + b_2)\mathbf{j} + (a_3 + b_3)\mathbf{k}$	(2)
$t\mathbf{a} = ta_1\mathbf{i} + ta_2\mathbf{j} + ta_3\mathbf{k}$	(3)
$\|\mathbf{a}\| = \sqrt{a_1{}^2 + a_2{}^2 + a_3{}^2}$	(4)

Example 1

Let $\mathbf{a} = \mathbf{i} + 2\mathbf{j}$, $\mathbf{b} = \mathbf{j} + 2\mathbf{k}$, and $\mathbf{c} = 2\mathbf{i} + \mathbf{k}$.
(a) Find $2\mathbf{a} - 3\mathbf{b} + \mathbf{c}$. (b) Express $\mathbf{v} = 3\mathbf{i} - \mathbf{j} + 2\mathbf{k}$ as a linear combination of **a**, **b**, and **c**.

Solution

(a) $2\mathbf{a} - 3\mathbf{b} + \mathbf{c} = 2(\mathbf{i} + 2\mathbf{j}) - 3(\mathbf{j} + 2\mathbf{k}) + (2\mathbf{i} + \mathbf{k})$
$\phantom{(a) 2\mathbf{a} - 3\mathbf{b} + \mathbf{c}} = 4\mathbf{i} + \mathbf{j} - 5\mathbf{k}$

(b) We wish to find scalars r, s, and t such that

$$\mathbf{v} = r\mathbf{a} + s\mathbf{b} + t\mathbf{c},$$

or $\qquad 3\mathbf{i} - \mathbf{j} + 2\mathbf{k} = r(\mathbf{i} + 2\mathbf{j}) + s(\mathbf{j} + 2\mathbf{k}) + t(2\mathbf{i} + \mathbf{k}).$

Multiplying and rearranging terms, we have

$$3\mathbf{i} - \mathbf{j} + 2\mathbf{k} = (r + 2t)\mathbf{i} + (2r + s)\mathbf{j} + (2s + t)\mathbf{k}.$$

(Solution continued next page.)

This vector equation is equivalent to the following system of scalar equations.

$$r \qquad\quad + 2t = 3$$
$$2r + \quad s \qquad\quad = -1$$
$$2s + \quad t = 2$$

Solving this system gives $r = -\dfrac{5}{9}$, $s = \dfrac{1}{9}$, and $t = \dfrac{16}{9}$.

Therefore, $\mathbf{v} = -\dfrac{5}{9}\,\mathbf{a} + \dfrac{1}{9}\,\mathbf{b} + \dfrac{16}{9}\,\mathbf{c}$. ∎

Recall that the dot product of $\mathbf{a}$ and $\mathbf{b}$ is defined by

$$\mathbf{a} \cdot \mathbf{b} = \|\mathbf{a}\|\,\|\mathbf{b}\|\,\cos\theta,$$

where θ is the angle between $\mathbf{a}$ and $\mathbf{b}$. Using a slight modification of the discussion given on page 212, we can show that if $\mathbf{a} = a_1\mathbf{i} + a_2\mathbf{j} + a_3\mathbf{k}$ and $\mathbf{b} = b_1\mathbf{i} + b_2\mathbf{j} + b_3\mathbf{k}$, then:

$$\mathbf{a} \cdot \mathbf{b} = a_1b_1 + a_2b_2 + a_3b_3$$

We note that the formula

$$\cos\theta = \frac{\mathbf{a} \cdot \mathbf{b}}{\|\mathbf{a}\|\,\|\mathbf{b}\|}$$

enables us to find the angle between two nonzero vectors. In particular, $\mathbf{a}$ and $\mathbf{b}$ are orthogonal if and only if

$$\mathbf{a} \cdot \mathbf{b} = a_1b_1 + a_2b_2 + a_3b_3 = 0.$$

Expressing a vector $\mathbf{v}$ as a linear combination of $\mathbf{a}$, $\mathbf{b}$, and $\mathbf{c}$ (Example 1) becomes simpler if $\mathbf{a}$, $\mathbf{b}$, and $\mathbf{c}$ are orthogonal in pairs.

Example 2

Express $\mathbf{v} = 3\mathbf{i} - 5\mathbf{j} + 2\mathbf{k}$ as a linear combination of $\mathbf{a} = 2\mathbf{i} + 3\mathbf{j} + \mathbf{k}$, $\mathbf{b} = \mathbf{i} - \mathbf{j} + \mathbf{k}$, and $\mathbf{c} = 4\mathbf{i} - \mathbf{j} - 5\mathbf{k}$.

Solution

By taking dot products we can check that $\mathbf{a} \cdot \mathbf{b} = \mathbf{b} \cdot \mathbf{c} = \mathbf{c} \cdot \mathbf{a} = 0$. Thus $\mathbf{a}$, $\mathbf{b}$, and $\mathbf{c}$ are orthogonal.
To find r, s, and t in $\mathbf{v} = r\mathbf{a} + s\mathbf{b} + t\mathbf{c}$, dot multiply both sides of the equation in turn by $\mathbf{a}$, $\mathbf{b}$, and $\mathbf{c}$.

$$\mathbf{a} \cdot \mathbf{v} = r\mathbf{a} \cdot \mathbf{a} + s\mathbf{a} \cdot \mathbf{b} + t\mathbf{a} \cdot \mathbf{c}$$
$$= r(\mathbf{a} \cdot \mathbf{a}) + s(\mathbf{a} \cdot \mathbf{b}) + t(\mathbf{a} \cdot \mathbf{c})$$
$$= r(\mathbf{a} \cdot \mathbf{a}) + 0 + 0$$
$$r = \frac{\mathbf{a} \cdot \mathbf{v}}{\mathbf{a} \cdot \mathbf{a}} = \frac{-7}{14} = -\frac{1}{2}$$

$$\mathbf{b} \cdot \mathbf{v} = \mathbf{b} \cdot \mathbf{a} + s\mathbf{b} \cdot \mathbf{b} + t\mathbf{b} \cdot \mathbf{c}$$
$$= r(\mathbf{b} \cdot \mathbf{a}) + s(\mathbf{b} \cdot \mathbf{b}) + t(\mathbf{b} \cdot \mathbf{c})$$
$$= 0 + s(\mathbf{b} \cdot \mathbf{b}) + 0$$
$$s = \frac{\mathbf{b} \cdot \mathbf{v}}{\mathbf{b} \cdot \mathbf{b}} = \frac{10}{3}$$
$$\mathbf{c} \cdot \mathbf{v} = r\mathbf{c} \cdot \mathbf{a} + s\mathbf{c} \cdot \mathbf{b} + t\mathbf{c} \cdot \mathbf{c}$$
$$= r(\mathbf{c} \cdot \mathbf{a}) + s(\mathbf{c} \cdot \mathbf{b}) + t(\mathbf{c} \cdot \mathbf{c})$$
$$= 0 + 0 + t(\mathbf{c} \cdot \mathbf{c})$$
$$t = \frac{\mathbf{c} \cdot \mathbf{v}}{\mathbf{c} \cdot \mathbf{c}} = \frac{7}{42} = \frac{1}{6}$$

Therefore, $\mathbf{v} = -\dfrac{1}{2}\mathbf{a} + \dfrac{10}{3}\mathbf{b} + \dfrac{1}{6}\mathbf{c}$. ∎

Example 3 Find a nonzero vector that is orthogonal to both $\mathbf{a} = 2\mathbf{i} - \mathbf{j} - 3\mathbf{k}$ and $\mathbf{b} = \mathbf{i} + 2\mathbf{j} + 2\mathbf{k}$.

Solution We seek a vector $\mathbf{r} = x\mathbf{i} + y\mathbf{j} + z\mathbf{k}$ such that

$$\mathbf{a} \cdot \mathbf{r} = 2x + (-1)y + (-3)z = 0$$
$$\mathbf{b} \cdot \mathbf{r} = 1x + 2y + 2z = 0$$

We solve this system for x and y in terms of z.

$$2x - y = 3z$$
$$x + 2y = -2z$$
$$x = \frac{4}{5}z, \ y = -\frac{7}{5}z$$

In order to avoid fractions, set $z = 5$ and obtain

$$x = 4, y = -7, z = 5.$$

Therefore, $\mathbf{r} = 4\mathbf{i} - 7\mathbf{j} + 5\mathbf{k}$. ∎

The answer to Example 3 is not unique. Any nonzero scalar multiple of $\mathbf{r}$ has the required property.

EXERCISES 9-2

A 1. Let $\mathbf{a} = \mathbf{i} + 4\mathbf{j} - \mathbf{k}$ and $\mathbf{b} = \mathbf{i} - 2\mathbf{j} + 2\mathbf{k}$. (a) Find $\mathbf{a} + 2\mathbf{b}$ and $2\mathbf{b} - \mathbf{a}$. (b) Find $\|\mathbf{a}\|$, $\|\mathbf{b}\|$, $\|\mathbf{a} + \mathbf{b}\|$, and $\|\mathbf{a} - \mathbf{b}\|$.

2. Let $\mathbf{a} = \mathbf{i} + \mathbf{j} - 3\mathbf{k}$ and $\mathbf{b} = \mathbf{i} - 2\mathbf{j} + \mathbf{k}$. (a) Find $3\mathbf{a} - 2\mathbf{b}$ and $2\mathbf{a} - 3\mathbf{b}$. (b) Find $\|\mathbf{a}\|$, $\|\mathbf{b}\|$, $\|\mathbf{a} + \mathbf{b}\|$, and $\|\mathbf{a} - \mathbf{b}\|$.

In Exercises 3–6, (a) find the linear combination $3\mathbf{a} - 2\mathbf{b} + \mathbf{c}$. (b) Express $\mathbf{v}$ as a linear combination of $\mathbf{a}$, $\mathbf{b}$, and $\mathbf{c}$.

3. $\mathbf{a} = \mathbf{i} + \mathbf{j} + \mathbf{k}, \mathbf{b} = \mathbf{j} + \mathbf{k}, \mathbf{c} = \mathbf{k}; \mathbf{v} = -3\mathbf{i} - 2\mathbf{j} + 4\mathbf{k}$

4. $\mathbf{a} = \mathbf{i} + \mathbf{j}, \mathbf{b} = \mathbf{j} + \mathbf{k}, \mathbf{c} = \mathbf{i} + \mathbf{k}; \mathbf{v} = 2\mathbf{i} + 3\mathbf{j}$

5. $\mathbf{a} = \mathbf{i} + \mathbf{j}, \mathbf{b} = \mathbf{i} - \mathbf{j}, \mathbf{c} = \mathbf{i} + \mathbf{j} + \mathbf{k}; \mathbf{v} = 2\mathbf{i} + 4\mathbf{j} + 5\mathbf{k}$

6. $\mathbf{a} = \mathbf{i} - 2\mathbf{j} + 3\mathbf{k}, \mathbf{b} = \mathbf{j} - \mathbf{k}, \mathbf{c} = \mathbf{k}; \mathbf{v} = \mathbf{i} - 2\mathbf{j} + 3\mathbf{k}$

7. Which pairs of the following vectors are (a) orthogonal? (b) parallel? (Recall that two vectors are parallel if and only if one is a scalar multiple of the other.)

   $\mathbf{a} = 4\mathbf{i} - 2\mathbf{j} + 2\mathbf{k}$ $\mathbf{b} = 2\mathbf{i} + 3\mathbf{j} + \mathbf{k}$
   $\mathbf{c} = 2\mathbf{i} + 3\mathbf{j} - \mathbf{k}$ $\mathbf{d} = -2\mathbf{i} + \mathbf{j} - \mathbf{k}$

In Exercises 8–11, find the angle θ between $\mathbf{a}$ and $\mathbf{b}$. Use $\cos\theta = \dfrac{(\mathbf{a} \cdot \mathbf{b})}{\|\mathbf{a}\| \, \|\mathbf{b}\|}$.

8. $\mathbf{a} = \mathbf{i} + \mathbf{j} + \mathbf{k}, \mathbf{b} = 2\mathbf{i} + \mathbf{j} + 2\mathbf{k}$ 9. $\mathbf{a} = \mathbf{i} - 2\mathbf{j} + 3\mathbf{k}, \mathbf{b} = 2\mathbf{i} + \mathbf{j}$

10. $\mathbf{a} = 3\mathbf{i} + 2\mathbf{j} + 2\mathbf{k}, \mathbf{b} = \mathbf{i} - 2\mathbf{j} - \mathbf{k}$ 11. $\mathbf{a} = \mathbf{i} - \mathbf{j} + \mathbf{k}, \mathbf{b} = \mathbf{i} + \mathbf{j} - \mathbf{k}$

In Exercises 12 and 13 express $\mathbf{v}$ as a linear combination of $\mathbf{a}$, $\mathbf{b}$, and $\mathbf{c}$. Use the fact that $\mathbf{a}$, $\mathbf{b}$, and $\mathbf{c}$ are orthogonal in pairs.

B 12. $\mathbf{a} = 3\mathbf{i} + 2\mathbf{j} + \mathbf{k}, \mathbf{b} = \mathbf{i} - \mathbf{j} - \mathbf{k}, \mathbf{c} = \mathbf{i} - 4\mathbf{j} + 5\mathbf{k}; \mathbf{v} = \mathbf{i} + \mathbf{j} + 2\mathbf{k}$

13. $\mathbf{a} = 2\mathbf{i} + 2\mathbf{j} + \mathbf{k}, \mathbf{b} = \mathbf{i} - 2\mathbf{j} + 2\mathbf{k}, \mathbf{c} = 2\mathbf{i} - \mathbf{j} - 2\mathbf{k}; \mathbf{v} = 3\mathbf{i} - \mathbf{j} + 4\mathbf{k}$

In Exercises 14 and 15, find a vector that is orthogonal to both $\mathbf{a}$ and $\mathbf{b}$.

14. $\mathbf{a} = \mathbf{i} + \mathbf{j} + \mathbf{k}, \mathbf{b} = 2\mathbf{i} - 5\mathbf{j} + 3\mathbf{k}$

15. $\mathbf{a} = \mathbf{i} + \mathbf{j} - \mathbf{k}, \mathbf{b} = \mathbf{i} - 3\mathbf{j} - 3\mathbf{k}$

16. Let $\mathbf{a} = 2\mathbf{i} - 3\mathbf{j} + \mathbf{k}$ and $\mathbf{b} = \mathbf{i} + 2\mathbf{j} - 4\mathbf{k}$. Let $\mathbf{v} = 18\mathbf{i} - 13\mathbf{j} + 16\mathbf{k}$. Find a vector $\mathbf{c}$ such that $\mathbf{v} = 2\mathbf{a} - \mathbf{b} + 5\mathbf{c}$.

17. Let $\mathbf{a} = \mathbf{i} + 4\mathbf{j} - \mathbf{k}$ and $\mathbf{b} = 3\mathbf{i} - 2\mathbf{j} + 4\mathbf{k}$. Let $\mathbf{v} = 3\mathbf{i} + 13\mathbf{j} - 2\mathbf{k}$. Find a vector $\mathbf{c}$ such that $\mathbf{v} = 3\mathbf{a} - 2\mathbf{b} + 3\mathbf{c}$.

18. A vector makes equal acute angles with $\mathbf{i}$, $\mathbf{j}$, and $\mathbf{k}$. Find the measures of the angles to the nearest tenth of a degree.

C 19. Is the following statement true or false? Give a reason for your answer. Given any three vectors $\mathbf{a}$, $\mathbf{b}$, and $\mathbf{c}$ in space, there is a nonzero vector $\mathbf{v}$ such that $\mathbf{a} \cdot \mathbf{v} = \mathbf{b} \cdot \mathbf{v} = \mathbf{c} \cdot \mathbf{v}$.

20. Let $\mathbf{a}$, $\mathbf{b}$, and $\mathbf{c}$ be *unit* vectors that are orthogonal in pairs and let $\mathbf{v}$ be an arbitrary vector. Show that $\mathbf{v} = (\mathbf{a} \cdot \mathbf{v})\mathbf{a} + (\mathbf{b} \cdot \mathbf{v})\mathbf{b} + (\mathbf{c} \cdot \mathbf{v})\mathbf{c}$. (Hint: Let $\mathbf{v} = r\mathbf{a} + s\mathbf{b} + t\mathbf{c}$ and determine r, s, and t as in Example 2.)

Self Quiz

1. Is the triangle with vertices $(1, 0, -2)$, $(3, -1, 2)$, and $(4, 2, -3)$ isosceles, right, both, or neither?

2. Find an equation of the sphere with radius 6 and center at $(-5, 3, 0)$.

3. Find an equation of the sphere that has center at $(4, 7, -4)$ and passes through the origin.

4. Let $\mathbf{a} = 3\mathbf{i} - \mathbf{j} + 2\mathbf{k}$ and $\mathbf{b} = -\mathbf{i} + 3\mathbf{j} - \mathbf{k}$. Find $4\mathbf{a} - \mathbf{b}$, $\|\mathbf{a}\|$, $\|\mathbf{b}\|$, and $\|\mathbf{a} + \mathbf{b}\|$.

5. Find the angle between $\mathbf{a} = 5\mathbf{i} - 5\mathbf{k}$ and $\mathbf{b} = (\mathbf{i} - 7\mathbf{j} + 7\mathbf{k})$ to the nearest degree.

6. Find a nonzero vector orthogonal to $\mathbf{i} - \mathbf{j} + \mathbf{k}$ and to $3\mathbf{i} + \mathbf{j} - 2\mathbf{k}$.

LINES AND PLANES

9-3

Lines in Space

Objective: To apply vectors to lines in space.

We can describe any P by a vector. Let O be a fixed reference point, called the origin. We call $\overrightarrow{OP}$ the position vector of P. If P is given by $P(x, y, z)$ then

$$\overrightarrow{OP} = x\mathbf{i} + y\mathbf{j} + z\mathbf{k}.$$

Using position vectors, we can write the vector $\overrightarrow{P_0P_1}$ whose initial point is $P_0(x_0, y_0, z_0)$ and whose terminal point is $P_1(x_1, y_1, z_1)$ as

$$\overrightarrow{P_0P_1} = (x_1 - x_0)\mathbf{i} + (y_1 - y_0)\mathbf{j} + (z_1 - z_0)\mathbf{k},$$

since $\overrightarrow{OP_0} + \overrightarrow{P_0P_1} = \overrightarrow{OP_1}$, or $\overrightarrow{P_0P_1} = \overrightarrow{OP_1} - \overrightarrow{OP_0}$. For example, if $P_0(1, 2, 5)$ and $P_1(4, -1, 3)$ are the initial and terminal points of $\overrightarrow{P_0P_1}$, then

$$\overrightarrow{P_0P_1} = (4 - 1)\mathbf{i} + (-1 - 2)\mathbf{j} + (3 - 5)\mathbf{k} = 3\mathbf{i} - 3\mathbf{j} - 2\mathbf{k}.$$

A line in space can easily be described by a simple vector equation. Suppose that a line l contains P_0 and is parallel to $\mathbf{m}$. (See Figure 9-5 on the following page.) Then P is on l if and only if $\overrightarrow{P_0P} = t\mathbf{m}$ for some real number t. (See Figure 9-6 on the following page.)

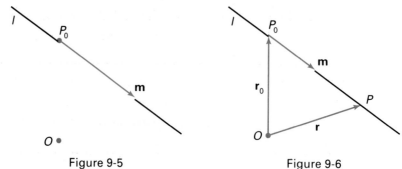

Figure 9-5   Figure 9-6

Thus $\overrightarrow{OP} = \overrightarrow{OP_0} + t\mathbf{m}$.

> If l contains P_0 and is parallel to $\mathbf{m}$, then
> $$\mathbf{r} = \mathbf{r}_0 + t\mathbf{m},$$
> where $\mathbf{r}$ is the position vector of P and $\mathbf{r}_0$ is the position vector of P_0.

This is a vector equation of l. In it, $\mathbf{r}_0$ and $\mathbf{m}$ are given constant vectors and $\mathbf{r}$ is a variable vector. As t varies, $\mathbf{r}$ traces out l.

Example 1   Find a vector equation of the line l that passes through the point $P_0(1, 4, 3)$ and is parallel to $3\mathbf{i} - 2\mathbf{j} + \mathbf{k}$.

Solution   Let $\mathbf{r}_0 = \overrightarrow{OP_0} = \mathbf{i} + 4\mathbf{j} + 3\mathbf{k}$ and let $\mathbf{m} = 3\mathbf{i} - 2\mathbf{j} + \mathbf{k}$. Then $\mathbf{r} = \mathbf{r}_0 + t\mathbf{m}$ becomes: $\mathbf{r} = (\mathbf{i} + 4\mathbf{j} + 3\mathbf{k}) + t(3\mathbf{i} - 2\mathbf{j} + \mathbf{k}) = (1 + 3t)\mathbf{i} + (4 - 2t)\mathbf{j} + (3 + t)\mathbf{k}$ ■

Next, we will find a vector equation for the line l determined by two points P_0 and P_1. Let $\mathbf{r}_0$ and $\mathbf{r}_1$ be the position vectors of P_0 and P_1, respectively. Let P with position vector $\mathbf{r}$ be any point on l (Figure 9-7).

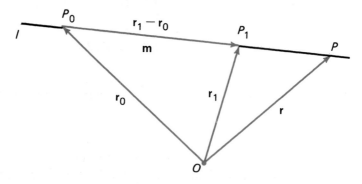

Figure 9-7

Since $\overrightarrow{P_0P_1} = \mathbf{r}_1 - \mathbf{r}_0$ is parallel to l, then $\mathbf{r} = \mathbf{r}_0 + t\mathbf{m}$ becomes
$$\mathbf{r} = \mathbf{r}_0 + t(\mathbf{r}_1 - \mathbf{r}_0).$$

If line l contains P_0 and P_1 and P is any point on the line, then
$$\mathbf{r} = (1 - t)\mathbf{r}_0 + t\mathbf{r}_1$$
where P_0, P_1, and P have position vectors $\mathbf{r}_0$, $\mathbf{r}_1$, and $\mathbf{r}$, respectively.

Example 2

(a) Find a vector equation of the line l determined by $P_0(1, 4, 0)$ and $P_1(2, 2, 3)$. Then (b) find the point where l intersects the xz-plane, and (c) draw the part of l lying in the first octant.

Solution

(a) Let $\mathbf{r}_0 = \overrightarrow{OP_0} = \mathbf{i} + 4\mathbf{j}$ and $\mathbf{r}_1 = \overrightarrow{OP_1} = 2\mathbf{i} + 2\mathbf{j} + 3\mathbf{k}$. Then $\mathbf{r} = (1 - t)\mathbf{r}_0 + t\mathbf{r}_1$ becomes
$$\mathbf{r} = (1 - t)(\mathbf{i} + 4\mathbf{j}) + t(2\mathbf{i} + 2\mathbf{j} + 3\mathbf{k})$$

or

$$\mathbf{r} = (1 + t)\mathbf{i} + (4 - 2t)\mathbf{j} + 3t\mathbf{k}. \qquad (*)$$

(b) The vector $\mathbf{r}$ positions a point in the xz-plane if and only if the coefficient of $\mathbf{j}$ is 0.

Therefore, $4 - 2t = 0$, or $t = 2$.

Substituting $t = 2$ into $(*)$, we have $\mathbf{r} = 3\mathbf{i} + 6\mathbf{k}$. Therefore, the desired point is $(3, 0, 6)$.

(c) $P_1(2, 2, 3)$ is in the first octant. Since l intersects the xz-plane at $(3, 0, 6)$ and the xy-plane at $P_0(1, 4, 0)$, these are the exit points in the first octant for l.

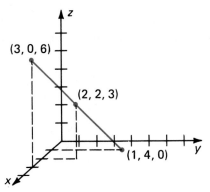

The distance d between a point P_1 and a line l is the shortest distance between P_1 and any point P on l. That is, if $\overrightarrow{P_1P}$ is perpendicular to l, then $\|\overrightarrow{P_1P}\| = d$.

Example 3   Find the distance between the line l whose vector equation is

$$\mathbf{r} = (2 + 3t)\mathbf{i} + (1 - t)\mathbf{j} + (1 + 2t)\mathbf{k}$$

and the point $P_1(7, -6, 4)$.

Solution   The equation $\mathbf{r} = \mathbf{r}_0 + t\mathbf{m}$ for l can be written as

$$\mathbf{r} = 2\mathbf{i} + \mathbf{j} + \mathbf{k} + t(3\mathbf{i} - \mathbf{j} + 2\mathbf{k}).$$

Sketch a vector diagram showing $P_0(2, 1, 1)$, $P_1(7, -6, 4)$, and P such that $\overrightarrow{P_1P}$ is perpendicular to $\mathbf{m} = 3\mathbf{i} - \mathbf{j} + 2\mathbf{k}$.

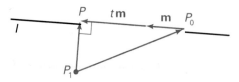

From the figure, $\overrightarrow{P_1P} = \overrightarrow{P_1P_0} + t\mathbf{m}$.

$$\overrightarrow{P_1P} = (2 - 7)\mathbf{i} + (1 - (-6))\mathbf{j} + (1 - 4)\mathbf{k} + t(3\mathbf{i} - \mathbf{j} + 2\mathbf{k})$$
$$= (-5 + 3t)\mathbf{i} + (7 - t)\mathbf{j} + (-3 + 2t)\mathbf{k}$$

Since $\overrightarrow{P_1P}$ is perpendicular to $\mathbf{m}$,

$$[(-5 + 3t)\mathbf{i} + (7 - t)\mathbf{j} + (-3 + 2t)\mathbf{k}] \cdot [3\mathbf{i} - \mathbf{j} + 2\mathbf{k}] = 0.$$

Solving this equation for t, we find that $t = 2$. Substituting 2 for t in the equation for $\overrightarrow{P_1P}$, we obtain

$$\overrightarrow{P_1P} = \mathbf{i} + 5\mathbf{j} + \mathbf{k}.$$

Hence, $\|\overrightarrow{P_1P}\| = \sqrt{1^2 + 5^2 + 1^2} = 3\sqrt{3}$.   ∎

Figure 9-8 indicates how $\mathbf{r} = (1 - t)\mathbf{r}_0 + t\mathbf{r}_1$ varies as t varies. Notice that when $t = 0$, $\mathbf{r} = \mathbf{r}_0$. When $t = 1$, $\mathbf{r} = \mathbf{r}_1$. Thus, $\mathbf{r} = (1 - t)\mathbf{r}_0 + t\mathbf{r}_1$ traces out $\overrightarrow{P_0P_1}$ if $0 \le t \le 1$. In particular, when $t = \dfrac{1}{2}$, then the midpoint of $\overrightarrow{P_0P_1}$ has position vector $\mathbf{r} = \dfrac{1}{2}(\mathbf{r}_0 + \mathbf{r}_1)$.

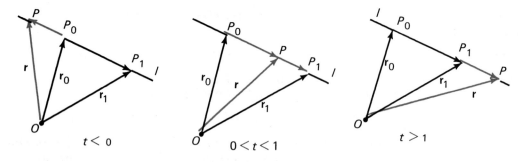

Figure 9-8

EXERCISES 9-3

In Exercises 1–4, find a vector equation of the line that contains

A 1. $(1, 2, 0)$ and is parallel to $\mathbf{i} - \mathbf{j} + 3\mathbf{k}$.   2. $(1, 1, 1)$ and is parallel to $-\mathbf{i} - \mathbf{j} - \mathbf{k}$.

   3. the origin and $(2, 3, -1)$.   4. $(1, -2, 4)$ and $(2, 1, 3)$.

In Exercises 5 and 6, find the point where the line intersects the given plane.

   5. $\mathbf{r} = (3 + t)\mathbf{i} + 2t\mathbf{j} + (2 - t)\mathbf{k}$;   6. $\mathbf{r} = (3 + t)\mathbf{i} + (1 + t)\mathbf{j} + 2t\mathbf{k}$;
     xy-plane xz-plane

In Exercises 7 and 8, find the distance from the origin to the line l.

   7. $\mathbf{r} = (3 + t)\mathbf{i} + (4 - 2t)\mathbf{j} + (-2 + 2t)\mathbf{k}$ 8. $\mathbf{r} = (2 + t)\mathbf{i} + (4 + 2t)\mathbf{j} + (-2 - t)\mathbf{k}$

In Exercises 9 and 10, find the distance from the point P_1 to the line l.

   9. $P_1(2, 0, 1)$, l: $\mathbf{r} = (3 + t)\mathbf{i} + (4 + t)\mathbf{j} + t\mathbf{k}$

  10. $P_1(2, 2, 1)$, l: $\mathbf{r} = (3 + t)\mathbf{i} + (4 - 2t)\mathbf{j} + (1 - t)\mathbf{k}$

In Exercises 11 and 12, find a vector equation of the line that contains

B 11. $(0, -1, 3)$ and is parallel to the line through $(2, 0, 1)$ and $(3, 2, 0)$.

  12. $(1, 2, 3)$ and is parallel to the line $\mathbf{r} = (1 + t)\mathbf{i} + 2t\mathbf{j} + (1 - t)\mathbf{k}$.

In Exercises 13 and 14, show that l_1 and l_2 are the same line. (Hint: Find two points on l_1 and show that they are on l_2.)

  13. l_1: $\mathbf{r} = (2 + t)\mathbf{i} + (-1 - 2t)\mathbf{j} + (1 - t)\mathbf{k}$
      l_2: $\mathbf{r} = (1 - s)\mathbf{i} + (1 + 2s)\mathbf{j} + (2 + s)\mathbf{k}$

  14. l_1: $\mathbf{r} = (5 - 2t)\mathbf{i} + 3t\mathbf{j} + (2 - t)\mathbf{k}$
      l_2: $\mathbf{r} = (1 + 2s)\mathbf{i} + (6 - 3s)\mathbf{j} + s\mathbf{k}$

In Exercises 15–18, find the points, if any, where the given line intersects the given sphere.

Example $\mathbf{r} = (1 + t)\mathbf{i} + 2\mathbf{j} - t\mathbf{k}$, $x^2 + y^2 + z^2 - 4y - 4z - 33 = 0$

Solution   Let $\mathbf{r} = x\mathbf{i} + y\mathbf{j} + z\mathbf{k}$. Then $x = 1 + t$, $y = 2$, $z = -t$.
                     Substitute into the equation of the sphere:

$$(1 + t)^2 + 2^2 + (-t)^2 - 4 \cdot 2 - 4(-t) - 33 = 0$$

                     This equation reduces to $t^2 + 3t - 18 = 0$, or $(t + 6)(t - 3) = 0$. Thus
                     $t = -6$ or $t = 3$, and the points of intersection are $(-5, 2, 6)$ and
                     $(4, 2, -3)$.

  15. $\mathbf{r} = 2t\mathbf{i} + t\mathbf{j} - t\mathbf{k}$, $x^2 + y^2 + z^2 - 4x + 2y - 12 = 0$

  16. $\mathbf{r} = (1 + t)\mathbf{i} + 2t\mathbf{k}$, $x^2 + y^2 + z^2 + 2x - 4y + 3z - 18 = 0$

17. $\mathbf{r} = \mathbf{i} + 2t\mathbf{j} + (2 - t)\mathbf{k}, x^2 + y^2 + z^2 - 6x - 2y + 2z + 2 = 0$

18. $\mathbf{r} = (2 + t)\mathbf{i} + t\mathbf{k}, x^2 + y^2 + z^2 - 2y = 0$

In Exercises 19 and 20, find a vector equation of the line that

19. passes through $(1, 2, 0)$ and makes equal angles with $\mathbf{i}$, $\mathbf{j}$, and $\mathbf{k}$.

20. passes through the origin and makes equal angles with $\mathbf{i} + \mathbf{j}$, $\mathbf{j} + \mathbf{k}$, and $\mathbf{i} + \mathbf{k}$.

C 21. Let $\mathbf{r} = x\mathbf{i} + y\mathbf{j} + z\mathbf{k}$, $\mathbf{r}_0 = x_0\mathbf{i} + y_0\mathbf{j} + z_0\mathbf{k}$, and $\mathbf{m} = d\mathbf{i} + e\mathbf{j} + f\mathbf{k}$. Show that the vector equation $\mathbf{r} = \mathbf{r}_0 + t\mathbf{m}$ of the line l is equivalent to

$$x = x_0 + dt, \ y = y_0 + et, \ z = z_0 + ft.$$

These are the **scalar parametric equations** of l.

22. Show that if the line l described in Exercise 21 is not parallel to any coordinate axis, then it can be described by equations of the form

$$\frac{x - x_0}{d} = \frac{y - y_0}{e} = \frac{z - z_0}{f}.$$

These are the **symmetric equations** of l.

9-4   Planes

Objective: To apply vectors to planes in space.

Let P_0 be a point and $\mathbf{n}$ be a nonzero vector. The set of all points P such that $\overrightarrow{P_0P}$ is perpendicular to $\mathbf{n}$, that is, such that $\mathbf{n} \cdot \overrightarrow{P_0P} = 0$ (Figure 9-9), is called the *plane through P with **normal vector** $\mathbf{n}$*. (The term normal vector is used to describe a vector that is perpendicular to a plane.) Letting $\mathbf{r}_0 = \overrightarrow{OP_0}$ and $\mathbf{r} = \overrightarrow{OP}$, we have the following as a vector equation of Q.

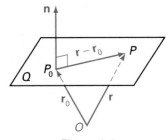

Figure 9-9

$$\mathbf{n} \cdot (\mathbf{r} - \mathbf{r}_0) = 0$$

Now introduce a coordinate system and let $\mathbf{n} = a\mathbf{i} + b\mathbf{j} + c\mathbf{k}$, $\mathbf{r}_0 = x_0\mathbf{i} + y_0\mathbf{j} + z_0\mathbf{k}$, and $\mathbf{r} = x\mathbf{i} + y\mathbf{j} + z\mathbf{k}$. Then

$$\mathbf{n} \cdot (\mathbf{r} - \mathbf{r}_0) = 0$$

becomes

$$(a\mathbf{i} + b\mathbf{j} + c\mathbf{k}) \cdot [(x - x_0)\mathbf{i} + (y - y_0)\mathbf{j} + (z - z_0)\mathbf{k}] = 0.$$

Therefore, a scalar equation of Q is given by:

$$a(x - x_0) + b(y - y_0) + c(z - z_0) = 0$$

This equation also can be written in the form

$$ax + by + cz + d = 0,$$

where $d = -(ax_0 + by_0 + cz_0)$.

It can be shown (Exercise 24, page 337) that *the graph of every equation of this form is a plane having* $a\mathbf{i} + b\mathbf{j} + c\mathbf{k}$ *as a normal vector* provided a, b, and c are not all 0.

In drawing a plane it often is sufficient to show only the part of the plane that lies in the first octant. We do this by drawing its **traces**. These are the lines in which the given plane intersects the coordinate planes. It often is helpful to plot the **intercepts** of the given plane. These are the points where the coordinate axes intersect the given plane.

Example 1   Graph the planes whose equations are given.

(a) $4x + 3y + 2z = 8$ (b) $z = x + y - 1$ (c) $2x + \dfrac{3}{2}y = 6$

Solution

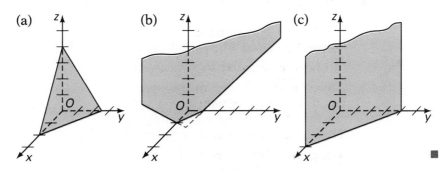

Note that in space the graph of $2x + \dfrac{3}{2}y = 6$ in Example 1 is a *plane*, *not* a line. The absence of z simply means that the plane is parallel to the z-axis.

Example 2   Find a vector equation of the line l containing $(0, 0, 3)$ and perpendicular to the plane $4x - 2y + z - 5 = 0$.

Solution   A vector perpendicular to the plane and therefore parallel to l is
$$\mathbf{m} = 4\mathbf{i} - 2\mathbf{j} + \mathbf{k}.$$
The position vector of $(0, 0, 3)$ is $\mathbf{r}_0 = 3\mathbf{k}$. Substitute these values into $\mathbf{r} = \mathbf{r}_0 + t\mathbf{m}$ to obtain
$$\mathbf{r} = 3\mathbf{k} + t(4\mathbf{i} - 2\mathbf{j} + \mathbf{k}), \text{ or } \mathbf{r} = 4t\mathbf{i} - 2t\mathbf{j} + (t + 3)\mathbf{k}. \quad \blacksquare$$

Example 3   Find an equation of the plane Q that contains the point $(2, -1, 1)$ and is perpendicular to the line $\mathbf{r} = (2 + t)\mathbf{i} - 3t\mathbf{j} + (1 + 2t)\mathbf{k}$.

Solution   Writing the equation of the line as
$$\mathbf{r} = (2\mathbf{i} + \mathbf{k}) + t(\mathbf{i} - 3\mathbf{j} + 2\mathbf{k}),$$
we see that $\mathbf{i} - 3\mathbf{j} + 2\mathbf{k}$ is parallel to the line and hence perpendicular to Q. Thus, $a(x - x_0) + b(y - y_0) + c(z - z_0) = 0$ becomes
$$1(x - 2) - 3(y + 1) + 2(z - 1) = 0,$$
or
$$x - 3y + 2z - 7 = 0. \quad \blacksquare$$

Example 4   Find an equation of the plane that passes through $(1, -1, -3)$, $(0, 3, 2)$, and $(0, 0, -2)$.

Solution   Substitute the coordinates of the given points into $ax + by + cz + d = 0$ to obtain
$$a - b - 3c + d = 0$$
$$3b + 2c + d = 0$$
$$-2c + d = 0$$
Solve this system for c, b, and a in terms of d.
$$c = \frac{d}{2}$$
$$b = \frac{1}{3}(-2c - d) = \frac{1}{3}(-d - d) = -\frac{2d}{3}$$
$$a = b + 3c - d = -\frac{2d}{3} + \frac{3d}{2} - d = -\frac{d}{6}$$
To avoid fractions, set $d = -6$ and obtain $a = 1, b = 4, c = -3$. Therefore, an equation of the plane is
$$x + 4y - 3z - 6 = 0. \quad \blacksquare$$

Next we will show that the distance between the point $P_1(x_1, y_1, z_1)$ and the plane $ax + by + cz + d = 0$ is given by:

$$D = \frac{|ax_1 + by_1 + cz_1 + d|}{\sqrt{a^2 + b^2 + c^2}}$$

First, choose any point $P_0(x_0, y_0, z_0)$ of the plane and let $\mathbf{n} = a\mathbf{i} + b\mathbf{j} + c\mathbf{k}$. We see from Figure 9-10 that

$$D = \|\overrightarrow{P_0P_1}\| \, |\cos \theta| = \frac{|\mathbf{n} \cdot \overrightarrow{P_0P_1}|}{\|\mathbf{n}\|}.$$

Since
$$\overrightarrow{P_0P_1} = (x_1 - x_0)\mathbf{i} + (y_1 - y_0)\mathbf{j} + (z_1 - z_0)\mathbf{k},$$
this becomes

$$\begin{aligned} D &= \frac{|a(x_1 - x_0) + b(y_1 - y_0) + c(z_1 - z_0)|}{\sqrt{a^2 + b^2 + c^2}} \\ &= \frac{|ax_1 + by_1 + cz_1 - (ax_0 + by_0 + cz_0)|}{\sqrt{a^2 + b^2 + c^2}} \end{aligned}$$

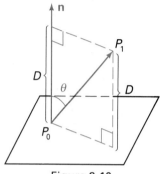

Figure 9-10

Since the quantity in parentheses is $-d$, the formula is established.

Example 5   (a) Verify that the planes Q_1: $6x - 2y + 4z = 8$ and Q_2: $3x - y + 2z = 13$ are parallel. (b) Find the distance between them.

Solution   (a) $\mathbf{n}_1 = 6\mathbf{i} - 2\mathbf{j} + 4\mathbf{k}$ is normal to Q_1 and $\mathbf{n}_2 = 3\mathbf{i} - \mathbf{j} + 2\mathbf{k}$ is normal to Q_2. Since $\mathbf{n}_1 = 2\mathbf{n}_2$, $\mathbf{n}_1$ and $\mathbf{n}_2$ are parallel. This implies that Q_1 and Q_2 are parallel.
   (b) Choose $P_1(0, 0, 2)$ on Q_1. The distance we seek is the distance D between P_1 and Q_2.

Therefore, $D = \dfrac{|3 \cdot 0 - (-1)0 + 2 \cdot 2 - 13|}{\sqrt{3^2 + (-1)^2 + 2^2}} = \dfrac{9}{\sqrt{14}}.$ ∎

EXERCISES 9-4

In Exercises 1–4, (a) find an equation of the plane through P_0 that is perpendicular to the given vector or line. (b) Then draw the first-octant part of the plane.

A 1. $P_0(0, 0, 2)$; $2\mathbf{i} + \mathbf{j} + 2\mathbf{k}$ 2. $P_0(4, 3, 2)$; $2\mathbf{j} + 3\mathbf{k}$

3. $P_0(2, 3, 4)$; $\mathbf{r} = (1 + t)\mathbf{i} + 2t\mathbf{k}$ 4. $P_0(1, 1, 1)$; $\mathbf{r} = (-1 + 2t)\mathbf{i} + t\mathbf{j} + 3t\mathbf{k}$

In Exercises 5–8, find a vector equation of the line through P_1 that is perpendicular to Q.

5. $P_1(3, 2, 1)$; $Q: 2x + y - 2z + 6 = 0$

6. $P_1(2, 0, 1)$; $Q: x - 2y + 3z = 0$

7. $P_1(0, -2, 2)$; $Q: 3x + 2y - 6z - 4 = 0$

8. $P_1(1, 2, 3)$; $Q: 3x - 4z = 5$

9–12. Find the distance between P_1 and Q in Exercises 5–8.

In Exercises 13 and 14, find an equation of the plane containing

13. $(1, 3, 2)$, $(0, 1, -1)$, and $(0, 1, 1)$.

14. $(1, 2, 3,)$, $(0, 1, 2)$, and $(0, 0, -1)$.

In Exercises 15 and 16, (a) verify that the given planes are parallel and then (b) find the distance between them.

15. $x - 2y + 2z + 6 = 0$; $2x - 4y + 4z - 5 = 0$

16. $2x + 4y - 5z = 0$; $2x + 4y - 5z + 10 = 0$

In Exercises 17 and 18, find the point where the line *l* intersects the plane Q. (Hint: See the example preceding Exercise 15, Section 9-3, page 331.

B 17. $l: \mathbf{r} = (3 - t)\mathbf{i} - 2t\mathbf{j} + (1 + t)\mathbf{k}$
$Q: 2x + y + 3z = 6$

18. $l: \mathbf{r} = 2t\mathbf{i} + (2 - t)\mathbf{j} + \mathbf{k}$
$Q: x + y - 3z = 1$

In Exercises 19 and 20, show that the line *l* is parallel to the plane Q. (Hint: Show that *l* and Q have no point in common.)

19. $l: \mathbf{r} = (2 + t)\mathbf{i} + 4t\mathbf{j} + (1 + 2t)\mathbf{k}$
$Q: 2x - y + z + 3 = 0$

20. $l: \mathbf{r} = (3 - 2t)\mathbf{i} + (1 + t)\mathbf{j} + 3t\mathbf{k}$
$Q: x - y + z = 0$

In Exercises 21 and 22, find a vector equation of the line of intersection of the given planes. (Hint: First find two points that belong to both planes; that is, find two simultaneous solutions of the given equations.)

21. $x + y - z = 3$; $x - y + 3z = 1$

22. $x + 2y - z = 1$; $x - 4y + z = 5$

23. Let *l* be a line, P be a point not on *l*, and Q be a plane that does not contain P and is neither parallel nor perpendicular to *l*. How many planes are there (none, one, or infinitely many) that
(a) contain P and are perpendicular to *l*?
(b) contain P and are parallel to *l*?
(c) contain P and are perpendicular to Q?
(d) contain P and are parallel to Q?
(e) contain *l* and are perpendicular to Q?
(f) contain *l* and are parallel to Q?

C 24. Show that the graph of $ax + by + cz + d = 0$ (a, b, and c not all 0) is a plane. (Hint: First show that there is a solution (x_0, y_0, z_0) of the given equation. Then show that the given equation is equivalent to $a(x - x_0) + b(y - y_0) + c(z - z_0) = 0$ and hence to one of the form $\mathbf{n} \cdot (\mathbf{r} - \mathbf{r}_0) = 0$.)

The cross product of two vectors

$$\mathbf{t} = t_1\mathbf{i} + t_2\mathbf{j} + t_3\mathbf{k} \text{ and } \mathbf{v} = v_1\mathbf{i} + v_2\mathbf{j} + v_3\mathbf{k}$$

is given by

$$\mathbf{t} \times \mathbf{v} = (t_2v_3 - t_3v_2)\mathbf{i} + (t_3v_1 - t_1v_3)\mathbf{j} + (t_1v_2 - t_2v_1)\mathbf{k}.$$

1. Show that the cross product of two parallel vectors is **0**.

2. Show that if **a** and **b** are vectors then $\mathbf{a} \times \mathbf{b}$ is perpendicular to each of the vectors **a** and **b**.

Self Quiz

9-3 | 9-4

1. Find a vector equation of the line passing through $(-1, 3, -2)$ and $(5, 4, -7)$.

2. Find the distance from the origin to the line defined by $\mathbf{r} = (1 + t)\mathbf{i} + (-5 + 2t)\mathbf{j} + (3 - t)\mathbf{k}$.

3. Find the point where the line $\mathbf{r} = (6 - 2t)\mathbf{i} + (3 + 4t)\mathbf{j} + (-2 + 5t)\mathbf{k}$ intersects the xy-plane.

4. Find a scalar equation of the plane which passes through $(-5, 4, 9)$ and is perpendicular to the line $\mathbf{r} = (1 + 3t)\mathbf{i} - t\mathbf{j} + 8t\mathbf{k}$.

5. Find an equation of the plane that passes through $(3, -5, 0)$, $(3, -1, 2)$, and $(0, 1, 1)$.

6. Find a vector equation of the line which passes through $(5, 2, -1)$ and is perpendicular to the plane $3x - 4y + z - 2 = 0$.

ADDITIONAL PROBLEMS

1. Find the distance between the point $(2, -3, 1)$ and the plane $7x - 4y + 4z - 3 = 0$.

2. (a) Show that
   $$l_1: \mathbf{r} = (2 + 5t)\mathbf{i} + (1 + 4t)\mathbf{j} + (-2 + t)\mathbf{k} \text{ and}$$
   $$l_2: \mathbf{r} = (2 + 2t)\mathbf{i} + (1 + t)\mathbf{j} + (-2 - 2t)\mathbf{k}$$
   intersect in $P(2, 1, -2)$.
   (b) Show that $\mathbf{n} = 3\mathbf{i} - 4\mathbf{j} + \mathbf{k}$ is perpendicular to both l_1 and l_2.
   (c) Use the results of (a) and (b) to find an equation of the plane that contains l_1 and l_2.

3. Express $\mathbf{v} = -2\mathbf{i} + 3\mathbf{j} + \mathbf{k}$ as a linear combination of $\mathbf{a} = \mathbf{i} + \mathbf{j}$, $\mathbf{b} = \mathbf{j} - \mathbf{k}$, and $\mathbf{c} = \mathbf{i} - \mathbf{k}$.

4. Find the center and radius of the sphere
   $$x^2 + y^2 + z^2 - 4x + 28y + 10z = 0.$$

5. Find the angle between $2\mathbf{i} + 2\mathbf{j}$ and $\mathbf{j} - \mathbf{k}$.

6. Find a vector equation of the line containing $(3, 7, -2)$ and $(5, -1, 4)$.

7. Find an equation of the sphere with center at $(-1, 0, 2)$ and passing through the point $(5, 7, -4)$.

8. Show that
   $$\frac{\|\mathbf{a} + \mathbf{b}\|^2 - \|\mathbf{a} - \mathbf{b}\|^2}{4} = \mathbf{a} \cdot \mathbf{b}.$$
   (Hint: Recall that $\|\mathbf{v}\|^2 = \mathbf{v} \cdot \mathbf{v}$.)

9. Find the distance between the parallel planes $12x - 4y + 3z - 12 = 0$ and $12x - 4y + 3z - 4 = 0$.

10. Find the point where the line $\mathbf{r} = (4 - t)\mathbf{i} + 3t\mathbf{j} - 2t\mathbf{k}$ intersects the plane $3x - y + 7z - 2 = 0$.

11. Find an equation of the line that passes through $(2, -3, 1)$ and is parallel to the line $\mathbf{r} = (7 - t)\mathbf{i} - 4t\mathbf{j} + (-1 + 6t)\mathbf{k}$.

12. Find the point of intersection of the lines $\mathbf{r} = 3t\mathbf{i} + (-4 + t)\mathbf{j} - 2t\mathbf{k}$ and $\mathbf{r} = (-1 + 2t)\mathbf{i} + 5t\mathbf{j} + (1 - t)\mathbf{k}$.

13. Find an equation of the plane that passes through $(7, 1, -5)$ and is perpendicular to the line $\mathbf{r} = (-2 + 3t)\mathbf{i} - 8t\mathbf{j} + (-3 + 9t)\mathbf{k}$.

14. Find a vector equation of the line of intersection of the planes
    $$x + 3y + z = 5 \text{ and } x - y - z = 7.$$
    (Hint: First find two points common to the two planes.)

CHAPTER SUMMARY

1. A rectangular coordinate system for three-dimensional space is formed by adding a z-axis perpendicular to the xy-system at the origin. The plane determined by the x-axis and the z-axis is called the xz-plane. Similarly, the planes determined by the x- and y-axes and the y- and z-axes are called the xy- and yz-planes, respectively. The distance between $P_1(x_1, y_1, z_1)$ and $P_2(x_2, y_2, z_2)$ is given by

    $$P_1P_2 = \sqrt{(x_2 - x_1)^2 + (y_2 - y_1)^2 + (z_2 - z_1)^2}.$$

    The equation for a sphere with center (h, k, l) and radius r is

    $$(x - h)^2 + (y - k)^2 + (z - l)^2 = r^2.$$

2. A vector $\mathbf{a}$ in space may be represented as a linear combination of the three unit vectors $\mathbf{i}$, $\mathbf{j}$, and $\mathbf{k}$; that is,

    $$\mathbf{a} = a_1\mathbf{i} + a_2\mathbf{j} + a_3\mathbf{k}.$$

    The norm of $\mathbf{a}$ is given by

    $$\|\mathbf{a}\| = \sqrt{a_1^2 + a_2^2 + a_3^2}.$$

    The dot product of $\mathbf{a}$ and $\mathbf{b}$ ($\mathbf{b} = b_1\mathbf{i} + b_2\mathbf{j} + b_3\mathbf{k}$) is given by

    $$\mathbf{a} \cdot \mathbf{b} = a_1b_1 + a_2b_2 + a_3b_3.$$

    The formula $\cos \theta = \dfrac{\mathbf{a} \cdot \mathbf{b}}{\|\mathbf{a}\|\,\|\mathbf{b}\|}$ enables us to find the angle between two nonzero vectors. Two vectors $\mathbf{a}$ and $\mathbf{b}$ are orthogonal if and only if $\mathbf{a} \cdot \mathbf{b} = 0$.

3. Given a point P_0 and a vector $\mathbf{m}$, a vector equation for the line l through P_0 and parallel to $\mathbf{m}$ is given by

    $$\mathbf{r} = \mathbf{r}_0 + t\mathbf{m},$$

    where $\mathbf{r} = \overrightarrow{OP}$ (where P is on l), $\mathbf{r}_0 = \overrightarrow{OP_0}$, and t is a scalar variable. A vector equation for a line determined by two points P_0 and P_1 is

    $$\mathbf{r} = (1 - t)\mathbf{r}_0 + t\mathbf{r}_1,$$

    where $\mathbf{r}_0 = \overrightarrow{OP_0}$, $\mathbf{r}_1 = \overrightarrow{OP_1}$, and t is a scalar variable.

4. Through a given point P_0 there is only one plane Q perpendicular to a nonzero vector $\mathbf{n}$. A vector equation for Q is $\mathbf{n} \cdot (\mathbf{r} - \mathbf{r}_0) = 0$, where $\mathbf{r} = \overrightarrow{OP}$ and $\mathbf{r}_0 = \overrightarrow{OP_0}$. If $\mathbf{n} = a\mathbf{i} + b\mathbf{j} + c\mathbf{k}$, $\mathbf{r}_0 = x_0\mathbf{i} + y_0\mathbf{j} + z_0\mathbf{k}$, and $\mathbf{r} = x\mathbf{i} + y\mathbf{j} + z\mathbf{k}$, a scalar equation for Q is

    $$a(x - x_0) + b(y - y_0) + c(z - z_0) = 0.$$

    This equation can be written as $ax + by + cz + d = 0$, where $d = -(ax_0 + by_0 + cz_0)$. The graph of every equation of this form is a plane having $a\mathbf{i} + b\mathbf{j} + c\mathbf{k}$ as a normal vector provided a, b, and c are not all zero. The distance D between the plane $ax + by + cz + d = 0$ and $P_1(x_1, y_1, z_1)$ is given by

$$D = \frac{|ax_1 + by_1 + cz_1 + d|}{\sqrt{a^2 + b^2 + c^2}}.$$

CHAPTER TEST

Exercises 1–3 refer to the sphere with the equation $x^2 + y^2 + z^2 - 6x + 4y - 10z = 11$.

9-1
1. Find the center and radius of the given sphere.

2. Find the distance from the center of the given sphere to the origin.

3. Find the distance from the center of the given sphere to the xz-plane.

Exercises 4–6 refer to the vectors $\mathbf{c} = 5\mathbf{i} + 7\mathbf{j} + \mathbf{k}$ and $\mathbf{d} = 6\mathbf{j} + 8\mathbf{k}$.

9-2
4. Find the norm of $\mathbf{c} - \mathbf{d}$.

5. Find the angle between $\mathbf{c}$ and $\mathbf{d}$.

6. Find a vector orthogonal to both $\mathbf{c}$ and $\mathbf{d}$.

Exercises 7–9 refer to the line l that contains the point $(2, 2, 3)$ and is parallel to $\mathbf{i} - \mathbf{j} - 2\mathbf{k}$.

9-3
7. Find a vector equation for l.

8. Find the distance from l to the origin.

9. Find the point where l intersects the xz-plane.

Exercises 10–12 refer to the plane that contains the point $(6, 6, -4)$ and is perpendicular to $\mathbf{i} + \mathbf{j} + 2\mathbf{k}$.

9-4
10. Give an equation for the plane in (a) vector form and (b) scalar form.

11. Draw the first-octant part of the plane.

12. Find the distance from the point $(0, 0, 0)$ to the plane.

The Distance Between a Point and a Line

The distance formula can be used to find the distance D between a point $Q(x_1, y_1, z_1)$ and a point $P(x, y, z)$ on a line l when x, y, and z are given functions of a variable, such as t. The distance d between a point Q and a line l is defined to be the least value of D.

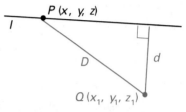

You can approximate d by finding when D stops decreasing and starts increasing. The following program can be used to find D for various values of t. In the program, $x = 2 + 3t$, $y = 1 - t$, and $z = 1 + 2t$.

```
10 INPUT X1, Y1, Z1, T1, T2
20 FOR T = T1 TO T2
30 X = 2 + 3 * T
40 Y = 1 − T
50 Z = 1 + 2 * T
60 D = SQR ( (X − X1) ↑ 2 + (Y − Y1) ↑ 2 + (Z − Z1) ↑ 2)
70 PRINT "T =" ; T ; " " ; "D = " ; INT (10000 * D + .5) / 10000
80 NEXT T
90 END
```

EXERCISES

Let $Q(x_1, y_1, z_1)$ be $Q(6, 1, 2)$ and let T1 $= -10$ and T2 $= 10$.

1. Run the program using these values of X1, Y1, Z1, T1, and T2.
2. (a) From Exercise 1, for what T is D least?
   (b) Find the coordinates $P(x, y, z)$ for the value of T from part (a).
   (c) Find the distance between $P(x, y, z)$ in part (b) and $Q(6, 1, 2)$.
3. (a) Use the method of Example 3 of Section 9-3 to find d.
   (b) Compare your answer in part (a) with your answer in Exercise 2(c).

Designing a Roller Coaster Ride

If you have ridden a roller coaster, you probably were not thinking about the work that went into making the ride. After all, you were sometimes traveling at about 70 mi/h and experiencing forces about three times that of gravity.

Roller coaster engineers and designers, however, must plan carefully to insure the safety and correct functioning of the ride. The planning of such a "space curve" requires much mathematics and consideration of many physical forces. A roller coaster is towed by a chain drive to the top of the first peak. From that point on, gravity takes over. The ups and downs of the course must be planned so that the car does not get stuck in a valley or career past the loading platform.

To build a safe yet exciting ride, planners design curves that are banked. This banking turns the force due to gravity and the centrifugal force tending to pull the car outward into a resultant force directed toward the track. The formulas needed to design the proper bank depend on the notion of a space curve, vectors, and the notion of the curvature of a graph.

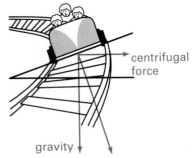

Many roller coaster rides today are built so that riders can travel in one or more vertical loops. These loops are built in such a way that riders slow down as they travel up the loop, speed up at the top, and then slow down as they come out of the loop. A loop that produces this effect is not a circle, but rather a fish-shaped or teardrop curve, called a clothoid. This mathematical curve was first discovered in the eighteenth century. However, its application to roller coaster design was not appreciated until the 1970s.

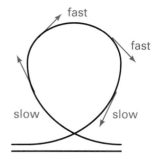

Guide to Review and Reference

Chapter 1

1. The terminal side of an angle in standard position passes through $(-5, 2\sqrt{6})$. Find the values of the six trigonometric functions of the angle.

2. Find $\csc 135° \cdot \cos(-45°) - \sec 240°$.

3. When the angle of elevation of the sun is 78°, a tree casts a 13 m shadow. How tall is the tree to the nearest meter?

4. If $\cos \theta = \dfrac{1}{3}$ and $270° < \theta < 360°$, find $\sin \theta$ and $\tan \theta$.

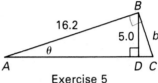

5. Using tables or a calculator, find θ and b if $\angle ABC = 90°$, $AB = 16.2$, and $BD = 5.0$.

Exercise 5

6. Using tables or a calculator, find $\tan 72°20' + \sec 132.7°$ to four decimal places.

7. (a) Convert to radian measure: $-30°$, $18°$

   (b) Convert to degree measure: $\dfrac{3\pi}{4}$, 4

8. Find the length of an arc of a circle of radius 15 cm that is intercepted by a central angle of 165°.

Chapter 2

1. Find the exact value of:

   (a) $\cos \dfrac{17\pi}{3}$ (b) t if $\sec t = -2$ and $\pi < t < 2\pi$

2. A particle starts at $(-2, 0)$ and moves counterclockwise around the circle $x^2 + y^2 = 4$ with angular speed 40°/min. Where will the particle be 6 minutes later?

3. Determine whether the function $f(x) = 3x^2 - 7$ is even, odd, or neither.

4. Graph $y = 1 + \cos x$ for $-\dfrac{\pi}{2} \le x \le 2\pi$.

5. Evaluate: (a) $\text{Tan}^{-1}(-1)$ (b) $\sec\left(\text{Sin}^{-1}\dfrac{3}{5}\right)$ (c) $\text{Cos}^{-1}\left(\cos\dfrac{5\pi}{6}\right)$

6. Graph $f(x) = -\cot x$ for $-\pi < x < \pi$. Then graph $g(x) = \tan x$ on the same axes. Find all points of intersection.

Chapter 3

1. If $\cos x = -0.6$ and $\sin x > 0$, find:

   (a) $\cos\left(x + \dfrac{\pi}{6}\right)$ (b) $\sin\left(x - \dfrac{\pi}{2}\right)$ (c) $\tan\left(x - \dfrac{\pi}{3}\right)$

2. Prove the identity: $\sin 2x \cdot \tan x + \cos 2x = 1$

3. Solve $2\cos^4 x - 5\cos^2 x + 2 = 0$ for $0 \le x \le \pi$.

4. Express $\dfrac{1 + \tan\theta\,(\csc\theta - \cot\theta\sec\theta)}{\cot\theta}$ as a single trigonometric function.

5. Prove: $\dfrac{\sec^2 x}{\sec x - 1} = \dfrac{\sec x + 1}{\sin^2 x}$ 6. Find $\tan 15°$ using identities.

Chapter 4

1. In $\triangle TUB$, $TU = 13$, $UB = 8$, and $\angle U = 60°$. Find BT. Leave your answer in simplest radical form.

2. In $\triangle VAT$, $VA = 300$, $\angle V = 75°$, and $\angle A = 45°$. Find TV. Leave your answer in simplest radical form.

3. The measure of $\angle A$ in $\triangle JAR$ may be found by using either the law of sines or the law of cosines. Find A using both methods.

4. Find the area of $\triangle JAR$.

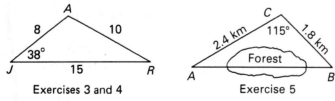

Exercises 3 and 4   Exercise 5

5. Two road crews are clearing a trail through a forest from points A and B. The crews cannot see each other. However, they both can observe a tower at point C. A lookout at the tower reports that $\angle ACB$ is 115°. If the distances to the tower are as marked, at what angles A and B should the crews begin their work?

6. In $\triangle CAN$, $\angle C = 35°$ and $CN = 20$. For what values of NA is it true that there is a unique triangle for the given information?

Chapter 5

For each of the given functions (a) graph the function for $-2\pi \le x \le 2\pi$, and (b) give the period and amplitude.

1. $y = 5 \cos \dfrac{x}{2} + 2$

2. $y = \sin\left(2x + \dfrac{\pi}{3}\right)$

3. Find an equation for the sinusoid shown.

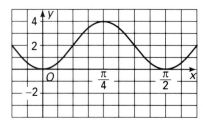

4. Find all solutions of $-\cos\theta + \sqrt{3}\sin\theta = 1$ in the interval $0° \le \theta < 360°$.

5. An object hanging at rest is given an initial upward velocity of 6π cm/sec. Thereafter, it oscillates with amplitude 3 cm. Find an equation giving its position s as a function of time t.

6. Graph $y = \sin x + \cos 2x$ by plotting points for $0 \le x \le 2\pi$.

Chapter 6

1. Find a unit vector orthogonal to $-15\mathbf{i} - 20\mathbf{j}$.

2. Find the angle between $\mathbf{i} + 3\mathbf{j}$ and $\mathbf{i} - 3\mathbf{j}$.

3. A particle is acted on by the following forces measured counterclockwise from the horizontal:
   $\mathbf{F}_1$: 10 N at 60° $\mathbf{F}_2$: 15 N at 0° $\mathbf{F}_3$: 4 N at 120°
   Find the magnitude and direction of the additional force that is needed to keep the particle stationary.

4. A 100 kg box rests on a ramp with a 30° slope. What frictional force is needed to keep the box stationary on the ramp?

5. A ship sails due east for 300 km and then due south for 50 km. What are its distance and bearing from its starting point?

Chapter 7

1. Graph $r = 3 - 2 \sin \theta$ using polar coordinates.

2. Find a polar equation for $x + \sqrt{2}y = 0$.

3. Simplify $\dfrac{i + (2 - i)(3 + 2i)}{(3 + i)}$.

4. Convert the complex number $\left(\sqrt{2}, -i\sqrt{2}\right)$ to polar form and find its reciprocal in polar form.

5. Find $\left(\sqrt{3} + i\right)^{-6}$. Express your answer in $x + yi$ form.

6. Express the three cube roots of -1 in both $x + yi$ form and polar form.

Chapter 8

1. Find the 10th term of the geometric sequence
   $2, -2\sqrt{3}, 6, -6\sqrt{3}, \ldots$.

2. Find the sum of the infinite geometric series
   $72 - 24 + 8 - \dfrac{8}{3} + \cdots$.

3. Approximate each of the following by using the first four terms of the appropriate Taylor series.

   (a) $\sin 2.6$ (b) $\cos \dfrac{2\pi}{3}$ (c) $e^{0.25}$ (d) $\cosh 1.8$ (e) $\sinh \dfrac{\pi}{4}$

4. Find the value of each of the following.

   (a) $e^{\frac{3\pi i}{4}}$ (b) $e^{-\frac{\pi i}{2}}$

Chapter 9

1. Find a vector orthogonal to both $\mathbf{i} + \mathbf{j} - 2\mathbf{k}$ and $2\mathbf{i} - \mathbf{j}$.

2. How far is the center of the sphere $x^2 + y^2 + z^2 - 2y + 6z = 20$ from the point where the plane $x + 2y + z = 12$ intersects the y-axis?

3. Give vector and scalar equations for the plane containing $(-1, 0, 3)$ and perpendicular to $2\mathbf{j} - \mathbf{k}$.

4. Give the vector equation for a line through $(1, 2, -4)$ perpendicular to the plane $y - z = 0$.

5. Find the distance between the point and the plane in Exercise 4.

Comprehensive Test B

Chapter 1

1. The terminal side of an angle θ in standard position passes through $(8, -\sqrt{17})$. What is $\csc \theta$?

   (a) $-\dfrac{9\sqrt{17}}{17}$ (b) $-\dfrac{9\sqrt{17}}{8}$ (c) $-\dfrac{\sqrt{17}}{8}$ (d) $\dfrac{\sqrt{17}}{8}$

2. Evaluate $\tan 120° \cdot \cos 210° - \csc 315° \cdot \sin 45°$.

   (a) 0   (b) $\dfrac{1}{2}$ (c) 2   (d) $2\dfrac{1}{2}$

3. Find x to the nearest tenth.

   (a) 5.1   (b) 16.0
   (c) 40.1   (d) 43.9

   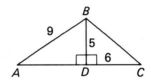

4. Find $\angle ABC$ to the nearest degree.

   (a) 93°   (b) 100°   (c) 106°   (d) 139°

5. Find θ if $\cos \theta = -0.5$ and $-180° < \theta < 180°$.

   (a) 60° or $-60°$ (b) 150° or $-150°$ (c) 120° or $-120°$ (d) $-120°$ or $-60°$

6. The angle of elevation of the sun is 29°10′. Find the height of a tree with a shadow 210 feet long.

   (a) 61 ft   (b) 102 ft   (c) 117 ft   (d) 183 ft

7. If $\csc 595° = \csc x$ for $-90° < x < 90°$, find x.

   (a) 35°   (b) 55°   (c) $-35°$ (d) $-55°$

8. Convert 210° to radian measure.

   (a) $\dfrac{5\pi}{6}$ (b) $\dfrac{3\pi}{4}$ (c) $\dfrac{7\pi}{6}$ (d) $\dfrac{7\pi}{12}$

9. Convert $\dfrac{7\pi}{4}$ to degree measure.

   (a) 315°   (b) 300°   (c) 270°   (d) 135°

10. A circle has radius 4. Find the area of a sector determined by a 60° central angle.

    (a) $\dfrac{4\pi}{3}$ (b) $\dfrac{8\pi}{3}$ (c) $\dfrac{16\pi}{3}$ (d) $\dfrac{2\pi}{3}$

Chapter 2

1. Find $\tan \pi$.
   (a) 0   (b) 0.0549   (c) 1   (d) undefined

2. Find $\tan t$ if $\cos t = -\dfrac{5}{13}$ and $\dfrac{\pi}{2} < t < \pi$.

   (a) $\dfrac{12}{13}$ (b) $\dfrac{-5}{12}$ (c) $\dfrac{12}{5}$ (d) $\dfrac{-12}{5}$

3. Which of the following is *not* an even function?
   (a) $f(x) = \cos x$ (b) $f(x) = |\sin x|$
   (c) $f(x) = \sin |x|$ (d) $f(x) = \tan x$

4. Find the vertical asymptote of $f(x) = \dfrac{x + 5}{x - 3}$.

   (a) $x = 3$ (b) $y = 5$ (c) $y = 3$ (d) $x = -5$

5. Evaluate $\mathrm{Tan}^{-1}\left(\sin \dfrac{\pi}{2}\right)$.

   (a) undefined   (b) $\dfrac{\pi}{2}$ (c) $\dfrac{\pi}{4}$ (d) $-\dfrac{\pi}{4}$

6. Evaluate $\mathrm{Sin}^{-1}\left(\sin \dfrac{\pi}{3}\right)$.

   (a) $\sin \dfrac{\pi}{3}$ (b) $\dfrac{\pi}{3}$ (c) $\sin\left(\mathrm{Sin}^{-1}\dfrac{\pi}{3}\right)$ (d) $\dfrac{2\pi}{3}$

7. Which graph never intersects the x-axis?
   (a) $y = |\cos x|$ (b) $y = \sec x$ (c) $y = \tan x$ (d) $y = \cot x$

8. Find the range of $y = \mathrm{Tan}^{-1}(x)$.

   (a) $0 < y < \dfrac{\pi}{2}$ (b) $-\dfrac{\pi}{2} < y < \dfrac{\pi}{2}$
   (c) $0 < y < \pi$ (d) $-\pi < y < \pi$

9. All of the following are undefined except:
   (a) $\csc\left(-\dfrac{\pi}{2}\right)$ (b) $\sec\left(\dfrac{\pi}{2}\right)$ (c) $\tan\left(\dfrac{\pi}{2}\right)$ (d) $\mathrm{Sin}^{-1}\left(\dfrac{\pi}{2}\right)$

10. A point on a circle of radius 10 m has a linear speed of 15 m/s. Find its angular speed.
    (a) 1.5 radians/s   (b) $0.\overline{6}$ radians/s
    (c) $\dfrac{15}{2\pi}$ radians/s   (d) $\dfrac{4\pi}{3}$ radians/s

Chapter 3

1. For all angles for which it is defined, $\dfrac{\cos \theta}{1 + \sin \theta} =$

   (a) $\sec \theta + \tan \theta$
   (b) $\sec \theta - \tan \theta$
   (c) $\tan \theta - \sec \theta$
   (d) $\sec \theta \tan \theta$

2. Express $\cos \theta + \sin \theta \tan \theta$ as a single trigonometric function.

   (a) $\csc \theta$
   (b) $\cos \theta$
   (c) $\sec \theta$
   (d) $\sin \theta$

3. If $\cos x = \dfrac{3}{5}$ and $\sin x < 0$, evaluate $\cos \left(x + \dfrac{\pi}{3} \right)$.

   (a) $\dfrac{3}{5} - \dfrac{2\sqrt{3}}{5}$
   (b) $\dfrac{3}{10} - \dfrac{2\sqrt{3}}{5}$
   (c) $\dfrac{3\sqrt{3}}{5} - \dfrac{2}{5}$
   (d) $\dfrac{3}{10} + \dfrac{2\sqrt{3}}{5}$

4. For all angles for which it is defined, $(\csc x \tan x - \cos x) \cot x =$

   (a) $\csc x - \sin x$
   (b) $1 - \cot x$
   (c) $\cos x$
   (d) $\sin x$

5. If $\tan x = 0.5$, find $\tan 2x$.

   (a) 0.25
   (b) 0.75
   (c) 1
   (d) $1.\overline{3}$

6. If $-180° \le \theta \le 180°$, for how many values of θ will $\dfrac{1 - \sin \theta}{2} = \cos^2 \theta$?

   (a) 0
   (b) 1
   (c) 2
   (d) 3

7. If $\sin \left(x + \dfrac{\pi}{4} \right) = \dfrac{\sqrt{2}}{3}$, evaluate $\sin x + \cos x$.

   (a) $\dfrac{1}{3}$
   (b) $\dfrac{1}{2}$
   (c) $\dfrac{2}{3}$
   (d) 1

8. The sum of the solutions to $2 \sin^3 \theta - \sin \theta = 0$, $\dfrac{\pi}{2} < \theta < \dfrac{3\pi}{2}$, is:

   (a) 3π
   (b) 2π
   (c) $\dfrac{7\pi}{4}$
   (d) $\sqrt{2}$

Chapter 4

1. According to the law of sines, in any $\triangle ABC$, $AB \cdot \sin A =$

   (a) $AC \cdot \sin B$
   (b) $BC \cdot \sin B$
   (c) $AC \cdot \sin C$
   (d) $BC \cdot \sin C$

2. State the law of cosines for $\triangle FGH$.

   (a) $(FG)^2 = (FH)^2 + (GH)^2 + 2(FH)(GH) \cos H$
   (b) $(FG)^2 = (FH)^2 + (GH)^2 - 2(FH)(GH) \cos H$
   (c) $(FG)^2 = (FH)^2 - (GH)^2 + 2(FH)(FG) \cos F$
   (d) $(FG)^2 = (FH)^2 - (GH)^2 - 2(FH)(GH) \cos H$

3. Two observation posts 3.74 km apart report the angle of elevation of an aircraft to be 37° and 23°. If the aircraft is between the two observation posts, how far is it from the nearer one?

   (a) 0.60 km   (b) 1.69 km   (c) 1.74 km   (d) 2.60 km

4. In $\triangle DEF$, if $DE = 6.2$, $EF = 7.1$, and $\angle E = 28°$, how many possible measures are there for $\angle F$?

   (a) none   (b) one   (c) two   (d) three

5. Two sides of a triangle have lengths 10 and 15. They meet at an angle of 75°. Find the length of the third side.

   (a) 12   (b) 14   (c) 16   (d) 18

6. Find $\cos \theta$.

   (a) $\dfrac{5}{8}$ (b) $\dfrac{13}{20}$

   (c) $-\dfrac{5}{16}$ (d) $\dfrac{5}{16}$

   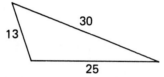

7. Find the area of the triangle shown.

   (a) 105   (b) 120

   (c) 135   (d) 160

8. In $\triangle RAT$, $\angle R = 35°$, $RA = 25$ m, and $AT = 15$ m. The following answers are suggested for RT to the nearest meter.

   (I) 34 m   (II) 25 m   (III) 16 m

   Which is (are) correct?

   (a) I and III   (b) II and III   (c) II only   (d) I only

Chapter 5

1. Which equation has the following as its graph?

   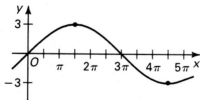

   (a) $y = \dfrac{1}{3} \sin 3x$ (b) $y = 3 \sin 3x$ (c) $y = \dfrac{1}{3} \sin \dfrac{1}{3}x$ (d) $y = 3 \sin \dfrac{1}{3}x$

2. Which of the following coincides with the graph of $y = \cos 2x$?

   (a) $y = \sin\left(2x - \dfrac{\pi}{2}\right)$ (b) $y = \sin 2\left(x + \dfrac{\pi}{2}\right)$

   (c) $y = 2 \cos x$ (d) $y = \sin\left(2x + \dfrac{\pi}{2}\right)$

3. Which of the following has a period of 8?

   (a) $y = 8 \sin x$

   (b) $y = \sin 8x$

   (c) $y = \sin 4\pi x$

   (d) $y = \sin \dfrac{\pi x}{4}$

4. $\cos 50° - \sin 50° =$

   (a) $\sqrt{2} \cos 5°$ (b) $\sqrt{2} \cos 95°$ (c) $2 \cos 5°$ (d) $2 \cos 95°$

5. An object on the end of a spring is pulled 8 cm below its equilibrium point at time $t = 0$ and then released. If the object makes a complete up-and-down movement every 6 s, which of the following describes its motion?

   (a) $y = 8 \sin \left(3\pi t - \dfrac{\pi}{2} \right)$

   (b) $y = 8 \sin \left(6t - \dfrac{\pi}{2} \right)$

   (c) $y = 8 \cos \left(\dfrac{\pi t}{3} + \pi \right)$

   (d) $y = 6 \cos \left(\dfrac{\pi}{4}t - \pi \right)$

6. Which equation has the following as its graph?

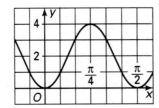

   (a) $y = 2 \sin \left(2x - \dfrac{\pi}{2} \right) + 2$

   (b) $y = 4 \sin 2x$

   (c) $y = 2 \cos 4\left(x - \dfrac{\pi}{4} \right) + 2$

   (d) $y = 4 \cos (x - \pi)$

Chapter 6

1. If $(si + tj) + (ti + 2sj) = 6i - 4j$. Find $s - t$.

   (a) 6   (b) -6 (c) 26   (d) -26

2. A boat has a heading of 18° when crossing a stream that flows due south. If the boat reaches shore directly opposite its launching point and has a speed in still water of 14.0 km/h, what is the speed of the current?

   (a) 4.3 km/h   (b) 4.5 km/h   (c) 13.3 km/h   (d) 14.7 km/h

3. Two forces act on a point. If the magnitudes are 12 N and 18 N and the angles of application (measured counterclockwise from the horizontal) are 0° and 165°, respectively, what is the magnitude of the third force needed to keep the point stationary?

   (a) 13 N (b) 7 N (c) 30 N (d) 285 N

4. Find the angle between **i** + 3**j** and 4**i** − 3**j**.

   (a) 72° (b) 98° (c) 102° (d) 108°

5. (2**i** + 3**j**) · (3**i** − 4**j**) =

   (a) 1 (b) −6 (c) 6**i** − 12**j** (d) 18**i** + **j**

6. Find the work done by a force **F** = 3**i** − 4**j** in moving a point from (5, 2) to (10, −10). Assume force is in newtons, and the distance is in meters.

   (a) 33 J (b) 47 J (c) 60 J (d) 63 J

7. Which vector is not orthogonal to 2**i** − 4**j**?

   (a) 4**i** + 2**j** (b) −2**i** − **j** (c) −4**i** + 2**j** (d) −4**i** − 2**j**

Chapter 7

1. Which of the following is the graph of $r = 2 \cos 2\theta$?

   (a)
   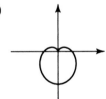

   (b)
   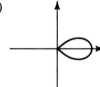

   (c)
   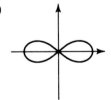

   (d)
   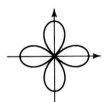

2. Simplify $\dfrac{(3 + i)(1 - 2i)}{2 + i}$.

   (a) $1 - 3i$ (b) $\dfrac{7}{5} - \dfrac{11}{5}i$ (c) $\dfrac{5}{3} - 5i$ (d) $\dfrac{7}{3} - \dfrac{11}{3}i$

3. Find the reciprocal of $1 - i\sqrt{3}$ in polar form.

   (a) $2 (\cos 120° + i \sin 120°)$ (b) $-2 (\cos 120° + i \sin 120°)$

   (c) $-\dfrac{1}{2} (\cos 240° + i \sin 240°)$ (d) $-\dfrac{1}{2} (\cos 120° + i \sin 120°)$

4. Find $\left(\sqrt{3} - i\right)^{-12}$.

   (a) $\dfrac{1}{4096}$

   (b) $\dfrac{1}{4096}(\cos 30° + i \sin 30°)$

   (c) $24 (\cos (-30°) + i \sin (-30°))$

   (d) $729 - 12i$

5. Which of the following is not a cube root of $4\sqrt{2} + \left(4\sqrt{2}\right)i$?

   (a) $-\sqrt{2} + i\sqrt{2}$

   (b) $2 (\cos 315° + i \sin 315°)$

   (c) $2 (\cos 255° + i \sin 255°)$

   (d) $2 (\cos 15° + i \sin 15°)$

Chapter 8

1. Find the sum of the first seven terms of the geometric series with $a = 24$, and $r = -\dfrac{1}{2}$.

   (a) $\dfrac{129}{8}$
   (b) $\dfrac{63}{4}$
   (c) $\dfrac{3}{8}$
   (d) $\dfrac{387}{8}$

2. Solve for x: $\ln x + \ln 1 = e^{\ln 4}$

   (a) 4
   (b) e^4
   (c) 3
   (d) $e^4 - 1$

3. Find $e^{\left(2 + \frac{3\pi}{2}i\right)}$.

   (a) $-e^2$
   (b) $-ie^2$
   (c) $-2i$
   (d) ie^2

4. $x^2 - \dfrac{x^6}{3!} + \dfrac{x^{10}}{5!} - \dfrac{x^{14}}{7!} + \cdots$ is a Taylor series for:

   (a) $\sinh x^2$
   (b) e^{x^2}
   (c) $\cos |x|^2$
   (d) $\sin |x|^2$

5. An approximation for sin 3 obtained by evaluating the first four terms of the Taylor series for $\sin x$ is:

   (a) 0.434
   (b) 0.910
   (c) 0.091
   (d) 0.043

6. If $\sinh x = \dfrac{-12}{13}$, $\cosh x = $

   (a) $\dfrac{\sqrt{313}}{13}$
   (b) $\dfrac{5}{13}$
   (c) $\pm\dfrac{5}{13}$
   (d) $\pm\dfrac{\sqrt{313}}{13}$

Chapter 9

1. Find the distance between $(2, 5, -6)$ and $(6, 2, 6)$.

   (a) 13
   (b) $\sqrt{19}$
   (c) 5
   (d) $\sqrt[3]{169}$

2. Which vector is *not* orthogonal to $\mathbf{i} + 2\mathbf{j} - \mathbf{k}$?

   (a) $5\mathbf{i} + 2\mathbf{j} + 9\mathbf{k}$
   (b) $\mathbf{i} - 2\mathbf{j} + \mathbf{k}$
   (c) $7\mathbf{i} + 7\mathbf{k}$
   (d) $\mathbf{i} - 2\mathbf{j} - 3\mathbf{k}$

3. Give a vector equation for a line through $(2, 0, 5)$ and $(3, -1, 2)$.

(a) $2\mathbf{i} + 5\mathbf{k} + t(\mathbf{i} - \mathbf{j} + 3\mathbf{k})$ (b) $\mathbf{i} + \mathbf{j} + 3t\mathbf{k}$

(c) $2\mathbf{i} + 5\mathbf{k} + t(\mathbf{i} - \mathbf{j} - 3\mathbf{k})$ (d) $t(-\mathbf{i} + \mathbf{j} + 3\mathbf{k})$

4. Find the distance between $P(2, 1, -3)$ and the line whose equation is $r = (-1 + t)\mathbf{i} + 2t\mathbf{j} + (5 - t)\mathbf{k}$.

(a) $\dfrac{5}{6}\sqrt{66}$ (b) 72

(c) $6\sqrt{2}$ (d) $2\sqrt{19}$

5. Give a scalar equation of the plane through the origin and perpendicular to $3\mathbf{i} + 4\mathbf{j} - 12\mathbf{k}$.

(a) $-3x + 4y - 12z = 0$ (b) $3x - 4y + 12z = 0$

(c) $3x + 4y + 12z = 0$ (d) $3x + 4y - 12z = 0$

6. Give a scalar equation of the plane through $(1, 2, 1)$, $(0, 0, 4)$, and $(1, 5, -4)$.

(a) $4x + 2y - 3z = 5$ (b) $11x + 2y + 5z = 20$

(c) $-x + 5y + 3z = 12$ (d) $12x + 3y + 6z = 24$

Spherical Trigonometry

A plane can intersect a sphere in either a point or a circle. If a plane passes through the center of the sphere, the circle of intersection is called a **great circle**. Given any two points on a sphere (not diametrically opposite), there is a unique great circle containing them. The shorter arc of this circle is the shortest path on the sphere connecting the points.

If we assume that Earth is a sphere (which it nearly is), then the **equator** is a great circle as are all the circles on Earth containing the North and South Poles. Circles on Earth parallel to the equator are called **parallels**; circles containing the North and South Poles are called **meridians**. The figure at the left below illustrates parallels (red) and meridians (black).

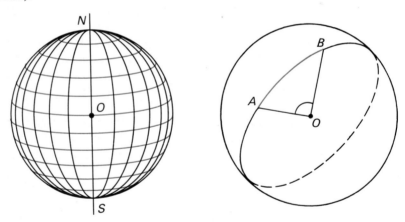

If *A* and *B* are any two points (not diametrically opposite) on a sphere with center *O*, we denote by $\overset{\frown}{AB}$ the shorter great-circle arc connecting them. The measure of this arc is defined to be the measure of ∠ *AOB* in degrees. (See the figure at the right above.) Thus, the measure of a great-circle arc with one end at the North Pole and the other on the equator is 90°.

Example 1   Find the length in kilometers of a 40° great-circle arc on Earth. (The radius of Earth is 6368 km.)

Solution   We use the fact that an arc subtending an angle of *θ* radians on a circle

356

of radius r has length $r\theta$. Since $r = 6368$, the required length is

$$6368 \cdot \frac{40\pi}{180} = 4446 \text{ km.} \quad \blacksquare$$

We can locate a point on Earth by giving its *latitude* and its *longitude*. The **latitude** of a point P is defined to be its angular distance from the equator, measured in degrees. Thus, the latitude of a point is an angle ϕ, $0° \leq \phi \leq 90°$. We write N after this degree measure if P lies above the equator and S if it lies below the equator. To specify the longitude of a point, the meridian through Greenwich, England is chosen and is called the **prime meridian**. The **longitude** of P is the angular distance between the meridian on which P lies and the prime meridian. Thus, the longitude of P is an angle λ, $0° \leq \lambda \leq 180°$. The longitude is designated E or W according as this distance is measured to the east or west of the prime meridian. For example, the latitude and longitude of Chicago, Illinois is 41°50′ N, 87°40′ W and the latitude and longitude of Sydney, Australia, is 34°0′ S, 151°0′ E.

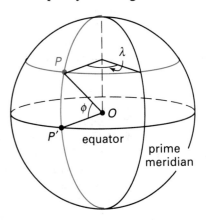

Two great-circle arcs having a common endpoint form a **spherical angle**. We assign to this angle the measure of the planar angle formed by the rays tangent to the arcs at their common endpoint. In the figure at the left below, $\widehat{AB}$ and $\widehat{AC}$ form a spherical angle, $\angle A$. A **spherical triangle** is formed by three great-circle arcs $\widehat{AB}$, $\widehat{BC}$, and $\widehat{CA}$ as shown in the figure at the right below. Spherical triangles on Earth are called **terrestrial triangles**. Since we can measure great-circle arc lengths in degrees, all six parts of spherical triangle ABC are measured in degrees.

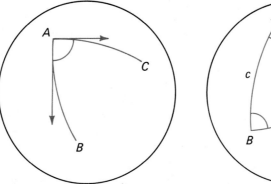
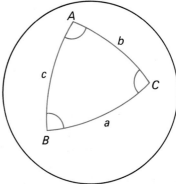

Example 2 Find the measure of ∠ N in spherical triangle *NAB*, in which *N* is at the North Pole, (a) *A* is at Houston (29°45′ N, 95°22′ W), and *B* is at Boston (42°21′ N, 71°0′ W). (b) *A* is at Chicago (41°50′ N, 87°40′ W) and *B* is at Rome (41°50′ N, 12°15′ E).

Solution (a) Since one vertex is at the North Pole and both longitudes are West, we can find the measure of ∠ N by subtracting:

$$\angle N = 95°22' - 71°0'$$
$$= 24°22'$$

(b) Since one vertex is at the North Pole, Chicago has west longitude, and Rome has east longitude, we can find the measure of ∠ N by adding:

$$\angle N = 87°40' + 12°15'$$
$$= 99°55'. \quad \blacksquare$$

One can prove theorems about spherical triangles. In many ways these theorems parallel familiar theorems about plane triangles. For example, the "Pythagorean Theorem" of spherical trigonometry gives a relationship among the three sides of a *spherical* right triangle (with ∠ C = 90°):

$$\cos c = \cos a \cos b.$$

Most terrestrial triangles used in air and sea travel are not right triangles. For such oblique triangles we have the following:

The Law of Sines

$$\frac{\sin a}{\sin A} = \frac{\sin b}{\sin B} = \frac{\sin c}{\sin C}$$

The Law of Cosines for Sides

$$\cos a = \cos b \cos c + \sin b \sin c \cos A$$

with analogous formulas for cos *b* and cos *c*:

$$\cos b = \cos a \cos c + \sin a \sin c \cos B$$
$$\cos c = \cos a \cos b + \sin a \sin b \cos C$$

The Law of Cosines for Angles

$$\cos A = -\cos B \cos C + \sin B \sin C \cos a$$

with analogous formulas for cos *B* and cos *C*:

$$\cos B = -\cos A \cos C + \sin A \sin C \cos b$$
$$\cos C = -\cos A \cos B + \sin A \sin B \cos c$$

Example 3 In spherical triangle ABC, $\angle A = 72°30'$, $\angle B = 124°10'$, and $\angle C = 37°50'$. Find a to the nearest $10'$.

Solution Since three angle measures are given and a is to be found, use the law of cosines for angles. From $\cos A = -\cos B \cos C + \sin B \sin C \cos a$, we obtain:

$$\cos a = \frac{\cos A + \cos B \cos C}{\sin B \sin C}$$

$$= -0.2815$$

Using a calculator, we find that:

$$a = 106.35° = 106° + 0.35 \cdot 60' = 106°20' \quad \blacksquare$$

EXERCISES

Spherical triangle *ABN* has vertex *N* at the North Pole and vertices *A* and *B* at the given cities. Use the table on page 362 to determine the measure of $\angle N$.

1. Boston and Chicago

2. Houston and Rome

3. Rio de Janiero and Dar-es-Salaam

4. Paris and Sydney

Find the required side or angle of the spherical triangle with the given parts. Give answers to the nearest ten minutes or nearest tenth of a degree, as indicated by the given data.

5. $a = 63°20'$; $b = 114°40'$; $c = 98°30'$; $\angle B = ?$

6. $\angle A = 123°10'$; $\angle B = 72°30'$; $\angle C = 63°20'$; $c = ?$

7. $\angle A = 118°30'$; $b = 39°0'$; $c = 87°30'$; $a = ?$

8. $a = 30.0°$; $b = 40.0°$; $c = 87°30'$; $\angle B = ?$

9. $\angle A = 84.5°$; $\angle B = 84.5°$; $\angle C = 53.8°$; $b = ?$

10. $\angle A = 42°10'$; $\angle B = 142°20'$; $c = 116°20'$; $\angle C = ?$

11. $a = 110.6°$; $\angle B = 63.3°$; $c = 37.5°$; $\angle A = ?$

12. $a = 52.5°$; $\angle B = 102.9°$; $\angle C = 94.6°$; $c = ?$

13. A ship traveling on a great circle crosses the equator into the Northern Hemisphere at an angle of 30°. What will be its greatest northern latitude? (Hint: At this point its track is perpendicular to the meridian.)

An important problem in navigation is finding the distance between two points, A and B, on Earth and planning the great-circle course between them. To solve such a problem we use the **terrestrial triangle** that has A, B, and the North Pole N as vertices. Using the latitudes of A and B, we can find the sides b and a. The angle at N can be found from the longitudes of A and B. Thus, finding the great-circle distance d from A to B is just a matter of applying the law of cosines for sides. In navigation, distances are usually given in nautical miles. A **nautical mile** is the length of a one-minute great-circle arc. It equals about 1.15 land miles or 1.85 kilometers.

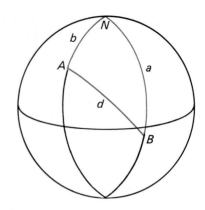

Example 4 Find the great-circle distance between Chicago (41°50′ N, 87°40′ W) and Rome (41°50′ N, 12°15′ E).

Solution

Step 1 Let C, R, and N denote Chicago, Rome, and the North Pole, respectively. Apply the law of cosines for sides to the terrestrial triangle CRN to find $\widehat{CR}$. Since both C and R have the same latitude,

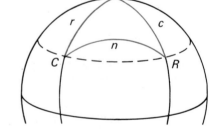

$$c = r = 90° - 41°50′$$
$$= 48°10′$$

Since C has west longitude and R has east longitude

$$N = 87°40′ + 12°15′$$
$$= 99°55′$$

By the law of cosines for sides,

$$\cos n = \cos c \cos r + \sin c \sin r \cos N$$
$$= \cos 48°10′ \cos 48°10′ + \sin^2 48°10′ \cos 99°55′ = 0.3492.$$

Hence $n = 69°34′$.

Step 2 To convert to nautical miles, express n in minutes.

$$n = (69 \cdot 60)′ + 34′ = 4174′$$

Therefore, the great-circle distance from Chicago to Rome is 4174 nautical miles. ■

Example 5 Find the distance between Chicago and Rome along the parallel on which they lie. (Use 3438 nautical miles as the radius of Earth.)

Solution

Use the arclength formula, $s = r\theta$, where r and θ are as shown in the diagram at the right and θ is measured in radians. Since C has latitude 41°50′, $\angle COO' = 48°10′$. From $\triangle COO'$ with $OC = 3438$,

$$r = 3438 \sin 48°10′ = 2562.$$

From Example 4, $N = \theta = 99°55′$, or 99.9167°. Hence,

$$s = 2562 \cdot 99.9167° \cdot \frac{\pi}{180} = 4468.$$

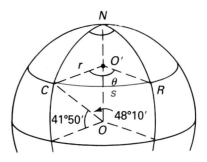

The parallel of latitude distance is 4468 nautical miles. ∎

Examples 4 and 5 illustrate the advantage of using a great-circle track, especially on long trips. Many maps show parallels of latitude as horizontal lines. The figure below shows how this distortion disguises the true situation.

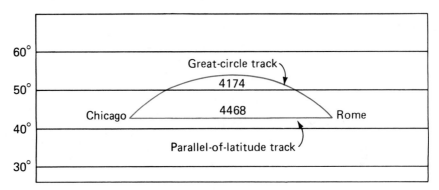

The **course** of a plane or ship is the angle θ, $0° \leq \theta < 360°$, that its path makes with the meridian, measured clockwise from north. The figure at the right shows the **initial course**, θ_1, and the **course on arrival**, θ_2, of the great-circle track from Los Angeles to Jakarta, Indonesia.

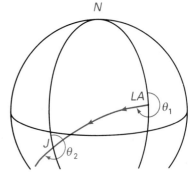

Example 6

A long-range jet is to fly the great-circle track from Los Angeles, 33°54′ N, 118°24′ W to Jakarta, 6°18′ S, 104°48′ E. How far will the jet travel, and what will be its initial course and its course on arrival?

(Solution on next page.)

Solution   Let A, B, and C denote Los Angeles, Jakarta, and the North Pole, respectively. Then we have:

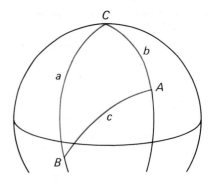

$$a = 90° + 6°18' = 96°18'$$
$$b = 90° - 33°54' = 56°6'$$
$$\angle C = 360° - (118°24' + 104°48')$$
$$= 136°48'$$

$$\cos c = \cos a \cos b + \sin a \sin b \cos C$$
$$= -0.6626$$
$$c = 131°30'$$

Therefore, the plane will travel $131 \times 60 + 30 = 7890$ nautical miles.

$$\cos A = \frac{\cos a - \cos b \cos c}{\sin b \sin c}$$
$$= 0.4180$$
$$\angle A = 65°18'$$

Therefore, the initial course is $360° - 65°18' = 294°42'$.

$$\cos B = \frac{\cos b - \cos a \cos c}{\sin b \sin c}$$
$$= 0.6516$$
$$\angle B = 49°20'$$

Therefore, the course at arrival is $180° + 49°20' = 229°20'$. Note that although the plane is flying to the Southern Hemisphere, its initial course is north of west. ∎

The table below gives latitudes and longitudes of some major cities.

	Latitude	Longitude
Boston	42°21′ N	71°0′ W
Chicago	41°50′ N	87°40′ W
Dar-es-Salaam	6°47′ S	39°15′ E
Houston	29°45′ N	95°22′ W
Mexico City	19°26′ N	99°7′ W
Paris	48°52′ N	2°20′ E
Rio de Janiero	23°0′ S	43°20′ W
Rome	41°50′ N	12°15′ E
San Francisco	37°45′ N	122°27′ W
Sydney	34°0′ S	151°0′ E

EXERCISES

Note: All tracks are great-circle arcs unless otherwise specified. Give your answers to the nearest nautical mile or to the nearest ten minutes.

14. How many nautical miles is Houston from
    (a) the equator?
    (b) the North Pole?
    (c) the South Pole?

15. How many nautical miles is Sydney from
    (a) the equator?
    (b) the North Pole?
    (c) the South Pole?

For Exercises 16–19, note that the relevant terrestrial triangle is a right triangle, so that, for example, the law of cosines for sides reduces to $\cos a = \cos b \cos c$.

16. A ship leaves San Francisco with initial heading 270° and follows a great-circle track. What will the ship's position be after it has traveled 2400 nautical miles? What will its course be then?

17. A plane leaves Houston with an initial course 90° and flies a great-circle track. Find the longitude of the point where the plane crosses the equator. Find its course at that point.

18. A plane leaves Boston with initial course 63°30′. Find the latitude and longitude of the northernmost point of its track. (Hint: At this point the track is perpendicular to the meridian.)

19. What is the position of the northernmost point of the great-circle track from Chicago to Rome? (Hint: The triangle in Example 4 is isosceles.)

In Exercises 20–25, a great-circle track is described. Find (a) its length, (b) the initial course, and (c) the course on arrival.

20. Boston to Rome

21. Paris to Mexico City

22. Boston to Dar-es-Salaam

23. San Francisco to Sydney

24. Houston to Rio de Janiero

25. Rome to Rio de Janiero

Computations with Logarithms

The properties of the function $y = \log_{10} x$ may be useful in making numerical computations. This function is called the **logarithmic function to the base 10** and the y-values are called **logarithms** to the base 10, or **common logarithms**, denoted $\log x$.

In order to make computations with logarithms, it is necessary to apply the four properties listed on page 291 to express the logarithm of a product, quotient, power, root, or a combination of these, in expanded form.

Example 1

Expand $\log \dfrac{(17)^3\sqrt{29}}{13}$.

Solution

Rewrite the expression with fractional exponents: $\dfrac{(17)^3(29)^{\frac{1}{2}}}{13}$.

$$\log \frac{(17)^3\sqrt{29}}{13} = \log \frac{(17)^3(29)^{\frac{1}{2}}}{13} = \log (17)^3(29)^{\frac{1}{2}} - \log 13$$

$$= \log (17)^3 + \log (29)^{\frac{1}{2}} - \log 13$$

$$= 3 \log 17 + \frac{1}{2} \log 29 - \log 13 \quad \blacksquare$$

Table 5 on page 408 gives values to four significant digits of logarithms for certain numbers from 1 to 10. In order to extend the use of this table to numbers between 0 and 1 or to numbers greater than 10, it is helpful first to express such numbers in scientific notation. You should recall that scientific notation involves the product of a number between 1 and 10 and a power of 10. For example, in scientific notation,

$$0.452 = 4.52 \times 10^{-1} \qquad 45.2 = 4.52 \times 10^{1} \qquad 452 = 4.52 \times 10^{2}$$

To save space in the entries in Table 5, all decimal points have been omitted. A decimal point is understood to belong between each pair of digits in the left-hand column of the table, and before each four-digit entry in the body of the table. For example, from Table 5, $\log 4.52 = 0.6551$.

To make computations with logarithms, you need to find the logarithm of a given number and the number whose logarithm is given.

Example 2 Find (a) log 45.2 and (b) log 0.0452.

Solution (a) log 45.2 = log (4.52 × 10¹)
$$= \log 4.52 + \log 10^1 = 0.6551 + 1 = 1.6551$$
(b) We have $0.0452 = 4.52 \times 10^{-2}$. Then
$$\log 0.0452 = \log 4.52 + \log 10^{-2} = 0.6551 + (-2)$$
$$= 0.6551 + 8 - 10 = 8.6551 - 10 \quad ■$$

    In general, $\log_{10} x$, for any positive real number x, can be thought of as consisting of two parts. One part is a nonnegative decimal between 0 and 1 that represents the logarithm of a number between 1 and 10, and is called the **mantissa** of the logarithm. The other part is an integer that represents the logarithm of a power of 10, and is called the **characteristic** of the logarithm. The use of the characteristic -10 in Example 2 is customary. However, it may sometimes be more convenient to use another form. In such a case we could use $1 - 3$, $18 - 20$, or any other pair of nonnegative integers b and d such that $b - d = -2$.

Example 3 Find N if log $N = 3.7324$.

Solution Write log N as $3 + 0.7324$. From Table 5, 0.7324 is log 5.40. Since log N has characteristic 3, $N = 5.40 \times 10^3 = 5400$. ■

    The laws of logarithms listed on page 291 can be used to compute products, quotients, powers, roots, or any combination of these.

Example 4 Compute $\dfrac{(2.24)^2(8.2)}{(2.12)^3}$.

Solution Let $N = \dfrac{(2.24)^2(8.2)}{(2.12)^3}$. Then by law 4, $\log N = \log \dfrac{(2.24)^2(8.2)}{(2.12)^3}$.

log N = 2 log 2.24 + log 8.2 − 3 log 2.12   by laws 1, 2, and 3

From Table 5, log 2.24 = 0.3502, log 8.2 = 0.9138, and log 2.12 = 0.3263.

Therefore, log N = 0.6353 and N = 4.32. ■

Example 5 Compute $\sqrt[7]{0.862}$.

Solution Let $N = \sqrt[7]{0.862} = (0.862)^{\frac{1}{7}}$. Then, $\log N = \dfrac{1}{7} \log 0.862$.

From Table 5, log 0.862 = $-1 + 0.9355 = 6.9355 - 7$.

Hence, $\log N = \dfrac{1}{7}(6.9355 - 7) = 0.9908 - 1$ and $N = 0.979$. ■

    When sums or differences are involved, logarithms can help to compute the terms involving products, quotients, powers, or roots.

Example 6   Compute $(4.13)^3 - \sqrt[4]{(20.5)^3}$.

Solution   Let $N = (4.13)^3$ and $M = \sqrt[4]{(20.5)^3}$.

$$\log N = 3 \log 4.13$$
$$\log N = 3(0.6160)$$
$$\log N = 1.8480$$
$$N = 70.47$$

$$\log M = \frac{3}{4} \log 20.5$$
$$\log M = \frac{3}{4}(1.3118)$$
$$\log M = 0.9839$$
$$M = 9.64$$

Therefore, $N - M = 70.47 - 9.64 = 60.83$. ■

If any product, quotient, power, or root involves negative numbers, we can still use logarithms for computations on the absolute values of these numbers, and then make appropriate adjustments for signs.

Example 7   In $\triangle ABC$, $b = 2780$, $\angle A = 34°$, and $\angle B = 14°$. Find a.

Solution   Using the law of sines:

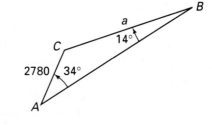

$$\frac{a}{\sin A} = \frac{b}{\sin B}.$$

Hence, $a = \dfrac{2780 \sin 34°}{\sin 14°}$.

$$\log a = \log 2780 + \log \sin 34° - \log \sin 14°$$
$$= \log 2780 + \log 0.5592 - \log 0.2419$$

From Table 5, $\log a = 3.4440 + (9.7474 - 10) - (9.3838 - 10) = 3.8076$.

Therefore, $a = 6420$. ■

Table 6 (p. 410) gives logarithmic values of trigonometric functions.

EXERCISES

Expand the given logarithm.

1. $\log (127)(42)$

2. $\log (5.8)(11.4)$

3. $\log (17)^{15}$

4. $\log \sqrt[5]{47}$

5. $\log \dfrac{59.7}{23.2}$

6. $\log \dfrac{10.01}{2.032}$

7. $\log \dfrac{(14.1)(2.5)^2}{3.2}$

8. $\log \dfrac{(11.2)^2 \sqrt{71}}{\sqrt[3]{85}}$

9. $\log \dfrac{[(10.2)^2(5.1)]^{\frac{1}{3}}}{(4.1)(6.7)^3}$

Find log *x*.

10. $x = 6.81$

11. $x = 5.22$

12. $x = 40$

13. $x = 57$

14. $x = 61.3$

15. $x = 22.73$

16. $x = 7490$

17. $x = 1547$

18. $x = 0.023$

19. $x = 0.3603$

20. $x = 0.00392$

21. $x = 0.0009396$

Find *x* if log *x* = *y*.

22. $y = 1.3096$

23. $y = 2.6561$

24. $y = 9.7451 - 10$

25. $y = 2.5551$

26. $y = 4.8762 - 5$

27. $y = 8.4843 - 10$

28. $y = 9.9800 - 10$

29. $y = 0.5520$

30. $y = 3.1887$

31. $y = 8.5482 - 10$

32. $y = 1.7787 - 4$

33. $y = 5.1220$

Use logarithms to find the value of each expression.

34. $\dfrac{(22.4)^2(12.4)}{(8.54)}$

35. $\dfrac{(16.5)(5.92)^2}{11.7}$

36. $\sqrt[6]{0.0598}$

37. $3.251\sqrt{\dfrac{(64.9)(21.2)}{15.7}}$

38. $\sqrt[3]{\dfrac{(51.2)(7.81)}{2.93}}$

39. $\sqrt{\dfrac{258}{(2.17)^2(9.32)}}$

40. $\sqrt[5]{\dfrac{(6)(3.91)}{(2)(3.02)}}$

41. $\sqrt[4]{\dfrac{(3.52)(12.8)}{(5.14)(1.97)}}$

42. $\sqrt[8]{\sqrt[3]{(16.4)^2}}$

43. $\sqrt[7]{\sqrt{2.95^3}}$

44. $\dfrac{3}{\sqrt[4]{0.725}}$

45. $\dfrac{(12.2)^3(4193)(5.271)}{17.83\sqrt[3]{0.934}}$

Solve the following oblique triangles, using logarithms as convenient. In each case, make a sketch.

46. $\angle C = 76°; \angle B = 41°, a = 7.08$

47. $c = 27, \angle A = 25°30'; \angle B = 95°30'$

48. $\angle A = 71°; \angle C = 77°, b = 113$

49. $b = 128, c = 57.1; \angle B = 110°$

Extra Practice

For use after Chapter 1

Draw an angle in standard position having the given measure.

1. $60°$
2. $-60°$
3. $-390°$
4. $765°$
5. $-135°$
6. $1110°$
7. $-270°$
8. $200°$

Let θ be an angle in standard position whose terminal side passes through the given point. Find (a) $\sin \theta$ and $\cos \theta$ to four decimal places and (b) θ to the nearest degree.

9. $(9, 12)$
10. $(2, 2\sqrt{3})$
11. $(3, 6\sqrt{2})$
12. $(5, 2\sqrt{6})$
13. $(4, 2\sqrt{5})$
14. $(2\sqrt{2}, 2\sqrt{2})$
15. $(2\sqrt{5}, 4\sqrt{5})$
16. $(\sqrt{13}, 2\sqrt{3})$

For the given angle θ, find $\sin \theta$ and $\cos \theta$ to four significant digits.

17. $23°40'$
18. $40°50'$
19. $65°10'$
20. $78°40'$

Find a value of θ that satisfies the given equation. Round your answer to the nearest minute or the nearest hundredth of a degree.

21. $\cos \theta = 0.2391$
22. $\sin \theta = 0.4950$
23. $\cos \theta = 0.5422$

Find the values of the six trigonometric functions for an angle θ in standard position whose terminal side passes through the given point. Simplify radicals and leave your answers in ratio form.

24. $(8, 6)$
25. $(12, 5)$
26. $(2, \sqrt{5})$
27. $(20, 21)$
28. $(2, 2\sqrt{3})$
29. $(\sqrt{2}, \sqrt{2})$
30. $(4, 5)$
31. $(3, 6)$

Solve $\triangle ABC$ in which $\angle C = 90°$.

32. $c = 65.0, a = 36.0$
33. $c = 120, b = 65$
34. $c = 42, \angle A = 50°$
35. $a = 35, b = 21$
36. $a = 110, \angle B = 40°$
37. $b = 72.0, \angle B = 36°$
38. $c = 140, \angle A = 29°$
39. $a = 18.0, b = 24.0$

Find exact values of the six trigonometric functions for each angle. Leave your answers in simplest form.

40. $120°$
41. $225°$
42. $315°$
43. $270°$

Use a reference triangle to find each of the following.

44. $\cos (-42°)$
45. $\tan 200°$
46. $\sec 115°30'$
47. $\sin 280.5°$
48. $\csc (-50°)$
49. $\cot 185.5°$

Convert the following to radian measure. Leave your answers in terms of π.

50. $225°$ **51.** $240°$ **52.** $-100°$ **53.** $20°$ **54.** $-60°$

Convert the following to degree measure.

55. $\dfrac{5\pi}{12}$ **56.** $\dfrac{3\pi}{4}$ **57.** $-\dfrac{\pi}{12}$ **58.** $\dfrac{7\pi}{6}$ **59.** 2π

For use after Chapter 2

Find the exact values of sin t, cos t, tan t, sec t, csc t, and cot t for the given values of t.

1. $\dfrac{4\pi}{3}$ **2.** $\dfrac{\pi}{2}$ **3.** $-\dfrac{11\pi}{6}$ **4.** $\dfrac{7\pi}{4}$

5. $-\dfrac{7\pi}{3}$ **6.** $-\dfrac{7\pi}{6}$ **7.** $-\dfrac{\pi}{3}$ **8.** $\dfrac{13\pi}{6}$

Suppose that a point P travels uniformly at an angular speed of ω in a counterclockwise direction around a circular path of radius a with center at the origin. Find the coordinates of the final position of P if P starts at the point $(a, 0)$ and moves for t units of time.

9. $a = 3$ cm; $\omega = \dfrac{5\pi}{6}$ radians/s; $t = 12$ s

10. $a = 4$ cm; $\omega = \dfrac{3\pi}{4}$ radians/s; $t = 4$ s

11. $a = 2$ cm; $\omega = \dfrac{5\pi}{3}$ radians/s; $t = 9$ s

12. $a = 3$ cm; $\omega = 6$ rpm; $t = 24$ min

13. $a = 4$ cm; $\omega = 4$ rpm; $t = 2$ min

14. $a = 12$ cm; $\omega = \dfrac{4\pi}{3}$ radians/s; $t = 18$ s

15. $a = 4$ cm; $\omega = 8$ rpm; $t = 15$ s

16. $a = 6$ cm; $\omega = 3.5$ rpm; $t = 2$ min

Graph each function over the given interval.

17. $y = \cos x$; $-2\pi \le x \le 0$ **18.** $y = \sin x$; $-2\pi \le x \le 0$

19. $y = -\cos x$; $-2\pi \le x \le 0$ **20.** $y = -\sin x$; $-2\pi \le x \le 0$

21. $y = \tan x$; $-2\pi \le x \le 0$ **22.** $y = 1 + \csc x$; $0 \le x \le 2\pi$

23. $y = |\sec x|$; $0 \le x \le 2\pi$ **24.** $y = |\cot x|$; $0 \le x \le 2\pi$

Evaluate.

25. $\text{Cos}^{-1} \dfrac{\sqrt{2}}{2}$

26. $\text{Sin}^{-1} 0$

27. $\text{Tan}^{-1} \dfrac{\sqrt{3}}{3}$

28. $\sin \left(\text{Cos}^{-1} \left(-\dfrac{\sqrt{3}}{2} \right) \right)$

29. $\cos (\text{Sin}^{-1} 1)$

30. $\text{Sin}^{-1} \left(\tan \dfrac{\pi}{4} \right)$

31. $\text{Cos}^{-1} \left(\tan \dfrac{\pi}{4} \right)$

32. $\tan \left(\text{Sin}^{-1} \dfrac{\sqrt{3}}{2} \right)$

33. $\text{Cos}^{-1} \left(\cos \left(-\dfrac{\pi}{2} \right) \right)$

For use after Chapter 3

Express the following in terms of the given trigonometric function. Give your answers in simplest form.

1. $\cot^2 x - 1$; $\sin x$

2. $\cos^2 x \csc x$; $\sin x$

3. $\csc x(1 - \sin x)(1 + \sin x)$; $\cos x$

4. $\csc^2 x + \cot^2 x$; $\cos x$

Express each of the following in terms of a single trigonometric function.

5. $\cot x(\sin x - \csc x)$

6. $\dfrac{1 + \tan^2 x}{\csc x \sec x}$

7. $\dfrac{\cos x (\sin x + \cos x \tan x)}{1 - \sin^2 x}$

8. $\dfrac{(\csc x - \cot x)(\csc x + \cot x)}{\sin x}$

Prove each identity.

9. $\csc^2 x(1 - \sin^2 x) = \cot^2 x$

10. $(\csc x + 1)(\sec x - \tan x) = \cot x$

11. $(\tan x - \sin x)(1 + \sec x) = \sin x \tan^2 x$

12. $(1 - \sin x)(\sec x + \tan x) = \cos x$

13. $\tan^2 x + \csc^2 x = \csc^2 x \sec^2 x - 1$

14. $\dfrac{1 - \cos^2 x}{\csc^2 x - 1} = \sin^4 x \sec^2 x$

15. $\dfrac{\sin x}{1 - \cos x} - \dfrac{\sin x}{1 + \cos x} = 2 \cot x$

16. $(\cot x + \cos x)(1 - \sin x) = \cot x (1 - \sin^2 x)$

Find each of the following if $\sin x_1 = -\dfrac{4}{5}$, $\sin x_2 = \dfrac{5}{13}$, $\pi < x_1 < \dfrac{3\pi}{2}$, and $0 < x_2 < \dfrac{\pi}{2}$.

17. $\sin (x_1 + x_2)$

18. $\sin (x_1 - x_2)$

19. $\cos (x_1 - x_2)$

20. $\cos (x_1 + x_2)$

Find each of the following if $\cos x_1 = -\dfrac{3}{5}$, $\cos x_2 = \dfrac{5}{13}$, $\pi < x_1 < \dfrac{3\pi}{2}$,

and $\dfrac{3\pi}{2} < x_2 < 2\pi$.

21. $\sin(x_1 + x_2)$ **22.** $\cos(x_1 + x_2)$ **23.** $\sin(x_1 - x_2)$ **24.** $\cos(x_1 - x_2)$

For each angle θ satisfying the given condition in the given quadrant, find
(a) $\sin 2\theta$ and (b) $\cos 2\theta$.

25. $\cos \theta = \dfrac{5}{13}$; I   **26.** $\sin \theta = \dfrac{12}{13}$; II   **27.** $\cos \theta = \dfrac{3}{5}$; IV

Use a half-angle formula to evaluate each expression.

28. $\sin \dfrac{x}{2}$, if $\cos x = -\dfrac{17}{25}$ and $0 < x < \pi$

29. $\cos \dfrac{x}{2}$, if $\cos x = -\dfrac{19}{36}$ and $\pi < x < 2\pi$

30. $\tan \dfrac{x}{2}$, if $\sin x = \dfrac{7}{25}$ and $0 < x < 2\pi$

31. $\tan \dfrac{x}{2}$, if $\cos x = \dfrac{24}{25}$ and $0 < x < \pi$

32. $\sin \dfrac{x}{2}$, if $\cos x = -\dfrac{24}{25}$ and $\pi < x < 2\pi$

Solve each equation for $0 \le x \le 2\pi$.

33. $\tan 2x = 2 \cos x$ **34.** $2 \sin x \sec x = \sec x$

35. $\tan x = \sin 2x$ **36.** $3 \csc x - \sin x = 2$

Give the general solution for each equation.

37. $\sin x \cos x = 0$ **38.** $\cos 2x + \sin x = 1$

39. $-\sin^2 x + 2 = \sin^4 x$ **40.** $\tan 2x + \sec^2 2x = 1$

For use after Chapter 4

Complete Exercises 1–6 for $\triangle ABC$ with the given parts.

1. $a = 4, b = 7, c = 9, \angle B = ?$ **2.** $a = 12, b = 9, \angle C = 45°, c = ?$

3. $b = 4, c = 5, \angle A = 60°, a = ?$ **4.** $a = 21, b = 13, c = 24, \angle A = ?$

5. $a = 13, c = 17, \angle B = 60°, b = ?$ **6.** $a = 5, b = 12, c = 16, \angle C = ?$

7. Find the lengths of the diagonals of a parallelogram with sides of lengths 15 and 40 and one angle of measure 120°30'.

8. Find the lengths of the diagonals of a parallelogram with sides of lengths 5 and 11 and one angle of measure 80°.

Solve each △ABC having the given angles and the given side. Give lengths to the nearest tenth.

9. $\angle A = 40°, \angle C = 62°, b = 7$

10. $\angle B = 112°, \angle C = 40°20', a = 20$

11. $\angle A = 110°20', \angle B = 38°, c = 13$

12. $\angle A = 47.5°, \angle C = 100°, b = 32$

13. $\angle B = 71°10', \angle C = 54°, a = 22$

14. $\angle B = 65°50', \angle C = 60°10', a = 18$

15. $\angle A = 42°40', \angle B = 63°, c = 17$

16. $\angle A = 82°10', \angle C = 43°50', b = 38$

For each △ABC, state all possible values for ∠B to the nearest 10' or tenth of a degree. Sketch a diagram to illustrate each solution. If no triangle can be formed, so state.

17. $\angle A = 48°, a = 12, b = 17$

18. $\angle A = 30°, a = 21, b = 42$

19. $\angle A = 22°, a = 20, b = 21$

20. $\angle A = 71°10', a = 12, b = 11$

21. $\angle A = 30°, a = 67.5, b = 135$

22. $\angle A = 64.5°, a = 28, b = 30.6$

23. $\angle A = 55°40', a = 6, b = 9$

24. $\angle A = 38°30', a = 57, b = 43$

Find the area of △ABC with the given sides and angles. Estimate your answer to the nearest tenth.

25. $a = 9, b = 14, c = 21$

26. $\angle A = 51°20', \angle B = 43°, c = 18$

27. $c = 17, b = 28, \angle A = 84.5°$

28. $a = 13, c = 15, \angle C = 72°$

29. $\angle B = 18°20', \angle C = 56°, a = 5$

30. $a = 22, b = 17.5, c = 34$

31. $a = 7, b = 8, \angle C = 101°$

32. $b = 108, c = 74, \angle B = 82°$

For use after Chapter 5

Graph each function for $0 \le x \le 2\pi$.

1. $y = -3 \sin x$

2. $y = 4 \sin x$

3. $y = 2 \cos x$

4. $y = 2 \sin 3x$

5. $y = \cos \frac{1}{2} x$

6. $y = -3 \sin 2x$

7. $y = |\cos 2x|$

8. $y = \frac{1}{2} \cos \frac{3}{2} x$

9. $y = \sin \pi x$

Graph all three functions in each exercise on one set of axes. Label the graph of each function.

10. (a) $y = \sin x$ (b) $y = \sin\left(x - \dfrac{\pi}{2}\right)$ (c) $y = \sin\left(x + \dfrac{\pi}{2}\right)$

11. (a) $y = \cos x$ (b) $y = \cos(x + \pi)$ (c) $y = \cos x + 2$

12. (a) $y = 2 \sin x$ (b) $y = 2 \sin x - 2$ (c) $y = 2 \sin x + 2$

13. (a) $y = \cos x$ (b) $y = \sin\left(x - \dfrac{\pi}{2}\right)$ (c) $y = -\cos(x - \pi)$

Graph each of the following.

14. $y = 2 \sin(x + \pi) + 1$

15. $y = 3 \cos\left(2x + \dfrac{\pi}{2}\right) - 1$

16. $y = \dfrac{1}{2} \sin\left(x - \dfrac{\pi}{2}\right) + 2$

17. $y = 4 \cos(2x - \pi) + \dfrac{1}{2}$

Rewrite each expression in the form $C \cos(\theta - \phi)$ for $-180° \le \phi < 180°$. Find ϕ to the nearest degree.

18. $3 \cos \theta - 4 \sin \theta$

19. $\cos \theta + \sin \theta$

20. $5 \cos \theta - \sqrt{2} \sin \theta$

21. $\sqrt{7} \cos \theta - 3 \sin \theta$

22. $3 \cos \theta - 3 \sin \theta$

23. $-3 \cos \theta + 4 \sin \theta$

24. $\sqrt{2} \cos \theta + \sqrt{2} \sin \theta$

25. $2 \cos \theta + \dfrac{3}{2} \sin \theta$

For each of the following cases of simple harmonic motion, (a) write an equation for the position y of the object as a function of time t in the form $y = a \sin(\omega t + \beta)$, (b) graph the function, and (c) find the frequency of the oscillation. Neglect the effects of friction and air resistance.

26. An object attached to a spring is pulled downward 3 cm from its equilibrium position and released. It makes one complete oscillation every 3 seconds.

27. An object attached to a spring is pulled downward 7 cm from its equilibrium position and released. It makes one complete oscillation every 2 seconds.

28. The water line of a buoy starts from a position 20 cm below the surface of the water and bobs up to a maximum position 20 cm above the surface in 1 second.

29. The buoy in Exercise 28 covers the same vertical range in the same amount of time, but starts from a position 10 cm below the surface of the water and bobs up first.

30–33. Express the functions in Exercises 26–29 in the form $y = a \cos(\omega t + \beta)$.

Graph each function for $0 \le x \le 2\pi$.

34. $y = \sin x + \cos 2x$

35. $y = \sin 2x - \cos x$

36. $y = -\sin x - \cos x$

37. $y = 3 \sin x - \sin 3x$

38. $y = \cos x + \cos \dfrac{x}{2}$

39. $y = \sin x - \sin \dfrac{x}{2}$

40. $y = \cos x + \sin \dfrac{x}{2}$

41. $y = 2 \cos x - \cos 2x$

42. $y = 2 \cos x + \cos 2x$

For use after Chapter 6

Draw the vectors **a**, **b**, **a** + **b**, and **a** − **b** with their initial points at the origin. Then find the norm of each of the four vectors.

1. $\mathbf{a} = -\mathbf{i}, \mathbf{b} = \mathbf{j}$

2. $\mathbf{a} = 6\mathbf{i} + 2\mathbf{j}, \mathbf{b} = 3\mathbf{j}$

3. $\mathbf{a} = 3\mathbf{i} - 2\mathbf{j}, \mathbf{b} = -2\mathbf{i} + \mathbf{j}$

4. $\mathbf{a} = -2\mathbf{i} + 5\mathbf{j}, \mathbf{b} = \mathbf{i} - 4\mathbf{j}$

For Exercises 5–8, find the specified linear combinations of $\mathbf{u} = 2\mathbf{i} - \mathbf{j}$ and $\mathbf{v} = -\mathbf{i} + 5\mathbf{j}$.

5. $2\mathbf{u} + 3\mathbf{v}$

6. $\dfrac{1}{2}\mathbf{u} + \mathbf{v}$

7. $\dfrac{2}{3}\mathbf{u} + \dfrac{1}{3}\mathbf{v}$

8. $-3\mathbf{u} + \dfrac{1}{3}\mathbf{v}$

9. A ship sails due south for 200 km, then due east for 50 km. What are the distance and bearing of the starting point from the ship?

10. A plane flies 620 km due north and then 225 km on a heading of 80°. What are its distance and bearing from its starting point?

11. A rescue ship has left port and travels 15 km on a heading of 40°. It hears a distress call from a ship 60 km from port, bearing 240°. What heading should the rescue ship take to reach the ship in distress and how far must it travel?

12. A plane heads due west with an air speed of 380 km/h. A 60 km/h wind is blowing from 210°. Find the plane's true course and ground speed.

13. If the plane in Exercise 12 was heading due south while everything else remained the same, what would be its true course and ground speed?

14. The speed in still water of a boat is 30 km/h. The boat heads directly east across a 75 m wide river flowing due south at 4.5 km/h.
(a) How far downstream will the boat land?
(b) What heading should the operator have used in order to land directly opposite the starting point?

15. Two planes take off in still air from the same airport at the same time. The speed of one plane is 480 km/h and its heading is 160°. The speed of the other plane is 430 km/h and its heading is 200°. How far apart are they at the end of one hour and what is the bearing of each from the other at that time?

16. A plane is flying on a heading of 300° at an air speed of 500 km/h. Its true course is 330° and the wind is blowing from 245°. What is the speed of the wind?

17. A 200-kg crate is on a plane inclined at 20° with the horizontal. Find the components of the gravitational force on the crate parallel and perpendicular to the plane.

18. A 180-kg motor slides with constant velocity down a ramp inclined at 23° with the horizontal. Find the frictional force on the motor.

19. A 275-kg object is suspended by two cables making 30° and 150° angles with the horizontal. Find the tension in each cable.

20. A 55-kg box is resting on a ramp that makes an angle of 20° with the horizontal. What is the frictional force that will keep the box from sliding down the ramp?

21. A 300-kg object is suspended by cables making angles of 30° and 130°, respectively, with the horizontal. Find the tension in each cable.

22. A crate is resting on a ramp that makes an angle of 22° with the horizontal. A frictional force of 480 N is required to keep the crate from sliding down the ramp. Find the mass of the crate.

The given forces act on a particle *P*. Find the magnitude and direction of the additional force that will keep *P* stationary. (Angles are measured counterclockwise from the horizontal.)

23. $\mathbf{F}_1$: 12 N, 5°
    $\mathbf{F}_2$: 18 N, 230°
    $\mathbf{F}_3$: 16 N, 150°

24. $\mathbf{F}_1$: 30 N, 20°
    $\mathbf{F}_2$: 20 N, 250°
    $\mathbf{F}_3$: 24 N, 130°

Find $\mathbf{a} \cdot \mathbf{b}$. Are the vectors orthogonal? Are they parallel?

25. $\mathbf{a} = -\mathbf{i} - \mathbf{j}$; $\mathbf{b} = 4\mathbf{i}$

26. $\mathbf{a} = 2\mathbf{i} - 3\mathbf{j}$; $\mathbf{b} = 3\mathbf{i} + 2\mathbf{j}$

27. $\mathbf{a} = -\mathbf{i} + 3\mathbf{j}$; $\mathbf{b} = 2\mathbf{i} - 6\mathbf{j}$

28. $\mathbf{a} = 3\mathbf{i} - \mathbf{j}$; $\mathbf{b} = 2\mathbf{i} + 2\mathbf{j}$

Resolve **v** into components parallel to **a** and to **b**.

29. $\mathbf{v} = \mathbf{i} + 3\mathbf{j}$; $\mathbf{a} = 4\mathbf{i} - 3\mathbf{j}$; $\mathbf{b} = 3\mathbf{i} + 4\mathbf{j}$

30. $\mathbf{v} = 2\mathbf{i} - 4\mathbf{j}$; $\mathbf{a} = -3\mathbf{i} + 2\mathbf{j}$; $\mathbf{b} = 2\mathbf{i} + 3\mathbf{j}$

31. $\mathbf{v} = -\mathbf{i} + 2\mathbf{j}$; $\mathbf{a} = -3\mathbf{i} + 4\mathbf{j}$; $\mathbf{b} = 4\mathbf{i} + 3\mathbf{j}$

32. $\mathbf{v} = 2\mathbf{i} + 5\mathbf{j}$; $\mathbf{a} = 5\mathbf{i} - \mathbf{j}$; $\mathbf{b} = \mathbf{i} + 5\mathbf{j}$

In Exercises 33–40, forces are measured in newtons and distances in meters. Give answers in joules.

33. Find the work done by the force $\mathbf{F} = 3\mathbf{i} + 2\mathbf{j}$ in moving a particle from $A(3, 0)$ to $B(0, 5)$.

34. Find the work done by the force $\mathbf{F} = 3\mathbf{j}$ in moving an object from $A(2, 6)$ to $B(3, -3)$.

35. How much work is done by force $\mathbf{F} = -4\mathbf{i} + 6\mathbf{j}$ in moving an object (a) from $A(4, 0)$ to $B(2, 9)$ to $C(8, 0)$? (b) from $A(4, 0)$ to $C(8, 0)$?

36. What is the combined work done by forces $\mathbf{F} = 3\mathbf{i} - 5\mathbf{j}$ and $\mathbf{G} = -2\mathbf{i} + 6\mathbf{j}$ in moving an object from $A(-3, 4)$ to $B(3, -6)$? What is the work done by $\mathbf{F} + \mathbf{G}$ in moving an object from A to B?

37. The mass of a loaded helicopter is 1500 kg. How much energy does its engine expend in ascending vertically for 75 m?

38. How much energy does a 60-kg climber expend in climbing a rock formation that is 25 m high and nearly vertical?

39. A crane loads 7500-kg containers onto the deck of a ship. How much energy does the crane expend in lifting a container 35 m off the ground?

40. How much energy does a 75-kg person expend in climbing an 11.3° grade for a horizontal distance of 6 km?

For use after Chapter 7

Find a polar equation for each of the following.

1. $y = 3$
2. $x = -3$
3. $x^2 + y^2 = 9$
4. $x - y = 1$

Graph the following polar equations.

5. $r \sin \theta = 2$
6. $r = 4 \sin \theta$
7. $\theta = \dfrac{3\pi}{2}$
8. $\theta + \dfrac{\pi}{3} = 0$

Graph the following polar equations.

9. $r = 1 + \sin 3\theta$ (three-leaved rose)
10. $r = 4 \cos \theta$ (circle)
11. $r = 2 \sin \theta$ (circle)
12. $r = 1 + \sin 2\theta$ (lemniscate)

13. $r = 2 \sin 2\theta$ (four-leaved rose)　　　　**14.** $r = 3 \cos 3\theta$ (three-leaved rose)

15. $r = 1 - 2 \cos \theta$ (limaçon with small loop)

16. $r = 2\theta, \theta \geq 0$ (spiral)

Determine the type of each conic.

17. $r = \dfrac{4}{1 + \cos \theta}$ 　　　　　　　　**18.** $r = \dfrac{4}{2 - \cos \theta}$

19. $r = \dfrac{6}{1 + 3 \sin \theta}$ 　　　　　　　　**20.** $r = \dfrac{9}{3 - 2 \sin \theta}$

21–24. Sketch the graphs of the conics in Exercises 17–20.

Find (a) $w + z$, (b) wz, and (c) $\dfrac{w}{z}$ in the form $x + yi$.

25. $w = 2 + i, z = 3 - i$ 　　　　　　**26.** $w = i, z = 5 - i$

27. $w = 2 - 2i, z = 3 + 3i$ 　　　　**28.** $w = 5 - i, z = 5 + i$

Find (a) the conjugate, (b) the modulus, and (c) the reciprocal of z, and (d) z^2.

29. $z = 3 - i$ 　　　　　　　　　　　**30.** $z = \sqrt{2} + \sqrt{2}i$

31. $z = \dfrac{4}{5} + \dfrac{3}{5}i$ 　　　　　　　　**32.** $z = \dfrac{\sqrt{3}}{3} + \dfrac{1}{3}i$

Graph the given complex number.

33. $\cos 45° + i \sin 45°$ 　　　　　　**34.** $4(\cos 60° + i \sin 60°)$

35. $2(\cos 120° + i \sin 120°)$ 　　　**36.** $3(\cos 30° + i \sin 30°)$

Write each complex number in polar form. If necessary, express angles to the nearest tenth of a degree.

37. $-1 + i$ 　　　　**38.** $2i$ 　　　　　　**39.** $3 - i\sqrt{3}$

40. $1 + 4i$ 　　　　**41.** $-2 - 2i\sqrt{3}$ 　　　**42.** $5 - 5i$

For each of the following, find z_1z_2.

43. $z_1 = \sqrt{2}(\cos 315° + i \sin 315°), z_2 = 2(\cos 30° + i \sin 30°)$

44. $z_1 = 2(\cos 270° + i \sin 270°), z_2 = 3(\cos 30° + i \sin 30°)$

45. $z_1 = \sqrt{2}(\cos 60° + i \sin 60°), z_2 = 3(\cos 30° + i \sin 30°)$

46. $z_1 = \sqrt{3}(\cos 90° + i \sin 90°), z_2 = \sqrt{2}(\cos 45° + i \sin 45°)$

47–50. In Exercises 43–46, find $\dfrac{z_1}{z_2}$.

Find the reciprocal of *z* in (a) polar form and (b) the form *x* + *yi*.

51. $z = \cos 300° + i \sin 300°$ **52.** $z = \cos 60° + i \sin 60°$

53. $z = 1 - i$ **54.** $z = \sqrt{2} + i$

Use De Moivre's theorem to express each of the following in the form *x* + *yi*.

55. $[3(\cos 45° + i \sin 45°)]^3$ **56.** $[2(\cos 60° + i \sin 60°)]^5$

57. $\left[\frac{1}{2}(\cos 72° + i \sin 72°)\right]^5$ **58.** $(3 + 3i)^4$

59. $[2(\cos 15° + i \sin 15°)]^4 \left[\frac{1}{2}(\cos 20° + i \sin 20°)\right]^6$

60. $\dfrac{(\cos 27° + i \sin 27°)^6}{(\cos 18° + i \sin 18°)^4}$

61. $\dfrac{(\sqrt{3} + i)^3}{(1 - i)^2}$ **62.** $\dfrac{(1 + i)^4}{(2 - 2i)^3}$

Find the indicated roots in the form specified, and sketch their graphs along with that of the given number.

63. The cube roots of $1 + i$ in the form *x* + *yi*

64. The sixth roots of *i* in the form *x* + *yi*

65. The fifth roots of $1 - i$ in the form *x* + *yi*

66. The seventh roots of $\sqrt{3} - i$ in polar form

67. The fourth roots of $-i$ in polar form

68. The cube roots of $\sqrt{3} + i$ in polar form

69. The fourth roots of $1 - i$ in polar form

70. The fifth roots of $\sqrt{2} + i$ in polar form

For use after Chapter 8

1. Use the first three terms of the Taylor series for sin x to find the following.

   (a) $\sin \dfrac{\pi}{3} \approx \sin 1.05$

   (b) $\sin \dfrac{\pi}{4} \approx \sin 0.79$

   (c) Compare the values you obtained in (a) and (b) with the known values of $\sin \dfrac{\pi}{3}$ and $\sin \dfrac{\pi}{4}$.

2. Use the first three terms of the Taylor series for cos x to find the following.

   (a) $\cos \dfrac{\pi}{6} \approx \cos 0.52$

   (b) $\cos \dfrac{2\pi}{3} \approx \cos 2.09$

   (c) Compare the values you obtained in (a) and (b) with the known values of $\cos \dfrac{\pi}{6}$ and $\cos \dfrac{2\pi}{3}$.

3. Use the first four terms of the Taylor series for sin x to evaluate each of the following to three decimal places.

   (a) $\sin 2.2$

   (b) $\sin (-0.94)$

   (c) Compare the answers to (a) and (b). Give a reason for their relationship.

4. Use the first four terms of the Taylor series for cos x to evaluate each of the following to three decimal places.

   (a) $\cos 2.2$

   (b) $\cos (-0.94)$

   (c) Compare the answers to (a) and (b). Give a reason for their relationship.

Find the value of the expression using the given equation.

5. $\cosh 1$; $\cosh x = 1 + \dfrac{x^2}{2!} + \dfrac{x^4}{4!} + \dfrac{x^6}{6!} + \cdots$

6. $\sinh 1$; $\sinh x = x + \dfrac{x^3}{3!} + \dfrac{x^5}{5!} + \dfrac{x^7}{7!} + \cdots$

7. $e^{-\pi i}$; $e^{ix} = \cos x + i \sin x$

Find the value of the following.

8. $e^{\frac{3\pi i}{2}}$

9. $e^{\frac{\pi i}{4}}$

10. Sketch the graph of the function f defined by $f(x) = |x|$, $-\pi \le x \le \pi$ and $f(x + 2\pi) = f(x)$.

For use after Chapter 9

In Exercises 1 and 2, (a) plot the given points and draw the triangle that they determine. (b) Use the distance formula to check whether the triangle is isosceles, right, both or neither.

1. $(2, 4, 2), (4, 5, 4), (4, 6, 1)$

2. $(4, 5, -6), (3, 6, -2), (2, 4, -4)$

Find an equation of the sphere having the given center and radius. Give your answer in the form $x^2 + y^2 + z^2 + ax + by + cz + d = 0$.

3. center $(0, 0, 0)$, radius 3

4. center $(1, 0, 1)$, radius 2

5. center $(2, 0, 2)$, radius 5

6. center $(1, 2, 3)$, radius 4

Find the center and radius of the given sphere.

7. $x^2 + y^2 + z^2 = 16$

8. $x^2 + y^2 + z^2 - 2x - 2y = 7$

9. $x^2 + y^2 + z^2 + 2y - 4z + 1 = 0$

10. $x^2 + y^2 + z^2 + 2z = 24$

In Exercises 11 and 12, $\mathbf{a} = \mathbf{i} - 2\mathbf{j} + \mathbf{k}$ and $\mathbf{b} = 3\mathbf{j} - 2\mathbf{k}$. (a) Find the given linear combinations. (b) Find $\|\mathbf{a}\|, \|\mathbf{b}\|, \|\mathbf{a} + \mathbf{b}\|$, and $\|\mathbf{a} - \mathbf{b}\|$.

11. $4\mathbf{a} - 2\mathbf{b}, 2\mathbf{a} + 4\mathbf{b}$

12. $3\mathbf{a} - 5\mathbf{b}, 5\mathbf{a} + 3\mathbf{b}$

In Exercises 13–16, find the angle between $\mathbf{a}$ and $\mathbf{b}$.

13. $\mathbf{a} = \mathbf{i} + \mathbf{j} - \mathbf{k}, \mathbf{b} = 2\mathbf{i} + \mathbf{j} + \mathbf{k}$

14. $\mathbf{a} = 2\mathbf{i} + 2\mathbf{j} - \mathbf{k}, \mathbf{b} = 3\mathbf{i} + \mathbf{j} - \mathbf{k}$

15. $\mathbf{a} = \mathbf{i} - \mathbf{j} - \mathbf{k}, \mathbf{b} = \mathbf{i} + 2\mathbf{j} + \mathbf{k}$

16. $\mathbf{a} = 2\mathbf{i} - \mathbf{j} + \mathbf{k}, \mathbf{b} = 3\mathbf{i} - 3\mathbf{j} + \mathbf{k}$

In Exercises 17 and 18, express $\mathbf{v}$ as a linear combination of $\mathbf{a}, \mathbf{b}$, and $\mathbf{c}$. Use the fact that $\mathbf{a}, \mathbf{b}$, and $\mathbf{c}$ are orthogonal in pairs.

17. $\mathbf{a} = 2\mathbf{i} - 2\mathbf{j} + \mathbf{k}, \mathbf{b} = \mathbf{i} + 2\mathbf{j} + 2\mathbf{k}, \mathbf{c} = 2\mathbf{i} + \mathbf{j} - 2\mathbf{k};$
    $\mathbf{v} = \mathbf{i} + \mathbf{j} + 2\mathbf{k}$

18. $\mathbf{a} = 4\mathbf{i} - 2\mathbf{j} + 2\mathbf{k}, \mathbf{b} = 2\mathbf{i} + 3\mathbf{j} - \mathbf{k}, \mathbf{c} = \mathbf{i} - 2\mathbf{j} - 4\mathbf{k};$
    $\mathbf{v} = 3\mathbf{i} + \mathbf{j} + \mathbf{k}$

Find a vector equation of the line having the stated properties.

19. Passes through $(1, 0, 1)$ and is parallel to $\mathbf{i} + \mathbf{j} - 2\mathbf{k}$

20. Passes through the origin and $(2, 2, 4)$

21. Passes through $(1, 0, 3)$ and makes equal angles with $\mathbf{i}$, $\mathbf{j}$, and $\mathbf{k}$

22. Passes through $(1, 2, 3)$ and is parallel to the line through $(1, 1, 3)$ and $(2, 0, 1)$

Find the point where the line *l* intersects the given plane.

23. $\mathbf{r} = (1 + t)\mathbf{i} + 3t\mathbf{j} + (1 - t)\mathbf{k}$; xz-plane

24. $\mathbf{r} = (1 - t)\mathbf{i} + (2 + t)\mathbf{j} + 2t\mathbf{k}$; xy-plane

Find the distance from the origin to the line *l*.

25. $\mathbf{r} = (4 - t)\mathbf{i} + (3 + t)\mathbf{j} + (1 - t)\mathbf{k}$

26. $\mathbf{r} = (3 + 2t)\mathbf{i} + (-3 + t)\mathbf{j} + (2 + 2t)\mathbf{k}$

In Exercises 27 and 28, (a) find an equation of the plane through P_0 that is perpendicular to the given vector or line. (b) Then draw the first-octant part of the plane.

27. $P_0(1, 2, 1)$; $3\mathbf{i} + \mathbf{j} + 2\mathbf{k}$

28. $P_0(2, 0, 1)$; $\mathbf{r} = (2 + t)\mathbf{i} + (-1 + t)\mathbf{j} + t\mathbf{k}$

Find a vector equation of the line through P_1 that is perpendicular to Q.

29. $P_1(2, 2, 0)$; Q: $3x - y + 2z = 2$

30. $P_1(1, 3, 1)$; Q: $2x + 3y - z - 2 = 0$

31–32. Find the distance between P_1 and Q in Exercises 29 and 30.

Find an equation of the plane that passes through the given points.

33. $(4, 1, 2), (5, 2, 3), (-3, 3, 1)$ 34. $(0, 0, 1), (3, 1, 2), (2, 0, 2)$

In Exercises 35 and 36, (a) verify that the given planes are parallel and (b) find the distance between them.

35. $2x + y - 3z = 4$; $-2x - y + 3z = 6$

36. $x - y + 2z - 3 = 0$; $2x - 2y + 4z - 5 = 0$

Algebra and Geometry Review

The Pythagorean Theorem and Radicals (Chapter 1)

Example   (a) Simplify $\sqrt{108}$.

(b) The hypotenuse of a right triangle has length 15 and the length of another side is 10. Find the length of the third side.

(c) Simplify $\dfrac{1}{\sqrt{18}}$.

Solution   (a)
$$\sqrt{108} = \sqrt{9 \cdot 4 \cdot 3}$$
$$= \sqrt{9} \cdot \sqrt{4} \cdot \sqrt{3} = 3 \cdot 2\sqrt{3} = 6\sqrt{3}$$

(b) From the Pythagorean theorem with b as the length of the third side,
$$10^2 + b^2 = 15^2$$
$$b^2 = 225 - 100$$
$$= 125$$
$$b = \sqrt{125} = \sqrt{5^2 \cdot 5} = 5\sqrt{5}$$

(c) Multiply numerator and denominator by $\sqrt{18}$.
$$\frac{1}{\sqrt{18}} = \frac{1}{\sqrt{18}} \cdot \frac{\sqrt{18}}{\sqrt{18}} = \frac{\sqrt{18}}{18}$$
$$= \frac{\sqrt{9}\,\sqrt{2}}{18} = \frac{3\sqrt{2}}{18}$$
$$= \frac{\sqrt{2}}{6}$$

In $\triangle ABC$, $\angle C = 90°$.

1. $a = 13$; $b = 11$; $c = ?$ 2. $a = 42$; $b = 25$; $c = ?$

3. $a = 28$; $c = 46$; $b = ?$ 4. $c = 108$; $b = 39$; $a = ?$

5. $a = 52$; $b = 64$; $c = ?$ 6. $a = 47$; $c = 91$; $b = ?$

Simplify.

7. $\sqrt{75}$ 8. $\sqrt{288}$ 9. $\sqrt{45}$

10. $\sqrt{112}$ 11. $\sqrt{96}$ 12. $\dfrac{1}{\sqrt{3}}$

13. $\dfrac{\sqrt{6}}{\sqrt{3}}$ 14. $\dfrac{\sqrt{12}}{\sqrt{3}}$ 15. $\dfrac{2}{\sqrt{32}}$

Parallel Lines (Chapter 1)

Example   Lines *a* and *b* are parallel.
Find θ.

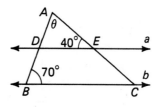

Solution   Since *a* and *b* are parallel,
$\angle DBC$ and $\angle ADE$ are congruent corresponding angles.
Hence, $\angle ADE = 70°$. Since the
sum of the measures of the
angles of a triangle is 180°,

$$\angle ADE + \angle DEA + \theta = 180°$$
$$70° + 40° + \theta = 180°$$
$$\theta = 70°$$

Lines *a* and *b* are parallel. Find θ.

1.

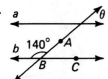

2.

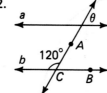

3.

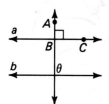

4.

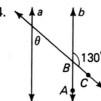

5.

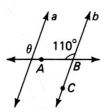

6.

7.

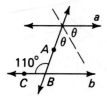

8.

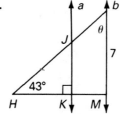

Special Right Triangles (Chapter 1)

Example   Find the value of x. Leave answers in simplest radical form.

(a)   (b)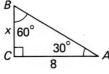

Solution   (a) The triangle is a right isosceles triangle, a 45°-45°-90° triangle. By the Pythagorean theorem:

$$x^2 + x^2 = 8^2$$
$$x^2 = 32$$
$$x = \sqrt{32} = \sqrt{16 \cdot 2} = 4\sqrt{2}$$

(b) The triangle is a 30°-60°-90° triangle. From geometry, $AB = 2BC$ and $AC = BC \sqrt{3}$.

$$8 = x \sqrt{3}$$
$$x = \frac{8}{\sqrt{3}} = \frac{8}{\sqrt{3}} \cdot \frac{\sqrt{3}}{\sqrt{3}} = \frac{8\sqrt{3}}{3}$$

Find the value of x. Leave answers in simplest radical form.

1.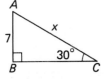

2.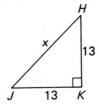

3.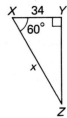

4.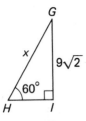

5.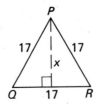

6.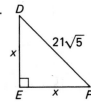

7.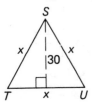

8.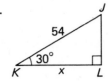

9.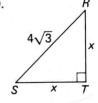

Graphing Equations (Chapter 2)

Example   Graph $y = x^2 - 4x + 3$.

Solution   Set up a table of ordered pairs and graph them, or complete the square in x and look for a familiar form.

$$y = x^2 - 4x + 3 + 1 - 1$$
$$= (x - 2)^2 - 1$$

The graph is the graph of $y = x^2$ translated 2 units to the right and 1 unit down.

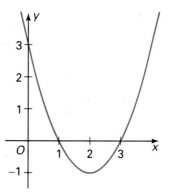

Graph each equation.

1. $y = 3x + 2$

2. $y = -2x - 2$

3. $y = x^2 - 2$

4. $y = 2x^2$

5. $y = \dfrac{4}{x}$

6. $y = -x^3$

7. $y = x^2 + x + 1$

8. $y = \dfrac{x^2}{2}$

9. $y = \dfrac{2}{x}$

Simplifying Rational Expressions (Chapter 3)

Example   Simplify $\dfrac{a^2}{a-2} - \dfrac{a+2}{a-2}$. Identify any restrictions on the variables.

Solution

$$\frac{a^2}{a-2} - \frac{a+2}{a-2} = \frac{a^2 - a - 2}{a-2}$$
$$= \frac{(a-2)(a+1)}{a-2}$$
$$= a + 1 \text{ where } a \neq 2$$

Simplify. Identify any restrictions on the variables.

1. $\dfrac{x^2 + 2x}{x^2 + 7x + 10}$

2. $\dfrac{y - 3}{y^2 - 6y + 9}$

3. $\dfrac{2z^2 - z - 3}{3z + 3}$

4. $\dfrac{u^2 - 5u}{2 - u} + \dfrac{12 - 3u}{2 - u}$

5. $(4x - 4) \div \dfrac{x^2 - 1}{x}$

6. $\dfrac{1 - t}{1 + t} + \dfrac{1 + t}{1 - t}$

7. $\dfrac{x^2 + x - 2}{x^2 + 2x - 3}$

8. $\dfrac{x^2 - 1}{x^2 + 3x + 2}$

9. $\dfrac{a^4 - 5a^2 + 4}{a^2 - a - 2}$

Solving Equations (Chapter 3)

Example   Solve: (a) $\dfrac{1.24}{x} = \dfrac{3.45}{0.1367}$ (b) $x^2 - 5x - 36 = 0$

Solution   (a) From the given proportion, $\dfrac{x}{1.24} = \dfrac{0.1367}{3.45}$.

Hence, $x = \dfrac{1.24 \cdot 0.1367}{3.45} = 0.0491$.

(b) Solve the given quadratic equation by factoring.

$$x^2 - 5x - 36 = (x - 9)(x + 4) = 0$$

Hence, $x - 9 = 0$ or $x + 4 = 0$. Thus, $x = 9$ or -4.

Solve each equation.

1. $x^2 + 5x + 4 = 0$

2. $3x^2 - 8x - 3 = 0$

3. $6x^2 - 13x + 6 = 0$

4. $x - 6 = \sqrt{x}$

5. $\dfrac{x}{x - 2} - 7 = \dfrac{2}{x - 2}$

6. $\dfrac{1}{x^2 - x} - \dfrac{3}{x} = -1$

7. $x^2 - 10x + 25 = 81$

8. $x^2 + 8x + 4 = -12$

9. $3 + \dfrac{10}{x^2 - 1} = \dfrac{5}{x - 1}$

Solve by completing the square.

10. $y^2 + 10y - 5 = 0$

11. $x^2 - 6x - 2 = 0$

12. $y^2 + 8y - 6 = 0$

13. $x^2 - 8x - 20 = 0$

14. $x^2 - 6x + 4 = 0$

15. $x^2 + 8x - 1 = 0$

16. $x^2 + 10x + 21 = 0$

17. $x^2 + 6x = 31$

18. $2x^2 + 5x = 3$

Area of a Polygon (Chapter 4)

Example   Find the area of an equilateral triangle ABC where sides have length 4.

Solution   Triangle ABC is equilateral. Draw $\overline{AD}$, the altitude to $\overline{BC}$. Triangle ADB is a 30°-60°-90° triangle and $AD = 2\sqrt{3}$.

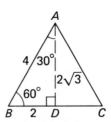

Area of $\triangle ADB = \dfrac{1}{2}bh = 2\sqrt{3}$

Area of $\triangle ABC = 2(2\sqrt{3}) = 4\sqrt{3}$

Find the area of each polygon.

1.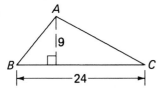

2.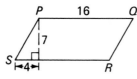

3.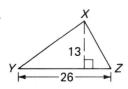

4.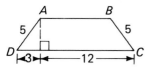

5.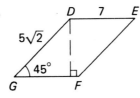

6.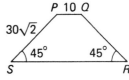

7.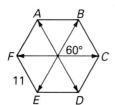

8.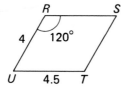

9.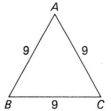

Conic Sections (Chapter 7)

Example   Identify the graph of each equation. Then graph the equation.
(a) $x^2 - 2x + y^2 - 4y = -1$ (b) $9x^2 + 16y^2 = 144$

Solution   (a) Complete the square in x and in y.

$$(x^2 - 2x + 1) + (y^2 - 4y + 4) = -1 + 1 + 4$$
$$(x - 1)^2 + (y - 2)^2 = 4 = 2^2$$

The graph is a circle of radius 2 and center $(1, 2)$ shown at the left below.

(b)
$$\frac{x^2}{16} + \frac{y^2}{9} = 1$$

The graph is an ellipse shown at the right below.

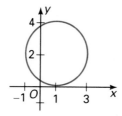
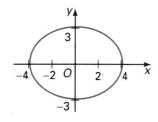

Identify the graph of each equation. Then graph the equation.

1. $y = x^2 + 2x + 3$

2. $9x^2 + 16y^2 = 144$

3. $x^2 + 2x + y^2 + 2y = 2$

4. $x^2 + 4y^2 + 2 = 6$

5. $x^2 + y = 2x + 3$

6. $y - x^2 = 12x + 4$

7. $25x^2 - 4y^2 - 25 = 75$

8. $x^2 - 6x + y^2 + 4y = 3$

Systems of Equations (Chapter 9)

Example   Solve: (a) $-5x + 3y = 25$
$$4x + 2y = 2$$

(b) $x + y + z = 3$
$$x + 2y - z = -1$$
$$-2x - y + 3z = 11$$

Solution   (a) To eliminate y from the system, multiply the first equation by -2. Multiply the second equation by 3. Then add to find x.

$$10x - 6y = -50$$
$$\underline{12x + 6y = 6}$$
$$22x \quad\quad = -44$$
$$x \quad\quad = -2$$

Now substitute -2 for x into one of the given equations.

$$-5(-2) + 3y = 25$$
$$3y = 15$$
$$y = 5$$

The solution is $(-2, 5)$.

(b) Use the method of Example 1 (a) to eliminate z from equations 1 and 2, then from equations 2 and 3.

$$x + y + z = 3$$
$$\underline{x + 2y - z = -1}$$
$$2x + 3y = 2$$
$$3x + 6y - 3z = -3$$
$$\underline{-2x - y + 3z = 11}$$
$$x + 5y = 8$$

Use the method of Example (a) to solve:

$$2x + 3y = 2$$
$$x + 5y = 8$$

From this system, $x = -2$ and $y = 2$, substitute these values of x and y into one of the original equations to obtain $z = 3$.
The solution is $(-2, 2, 3)$.

Solve each system of equations.

1. $3x - 2y = 8$
   $3x - 6y = -16$

2. $5x + 6y = 16$
   $6x - 5y = 7$

3. $7x + 10y = 17$
   $8x + 15y = 23$

4. $5x + 2y = -1$
   $4x + 3y = 9$

5. $3x - 7y = 9$
   $2x - 5y = 7$

6. $10x + 3y = 2$
   $6x - 7y = -12$

7. $x + y - z = 2$
   $4x - y + z = 0$
   $6x + y + z = 4$

8. $x + y - z = 4$
   $4x - y + z = 0$
   $6x + y + z = 8$

9. $x + y + z = 1$
   $x + 2y - z = -1$
   $2x - y - 3z = 8$

10. $-x - y + z = 7$
    $x + 2y + 3z = 7$
    $4x - 3y - 2z = -6$

11. $x + 2y + z = 11$
    $x + 3y + 3z = 19$
    $2x + y + z = 13$

12. $2x - y + 3z = 3$
    $x + y + 2z = 1$
    $x - 2y - z = -2$

Table 1   Trigonometric Functions of θ (θ in decimal degrees)

θ Degrees	θ Radians	$\sin\theta$	$\cos\theta$	$\tan\theta$	$\cot\theta$	$\sec\theta$	$\csc\theta$		
0.0	.0000	.0000	1.0000	.0000	undefined	1.000	undefined	1.5708	**90.0**
0.1	.0017	.0017	1.0000	.0017	573.0	1.000	573.0	1.5691	89.9
0.2	.0035	.0035	1.0000	.0035	286.5	1.000	286.5	1.5673	89.8
0.3	.0052	.0052	1.0000	.0052	191.0	1.000	191.0	1.5656	89.7
0.4	.0070	.0070	1.0000	.0070	143.2	1.000	143.2	1.5638	89.6
0.5	.0087	.0087	1.0000	.0087	114.6	1.000	114.6	1.5621	89.5
0.6	.0105	.0105	.9999	.0105	95.49	1.000	95.49	1.5603	89.4
0.7	.0122	.0122	.9999	.0122	81.85	1.000	81.85	1.5586	89.3
0.8	.0140	.0140	.9999	.0140	71.62	1.000	71.62	1.5568	89.2
0.9	.0157	.0157	.9999	.0157	63.66	1.000	63.66	1.5551	89.1
1.0	.0175	.0175	.9998	.0175	57.29	1.000	57.30	1.5533	**89.0**
1.1	.0192	.0192	.9998	.0192	52.08	1.000	52.09	1.5516	88.9
1.2	.0209	.0209	.9998	.0209	47.74	1.000	47.75	1.5499	88.8
1.3	.0227	.0227	.9997	.0227	44.07	1.000	44.08	1.5481	88.7
1.4	.0244	.0244	.9997	.0244	40.92	1.000	40.93	1.5464	88.6
1.5	.0262	.0262	.9997	.0262	38.19	1.000	38.20	1.5446	88.5
1.6	.0279	.0279	.9996	.0279	35.80	1.000	35.81	1.5429	88.4
1.7	.0297	.0297	.9996	.0297	33.69	1.000	33.71	1.5411	88.3
1.8	.0314	.0314	.9995	.0314	31.82	1.000	31.84	1.5394	88.2
1.9	.0332	.0332	.9995	.0332	30.14	1.001	30.16	1.5376	88.1
2.0	.0349	.0349	.9994	.0349	28.64	1.001	28.65	1.5359	**88.0**
2.1	.0367	.0366	.9993	.0367	27.27	1.001	27.29	1.5341	87.9
2.2	.0384	.0384	.9993	.0384	26.03	1.001	26.05	1.5324	87.8
2.3	.0401	.0401	.9992	.0402	24.90	1.001	24.92	1.5307	87.7
2.4	.0419	.0419	.9991	.0419	23.86	1.001	23.88	1.5289	87.6
2.5	.0436	.0436	.9990	.0437	22.90	1.001	22.93	1.5272	87.5
2.6	.0454	.0454	.9990	.0454	22.02	1.001	22.04	1.5254	87.4
2.7	.0471	.0471	.9989	.0472	21.20	1.001	21.23	1.5237	87.3
2.8	.0489	.0488	.9988	.0489	20.45	1.001	20.47	1.5219	87.2
2.9	.0506	.0506	.9987	.0507	19.74	1.001	19.77	1.5202	87.1
3.0	.0524	.0523	.9986	.0524	19.08	1.001	19.11	1.5184	**87.0**
3.1	.0541	.0541	.9985	.0542	18.46	1.001	18.49	1.5167	86.9
3.2	.0559	.0558	.9984	.0559	17.89	1.002	17.91	1.5149	86.8
3.3	.0576	.0576	.9983	.0577	17.34	1.002	17.37	1.5132	86.7
3.4	.0593	.0593	.9982	.0594	16.83	1.002	16.86	1.5115	86.6
3.5	.0611	.0610	.9981	.0612	16.35	1.002	16.38	1.5097	86.5
3.6	.0628	.0628	.9980	.0629	15.89	1.002	15.93	1.5080	86.4
3.7	.0646	.0645	.9979	.0647	15.46	1.002	15.50	1.5062	86.3
3.8	.0663	.0663	.9978	.0664	15.06	1.002	15.09	1.5045	86.2
3.9	.0681	.0680	.9977	.0682	14.67	1.002	14.70	1.5027	86.1
4.0	.0698	.0698	.9976	.0699	14.30	1.002	14.34	1.5010	**86.0**
4.1	.0716	.0715	.9974	.0717	13.95	1.003	13.99	1.4992	85.9
4.2	.0733	.0732	.9973	.0734	13.62	1.003	13.65	1.4975	85.8
4.3	.0750	.0750	.9972	.0752	13.30	1.003	13.34	1.4957	85.7
4.4	.0768	.0767	.9971	.0769	13.00	1.003	13.03	1.4940	85.6
4.5	.0785	.0785	.9969	.0787	12.71	1.003	12.75	1.4923	85.5
4.6	.0803	.0802	.9968	.0805	12.43	1.003	12.47	1.4905	85.4
4.7	.0820	.0819	.9966	.0822	12.16	1.003	12.20	1.4888	85.3
4.8	.0838	.0837	.9965	.0840	11.91	1.004	11.95	1.4870	85.2
4.9	.0855	.0854	.9963	.0857	11.66	1.004	11.71	1.4853	85.1
5.0	.0873	.0872	.9962	.0875	11.43	1.004	11.47	1.4835	**85.0**
5.1	.0890	.0889	.9960	.0892	11.20	1.004	11.25	1.4818	84.9
5.2	.0908	.0906	.9959	.0910	10.99	1.004	11.03	1.4800	84.8
5.3	.0925	.0924	.9957	.0928	10.78	1.004	10.83	1.4783	84.7
5.4	.0942	.0941	.9956	.0945	10.58	1.004	10.63	1.4765	84.6
5.5	.0960	.0958	.9954	.0963	10.39	1.005	10.43	1.4748	84.5
5.6	.0977	.0976	.9952	.0981	10.20	1.005	10.25	1.4731	84.4
5.7	.0995	.0993	.9951	.0998	10.02	1.005	10.07	1.4713	84.3
5.8	.1012	.1011	.9949	.1016	9.845	1.005	9.895	1.4696	84.2
5.9	.1030	.1028	.9947	.1033	9.677	1.005	9.728	1.4678	84.1
6.0	.1047	.1045	.9945	.1051	9.514	1.006	9.567	1.4661	**84.0**
		$\cos\theta$	$\sin\theta$	$\cot\theta$	$\tan\theta$	$\csc\theta$	$\sec\theta$	θ Radians	θ Degrees

Table 1 Trigonometric Functions of θ (θ in decimal degrees)

θ Degrees	θ Radians	sin θ	cos θ	tan θ	cot θ	sec θ	csc θ		
6.0	.1047	.1045	.9945	.1051	9.514	1.006	9.567	1.4661	**84.0**
6.1	.1065	.1063	.9943	.1069	9.357	1.006	9.411	1.4643	83.9
6.2	.1082	.1080	.9942	.1086	9.205	1.006	9.259	1.4626	83.8
6.3	.1100	.1097	.9940	.1104	9.058	1.006	9.113	1.4608	83.7
6.4	.1117	.1115	.9938	.1122	8.915	1.006	8.971	1.4591	83.6
6.5	.1134	.1132	.9936	.1139	8.777	1.006	8.834	1.4574	83.5
6.6	.1152	.1149	.9934	.1157	8.643	1.007	8.700	1.4556	83.4
6.7	.1169	.1167	.9932	.1175	8.513	1.007	8.571	1.4539	83.3
6.8	.1187	.1184	.9930	.1192	8.386	1.007	8.446	1.4521	83.2
6.9	.1204	.1201	.9928	.1210	8.264	1.007	8.324	1.4504	83.1
7.0	.1222	.1219	.9925	.1228	8.144	1.008	8.206	1.4486	**83.0**
7.1	.1239	.1236	.9923	.1246	8.028	1.008	8.091	1.4469	82.9
7.2	.1257	.1253	.9921	.1263	7.916	1.008	7.979	1.4451	82.8
7.3	.1274	.1271	.9919	.1281	7.806	1.008	7.870	1.4434	82.7
7.4	.1292	.1288	.9917	.1299	7.700	1.008	7.764	1.4416	82.6
7.5	.1309	.1305	.9914	.1317	7.596	1.009	7.661	1.4399	82.5
7.6	.1326	.1323	.9912	.1334	7.495	1.009	7.561	1.4382	82.4
7.7	.1344	.1340	.9910	.1352	7.396	1.009	7.463	1.4364	82.3
7.8	.1361	.1357	.9907	.1370	7.300	1.009	7.368	1.4347	82.2
7.9	.1379	.1374	.9905	.1388	7.207	1.010	7.276	1.4329	82.1
8.0	.1396	.1392	.9903	.1405	7.115	1.010	7.185	1.4312	**82.0**
8.1	.1414	.1409	.9900	.1423	7.026	1.010	7.097	1.4294	81.9
8.2	.1431	.1426	.9898	.1441	6.940	1.010	7.011	1.4277	81.8
8.3	.1449	.1444	.9895	.1459	6.855	1.011	6.927	1.4259	81.7
8.4	.1466	.1461	.9893	.1477	6.772	1.011	6.845	1.4242	81.6
8.5	.1484	.1478	.9890	.1495	6.691	1.011	6.765	1.4224	81.5
8.6	.1501	.1495	.9888	.1512	6.612	1.011	6.687	1.4207	81.4
8.7	.1518	.1513	.9885	.1530	6.535	1.012	6.611	1.4190	81.3
8.8	.1536	.1530	.9882	.1548	6.460	1.012	6.537	1.4172	81.2
8.9	.1553	.1547	.9880	.1566	6.386	1.012	6.464	1.4155	81.1
9.0	.1571	.1564	.9877	.1584	6.314	1.012	6.392	1.4137	**81.0**
9.1	.1588	.1582	.9874	.1602	6.243	1.013	6.323	1.4120	80.9
9.2	.1606	.1599	.9871	.1620	6.174	1.013	6.255	1.4102	80.8
9.3	.1623	.1616	.9869	.1638	6.107	1.013	6.188	1.4085	80.7
9.4	.1641	.1633	.9866	.1655	6.041	1.014	6.123	1.4067	80.6
9.5	.1658	.1650	.9863	.1673	5.976	1.014	6.059	1.4050	80.5
9.6	.1676	.1668	.9860	.1691	5.912	1.014	5.996	1.4032	80.4
9.7	.1693	.1685	.9857	.1709	5.850	1.015	5.935	1.4015	80.3
9.8	.1710	.1702	.9854	.1727	5.789	1.015	5.875	1.3998	80.2
9.9	.1728	.1719	.9851	.1745	5.730	1.015	5.816	1.3980	80.1
10.0	.1745	.1736	.9848	.1763	5.671	1.015	5.759	1.3963	**80.0**
10.1	.1763	.1754	.9845	.1781	5.614	1.016	5.702	1.3945	79.9
10.2	.1780	.1771	.9842	.1799	5.558	1.016	5.647	1.3928	79.8
10.3	.1798	.1788	.9839	.1817	5.503	1.016	5.593	1.3910	79.7
10.4	.1815	.1805	.9836	.1835	5.449	1.017	5.540	1.3893	79.6
10.5	.1833	.1822	.9833	.1853	5.396	1.017	5.487	1.3875	79.5
10.6	.1850	.1840	.9829	.1871	5.343	1.017	5.436	1.3858	79.4
10.7	.1868	.1857	.9826	.1890	5.292	1.018	5.386	1.3840	79.3
10.8	.1885	.1874	.9823	.1908	5.242	1.018	5.337	1.3823	79.2
10.9	.1902	.1891	.9820	.1926	5.193	1.018	5.288	1.3806	79.1
11.0	.1920	.1908	.9816	.1944	5.145	1.019	5.241	1.3788	**79.0**
11.1	.1937	.1925	.9813	.1962	5.097	1.019	5.194	1.3771	78.9
11.2	.1955	.1942	.9810	.1980	5.050	1.019	5.148	1.3753	78.8
11.3	.1972	.1959	.9806	.1998	5.005	1.020	5.103	1.3736	78.7
11.4	.1990	.1977	.9803	.2016	4.959	1.020	5.059	1.3718	78.6
11.5	.2007	.1994	.9799	.2035	4.915	1.020	5.016	1.3701	78.5
11.6	.2025	.2011	.9796	.2053	4.872	1.021	4.973	1.3683	78.4
11.7	.2042	.2028	.9792	.2071	4.829	1.021	4.931	1.3666	78.3
11.8	.2059	.2045	.9789	.2089	4.787	1.022	4.890	1.3648	78.2
11.9	.2077	.2062	.9785	.2107	4.745	1.022	4.850	1.3631	78.1
12.0	.2094	.2079	.9781	.2126	4.705	1.022	4.810	1.3614	**78.0**
		cos θ	sin θ	cot θ	tan θ	csc θ	sec θ	θ Radians	θ Degrees

Table 1 Trigonometric Functions of θ (θ in decimal degrees)

θ Degrees	θ Radians	sin θ	cos θ	tan θ	cot θ	sec θ	csc θ		
12.0	.2094	.2079	.9781	.2126	4.705	1.022	4.810	1.3614	**78.0**
12.1	.2112	.2096	.9778	.2144	4.665	1.023	4.771	1.3596	77.9
12.2	.2129	.2113	.9774	.2162	4.625	1.023	4.732	1.3579	77.8
12.3	.2147	.2130	.9770	.2180	4.586	1.023	4.694	1.3561	77.7
12.4	.2164	.2147	.9767	.2199	4.548	1.024	4.657	1.3544	77.6
12.5	.2182	.2164	.9763	.2217	4.511	1.024	4.620	1.3526	77.5
12.6	.2199	.2181	.9759	.2235	4.474	1.025	4.584	1.3509	77.4
12.7	.2217	.2198	.9755	.2254	4.437	1.025	4.549	1.3491	77.3
12.8	.2234	.2215	.9751	.2272	4.402	1.025	4.514	1.3474	77.2
12.9	.2251	.2233	.9748	.2290	4.366	1.026	4.479	1.3456	77.1
13.0	.2269	.2250	.9744	.2309	4.331	1.026	4.445	1.3439	**77.0**
13.1	.2286	.2267	.9740	.2327	4.297	1.027	4.412	1.3422	76.9
13.2	.2304	.2284	.9736	.2345	4.264	1.027	4.379	1.3404	76.8
13.3	.2321	.2300	.9732	.2364	4.230	1.028	4.347	1.3387	76.7
13.4	.2339	.2317	.9728	.2382	4.198	1.028	4.315	1.3369	76.6
13.5	.2356	.2334	.9724	.2401	4.165	1.028	4.284	1.3352	76.5
13.6	.2374	.2351	.9720	.2419	4.134	1.029	4.253	1.3334	76.4
13.7	.2391	.2368	.9715	.2438	4.102	1.029	4.222	1.3317	76.3
13.8	.2409	.2385	.9711	.2456	4.071	1.030	4.192	1.3299	76.2
13.9	.2426	.2402	.9707	.2475	4.041	1.030	4.163	1.3282	76.1
14.0	.2443	.2419	.9703	.2493	4.011	1.031	4.134	1.3265	**76.0**
14.1	.2461	.2436	.9699	.2512	3.981	1.031	4.105	1.3247	75.9
14.2	.2478	.2453	.9694	.2530	3.952	1.032	4.077	1.3230	75.8
14.3	.2496	.2470	.9690	.2549	3.923	1.032	4.049	1.3212	75.7
14.4	.2513	.2487	.9686	.2568	3.895	1.032	4.021	1.3195	75.6
14.5	.2531	.2504	.9681	.2586	3.867	1.033	3.994	1.3177	75.5
14.6	.2548	.2521	.9677	.2605	3.839	1.033	3.967	1.3160	75.4
14.7	.2566	.2538	.9673	.2623	3.812	1.034	3.941	1.3142	75.3
14.8	.2583	.2554	.9668	.2642	3.785	1.034	3.915	1.3125	75.2
14.9	.2601	.2571	.9664	.2661	3.758	1.035	3.889	1.3107	75.1
15.0	.2618	.2588	.9659	.2679	3.732	1.035	3.864	1.3090	**75.0**
15.1	.2635	.2605	.9655	.2698	3.706	1.036	3.839	1.3073	74.9
15.2	.2653	.2622	.9650	.2717	3.681	1.036	3.814	1.3055	74.8
15.3	.2670	.2639	.9646	.2736	3.655	1.037	3.790	1.3038	74.7
15.4	.2688	.2656	.9641	.2754	3.630	1.037	3.766	1.3020	74.6
15.5	.2705	.2672	.9636	.2773	3.606	1.038	3.742	1.3003	74.5
15.6	.2723	.2689	.9632	.2792	3.582	1.038	3.719	1.2985	74.4
15.7	.2740	.2706	.9627	.2811	3.558	1.039	3.695	1.2968	74.3
15.8	.2758	.2723	.9622	.2830	3.534	1.039	3.673	1.2950	74.2
15.9	.2775	.2740	.9617	.2849	3.511	1.040	3.650	1.2933	74.1
16.0	.2793	.2756	.9613	.2867	3.487	1.040	3.628	1.2915	**74.0**
16.1	.2810	.2773	.9608	.2886	3.465	1.041	3.606	1.2898	73.9
16.2	.2827	.2790	.9603	.2905	3.442	1.041	3.584	1.2881	73.8
16.3	.2845	.2807	.9598	.2924	3.420	1.042	3.563	1.2863	73.7
16.4	.2862	.2823	.9593	.2943	3.398	1.042	3.542	1.2846	73.6
16.5	.2880	.2840	.9588	.2962	3.376	1.043	3.521	1.2828	73.5
16.6	.2897	.2857	.9583	.2981	3.354	1.043	3.500	1.2811	73.4
16.7	.2915	.2874	.9578	.3000	3.333	1.044	3.480	1.2793	73.3
16.8	.2932	.2890	.9573	.3019	3.312	1.045	3.460	1.2776	73.2
16.9	.2950	.2907	.9568	.3038	3.291	1.045	3.440	1.2758	73.1
17.0	.2967	.2924	.9563	.3057	3.271	1.046	3.420	1.2741	**73.0**
17.1	.2985	.2940	.9558	.3076	3.251	1.046	3.401	1.2723	72.9
17.2	.3002	.2957	.9553	.3096	3.230	1.047	3.382	1.2706	72.8
17.3	.3019	.2974	.9548	.3115	3.211	1.047	3.363	1.2689	72.7
17.4	.3037	.2990	.9542	.3134	3.191	1.048	3.344	1.2671	72.6
17.5	.3054	.3007	.9537	.3153	3.172	1.049	3.326	1.2654	72.5
17.6	.3072	.3024	.9532	.3172	3.152	1.049	3.307	1.2636	72.4
17.7	.3089	.3040	.9527	.3191	3.133	1.050	3.289	1.2619	72.3
17.8	.3107	.3057	.9521	.3211	3.115	1.050	3.271	1.2601	72.2
17.9	.3124	.3074	.9516	.3230	3.096	1.051	3.254	1.2584	72.1
18.0	.3142	.3090	.9511	.3249	3.078	1.051	3.236	1.2566	**72.0**
		cos θ	sin θ	cot θ	tan θ	csc θ	sec θ	θ Radians	θ Degrees

Table 1 Trigonometric Functions of θ (θ in decimal degrees)

θ Degrees	θ Radians	sin θ	cos θ	tan θ	cot θ	sec θ	csc θ		
18.0	.3142	.3090	.9511	.3249	3.078	1.051	3.236	1.2566	**72.0**
18.1	.3159	.3107	.9505	.3269	3.060	1.052	3.219	1.2549	71.9
18.2	.3177	.3123	.9500	.3288	3.042	1.053	3.202	1.2531	71.8
18.3	.3194	.3140	.9494	.3307	3.024	1.053	3.185	1.2514	71.7
18.4	.3211	.3156	.9489	.3327	3.006	1.054	3.168	1.2497	71.6
18.5	.3229	.3173	.9483	.3346	2.989	1.054	3.152	1.2479	71.5
18.6	.3246	.3190	.9478	.3365	2.971	1.055	3.135	1.2462	71.4
18.7	.3264	.3206	.9472	.3385	2.954	1.056	3.119	1.2444	71.3
18.8	.3281	.3223	.9466	.3404	2.937	1.056	3.103	1.2427	71.2
18.9	.3299	.3239	.9461	.3424	2.921	1.057	3.087	1.2409	71.1
19.0	.3316	.3256	.9455	.3443	2.904	1.058	3.072	1.2392	**71.0**
19.1	.3334	.3272	.9449	.3463	2.888	1.058	3.056	1.2374	70.9
19.2	.3351	.3289	.9444	.3482	2.872	1.059	3.041	1.2357	70.8
19.3	.3368	.3305	.9438	.3502	2.856	1.060	3.026	1.2339	70.7
19.4	.3386	.3322	.9432	.3522	2.840	1.060	3.011	1.2322	70.6
19.5	.3403	.3338	.9426	.3541	2.824	1.061	2.996	1.2305	70.5
19.6	.3421	.3355	.9421	.3561	2.808	1.062	2.981	1.2287	70.4
19.7	.3438	.3371	.9415	.3581	2.793	1.062	2.967	1.2270	70.3
19.8	.3456	.3387	.9409	.3600	2.778	1.063	2.952	1.2252	70.2
19.9	.3473	.3404	.9403	.3620	2.762	1.064	2.938	1.2235	70.1
20.0	.3491	.3420	.9397	.3640	2.747	1.064	2.924	1.2217	**70.0**
20.1	.3508	.3437	.9391	.3659	2.733	1.065	2.910	1.2200	69.9
20.2	.3526	.3453	.9385	.3679	2.718	1.066	2.896	1.2182	69.8
20.3	.3543	.3469	.9379	.3699	2.703	1.066	2.882	1.2165	69.7
20.4	.3560	.3486	.9373	.3719	2.689	1.067	2.869	1.2147	69.6
20.5	.3578	.3502	.9367	.3739	2.675	1.068	2.855	1.2130	69.5
20.6	.3595	.3518	.9361	.3759	2.660	1.068	2.842	1.2113	69.4
20.7	.3613	.3535	.9354	.3779	2.646	1.069	2.829	1.2095	69.3
20.8	.3630	.3551	.9348	.3799	2.633	1.070	2.816	1.2078	69.2
20.9	.3648	.3567	.9342	.3819	2.619	1.070	2.803	1.2060	69.1
21.0	.3665	.3584	.9336	.3839	2.605	1.071	2.790	1.2043	**69.0**
21.1	.3683	.3600	.9330	.3859	2.592	1.072	2.778	1.2025	68.9
21.2	.3700	.3616	.9323	.3879	2.578	1.073	2.765	1.2008	68.8
21.3	.3718	.3633	.9317	.3899	2.565	1.073	2.753	1.1991	68.7
21.4	.3735	.3649	.9311	.3919	2.552	1.074	2.741	1.1973	68.6
21.5	.3752	.3665	.9304	.3939	2.539	1.075	2.729	1.1956	68.5
21.6	.3770	.3681	.9298	.3959	2.526	1.076	2.716	1.1938	68.4
21.7	.3787	.3697	.9291	.3979	2.513	1.076	2.705	1.1921	68.3
21.8	.3805	.3714	.9285	.4000	2.500	1.077	2.693	1.1903	68.2
21.9	.3822	.3730	.9278	.4020	2.488	1.078	2.681	1.1886	68.1
22.0	.3840	.3746	.9272	.4040	2.475	1.079	2.669	1.1868	**68.0**
22.1	.3857	.3762	.9265	.4061	2.463	1.079	2.658	1.1851	67.9
22.2	.3875	.3778	.9259	.4081	2.450	1.080	2.647	1.1833	67.8
22.3	.3892	.3795	.9252	.4101	2.438	1.081	2.635	1.1816	67.7
22.4	.3910	.3811	.9245	.4122	2.426	1.082	2.624	1.1798	67.6
22.5	.3927	.3827	.9239	.4142	2.414	1.082	2.613	1.1781	67.5
22.6	.3944	.3843	.9232	.4163	2.402	1.083	2.602	1.1764	67.4
22.7	.3962	.3859	.9225	.4183	2.391	1.084	2.591	1.1746	67.3
22.8	.3979	.3875	.9219	.4204	2.379	1.085	2.581	1.1729	67.2
22.9	.3997	.3891	.9212	.4224	2.367	1.086	2.570	1.1711	67.1
23.0	.4014	.3907	.9205	.4245	2.356	1.086	2.559	1.1694	**67.0**
23.1	.4032	.3923	.9198	.4265	2.344	1.087	2.549	1.1676	66.9
23.2	.4049	.3939	.9191	.4286	2.333	1.088	2.538	1.1659	66.8
23.3	.4067	.3955	.9184	.4307	2.322	1.089	2.528	1.1641	66.7
23.4	.4084	.3971	.9178	.4327	2.311	1.090	2.518	1.1624	66.6
23.5	.4102	.3987	.9171	.4348	2.300	1.090	2.508	1.1606	66.5
23.6	.4119	.4003	.9164	.4369	2.289	1.091	2.498	1.1589	66.4
23.7	.4136	.4019	.9157	.4390	2.278	1.092	2.488	1.1572	66.3
23.8	.4154	.4035	.9150	.4411	2.267	1.093	2.478	1.1554	66.2
23.9	.4171	.4051	.9143	.4431	2.257	1.094	2.468	1.1537	66.1
24.0	.4189	.4067	.9135	.4452	2.246	1.095	2.459	1.1519	**66.0**
		cos θ	sin θ	cot θ	tan θ	csc θ	sec θ	θ Radians	θ Degrees

Table 1 Trigonometric Functions of θ (θ in decimal degrees)

θ Degrees	θ Radians	$\sin \theta$	$\cos \theta$	$\tan \theta$	$\cot \theta$	$\sec \theta$	$\csc \theta$		
24.0	.4189	.4067	.9135	.4452	2.246	1.095	2.459	1.1519	**66.0**
24.1	.4206	.4083	.9128	.4473	2.236	1.095	2.449	1.1502	65.9
24.2	.4224	.4099	.9121	.4494	2.225	1.096	2.439	1.1484	65.8
24.3	.4241	.4115	.9114	.4515	2.215	1.097	2.430	1.1467	65.7
24.4	.4259	.4131	.9107	.4536	2.204	1.098	2.421	1.1449	65.6
24.5	.4276	.4147	.9100	.4557	2.194	1.099	2.411	1.1432	65.5
24.6	.4294	.4163	.9092	.4578	2.184	1.100	2.402	1.1414	65.4
24.7	.4311	.4179	.9085	.4599	2.174	1.101	2.393	1.1397	65.3
24.8	.4328	.4195	.9078	.4621	2.164	1.102	2.384	1.1380	65.2
24.9	.4346	.4210	.9070	.4642	2.154	1.102	2.375	1.1362	65.1
25.0	.4363	.4226	.9063	.4663	2.145	1.103	2.366	1.1345	**65.0**
25.1	.4381	.4242	.9056	.4684	2.135	1.104	2.357	1.1327	64.9
25.2	.4398	.4258	.9048	.4706	2.125	1.105	2.349	1.1310	64.8
25.3	.4416	.4274	.9041	.4727	2.116	1.106	2.340	1.1292	64.7
25.4	.4433	.4289	.9033	.4748	2.106	1.107	2.331	1.1275	64.6
25.5	.4451	.4305	.9026	.4770	2.097	1.108	2.323	1.1257	64.5
25.6	.4468	.4321	.9018	.4791	2.087	1.109	2.314	1.1240	64.4
25.7	.4485	.4337	.9011	.4813	2.078	1.110	2.306	1.1222	64.3
25.8	.4503	.4352	.9003	.4834	2.069	1.111	2.298	1.1205	64.2
25.9	.4520	.4368	.8996	.4856	2.059	1.112	2.289	1.1188	64.1
26.0	.4538	.4384	.8988	.4877	2.050	1.113	2.281	1.1170	**64.0**
26.1	.4555	.4399	.8980	.4899	2.041	1.114	2.273	1.1153	63.9
26.2	.4573	.4415	.8973	.4921	2.032	1.115	2.265	1.1135	63.8
26.3	.4590	.4431	.8965	.4942	2.023	1.115	2.257	1.1118	63.7
26.4	.4608	.4446	.8957	.4964	2.014	1.116	2.249	1.1100	63.6
26.5	.4625	.4462	.8949	.4986	2.006	1.117	2.241	1.1083	63.5
26.6	.4643	.4478	.8942	.5008	1.997	1.118	2.233	1.1065	63.4
26.7	.4660	.4493	.8934	.5029	1.988	1.119	2.226	1.1048	63.3
26.8	.4677	.4509	.8926	.5051	1.980	1.120	2.218	1.1030	63.2
26.9	.4695	.4524	.8918	.5073	1.971	1.121	2.210	1.1013	63.1
27.0	.4712	.4540	.8910	.5095	1.963	1.122	2.203	1.0996	**63.0**
27.1	.4730	.4555	.8902	.5117	1.954	1.123	2.195	1.0978	62.9
27.2	.4747	.4571	.8894	.5139	1.946	1.124	2.188	1.0961	62.8
27.3	.4765	.4586	.8886	.5161	1.937	1.125	2.180	1.0943	62.7
27.4	.4782	.4602	.8878	.5184	1.929	1.126	2.173	1.0926	62.6
27.5	.4800	.4617	.8870	.5206	1.921	1.127	2.166	1.0908	62.5
27.6	.4817	.4633	.8862	.5228	1.913	1.128	2.158	1.0891	62.4
27.7	.4835	.4648	.8854	.5250	1.905	1.129	2.151	1.0873	62.3
27.8	.4852	.4664	.8846	.5272	1.897	1.130	2.144	1.0856	62.2
27.9	.4869	.4679	.8838	.5295	1.889	1.132	2.137	1.0838	62.1
28.0	.4887	.4695	.8829	.5317	1.881	1.133	2.130	1.0821	**62.0**
28.1	.4904	.4710	.8821	.5340	1.873	1.134	2.123	1.0804	61.9
28.2	.4922	.4726	.8813	.5362	1.865	1.135	2.116	1.0786	61.8
28.3	.4939	.4741	.8805	.5384	1.857	1.136	2.109	1.0769	61.7
28.4	.4957	.4756	.8796	.5407	1.849	1.137	2.103	1.0751	61.6
28.5	.4974	.4772	.8788	.5430	1.842	1.138	2.096	1.0734	61.5
28.6	.4992	.4787	.8780	.5452	1.834	1.139	2.089	1.0716	61.4
28.7	.5009	.4802	.8771	.5475	1.827	1.140	2.082	1.0699	61.3
28.8	.5027	.4818	.8763	.5498	1.819	1.141	2.076	1.0681	61.2
28.9	.5044	.4833	.8755	.5520	1.811	1.142	2.069	1.0664	61.1
29.0	.5061	.4848	.8746	.5543	1.804	1.143	2.063	1.0647	**61.0**
29.1	.5079	.4863	.8738	.5566	1.797	1.144	2.056	1.0629	60.9
29.2	.5096	.4879	.8729	.5589	1.789	1.146	2.050	1.0612	60.8
29.3	.5114	.4894	.8721	.5612	1.782	1.147	2.043	1.0594	60.7
29.4	.5131	.4909	.8712	.5635	1.775	1.148	2.037	1.0577	60.6
29.5	.5149	.4924	.8704	.5658	1.767	1.149	2.031	1.0559	60.5
29.6	.5166	.4939	.8695	.5681	1.760	1.150	2.025	1.0542	60.4
29.7	.5184	.4955	.8686	.5704	1.753	1.151	2.018	1.0524	60.3
29.8	.5201	.4970	.8678	.5727	1.746	1.152	2.012	1.0507	60.2
29.9	.5219	.4985	.8669	.5750	1.739	1.154	2.006	1.0489	60.1
30.0	.5236	.5000	.8660	.5774	1.732	1.155	2.000	1.0472	**60.0**
		$\cos \theta$	$\sin \theta$	$\cot \theta$	$\tan \theta$	$\csc \theta$	$\sec \theta$	θ Radians	θ Degrees

Table 1 Trigonometric Functions of θ (θ in decimal degrees)

θ Degrees	θ Radians	sin θ	cos θ	tan θ	cot θ	sec θ	csc θ		
30.0	.5236	.5000	.8660	.5774	1.732	1.155	2.000	1.0472	**60.0**
30.1	.5253	.5015	.8652	.5797	1.725	1.156	1.994	1.0455	59.9
30.2	.5271	.5030	.8643	.5820	1.718	1.157	1.988	1.0437	59.8
30.3	.5288	.5045	.8634	.5844	1.711	1.158	1.982	1.0420	59.7
30.4	.5306	.5060	.8625	.5867	1.704	1.159	1.976	1.0402	59.6
30.5	.5323	.5075	.8616	.5890	1.698	1.161	1.970	1.0385	59.5
30.6	.5341	.5090	.8607	.5914	1.691	1.162	1.964	1.0367	59.4
30.7	.5358	.5105	.8599	.5938	1.684	1.163	1.959	1.0350	59.3
30.8	.5376	.5120	.8590	.5961	1.678	1.164	1.953	1.0332	59.2
30.9	.5393	.5135	.8581	.5985	1.671	1.165	1.947	1.0315	59.1
31.0	.5411	.5150	.8572	.6009	1.664	1.167	1.942	1.0297	**59.0**
31.1	.5428	.5165	.8563	.6032	1.658	1.168	1.936	1.0280	58.9
31.2	.5445	.5180	.8554	.6056	1.651	1.169	1.930	1.0263	58.8
31.3	.5463	.5195	.8545	.6080	1.645	1.170	1.925	1.0245	58.7
31.4	.5480	.5210	.8535	.6104	1.638	1.172	1.919	1.0228	58.6
31.5	.5498	.5225	.8526	.6128	1.632	1.173	1.914	1.0210	58.5
31.6	.5515	.5240	.8517	.6152	1.625	1.174	1.908	1.0193	58.4
31.7	.5533	.5255	.8508	.6176	1.619	1.175	1.903	1.0175	58.3
31.8	.5550	.5270	.8499	.6200	1.613	1.177	1.898	1.0158	58.2
31.9	.5568	.5284	.8490	.6224	1.607	1.178	1.892	1.0140	58.1
32.0	.5585	.5299	.8480	.6249	1.600	1.179	1.887	1.0123	**58.0**
32.1	.5603	.5314	.8471	.6273	1.594	1.180	1.882	1.0105	57.9
32.2	.5620	.5329	.8462	.6297	1.588	1.182	1.877	1.0088	57.8
32.3	.5637	.5344	.8453	.6322	1.582	1.183	1.871	1.0071	57.7
32.4	.5655	.5358	.8443	.6346	1.576	1.184	1.866	1.0053	57.6
32.5	.5672	.5373	.8434	.6371	1.570	1.186	1.861	1.0036	57.5
32.6	.5690	.5388	.8425	.6395	1.564	1.187	1.856	1.0018	57.4
32.7	.5707	.5402	.8415	.6420	1.558	1.188	1.851	1.0001	57.3
32.8	.5725	.5417	.8406	.6445	1.552	1.190	1.846	.9983	57.2
32.9	.5742	.5432	.8396	.6469	1.546	1.191	1.841	.9966	57.1
33.0	.5760	.5446	.8387	.6494	1.540	1.192	1.836	.9948	**57.0**
33.1	.5777	.5461	.8377	.6519	1.534	1.194	1.831	.9931	56.9
33.2	.5794	.5476	.8368	.6544	1.528	1.195	1.826	.9913	56.8
33.3	.5812	.5490	.8358	.6569	1.522	1.196	1.821	.9896	56.7
33.4	.5829	.5505	.8348	.6594	1.517	1.198	1.817	.9879	56.6
33.5	.5847	.5519	.8339	.6619	1.511	1.199	1.812	.9861	56.5
33.6	.5864	.5534	.8329	.6644	1.505	1.201	1.807	.9844	56.4
33.7	.5882	.5548	.8320	.6669	1.499	1.202	1.802	.9826	56.3
33.8	.5899	.5563	.8310	.6694	1.494	1.203	1.798	.9809	56.2
33.9	.5917	.5577	.8300	.6720	1.488	1.205	1.793	.9791	56.1
34.0	.5934	.5592	.8290	.6745	1.483	1.206	1.788	.9774	**56.0**
34.1	.5952	.5606	.8281	.6771	1.477	1.208	1.784	.9756	55.9
34.2	.5969	.5621	.8271	.6796	1.471	1.209	1.779	.9739	55.8
34.3	.5986	.5635	.8261	.6822	1.466	1.211	1.775	.9721	55.7
34.4	.6004	.5650	.8251	.6847	1.460	1.212	1.770	.9704	55.6
34.5	.6021	.5664	.8241	.6873	1.455	1.213	1.766	.9687	55.5
34.6	.6039	.5678	.8231	.6899	1.450	1.215	1.761	.9669	55.4
34.7	.6056	.5693	.8221	.6924	1.444	1.216	1.757	.9652	55.3
34.8	.6074	.5707	.8211	.6950	1.439	1.218	1.752	.9634	55.2
34.9	.6091	.5721	.8202	.6976	1.433	1.219	1.748	.9617	55.1
35.0	.6109	.5736	.8192	.7002	1.428	1.221	1.743	.9599	**55.0**
35.1	.6126	.5750	.8181	.7028	1.423	1.222	1.739	.9582	54.9
35.2	.6144	.5764	.8171	.7054	1.418	1.224	1.735	.9564	54.8
35.3	.6161	.5779	.8161	.7080	1.412	1.225	1.731	.9547	54.7
35.4	.6178	.5793	.8151	.7107	1.407	1.227	1.726	.9529	54.6
35.5	.6196	.5807	.8141	.7133	1.402	1.228	1.722	.9512	54.5
35.6	.6213	.5821	.8131	.7159	1.397	1.230	1.718	.9495	54.4
35.7	.6231	.5835	.8121	.7186	1.392	1.231	1.714	.9477	54.3
35.8	.6248	.5850	.8111	.7212	1.387	1.233	1.710	.9460	54.2
35.9	.6266	.5864	.8100	.7239	1.381	1.235	1.705	.9442	54.1
36.0	.6283	.5878	.8090	.7265	1.376	1.236	1.701	.9425	**54.0**
		cos θ	sin θ	cot θ	tan θ	csc θ	sec θ	θ Radians	θ Degrees

Table 1   Trigonometric Functions of θ (θ in decimal degrees)

θ Degrees	θ Radians	$\sin\theta$	$\cos\theta$	$\tan\theta$	$\cot\theta$	$\sec\theta$	$\csc\theta$		
36.0	.6283	.5878	.8090	.7265	1.376	1.236	1.701	.9425	**54.0**
36.1	.6301	.5892	.8080	.7292	1.371	1.238	1.697	.9407	53.9
36.2	.6318	.5906	.8070	.7319	1.366	1.239	1.693	.9390	53.8
36.3	.6336	.5920	.8059	.7346	1.361	1.241	1.689	.9372	53.7
36.4	.6353	.5934	.8049	.7373	1.356	1.242	1.685	.9355	53.6
36.5	.6370	.5948	.8039	.7400	1.351	1.244	1.681	.9338	53.5
36.6	.6388	.5962	.8028	.7427	1.347	1.246	1.677	.9320	53.4
36.7	.6405	.5976	.8018	.7454	1.342	1.247	1.673	.9303	53.3
36.8	.6423	.5990	.8007	.7481	1.337	1.249	1.669	.9285	53.2
36.9	.6440	.6004	.7997	.7508	1.332	1.250	1.666	.9268	53.1
37.0	.6458	.6018	.7986	.7536	1.327	1.252	1.662	.9250	**53.0**
37.1	.6475	.6032	.7976	.7563	1.322	1.254	1.658	.9233	52.9
37.2	.6493	.6046	.7965	.7590	1.317	1.255	1.654	.9215	52.8
37.3	.6510	.6060	.7955	.7618	1.313	1.257	1.650	.9198	52.7
37.4	.6528	.6074	.7944	.7646	1.308	1.259	1.646	.9180	52.6
37.5	.6545	.6088	.7934	.7673	1.303	1.260	1.643	.9163	52.5
37.6	.6562	.6101	.7923	.7701	1.299	1.262	1.639	.9146	52.4
37.7	.6580	.6115	.7912	.7729	1.294	1.264	1.635	.9128	52.3
37.8	.6597	.6129	.7902	.7757	1.289	1.266	1.632	.9111	52.2
37.9	.6615	.6143	.7891	.7785	1.285	1.267	1.628	.9093	52.1
38.0	.6632	.6157	.7880	.7813	1.280	1.269	1.624	.9076	**52.0**
38.1	.6650	.6170	.7869	.7841	1.275	1.271	1.621	.9058	51.9
38.2	.6667	.6184	.7859	.7869	1.271	1.272	1.617	.9041	51.8
38.3	.6685	.6198	.7848	.7898	1.266	1.274	1.613	.9023	51.7
38.4	.6702	.6211	.7837	.7926	1.262	1.276	1.610	.9006	51.6
38.5	.6720	.6225	.7826	.7954	1.257	1.278	1.606	.8988	51.5
38.6	.6737	.6239	.7815	.7983	1.253	1.280	1.603	.8971	51.4
38.7	.6754	.6252	.7804	.8012	1.248	1.281	1.599	.8954	51.3
38.8	.6772	.6266	.7793	.8040	1.244	1.283	1.596	.8936	51.2
38.9	.6789	.6280	.7782	.8069	1.239	1.285	1.592	.8919	51.1
39.0	.6807	.6293	.7771	.8098	1.235	1.287	1.589	.8901	**51.0**
39.1	.6824	.6307	.7760	.8127	1.230	1.289	1.586	.8884	50.9
39.2	.6842	.6320	.7749	.8156	1.226	1.290	1.582	.8866	50.8
39.3	.6859	.6334	.7738	.8185	1.222	1.292	1.579	.8849	50.7
39.4	.6877	.6347	.7727	.8214	1.217	1.294	1.575	.8831	50.6
39.5	.6894	.6361	.7716	.8243	1.213	1.296	1.572	.8814	50.5
39.6	.6912	.6374	.7705	.8273	1.209	1.298	1.569	.8796	50.4
39.7	.6929	.6388	.7694	.8302	1.205	1.300	1.566	.8779	50.3
39.8	.6946	.6401	.7683	.8332	1.200	1.302	1.562	.8762	50.2
39.9	.6964	.6414	.7672	.8361	1.196	1.304	1.559	.8744	50.1
40.0	.6981	.6428	.7660	.8391	1.192	1.305	1.556	.8727	**50.0**
40.1	.6999	.6441	.7649	.8421	1.188	1.307	1.552	.8709	49.9
40.2	.7016	.6455	.7638	.8451	1.183	1.309	1.549	.8692	49.8
40.3	.7034	.6468	.7627	.8481	1.179	1.311	1.546	.8674	49.7
40.4	.7051	.6481	.7615	.8511	1.175	1.313	1.543	.8657	49.6
40.5	.7069	.6494	.7604	.8541	1.171	1.315	1.540	.8639	49.5
40.6	.7086	.6508	.7593	.8571	1.167	1.317	1.537	.8622	49.4
40.7	.7103	.6521	.7581	.8601	1.163	1.319	1.534	.8604	49.3
40.8	.7121	.6534	.7570	.8632	1.159	1.321	1.530	.8587	49.2
40.9	.7138	.6547	.7559	.8662	1.154	1.323	1.527	.8570	49.1
41.0	.7156	.6561	.7547	.8693	1.150	1.325	1.524	.8552	**49.0**
41.1	.7173	.6574	.7536	.8724	1.146	1.327	1.521	.8535	48.9
41.2	.7191	.6587	.7524	.8754	1.142	1.329	1.518	.8517	48.8
41.3	.7208	.6600	.7513	.8785	1.138	1.331	1.515	.8500	48.7
41.4	.7226	.6613	.7501	.8816	1.134	1.333	1.512	.8482	48.6
41.5	.7243	.6626	.7490	.8847	1.130	1.335	1.509	.8465	48.5
41.6	.7261	.6639	.7478	.8878	1.126	1.337	1.506	.8447	48.4
41.7	.7278	.6652	.7466	.8910	1.122	1.339	1.503	.8430	48.3
41.8	.7295	.6665	.7455	.8941	1.118	1.341	1.500	.8412	48.2
41.9	.7313	.6678	.7443	.8972	1.115	1.344	1.497	.8395	48.1
42.0	.7330	.6691	.7431	.9004	1.111	1.346	1.494	.8378	**48.0**
		$\cos\theta$	$\sin\theta$	$\cot\theta$	$\tan\theta$	$\csc\theta$	$\sec\theta$	θ Radians	θ Degrees

Table 1 Trigonometric Functions of θ (θ in decimal degrees)

θ Degrees	θ Radians	$\sin\theta$	$\cos\theta$	$\tan\theta$	$\cot\theta$	$\sec\theta$	$\csc\theta$		
42.0	.7330	.6691	.7431	.9004	1.111	1.346	1.494	.8378	**48.0**
42.1	.7348	.6704	.7420	.9036	1.107	1.348	1.492	.8360	47.9
42.2	.7365	.6717	.7408	.9067	1.103	1.350	1.489	.8343	47.8
42.3	.7383	.6730	.7396	.9099	1.099	1.352	1.486	.8325	47.7
42.4	.7400	.6743	.7385	.9131	1.095	1.354	1.483	.8308	47.6
42.5	.7418	.6756	.7373	.9163	1.091	1.356	1.480	.8290	47.5
42.6	.7435	.6769	.7361	.9195	1.087	1.359	1.477	.8273	47.4
42.7	.7453	.6782	.7349	.9228	1.084	1.361	1.475	.8255	47.3
42.8	.7470	.6794	.7337	.9260	1.080	1.363	1.472	.8238	47.2
42.9	.7487	.6807	.7325	.9293	1.076	1.365	1.469	.8221	47.1
43.0	.7505	.6820	.7314	.9325	1.072	1.367	1.466	.8203	**47.0**
43.1	.7522	.6833	.7302	.9358	1.069	1.370	1.464	.8186	46.9
43.2	.7540	.6845	.7290	.9391	1.065	1.372	1.461	.8168	46.8
43.3	.7557	.6858	.7278	.9424	1.061	1.374	1.458	.8151	46.7
43.4	.7575	.6871	.7266	.9457	1.057	1.376	1.455	.8133	46.6
43.5	.7592	.6884	.7254	.9490	1.054	1.379	1.453	.8116	46.5
43.6	.7610	.6896	.7242	.9523	1.050	1.381	1.450	.8098	46.4
43.7	.7627	.6909	.7230	.9556	1.046	1.383	1.447	.8081	46.3
43.8	.7645	.6921	.7218	.9590	1.043	1.386	1.445	.8063	46.2
43.9	.7662	.6934	.7206	.9623	1.039	1.388	1.442	.8046	46.1
44.0	.7679	.6947	.7193	.9657	1.036	1.390	1.440	.8029	**46.0**
44.1	.7697	.6959	.7181	.9691	1.032	1.393	1.437	.8011	45.9
44.2	.7714	.6972	.7169	.9725	1.028	1.395	1.434	.7994	45.8
44.3	.7732	.6984	.7157	.9759	1.025	1.397	1.432	.7976	45.7
44.4	.7749	.6997	.7145	.9793	1.021	1.400	1.429	.7959	45.6
44.5	.7767	.7009	.7133	.9827	1.018	1.402	1.427	.7941	45.5
44.6	.7784	.7022	.7120	.9861	1.014	1.404	1.424	.7924	45.4
44.7	.7802	.7034	.7108	.9896	1.011	1.407	1.422	.7906	45.3
44.8	.7819	.7046	.7096	.9930	1.007	1.409	1.419	.7889	45.2
44.9	.7837	.7059	.7083	.9965	1.003	1.412	1.417	.7871	45.1
45.0	.7854	.7071	.7071	1.0000	1.000	1.414	1.414	.7854	**45.0**
		$\cos\theta$	$\sin\theta$	$\cot\theta$	$\tan\theta$	$\csc\theta$	$\sec\theta$	θ Radians	θ Degrees

Table 2 Trigonometric Functions of θ (θ in degrees and minutes)

m (θ) Degrees	Radians	sin θ	csc θ	tan θ	cot θ	sec θ	cos θ		
0° 00'	.0000	.0000	Undefined	.0000	Undefined	1.000	1.0000	1.5708	90° 00'
10'	.0029	.0029	343.8	.0029	343.8	1.000	1.0000	1.5679	50'
20'	.0058	.0058	171.9	.0058	171.9	1.000	1.0000	1.5650	40'
30'	.0087	.0087	114.6	.0087	114.6	1.000	1.0000	1.5621	30'
40'	.0116	.0116	85.95	.0116	85.94	1.000	.9999	1.5592	20'
50'	.0145	.0145	68.76	.0145	68.75	1.000	.9999	1.5563	10'
1° 00'	.0175	.0175	57.30	.0175	57.29	1.000	.9998	1.5533	89° 00'
10'	.0204	.0204	49.11	.0204	49.10	1.000	.9998	1.5504	50'
20'	.0233	.0233	42.98	.0233	42.96	1.000	.9997	1.5475	40'
30'	.0262	.0262	38.20	.0262	38.19	1.000	.9997	1.5446	30'
40'	.0291	.0291	34.38	.0291	34.37	1.000	.9996	1.5417	20'
50'	.0320	.0320	31.26	.0320	31.24	1.001	.9995	1.5388	10'
2° 00'	.0349	.0349	28.65	.0349	28.64	1.001	.9994	1.5359	88° 00'
10'	.0378	.0378	26.45	.0378	26.43	1.001	.9993	1.5330	50'
20'	.0407	.0407	24.56	.0407	24.54	1.001	.9992	1.5301	40'
30'	.0436	.0436	22.93	.0437	22.90	1.001	.9990	1.5272	30'
40'	.0465	.0465	21.49	.0466	21.47	1.001	.9989	1.5243	20'
50'	.0495	.0494	20.23	.0495	20.21	1.001	.9988	1.5213	10'
3° 00'	.0524	.0523	19.11	.0524	19.08	1.001	.9986	1.5184	87° 00'
10'	.0553	.0552	18.10	.0553	18.07	1.002	.9985	1.5155	50'
20'	.0582	.0581	17.20	.0582	17.17	1.002	.9983	1.5126	40'
30'	.0611	.0610	16.38	.0612	16.35	1.002	.9981	1.5097	30'
40'	.0640	.0640	15.64	.0641	15.60	1.002	.9980	1.5068	20'
50'	.0669	.0669	14.96	.0670	14.92	1.002	.9978	1.5039	10'
4° 00'	.0698	.0698	14.34	.0699	14.30	1.002	.9976	1.5010	86° 00'
10'	.0727	.0727	13.76	.0729	13.73	1.003	.9974	1.4981	50'
20'	.0756	.0756	13.23	.0758	13.20	1.003	.9971	1.4952	40'
30'	.0785	.0785	12.75	.0787	12.71	1.003	.9969	1.4923	30'
40'	.0814	.0814	12.29	.0816	12.25	1.003	.9967	1.4893	20'
50'	.0844	.0843	11.87	.0846	11.83	1.004	.9964	1.4864	10'
5° 00'	.0873	.0872	11.47	.0875	11.43	1.004	.9962	1.4835	85° 00'
10'	.0902	.0901	11.10	.0904	11.06	1.004	.9959	1.4806	50'
20'	.0931	.0929	10.76	.0934	10.71	1.004	.9957	1.4777	40'
30'	.0960	.0958	10.43	.0963	10.39	1.005	.9954	1.4748	30'
40'	.0989	.0987	10.13	.0992	10.08	1.005	.9951	1.4719	20'
50'	.1018	.1016	9.839	.1022	9.788	1.005	.9948	1.4690	10'
6° 00'	.1047	.1045	9.567	.1051	9.514	1.006	.9945	1.4661	84° 00'
10'	.1076	.1074	9.309	.1080	9.255	1.006	.9942	1.4632	50'
20'	.1105	.1103	9.065	.1110	9.010	1.006	.9939	1.4603	40'
30'	.1134	.1132	8.834	.1139	8.777	1.006	.9936	1.4573	30'
40'	.1164	.1161	8.614	.1169	8.556	1.007	.9932	1.4544	20'
50'	.1193	.1190	8.405	.1198	8.345	1.007	.9929	1.4515	10'
7° 00'	.1222	.1219	8.206	.1228	8.144	1.008	.9925	1.4486	83° 00'
10'	.1251	.1248	8.016	.1257	7.953	1.008	.9922	1.4457	50'
20'	.1280	.1276	7.834	.1287	7.770	1.008	.9918	1.4428	40'
30'	.1309	.1305	7.661	.1317	7.596	1.009	.9914	1.4399	30'
40'	.1338	.1334	7.496	.1346	7.429	1.009	.9911	1.4370	20'
50'	.1367	.1363	7.337	.1376	7.269	1.009	.9907	1.4341	10'
8° 00'	.1396	.1392	7.185	.1405	7.115	1.010	.9903	1.4312	82° 00'
10'	.1425	.1421	7.040	.1435	6.968	1.010	.9899	1.4283	50'
20'	.1454	.1449	6.900	.1465	6.827	1.011	.9894	1.4254	40'
30'	.1484	.1478	6.765	.1495	6.691	1.011	.9890	1.4224	30'
40'	.1513	.1507	6.636	.1524	6.561	1.012	.9886	1.4195	20'
50'	.1542	.1536	6.512	.1554	6.435	1.012	.9881	1.4166	10'
9° 00'	.1571	.1564	6.392	.1584	6.314	1.012	.9877	1.4137	81° 00'
		sec θ	cot θ	tan θ	csc θ	sin θ	Radians m (θ)	Degrees	

Table 2 Trigonometric Functions of θ (θ in degrees and minutes)

m (θ) Degrees	Radians	sin θ	csc θ	tan θ	cot θ	sec θ	cos θ		
9° 00′	.1571	.1564	6.392	.1584	6.314	1.012	.9877	1.4137	81° 00′
10′	.1600	.1593	6.277	.1614	6.197	1.013	.9872	1.4108	50′
20′	.1629	.1622	6.166	.1644	6.084	1.013	.9868	1.4079	40′
30′	.1658	.1650	6.059	.1673	5.976	1.014	.9863	1.4050	30′
40′	.1687	.1679	5.955	.1703	5.871	1.014	.9858	1.4021	20′
50′	.1716	.1708	5.855	.1733	5.769	1.015	.9853	1.3992	10′
10° 00′	.1745	.1736	5.759	.1763	5.671	1.015	.9848	1.3963	80° 00′
10′	.1774	.1765	5.665	.1793	5.576	1.016	.9843	1.3934	50′
20′	.1804	.1794	5.575	.1823	5.485	1.016	.9838	1.3904	40′
30′	.1833	.1822	5.487	.1853	5.396	1.017	.9833	1.3875	30′
40′	.1862	.1851	5.403	.1883	5.309	1.018	.9827	1.3846	20′
50′	.1891	.1880	5.320	.1914	5.226	1.018	.9822	1.3817	10′
11° 00′	.1920	.1908	5.241	.1944	5.145	1.019	.9816	1.3788	79° 00′
10′	.1949	.1937	5.164	.1974	5.066	1.019	.9811	1.3759	50′
20′	.1978	.1965	5.089	.2004	4.989	1.020	.9805	1.3730	40′
30′	.2007	.1994	5.016	.2035	4.915	1.020	.9799	1.3701	30′
40′	.2036	.2022	4.945	.2065	4.843	1.021	.9793	1.3672	20′
50′	.2065	.2051	4.876	.2095	4.773	1.022	.9787	1.3643	10′
12° 00′	.2094	.2079	4.810	.2126	4.705	1.022	.9781	1.3614	78° 00′
10′	.2123	.2108	4.745	.2156	4.638	1.023	.9775	1.3584	50′
20′	.2153	.2136	4.682	.2186	4.574	1.024	.9769	1.3555	40′
30′	.2182	.2164	4.620	.2217	4.511	1.024	.9763	1.3526	30′
40′	.2211	.2193	4.560	.2247	4.449	1.025	.9757	1.3497	20′
50′	.2240	.2221	4.502	.2278	4.390	1.026	.9750	1.3468	10′
13° 00′	.2269	.2250	4.445	.2309	4.331	1.026	.9744	1.3439	77° 00′
10′	.2298	.2278	4.390	.2339	4.275	1.027	.9737	1.3410	50′
20′	.2327	.2306	4.336	.2370	4.219	1.028	.9730	1.3381	40′
30′	.2356	.2334	4.284	.2401	4.165	1.028	.9724	1.3352	30′
40′	.2385	.2363	4.232	.2432	4.113	1.029	.9717	1.3323	20′
50′	.2414	.2391	4.182	.2462	4.061	1.030	.9710	1.3294	10′
14° 00′	.2443	.2419	4.134	.2493	4.011	1.031	.9703	1.3265	76° 00′
10′	.2473	.2447	4.086	.2524	3.962	1.031	.9696	1.3235	50′
20′	.2502	.2476	4.039	.2555	3.914	1.032	.9689	1.3206	40′
30′	.2531	.2504	3.994	.2586	3.867	1.033	.9681	1.3177	30′
40′	.2560	.2532	3.950	.2617	3.821	1.034	.9674	1.3148	20′
50′	.2589	.2560	3.906	.2648	3.776	1.034	.9667	1.3119	10′
15° 00′	.2618	.2588	3.864	.2679	3.732	1.035	.9659	1.3090	75° 00′
10′	.2647	.2616	3.822	.2711	3.689	1.036	.9652	1.3061	50′
20′	.2676	.2644	3.782	.2742	3.647	1.037	.9644	1.3032	40′
30′	.2705	.2672	3.742	.2773	3.606	1.038	.9636	1.3003	30′
40′	.2734	.2700	3.703	.2805	3.566	1.039	.9628	1.2974	20′
50′	.2763	.2728	3.665	.2836	3.526	1.039	.9621	1.2945	10′
16° 00′	.2793	.2756	3.628	.2867	3.487	1.040	.9613	1.2915	74° 00′
10′	.2822	.2784	3.592	.2899	3.450	1.041	.9605	1.2886	50′
20′	.2851	.2812	3.556	.2931	3.412	1.042	.9596	1.2857	40′
30′	.2880	.2840	3.521	.2962	3.376	1.043	.9588	1.2828	30′
40′	.2909	.2868	3.487	.2994	3.340	1.044	.9580	1.2799	20′
50′	.2938	.2896	3.453	.3026	3.305	1.045	.9572	1.2770	10′
17° 00′	.2967	.2924	3.420	.3057	3.271	1.046	.9563	1.2741	73° 00′
10′	.2996	.2952	3.388	.3089	3.237	1.047	.9555	1.2712	50′
20′	.3025	.2979	3.357	.3121	3.204	1.048	.9546	1.2683	40′
30′	.3054	.3007	3.326	.3153	3.172	1.049	.9537	1.2654	30′
40′	.3083	.3035	3.295	.3185	3.140	1.049	.9528	1.2625	20′
50′	.3113	.3062	3.265	.3217	3.108	1.050	.9520	1.2595	10′
18° 00′	.3142	.3090	3.236	.3249	3.078	1.051	.9511	1.2566	72° 00′
		cos θ	sec θ	cot θ	tan θ	csc θ	sin θ	Radians	Degrees m (θ)

Table 2   Trigonometric Functions of θ (θ in degrees and minutes)

$m(\theta)$ Degrees	Radians	$\sin\theta$	$\csc\theta$	$\tan\theta$	$\cot\theta$	$\sec\theta$	$\cos\theta$		
18° 00′	.3142	.3090	3.236	.3249	3.078	1.051	.9511	1.2566	72° 00′
10′	.3171	.3118	3.207	.3281	3.047	1.052	.9502	1.2537	50′
20′	.3200	.3145	3.179	.3314	3.018	1.053	.9492	1.2508	40′
30′	.3229	.3173	3.152	.3346	2.989	1.054	.9483	1.2479	30′
40′	.3258	.3201	3.124	.3378	2.960	1.056	.9474	1.2450	20′
50′	.3287	.3228	3.098	.3411	2.932	1.057	.9465	1.2421	10′
19° 00′	.3316	.3256	3.072	.3443	2.904	1.058	.9455	1.2392	71° 00′
10′	.3345	.3283	3.046	.3476	2.877	1.059	.9446	1.2363	50′
20′	.3374	.3311	3.021	.3508	2.850	1.060	.9436	1.2334	40′
30′	.3403	.3338	2.996	.3541	2.824	1.061	.9426	1.2305	30′
40′	.3432	.3365	2.971	.3574	2.798	1.062	.9417	1.2275	20′
50′	.3462	.3393	2.947	.3607	2.773	1.063	.9407	1.2246	10′
20° 00′	.3491	.3420	2.924	.3640	2.747	1.064	.9397	1.2217	70° 00′
10′	.3520	.3448	2.901	.3673	2.723	1.065	.9387	1.2188	50′
20′	.3549	.3475	2.878	.3706	2.699	1.066	.9377	1.2159	40′
30′	.3578	.3502	2.855	.3739	2.675	1.068	.9367	1.2130	30′
40′	.3607	.3529	2.833	.3772	2.651	1.069	.9356	1.2101	20′
50′	.3636	.3557	2.812	.3805	2.628	1.070	.9346	1.2072	10′
21° 00′	.3665	.3584	2.790	.3839	2.605	1.071	.9336	1.2043	69° 00′
10′	.3694	.3611	2.769	.3872	2.583	1.072	.9325	1.2014	50′
20′	.3723	.3638	2.749	.3906	2.560	1.074	.9315	1.1985	40′
30′	.3752	.3665	2.729	.3939	2.539	1.075	.9304	1.1956	30′
40′	.3782	.3692	2.709	.3973	2.517	1.076	.9293	1.1926	20′
50′	.3811	.3719	2.689	.4006	2.496	1.077	.9283	1.1897	10′
22° 00′	.3840	.3746	2.669	.4040	2.475	1.079	.9272	1.1868	68° 00′
10′	.3869	.3773	2.650	.4074	2.455	1.080	.9261	1.1839	50′
20′	.3898	.3800	2.632	.4108	2.434	1.081	.9250	1.1810	40′
30′	.3927	.3827	2.613	.4142	2.414	1.082	.9239	1.1781	30′
40′	.3956	.3854	2.595	.4176	2.394	1.084	.9228	1.1752	20′
50′	.3985	.3881	2.577	.4210	2.375	1.085	.9216	1.1723	10′
23° 00′	.4014	.3907	2.559	.4245	2.356	1.086	.9205	1.1694	67° 00′
10′	.4043	.3934	2.542	.4279	2.337	1.088	.9194	1.1665	50′
20′	.4072	.3961	2.525	.4314	2.318	1.089	.9182	1.1636	40′
30′	.4102	.3987	2.508	.4348	2.300	1.090	.9171	1.1606	30′
40′	.4131	.4014	2.491	.4383	2.282	1.092	.9159	1.1577	20′
50′	.4160	.4041	2.475	.4417	2.264	1.093	.9147	1.1548	10′
24° 00′	.4189	.4067	2.459	.4452	2.246	1.095	.9135	1.1519	66° 00′
10′	.4218	.4094	2.443	.4487	2.229	1.096	.9124	1.1490	50′
20′	.4247	.4120	2.427	.4522	2.211	1.097	.9112	1.1461	40′
30′	.4276	.4147	2.411	.4557	2.194	1.099	.9100	1.1432	30′
40′	.4305	.4173	2.396	.4592	2.177	1.100	.9088	1.1403	20′
50′	.4334	.4200	2.381	.4628	2.161	1.102	.9075	1.1374	10′
25° 00′	.4363	.4226	2.366	.4663	2.145	1.103	.9063	1.1345	65° 00′
10′	.4392	.4253	2.352	.4699	2.128	1.105	.9051	1.1316	50′
20′	.4422	.4279	2.337	.4734	2.112	1.106	.9038	1.1286	40′
30′	.4451	.4305	2.323	.4770	2.097	1.108	.9026	1.1257	30′
40′	.4480	.4331	2.309	.4806	2.081	1.109	.9013	1.1228	20′
50′	.4509	.4358	2.295	.4841	2.066	1.111	.9001	1.1199	10′
26° 00′	.4538	.4384	2.281	.4877	2.050	1.113	.8988	1.1170	64° 00′
10′	.4567	.4410	2.268	.4913	2.035	1.114	.8975	1.1141	50′
20′	.4596	.4436	2.254	.4950	2.020	1.116	.8962	1.1112	40′
30′	.4625	.4462	2.241	.4986	2.006	1.117	.8949	1.1083	30′
40′	.4654	.4488	2.228	.5022	1.991	1.119	.8936	1.1054	20′
50′	.4683	.4514	2.215	.5059	1.977	1.121	.8923	1.1025	10′
27° 00′	.4712	.4540	2.203	.5095	1.963	1.122	.8910	1.0996	63° 00′
		$\cos\theta$	$\sec\theta$	$\cot\theta$	$\tan\theta$	$\csc\theta$	$\sin\theta$	Radians	Degrees $m(\theta)$

Table 2 Trigonometric Functions of θ (θ in degrees and minutes)

m (θ) Degrees	Radians	sin θ	csc θ	tan θ	cot θ	sec θ	cos θ		
27° 00′	.4712	.4540	2.203	.5095	1.963	1.122	.8910	1.0996	63° 00′
10′	.4741	.4566	2.190	.5132	1.949	1.124	.8897	1.0966	50′
20′	.4771	.4592	2.178	.5169	1.935	1.126	.8884	1.0937	40′
30′	.4800	.4617	2.166	.5206	1.921	1.127	.8870	1.0908	30′
40′	.4829	.4643	2.154	.5243	1.907	1.129	.8857	1.0879	20′
50′	.4858	.4669	2.142	.5280	1.894	1.131	.8843	1.0850	10′
28° 00′	.4887	.4695	2.130	.5317	1.881	1.133	.8829	1.0821	62° 00′
10′	.4916	.4720	2.118	.5354	1.868	1.134	.8816	1.0792	50′
20′	.4945	.4746	2.107	.5392	1.855	1.136	.8802	1.0763	40′
30′	.4974	.4772	2.096	.5430	1.842	1.138	.8788	1.0734	30′
40′	.5003	.4797	2.085	.5467	1.829	1.140	.8774	1.0705	20′
50′	.5032	.4823	2.074	.5505	1.816	1.142	.8760	1.0676	10′
29° 00′	.5061	.4848	2.063	.5543	1.804	1.143	.8746	1.0647	61° 00′
10′	.5091	.4874	2.052	.5581	1.792	1.145	.8732	1.0617	50′
20′	.5120	.4899	2.041	.5619	1.780	1.147	.8718	1.0588	40′
30′	.5149	.4924	2.031	.5658	1.767	1.149	.8704	1.0559	30′
40′	.5178	.4950	2.020	.5696	1.756	1.151	.8689	1.0530	20′
50′	.5207	.4975	2.010	.5735	1.744	1.153	.8675	1.0501	10′
30° 00′	.5236	.5000	2.000	.5774	1.732	1.155	.8660	1.0472	60° 00′
10′	.5265	.5025	1.990	.5812	1.720	1.157	.8646	1.0443	50′
20′	.5294	.5050	1.980	.5851	1.709	1.159	.8631	1.0414	40′
30′	.5323	.5075	1.970	.5890	1.698	1.161	.8616	1.0385	30′
40′	.5352	.5100	1.961	.5930	1.686	1.163	.8601	1.0356	20′
50′	.5381	.5125	1.951	.5969	1.675	1.165	.8587	1.0327	10′
31° 00′	.5411	.5150	1.942	.6009	1.664	1.167	.8572	1.0297	59° 00′
10′	.5440	.5175	1.932	.6048	1.653	1.169	.8557	1.0268	50′
20′	.5469	.5200	1.923	.6088	1.643	1.171	.8542	1.0239	40′
30′	.5498	.5225	1.914	.6128	1.632	1.173	.8526	1.0210	30′
40′	.5527	.5250	1.905	.6168	1.621	1.175	.8511	1.0181	20′
50′	.5556	.5275	1.896	.6208	1.611	1.177	.8496	1.0152	10′
32° 00′	.5585	.5299	1.887	.6249	1.600	1.179	.8480	1.0123	58° 00′
10′	.5614	.5324	1.878	.6289	1.590	1.181	.8465	1.0094	50′
20′	.5643	.5348	1.870	.6330	1.580	1.184	.8450	1.0065	40′
30′	.5672	.5373	1.861	.6371	1.570	1.186	.8434	1.0036	30′
40′	.5701	.5398	1.853	.6412	1.560	1.188	.8418	1.0007	20′
50′	.5730	.5422	1.844	.6453	1.550	1.190	.8403	.9977	10′
33° 00′	.5760	.5446	1.836	.6494	1.540	1.192	.8387	.9948	57° 00′
10′	.5789	.5471	1.828	.6536	1.530	1.195	.8371	.9919	50′
20′	.5818	.5495	1.820	.6577	1.520	1.197	.8355	.9890	40′
30′	.5847	.5519	1.812	.6619	1.511	1.199	.8339	.9861	30′
40′	.5876	.5544	1.804	.6661	1.501	1.202	.8323	.9832	20′
50′	.5905	.5568	1.796	.6703	1.492	1.204	.8307	.9803	10′
34° 00′	.5934	.5592	1.788	.6745	1.483	1.206	.8290	.9774	56° 00′
10′	.5963	.5616	1.781	.6787	1.473	1.209	.8274	.9745	50′
20′	.5992	.5640	1.773	.6830	1.464	1.211	.8258	.9716	40′
30′	.6021	.5664	1.766	.6873	1.455	1.213	.8241	.9687	30′
40′	.6050	.5688	1.758	.6916	1.446	1.216	.8225	.9657	20′
50′	.6080	.5712	1.751	.6959	1.437	1.218	.8208	.9628	10′
35° 00′	.6109	.5736	1.743	.7002	1.428	1.221	.8192	.9599	55° 00′
10′	.6138	.5760	1.736	.7046	1.419	1.223	.8175	.9570	50′
20′	.6167	.5783	1.729	.7089	1.411	1.226	.8158	.9541	40′
30′	.6196	.5807	1.722	.7133	1.402	1.228	.8141	.9512	30′
40′	.6225	.5831	1.715	.7177	1.393	1.231	.8124	.9483	20′
50′	.6254	.5854	1.708	.7221	1.385	1.233	.8107	.9454	10′
36° 00′	.6283	.5878	1.701	.7265	1.376	1.236	.8090	.9425	54° 00′
		cos θ	sec θ	cot θ	tan θ	csc θ	sin θ	Radians m (θ)	Degrees

Table 2 Trigonometric Functions of θ (θ in degrees and minutes)

m (θ) Degrees	Radians	sin θ	csc θ	tan θ	cot θ	sec θ	cos θ		
36° 00′	.6283	.5878	1.701	.7265	1.376	1.236	.8090	.9425	54° 00′
10′	.6312	.5901	1.695	.7310	1.368	1.239	.8073	.9396	50′
20′	.6341	.5925	1.688	.7355	1.360	1.241	.8056	.9367	40′
30′	.6370	.5948	1.681	.7400	1.351	1.244	.8039	.9338	30′
40′	.6400	.5972	1.675	.7445	1.343	1.247	.8021	.9308	20′
50′	.6429	.5995	1.668	.7490	1.335	1.249	.8004	.9279	10′
37° 00′	.6458	.6018	1.662	.7536	1.327	1.252	.7986	.9250	53° 00′
10′	.6487	.6041	1.655	.7581	1.319	1.255	.7969	.9221	50′
20′	.6516	.6065	1.649	.7627	1.311	1.258	.7951	.9192	40′
30′	.6545	.6088	1.643	.7673	1.303	1.260	.7934	.9163	30′
40′	.6574	.6111	1.636	.7720	1.295	1.263	.7916	.9134	20′
50′	.6603	.6134	1.630	.7766	1.288	1.266	.7898	.9105	10′
38° 00′	.6632	.6157	1.624	.7813	1.280	1.269	.7880	.9076	52° 00′
10′	.6661	.6180	1.618	.7860	1.272	1.272	.7862	.9047	50′
20′	.6690	.6202	1.612	.7907	1.265	1.275	.7844	.9018	40′
30′	.6720	.6225	1.606	.7954	1.257	1.278	.7826	.8988	30′
40′	.6749	.6248	1.601	.8002	1.250	1.281	.7808	.8959	20′
50′	.6778	.6271	1.595	.8050	1.242	1.284	.7790	.8930	10′
39° 00′	.6807	.6293	1.589	.8098	1.235	1.287	.7771	.8901	51° 00′
10′	.6836	.6316	1.583	.8146	1.228	1.290	.7753	.8872	50′
20′	.6865	.6338	1.578	.8195	1.220	1.293	.7735	.8843	40′
30′	.6894	.6361	1.572	.8243	1.213	1.296	.7716	.8814	30′
40′	.6923	.6383	1.567	.8292	1.206	1.299	.7698	.8785	20′
50′	.6952	.6406	1.561	.8342	1.199	1.302	.7679	.8756	10′
40° 00′	.6981	.6428	1.556	.8391	1.192	1.305	.7660	.8727	50° 00′
10′	.7010	.6450	1.550	.8441	1.185	1.309	.7642	.8698	50′
20′	.7039	.6472	1.545	.8491	1.178	1.312	.7623	.8668	40′
30′	.7069	.6494	1.540	.8541	1.171	1.315	.7604	.8639	30′
40′	.7098	.6517	1.535	.8591	1.164	1.318	.7585	.8610	20′
50′	.7127	.6539	1.529	.8642	1.157	1.322	.7566	.8581	10′
41° 00′	.7156	.6561	1.524	.8693	1.150	1.325	.7547	.8552	49° 00′
10′	.7185	.6583	1.519	.8744	1.144	1.328	.7528	.8523	50′
20′	.7214	.6604	1.514	.8796	1.137	1.332	.7509	.8494	40′
30′	.7243	.6626	1.509	.8847	1.130	1.335	.7490	.8465	30′
40′	.7272	.6648	1.504	.8899	1.124	1.339	.7470	.8436	20′
50′	.7301	.6670	1.499	.8952	1.117	1.342	.7451	.8407	10′
42° 00′	.7330	.6691	1.494	.9004	1.111	1.346	.7431	.8378	48° 00′
10′	.7359	.6713	1.490	.9057	1.104	1.349	.7412	.8348	50′
20′	.7389	.6734	1.485	.9110	1.098	1.353	.7392	.8319	40′
30′	.7418	.6756	1.480	.9163	1.091	1.356	.7373	.8290	30′
40′	.7447	.6777	1.476	.9217	1.085	1.360	.7353	.8261	20′
50′	.7476	.6799	1.471	.9271	1.079	1.364	.7333	.8232	10′
43° 00′	.7505	.6820	1.466	.9325	1.072	1.367	.7314	.8203	47° 00′
10′	.7534	.6841	1.462	.9380	1.066	1.371	.7294	.8174	50′
20′	.7563	.6862	1.457	.9435	1.060	1.375	.7274	.8145	40′
30′	.7592	.6884	1.453	.9490	1.054	1.379	.7254	.8116	30′
40′	.7621	.6905	1.448	.9545	1.048	1.382	.7234	.8087	20′
50′	.7650	.6926	1.444	.9601	1.042	1.386	.7214	.8058	10′
44° 00′	.7679	.6947	1.440	.9657	1.036	1.390	.7193	.8029	46° 00′
10′	.7709	.6967	1.435	.9713	1.030	1.394	.7173	.7999	50′
20′	.7738	.6988	1.431	.9770	1.024	1.398	.7153	.7970	40′
30′	.7767	.7009	1.427	.9827	1.018	1.402	.7133	.7941	30′
40′	.7796	.7030	1.423	.9884	1.012	1.406	.7112	.7912	20′
50′	.7825	.7050	1.418	.9942	1.006	1.410	.7092	.7883	10′
45° 00′	.7854	.7071	1.414	1.000	1.000	1.414	.7071	.7854	45° 00′
		cos θ	sec θ	cot θ	tan θ	csc θ	sin θ	Radians	Degrees m (θ)

Table 3   Trigonometric Functions of θ (θ in radians)

Radians *m(θ)*	Degrees	sin θ	csc θ	tan θ	cot θ	sec θ	cos θ
0.00	0° 00′	0.0000	Undefined	0.0000	Undefined	1.000	1.000
.01	0° 34′	.0100	100.0	.0100	100.0	1.000	1.000
.02	1° 09′	.0200	50.00	.0200	49.99	1.000	0.9998
.03	1° 43′	.0300	33.34	.0300	33.32	1.000	0.9996
.04	2° 18′	.0400	25.01	.0400	24.99	1.001	0.9992
0.05	2° 52′	0.0500	20.01	0.0500	19.98	1.001	0.9988
.06	3° 26′	.0600	16.68	.0601	16.65	1.002	.9982
.07	4° 01′	.0699	14.30	.0701	14.26	1.002	.9976
.08	4° 35′	.0799	12.51	.0802	12.47	1.003	.9968
.09	5° 09′	.0899	11.13	.0902	11.08	1.004	.9960
0.10	5° 44′	0.0998	10.02	0.1003	9.967	1.005	0.9950
.11	6° 18′	.1098	9.109	.1104	9.054	1.006	.9940
.12	6° 53′	.1197	8.353	.1206	8.293	1.007	.9928
.13	7° 27′	.1296	7.714	.1307	7.649	1.009	.9916
.14	8° 01′	.1395	7.166	.1409	7.096	1.010	.9902
0.15	8° 36′	0.1494	6.692	0.1511	6.617	1.011	0.9888
.16	9° 10′	.1593	6.277	.1614	6.197	1.013	.9872
.17	9° 44′	.1692	5.911	.1717	5.826	1.015	.9856
.18	10° 19′	.1790	5.586	.1820	5.495	1.016	.9838
.19	10° 53′	.1889	5.295	.1923	5.200	1.018	.9820
0.20	11° 28′	0.1987	5.033	0.2027	4.933	1.020	0.9801
.21	12° 02′	.2085	4.797	.2131	4.692	1.022	.9780
.22	12° 36′	.2182	4.582	.2236	4.472	1.025	.9759
.23	13° 11′	.2280	4.386	.2341	4.271	1.027	.9737
.24	13° 45′	.2377	4.207	.2447	4.086	1.030	.9713
0.25	14° 19′	0.2474	4.042	0.2553	3.916	1.032	0.9689
.26	14° 54′	.2571	3.890	.2660	3.759	1.035	.9664
.27	15° 28′	.2667	3.749	.2768	3.613	1.038	.9638
.28	16° 03′	.2764	3.619	.2876	3.478	1.041	.9611
.29	16° 37′	.2860	3.497	.2984	3.351	1.044	.9582
0.30	17° 11′	0.2955	3.384	0.3093	3.233	1.047	0.9553
.31	17° 46′	.3051	3.278	.3203	3.122	1.050	.9523
.32	18° 20′	.3146	3.179	.3314	3.018	1.053	.9492
.33	18° 55′	.3240	3.086	.3425	2.920	1.057	.9460
.34	19° 29′	.3335	2.999	.3537	2.827	1.061	.9428
0.35	20° 03′	0.3429	2.916	0.3650	2.740	1.065	0.9394
.36	20° 38′	.3523	2.839	.3764	2.657	1.068	.9359
.37	21° 12′	.3616	2.765	.3879	2.578	1.073	.9323
.38	21° 46′	.3709	2.696	.3994	2.504	1.077	.9287
.39	22° 21′	.3802	2.630	.4111	2.433	1.081	.9249
0.40	22° 55′	0.3894	2.568	0.4228	2.365	1.086	0.9211
.41	23° 30′	.3986	2.509	.4346	2.301	1.090	.9171
.42	24° 04′	.4078	2.452	.4466	2.239	1.095	.9131
.43	24° 38′	.4169	2.399	.4586	2.180	1.100	.9090
.44	25° 13′	.4259	2.348	.4708	2.124	1.105	.9048
0.45	25° 47′	0.4350	2.299	0.4831	2.070	1.111	0.9004
.46	26° 21′	.4439	2.253	.4954	2.018	1.116	.8961
.47	26° 56′	.4529	2.208	.5080	1.969	1.122	.8916
.48	27° 30′	.4618	2.166	.5206	1.921	1.127	.8870
.49	28° 05′	.4706	2.125	.5334	1.875	1.133	.8823

Table 3 Trigonometric Functions of θ (θ in radians)

$m(\theta)$ Radians	Degrees	$\sin\theta$	$\csc\theta$	$\tan\theta$	$\cot\theta$	$\sec\theta$	$\cos\theta$
0.50	28° 39′	0.4794	2.086	0.5463	1.830	1.139	0.8776
.51	29° 13′	.4882	2.048	.5594	1.788	1.146	.8727
.52	29° 48′	.4969	2.013	.5726	1.747	1.152	.8678
.53	30° 22′	.5055	1.978	.5859	1.707	1.159	.8628
.54	30° 56′	.5141	1.945	.5994	1.668	1.166	.8577
0.55	31° 31′	0.5227	1.913	0.6131	1.631	1.173	0.8525
.56	32° 05′	.5312	1.883	.6269	1.595	1.180	.8473
.57	32° 40′	.5396	1.853	.6410	1.560	1.188	.8419
.58	33° 14′	.5480	1.825	.6552	1.526	1.196	.8365
.59	33° 48′	.5564	1.797	.6696	1.494	1.203	.8309
0.60	34° 23′	0.5646	1.771	0.6841	1.462	1.212	0.8253
.61	34° 57′	.5729	1.746	.6989	1.431	1.220	.8196
.62	35° 31′	.5810	1.721	.7139	1.401	1.229	.8139
.63	36° 06′	.5891	1.697	.7291	1.372	1.238	.8080
.64	36° 40′	.5972	1.674	.7445	1.343	1.247	.8021
0.65	37° 15′	0.6052	1.652	0.7602	1.315	1.256	0.7961
.66	37° 49′	.6131	1.631	.7761	1.288	1.266	.7900
.67	38° 23′	.6210	1.610	.7923	1.262	1.276	.7838
.68	38° 58′	.6288	1.590	.8087	1.237	1.286	.7776
.69	39° 32′	.6365	1.571	.8253	1.212	1.297	.7712
0.70	40° 06′	0.6442	1.552	0.8423	1.187	1.307	0.7648
.71	40° 41′	.6518	1.534	.8595	1.163	1.319	.7584
.72	41° 15′	.6594	1.517	.8771	1.140	1.330	.7518
.73	41° 50′	.6669	1.500	.8949	1.117	1.342	.7452
.74	42° 24′	.6743	1.483	.9131	1.095	1.354	.7385
0.75	42° 58′	0.6816	1.467	0.9316	1.073	1.367	0.7317
.76	43° 33′	.6889	1.452	.9505	1.052	1.380	.7248
.77	44° 07′	.6961	1.437	.9697	1.031	1.393	.7179
.78	44° 41′	.7033	1.422	.9893	1.011	1.407	.7109
.79	45° 16′	.7104	1.408	1.009	.9908	1.421	.7038
0.80	45° 50′	0.7174	1.394	1.030	0.9712	1.435	0.6967
.81	46° 25′	.7243	1.381	1.050	.9520	1.450	.6895
.82	46° 59′	.7311	1.368	1.072	.9331	1.466	.6822
.83	47° 33′	.7379	1.355	1.093	.9146	1.482	.6749
.84	48° 08′	.7446	1.343	1.116	.8964	1.498	.6675
0.85	48° 42′	0.7513	1.331	1.138	0.8785	1.515	0.6600
.86	49° 17′	.7578	1.320	1.162	.8609	1.533	.6524
.87	49° 51′	.7643	1.308	1.185	.8437	1.551	.6448
.88	50° 25′	.7707	1.297	1.210	.8267	1.569	.6372
.89	51° 00′	.7771	1.287	1.235	.8100	1.589	.6294
0.90	51° 34′	0.7833	1.277	1.260	0.7936	1.609	0.6216
.91	52° 08′	.7895	1.267	1.286	.7774	1.629	.6137
.92	52° 43′	.7956	1.257	1.313	.7615	1.651	.6058
.93	53° 17′	.8016	1.247	1.341	.7458	1.673	.5978
.94	53° 52′	.8076	1.238	1.369	.7303	1.696	.5898
0.95	54° 26′	0.8134	1.229	1.398	0.7151	1.719	0.5817
.96	55° 00′	.8192	1.221	1.428	.7001	1.744	.5735
.97	55° 35′	.8249	1.212	1.459	.6853	1.769	.5653
.98	56° 09′	.8305	1.204	1.491	.6707	1.795	.5570
.99	56° 43′	.8360	1.196	1.524	.6563	1.823	.5487

Table 3 Trigonometric Functions of θ (θ in radians)

Radians $m(\theta)$ Degrees		$\sin \theta$	$\csc \theta$	$\tan \theta$	$\cot \theta$	$\sec \theta$	$\cos \theta$
1.00	57° 18′	0.8415	1.188	1.557	0.6421	1.851	0.5403
1.01	57° 52′	.8468	1.181	1.592	.6281	1.880	.5319
1.02	58° 27′	.8521	1.174	1.628	.6142	1.911	.5234
1.03	59° 01′	.8573	1.166	1.665	.6005	1.942	.5148
1.04	59° 35′	.8624	1.160	1.704	.5870	1.975	.5062
1.05	60° 10′	0.8674	1.153	1.743	0.5736	2.010	0.4976
1.06	60° 44′	.8724	1.146	1.784	.5604	2.046	.4889
1.07	61° 18′	.8772	1.140	1.827	.5473	2.083	.4801
1.08	61° 53′	.8820	1.134	1.871	.5344	2.122	.4713
1.09	62° 27′	.8866	1.128	1.917	.5216	2.162	.4625
1.10	63° 02′	0.8912	1.122	1.965	0.5090	2.205	0.4536
1.11	63° 36′	.8957	1.116	2.014	.4964	2.249	.4447
1.12	64° 10′	.9001	1.111	2.066	.4840	2.295	.4357
1.13	64° 45′	.9044	1.106	2.120	.4718	2.344	.4267
1.14	65° 19′	.9086	1.101	2.176	.4596	2.395	.4176
1.15	65° 53′	0.9128	1.096	2.234	0.4475	2.448	0.4085
1.16	66° 28′	.9168	1.091	2.296	.4356	2.504	.3993
1.17	67° 02′	.9208	1.086	2.360	.4237	2.563	.3902
1.18	67° 37′	.9246	1.082	2.428	.4120	2.625	.3809
1.19	68° 11′	.9284	1.077	2.498	.4003	2.691	3717
1.20	68° 45′	0.9320	1.073	2.572	0.3888	2.760	0.3624
1.21	69° 20′	.9356	1.069	2.650	.3773	2.833	.3530
1.22	69° 54′	.9391	1.065	2.733	.3659	2.910	.3436
1.23	70° 28′	.9425	1.061	2.820	.3546	2.992	.3342
1.24	71° 03′	.9458	1.057	2.912	.3434	3.079	.3248
1.25	71° 37′	0.9490	1.054	3.010	0.3323	3.171	0.3153
1.26	72° 12′	.9521	1.050	3.113	.3212	3.270	.3058
1.27	72° 46′	.9551	1.047	3.224	.3102	3.375	.2963
1.28	73° 20′	.9580	1.044	3.341	.2993	3.488	.2867
1.29	73° 55′	.9608	1.041	3.467	.2884	3.609	.2771
1.30	74° 29′	0.9636	1.038	3.602	0.2776	3.738	0.2675
1.31	75° 03′	.9662	1.035	3.747	.2669	3.878	.2579
1.32	75° 38′	.9687	1.032	3.903	.2562	4.029	.2482
1.33	76° 12′	.9711	1.030	4.072	.2456	4.193	.2385
1.34	76° 47′	.9735	1.027	4.256	.2350	4.372	.2288
1.35	77° 21′	0.9757	1.025	4.455	0.2245	4.566	0.2190
1.36	77° 55′	.9779	1.023	4.673	.2140	4.779	.2092
1.37	78° 30′	.9799	1.021	4.913	.2035	5.014	.1994
1.38	79° 04′	.9819	1.018	5.177	.1931	5.273	.1896
1.39	79° 39′	.9837	1.017	5.471	.1828	5.561	.1798
1.40	80° 13′	0.9854	1.015	5.798	0.1725	5.883	0.1700
1.41	80° 47′	.9871	1.013	6.165	.1622	6.246	.1601
1.42	81° 22′	.9887	1.011	6.581	.1519	6.657	.1502
1.43	81° 56′	.9901	1.010	7.055	.1417	7.126	.1403
1.44	82° 30′	.9915	1.009	7.602	.1315	7.667	.1304
1.45	83° 05′	0.9927	1.007	8.238	0.1214	8.299	0.1205
1.46	83° 39′	.9939	1.006	8.989	.1113	9.044	.1106
1.47	84° 14′	.9949	1.005	9.887	.1011	9.938	.1006
1.48	84° 48′	.9959	1.004	10.98	.0911	11.03	.0907
1.49	85° 22′	.9967	1.003	12.35	.0810	12.39	.0807

Table 3 Trigonometric Functions of θ (θ in radians)

Radians $m(\theta)$	Degrees	sin θ	csc θ	tan θ	cot θ	sec θ	cos θ
1.50	85° 57′	0.9975	1.003	14.10	0.0709	14.14	0.0707
1.51	86° 31′	.9982	1.002	16.43	.0609	16.46	.0608
1.52	87° 05′	.9987	1.001	19.67	.0508	19.70	.0508
1.53	87° 40′	.9992	1.001	24.50	.0408	24.52	.0408
1.54	88° 14′	.9995	1.000	32.46	.0308	32.48	.0308
1.55	88° 49′	0.9998	1.000	48.08	0.0208	48.09	0.0208
1.56	89° 23′	.9999	1.000	92.62	.0108	92.63	.0108
1.57	89° 57′	1.000	1.000	1256	.0008	1256	.0008

Table 4 Squares and Square Roots

N	N^2	$\sqrt{N}$	$\sqrt{10N}$	N	N^2	$\sqrt{N}$	$\sqrt{10N}$
1.0	1.00	1.000	3.162	**5.5**	30.25	2.345	7.416
1.1	1.21	1.049	3.317	**5.6**	31.36	2.366	7.483
1.2	1.44	1.095	3.464	**5.7**	32.49	2.387	7.550
1.3	1.69	1.140	3.606	**5.8**	33.64	2.408	7.616
1.4	1.96	1.183	3.742	**5.9**	34.81	2.429	7.681
1.5	2.25	1.225	3.873	**6.0**	36.00	2.449	7.746
1.6	2.56	1.265	4.000	**6.1**	37.21	2.470	7.810
1.7	2.89	1.304	4.123	**6.2**	38.44	2.490	7.874
1.8	3.24	1.342	4.243	**6.3**	39.69	2.510	7.937
1.9	3.61	1.378	4.359	**6.4**	40.96	2.530	8.000
2.0	4.00	1.414	4.472	**6.5**	42.25	2.550	8.062
2.1	4.41	1.449	4.583	**6.6**	43.56	2.569	8.124
2.2	4.84	1.483	4.690	**6.7**	44.89	2.588	8.185
2.3	5.29	1.517	4.796	**6.8**	46.24	2.608	8.246
2.4	5.76	1.549	4.899	**6.9**	47.61	2.627	8.307
2.5	6.25	1.581	5.000	**7.0**	49.00	2.646	8.367
2.6	6.76	1.612	5.099	**7.1**	50.41	2.665	8.426
2.7	7.29	1.643	5.196	**7.2**	51.84	2.683	8.485
2.8	7.84	1.673	5.292	**7.3**	53.29	2.702	8.544
2.9	8.41	1.703	5.385	**7.4**	54.76	2.720	8.602
3.0	9.00	1.732	5.477	**7.5**	56.25	2.739	8.660
3.1	9.61	1.761	5.568	**7.6**	57.76	2.757	8.718
3.2	10.24	1.789	5.657	**7.7**	59.29	2.775	8.775
3.3	10.89	1.817	5.745	**7.8**	60.84	2.793	8.832
3.4	11.56	1.844	5.831	**7.9**	62.41	2.811	8.888
3.5	12.25	1.871	5.916	**8.0**	64.00	2.828	8.944
3.6	12.96	1.897	6.000	**8.1**	65.61	2.846	9.000
3.7	13.69	1.924	6.083	**8.2**	67.24	2.864	9.055
3.8	14.44	1.949	6.164	**8.3**	68.89	2.881	9.110
3.9	15.21	1.975	6.245	**8.4**	70.56	2.898	9.165
4.0	16.00	2.000	6.325	**8.5**	72.25	2.915	9.220
4.1	16.81	2.025	6.403	**8.6**	73.96	2.933	9.274
4.2	17.64	2.049	6.481	**8.7**	75.69	2.950	9.327
4.3	18.49	2.074	6.557	**8.8**	77.44	2.966	9.381
4.4	19.36	2.098	6.633	**8.9**	79.21	2.983	9.434
4.5	20.25	2.121	6.708	**9.0**	81.00	3.000	9.487
4.6	21.16	2.145	6.782	**9.1**	82.81	3.017	9.539
4.7	22.09	2.168	6.856	**9.2**	84.64	3.033	9.592
4.8	23.04	2.191	6.928	**9.3**	86.49	3.050	9.644
4.9	24.01	2.214	7.000	**9.4**	88.36	3.066	9.695
5.0	25.00	2.236	7.071	**9.5**	90.25	3.082	9.747
5.1	26.01	2.258	7.141	**9.6**	92.16	3.098	9.798
5.2	27.04	2.280	7.211	**9.7**	94.09	3.114	9.849
5.3	28.09	2.302	7.280	**9.8**	96.04	3.130	9.899
5.4	29.16	2.324	7.348	**9.9**	98.01	3.146	9.950
5.5	30.25	2.345	7.416	**10**	100.00	3.162	10.000

Table 5 Common Logarithms of Numbers *

N	0	1	2	3	4	5	6	7	8	9
10	0000	0043	0086	0128	0170	0212	0253	0294	0334	0374
11	0414	0453	0492	0531	0569	0607	0645	0682	0719	0755
12	0792	0828	0864	0899	0934	0969	1004	1038	1072	1106
13	1139	1173	1206	1239	1271	1303	1335	1367	1399	1430
14	1461	1492	1523	1553	1584	1614	1644	1673	1703	1732
15	1761	1790	1818	1847	1875	1903	1931	1959	1987	2014
16	2041	2068	2095	2122	2148	2175	2201	2227	2253	2279
17	2304	2330	2355	2380	2405	2430	2455	2480	2504	2529
18	2553	2577	2601	2625	2648	2672	2695	2718	2742	2765
19	2788	2810	2833	2856	2878	2900	2923	2945	2967	2989
20	3010	3032	3054	3075	3096	3118	3139	3160	3181	3201
21	3222	3243	3263	3284	3304	3324	3345	3365	3385	3404
22	3424	3444	3464	3483	3502	3522	3541	3560	3579	3598
23	3617	3636	3655	3674	3692	3711	3729	3747	3766	3784
24	3802	3820	3838	3856	3874	3892	3909	3927	3945	3962
25	3979	3997	4014	4031	4048	4065	4082	4099	4116	4133
26	4150	4166	4183	4200	4216	4232	4249	4265	4281	4298
27	4314	4330	4346	4362	4378	4393	4409	4425	4440	4456
28	4472	4487	4502	4518	4533	4548	4564	4579	4594	4609
29	4624	4639	4654	4669	4683	4698	4713	4728	4742	4757
30	4771	4786	4800	4814	4829	4843	4857	4871	4886	4900
31	4914	4928	4942	4955	4969	4983	4997	5011	5024	5038
32	5051	5065	5079	5092	5105	5119	5132	5145	5159	5172
33	5185	5198	5211	5224	5237	5250	5263	5276	5289	5302
34	5315	5328	5340	5353	5366	5378	5391	5403	5416	5428
35	5441	5453	5465	5478	5490	5502	5514	5527	5539	5551
36	5563	5575	5587	5599	5611	5623	5635	5647	5658	5670
37	5682	5694	5705	5717	5729	5740	5752	5763	5775	5786
38	5798	5809	5821	5832	5843	5855	5866	5877	5888	5899
39	5911	5922	5933	5944	5955	5966	5977	5988	5999	6010
40	6021	6031	6042	6053	6064	6075	6085	6096	6107	6117
41	6128	6138	6149	6160	6170	6180	6191	6201	6212	6222
42	6232	6243	6253	6263	6274	6284	6294	6304	6314	6325
43	6335	6345	6355	6365	6375	6385	6395	6405	6415	6425
44	6435	6444	6454	6464	6474	6484	6493	6503	6513	6522
45	6532	6542	6551	6561	6571	6580	6590	6599	6609	6618
46	6628	6637	6646	6656	6665	6675	6684	6693	6702	6712
47	6721	6730	6739	6749	6758	6767	6776	6785	6794	6803
48	6812	6821	6830	6839	6848	6857	6866	6875	6884	6893
49	6902	6911	6920	6928	6937	6946	6955	6964	6972	6981
50	6990	6998	7007	7016	7024	7033	7042	7050	7059	7067
51	7076	7084	7093	7101	7110	7118	7126	7135	7143	7152
52	7160	7168	7177	7185	7193	7202	7210	7218	7226	7235
53	7243	7251	7259	7267	7275	7284	7292	7300	7308	7316
54	7324	7332	7340	7348	7356	7364	7372	7380	7388	7396

*Mantissas; decimal points omitted. Characteristics are found by inspection.

Table 5   Common Logarithms of Numbers

N	0	1	2	3	4	5	6	7	8	9
55	7404	7412	7419	7427	7435	7443	7451	7459	7466	7474
56	7482	7490	7497	7505	7513	7520	7528	7536	7543	7551
57	7559	7566	7574	7582	7589	7597	7604	7612	7619	7627
58	7634	7642	7649	7657	7664	7672	7679	7686	7694	7701
59	7709	7716	7723	7731	7738	7745	7752	7760	7767	7774
60	7782	7789	7796	7803	7810	7818	7825	7832	7839	7846
61	7853	7860	7868	7875	7882	7889	7896	7903	7910	7917
62	7924	7931	7938	7945	7952	7959	7966	7973	7980	7987
63	7993	8000	8007	8014	8021	8028	8035	8041	8048	8055
64	8062	8069	8075	8082	8089	8096	8102	8109	8116	8122
65	8129	8136	8142	8149	8156	8162	8169	8176	8182	8189
66	8195	8202	8209	8215	8222	8228	8235	8241	8248	8254
67	8261	8267	8274	8280	8287	8293	8299	8306	8312	8319
68	8325	8331	8338	8344	8351	8357	8363	8370	8376	8382
69	8388	8395	8401	8407	8414	8420	8426	8432	8439	8445
70	8451	8457	8463	8470	8476	8482	8488	8494	8500	8506
71	8513	8519	8525	8531	8537	8543	8549	8555	8561	8567
72	8573	8579	8585	8591	8597	8603	8609	8615	8621	8627
73	8633	8639	8645	8651	8657	8663	8669	8675	8681	8686
74	8692	8698	8704	8710	8716	8722	8727	8733	8739	8745
75	8751	8756	8762	8768	8774	8779	8785	8791	8797	8802
76	8808	8814	8820	8825	8831	8837	8842	8848	8854	8859
77	8865	8871	8876	8882	8887	8893	8899	8904	8910	8915
78	8921	8927	8932	8938	8943	8949	8954	8960	8965	8971
79	8976	8982	8987	8993	8998	9004	9009	9015	9020	9025
80	9031	9036	9042	9047	9053	9058	9063	9069	9074	9079
81	9085	9090	9096	9101	9106	9112	9117	9122	9128	9133
82	9138	9143	9149	9154	9159	9165	9170	9175	9180	9186
83	9191	9196	9201	9206	9212	9217	9222	9227	9232	9238
84	9243	9248	9253	9258	9263	9269	9274	9279	9284	9289
85	9294	9299	9304	9309	9315	9320	9325	9330	9335	9340
86	9345	9350	9355	9360	9365	9370	9375	9380	9385	9390
87	9395	9400	9405	9410	9415	9420	9425	9430	9435	9440
88	9445	9450	9455	9460	9465	9469	9474	9479	9484	9489
89	9494	9499	9504	9509	9513	9518	9523	9528	9533	9538
90	9542	9547	9552	9557	9562	9566	9571	9576	9581	9586
91	9590	9595	9600	9605	9609	9614	9619	9624	9628	9633
92	9638	9643	9647	9652	9657	9661	9666	9671	9675	9680
93	9685	9689	9694	9699	9703	9708	9713	9717	9722	9727
94	9731	9736	9741	9745	9750	9754	9759	9763	9768	9773
95	9777	9782	9786	9791	9795	9800	9805	9809	9814	9818
96	9823	9827	9832	9836	9841	9845	9850	9854	9859	9863
97	9868	9872	9877	9881	9886	9890	9894	9899	9903	9908
98	9912	9917	9921	9926	9930	9934	9939	9943	9948	9952
99	9956	9961	9965	9969	9974	9978	9983	9987	9991	9996

Table 6   Four-place Logarithms of Trigonometric Functions*

Angle	L Sin	d	L Tan	cd	L Cot	d	L Cos	
0° 0′	——		——		——		10.0000	90° 0′
10′	7.4637		7.4637		12.5363	0	10.0000	50′
20′	7.7648	3011	7.7648	3011	12.2352	0	10.0000	40′
30′	7.9408	1760	7.9409	1761	12.0591	0	10.0000	30′
40′	8.0658	1250	8.0658	1249	11.9342	0	10.0000	20′
50′	8.1627	969	8.1627	969	11.8373	0	10.0000	10′
1° 0′	8.2419	792	8.2419	792	11.7581	1	9.9999	89° 0′
10′	8.3088	669	8.3089	670	11.6911	0	9.9999	50′
20′	8.3668	580	8.3669	580	11.6331	0	9.9999	40′
30′	8.4179	511	8.4181	512	11.5819	0	9.9999	30′
40′	8.4637	458	8.4638	457	11.5362	1	9.9998	20′
50′	8.5050	413	8.5053	415	11.4947	0	9.9998	10′
2° 0′	8.5428	378	8.5431	378	11.4569	1	9.9997	88° 0′
10′	8.5776	348	8.5779	348	11.4221	0	9.9997	50′
20′	8.6097	321	8.6101	322	11.3899	1	9.9996	40′
30′	8.6397	300	8.6401	300	11.3599	0	9.9996	30′
40′	8.6677	280	8.6682	281	11.3318	1	9.9995	20′
50′	8.6940	263	8.6945	263	11.3055	0	9.9995	10′
3° 0′	8.7188	248	8.7194	249	11.2806	1	9.9994	87° 0′
10′	8.7423	235	8.7429	235	11.2571	1	9.9993	50′
20′	8.7645	222	8.7652	223	11.2348	0	9.9993	40′
30′	8.7857	212	8.7865	213	11.2135	1	9.9992	30′
40′	8.8059	202	8.8067	202	11.1933	1	9.9991	20′
50′	8.8251	192	8.8261	194	11.1739	1	9.9990	10′
4° 0′	8.8436	185	8.8446	185	11.1554	1	9.9989	86° 0′
10′	8.8613	177	8.8624	178	11.1376	0	9.9989	50′
20′	8.8783	170	8.8795	171	11.1205	1	9.9988	40′
30′	8.8946	163	8.8960	165	11.1040	1	9.9987	30′
40′	8.9104	158	8.9118	158	11.0882	1	9.9986	20′
50′	8.9256	152	8.9272	154	11.0728	1	9.9985	10′
5° 0′	8.9403	147	8.9420	148	11.0580	2	9.9983	85° 0′
10′	8.9545	142	8.9563	143	11.0437	1	9.9982	50′
20′	8.9682	137	8.9701	138	11.0299	1	9.9981	40′
30′	8.9816	134	8.9836	135	11.0164	1	9.9980	30′
40′	8.9945	129	8.9966	130	11.0034	1	9.9979	20′
50′	9.0070	125	9.0093	127	10.9907	2	9.9977	10′
6° 0′	9.0192	122	9.0216	123	10.9784	1	9.9976	84° 0′
10′	9.0311	119	9.0336	120	10.9664	1	9.9975	50′
20′	9.0426	115	9.0453	117	10.9547	2	9.9973	40′
30′	9.0539	113	9.0567	114	10.9433	1	9.9972	30′
40′	9.0648	109	9.0678	111	10.9322	1	9.9971	20′
50′	9.0755	107	9.0786	108	10.9214	2	9.9969	10′
7° 0′	9.0859	104	9.0891	105	10.9109	1	9.9968	83° 0′
10′	9.0961	102	9.0995	104	10.9005	2	9.9966	50′
20′	9.1060	99	9.1096	101	10.8904	2	9.9964	40′
30′	9.1157	97	9.1194	98	10.8806	1	9.9963	30′
40′	9.1252	95	9.1291	97	10.8709	2	9.9961	20′
50′	9.1345	93	9.1385	94	10.8615	2	9.9959	10′
8° 0′	9.1436	91	9.1478	93	10.8522	1	9.9958	82° 0′
10′	9.1525	89	9.1569	91	10.8431	2	9.9956	50′
20′	9.1612	87	9.1658	89	10.8342	2	9.9954	40′
30′	9.1697	85	9.1745	87	10.8255	2	9.9952	30′
40′	9.1781	84	9.1831	86	10.8169	2	9.9950	20′
50′	9.1863	82	9.1915	84	10.8085	2	9.9948	10′
9° 0′	9.1943	80	9.1997	82	10.8003	2	9.9946	81° 0′
	L Cos	d	L Cot	cd	L Tan	d	L Sin	Angle

*These tables give the logarithms increased by 10. Hence in each case 10 should be subtracted.

Table 6 Four-place Logarithms of Trigonometric Functions

Angle	L Sin	d	L Tan	cd	L Cot	d	L Cos	Angle
9° 0′	9.1943	79	9.1997	81	10.8003	2	9.9946	81° 0′
10′	9.2022	78	9.2078	80	10.7922	2	9.9944	50′
20′	9.2100	76	9.2158	78	10.7842	2	9.9942	40′
30′	9.2176	75	9.2236	77	10.7764	2	9.9940	30′
40′	9.2251	73	9.2313	76	10.7687	2	9.9938	20′
50′	9.2324	73	9.2389	74	10.7611	2	9.9936	10′
10° 0′	9.2397	71	9.2463	73	10.7537	3	9.9934	80° 0′
10′	9.2468	70	9.2536	73	10.7464	2	9.9931	50′
20′	9.2538	68	9.2609	71	10.7391	2	9.9929	40′
30′	9.2606	68	9.2680	70	10.7320	3	9.9927	30′
40′	9.2674	66	9.2750	69	10.7250	2	9.9924	20′
50′	9.2740	66	9.2819	68	10.7181	3	9.9922	10′
11° 0′	9.2806	64	9.2887	66	10.7113	2	9.9919	79° 0′
10′	9.2870	64	9.2953	67	10.7047	3	9.9917	50′
20′	9.2934	63	9.3020	65	10.6980	2	9.9914	40′
30′	9.2997	61	9.3085	64	10.6915	3	9.9912	30′
40′	9.3058	61	9.3149	63	10.6851	2	9.9909	20′
50′	9.3119	60	9.3212	63	10.6788	3	9.9907	10′
12° 0′	9.3179	59	9.3275	61	10.6725	3	9.9904	78° 0′
10′	9.3238	58	9.3336	61	10.6664	2	9.9901	50′
20′	9.3296	57	9.3397	61	10.6603	3	9.9899	40′
30′	9.3353	57	9.3458	59	10.6542	3	9.9896	30′
40′	9.3410	56	9.3517	59	10.6483	3	9.9893	20′
50′	9.3466	55	9.3576	58	10.6424	3	9.9890	10′
13° 0′	9.3521	54	9.3634	57	10.6366	3	9.9887	77° 0′
10′	9.3575	54	9.3691	57	10.6309	3	9.9884	50′
20′	9.3629	53	9.3748	56	10.6252	3	9.9881	40′
30′	9.3682	52	9.3804	55	10.6196	3	9.9878	30′
40′	9.3734	52	9.3859	55	10.6141	3	9.9875	20′
50′	9.3786	51	9.3914	54	10.6086	3	9.9872	10′
14° 0′	9.3837	50	9.3968	53	10.6032	3	9.9869	76° 0′
10′	9.3887	50	9.4021	53	10.5979	3	9.9866	50′
20′	9.3937	49	9.4074	53	10.5926	4	9.9863	40′
30′	9.3986	49	9.4127	51	10.5873	3	9.9859	30′
40′	9.4035	48	9.4178	52	10.5822	3	9.9856	20′
50′	9.4083	47	9.4230	51	10.5770	4	9.9853	10′
15° 0′	9.4130	47	9.4281	50	10.5719	3	9.9849	75° 0′
10′	9.4177	46	9.4331	50	10.5669	3	9.9846	50′
20′	9.4223	46	9.4381	49	10.5619	4	9.9843	40′
30′	9.4269	45	9.4430	49	10.5570	3	9.9839	30′
40′	9.4314	45	9.4479	48	10.5521	4	9.9836	20′
50′	9.4359	44	9.4527	48	10.5473	4	9.9832	10′
16° 0′	9.4403	44	9.4575	47	10.5425	3	9.9828	74° 0′
10′	9.4447	44	9.4622	47	10.5378	4	9.9825	50′
20′	9.4491	42	9.4669	47	10.5331	4	9.9821	40′
30′	9.4533	43	9.4716	46	10.5284	3	9.9817	30′
40′	9.4576	42	9.4762	46	10.5238	4	9.9814	20′
50′	9.4618	41	9.4808	45	10.5192	4	9.9810	10′
17° 0′	9.4659	41	9.4853	45	10.5147	4	9.9806	73° 0′
10′	9.4700	41	9.4898	45	10.5102	4	9.9802	50′
20′	9.4741	40	9.4943	44	10.5057	4	9.9798	40′
30′	9.4781	40	9.4987	44	10.5013	4	9.9794	30′
40′	9.4821	40	9.5031	44	10.4969	4	9.9790	20′
50′	9.4861	39	9.5075	43	10.4925	4	9.9786	10′
18° 0′	9.4900		9.5118		10.4882		9.9782	72° 0′
	L Cos	d	L Cot	cd	L Tan	d	L Sin	Angle

Table 6 Four-place Logarithms of Trigonometric Functions

Angle	L Sin	d	L Tan	cd	L Cot	d	L Cos	Angle
18° 0′	9.4900	39	9.5118	43	10.4882	4	9.9782	72° 0′
10′	9.4939	38	9.5161	42	10.4839	4	9.9778	50′
20′	9.4977	38	9.5203	42	10.4797	4	9.9774	40′
30′	9.5015	37	9.5245	42	10.4755	5	9.9770	30′
40′	9.5052	38	9.5287	42	10.4713	4	9.9765	20′
50′	9.5090	36	9.5329	41	10.4671	4	9.9761	10′
19° 0′	9.5126	37	9.5370	41	10.4630	5	9.9757	71° 0′
10′	9.5163	36	9.5411	40	10.4589	4	9.9752	50′
20′	9.5199	36	9.5451	40	10.4549	5	9.9748	40′
30′	9.5235	35	9.5491	40	10.4509	4	9.9743	30′
40′	9.5270	36	9.5531	40	10.4469	5	9.9739	20′
50′	9.5306	35	9.5571	40	10.4429	4	9.9734	10′
20° 0′	9.5341	34	9.5611	39	10.4389	5	9.9730	70° 0′
10′	9.5375	34	9.5650	39	10.4350	4	9.9725	50′
20′	9.5409	34	9.5689	38	10.4311	5	9.9721	40′
30′	9.5443	34	9.5727	39	10.4273	5	9.9716	30′
40′	9.5477	33	9.5766	38	10.4234	5	9.9711	20′
50′	9.5510	33	9.5804	38	10.4196	4	9.9706	10′
21° 0′	9.5543	33	9.5842	37	10.4158	5	9.9702	69° 0′
10′	9.5576	33	9.5879	38	10.4121	5	9.9697	50′
20′	9.5609	32	9.5917	37	10.4083	5	9.9692	40′
30′	9.5641	32	9.5954	37	10.4046	5	9.9687	30′
40′	9.5673	31	9.5991	37	10.4009	5	9.9682	20′
50′	9.5704	32	9.6028	36	10.3972	5	9.9677	10′
22° 0′	9.5736	31	9.6064	36	10.3936	5	9.9672	68° 0′
10′	9.5767	31	9.6100	36	10.3900	6	9.9667	50′
20′	9.5798	30	9.6136	36	10.3864	5	9.9661	40′
30′	9.5828	31	9.6172	36	10.3828	5	9.9656	30′
40′	9.5859	30	9.6208	35	10.3792	5	9.9651	20′
50′	9.5889	30	9.6243	36	10.3757	6	9.9646	10′
23° 0′	9.5919	29	9.6279	35	10.3721	5	9.9640	67° 0′
10′	9.5948	30	9.6314	34	10.3686	6	9.9635	50′
20′	9.5978	29	9.6348	35	10.3652	5	9.9629	40′
30′	9.6007	29	9.6383	34	10.3617	6	9.9624	30′
40′	9.6036	29	9.6417	35	10.3583	5	9.9618	20′
50′	9.6065	28	9.6452	34	10.3548	6	9.9613	10′
24° 0′	9.6093	28	9.6486	34	10.3514	5	9.9607	66° 0′
10′	9.6121	28	9.6520	33	10.3480	6	9.9602	50′
20′	9.6149	28	9.6553	34	10.3447	6	9.9596	40′
30′	9.6177	28	9.6587	33	10.3413	6	9.9590	30′
40′	9.6205	27	9.6620	34	10.3380	5	9.9584	20′
50′	9.6232	27	9.6654	33	10.3346	6	9.9579	10′
25° 0′	9.6259	27	9.6687	33	10.3313	6	9.9573	65° 0′
10′	9.6286	27	9.6720	32	10.3280	6	9.9567	50′
20′	9.6313	27	9.6752	33	10.3248	6	9.9561	40′
30′	9.6340	26	9.6785	32	10.3215	6	9.9555	30′
40′	9.6366	26	9.6817	33	10.3183	6	9.9549	20′
50′	9.6392	26	9.6850	32	10.3150	6	9.9543	10′
26° 0′	9.6418	26	9.6882	32	10.3118	7	9.9537	64° 0′
10′	9.6444	26	9.6914	32	10.3086	6	9.9530	50′
20′	9.6470	25	9.6946	31	10.3054	6	9.9524	40′
30′	9.6495	26	9.6977	32	10.3023	6	9.9518	30′
40′	9.6521	25	9.7009	31	10.2991	7	9.9512	20′
50′	9.6546	24	9.7040	32	10.2960	6	9.9505	10′
27° 0′	9.6570		9.7072		10.2928		9.9499	63° 0′
	L Cos	d	L Cot	cd	L Tan	d	L Sin	Angle

Table 6 Four-place Logarithms of Trigonometric Functions

Angle	L Sin	d	L Tan	cd	L Cot	d	L Cos	Angle
27° 0'	9.6570		9.7072		10.2928		9.9499	63° 0'
10'	9.6595	25	9.7103	31	10.2897	7	9.9492	50'
20'	9.6620	25	9.7134	31	10.2866	6	9.9486	40'
30'	9.6644	24	9.7165	31	10.2835	7	9.9479	30'
40'	9.6668	24	9.7196	31	10.2804	6	9.9473	20'
50'	9.6692	24	9.7226	30	10.2774	7	9.9466	10'
28° 0'	9.6716	24	9.7257	31	10.2743	7	9.9459	62° 0'
10'	9.6740	24	9.7287	30	10.2713	6	9.9453	50'
20'	9.6763	23	9.7317	30	10.2683	7	9.9446	40'
30'	9.6787	24	9.7348	31	10.2652	7	9.9439	30'
40'	9.6810	23	9.7378	30	10.2622	7	9.9432	20'
50'	9.6833	23	9.7408	30	10.2592	7	9.9425	10'
29° 0'	9.6856	23	9.7438	30	10.2562	7	9.9418	61° 0'
10'	9.6878	22	9.7467	29	10.2533	7	9.9411	50'
20'	9.6901	23	9.7497	30	10.2503	7	9.9404	40'
30'	9.6923	22	9.7526	29	10.2474	7	9.9397	30'
40'	9.6946	23	9.7556	30	10.2444	7	9.9390	20'
50'	9.6968	22	9.7585	29	10.2415	7	9.9383	10'
30° 0'	9.6990	22	9.7614	29	10.2386	8	9.9375	60° 0'
10'	9.7012	22	9.7644	30	10.2356	7	9.9368	50'
20'	9.7033	21	9.7673	29	10.2327	7	9.9361	40'
30'	9.7055	22	9.7701	28	10.2299	8	9.9353	30'
40'	9.7076	21	9.7730	29	10.2270	7	9.9346	20'
50'	9.7097	21	9.7759	29	10.2241	8	9.9338	10'
31° 0'	9.7118	21	9.7788	29	10.2212	7	9.9331	59° 0'
10'	9.7139	21	9.7816	28	10.2184	8	9.9323	50'
20'	9.7160	21	9.7845	29	10.2155	8	9.9315	40'
30'	9.7181	21	9.7873	28	10.2127	7	9.9308	30'
40'	9.7201	20	9.7902	29	10.2098	8	9.9300	20'
50'	9.7222	21	9.7930	28	10.2070	8	9.9292	10'
32° 0'	9.7242	20	9.7958	28	10.2042	8	9.9284	58° 0'
10'	9.7262	20	9.7986	28	10.2014	8	9.9276	50'
20'	9.7282	20	9.8014	28	10.1986	8	9.9268	40'
30'	9.7302	20	9.8042	28	10.1958	8	9.9260	30'
40'	9.7322	20	9.8070	28	10.1930	8	9.9252	20'
50'	9.7342	20	9.8097	27	10.1903	8	9.9244	10'
33° 0'	9.7361	19	9.8125	28	10.1875	8	9.9236	57° 0'
10'	9.7380	19	9.8153	28	10.1847	8	9.9228	50'
20'	9.7400	20	9.8180	27	10.1820	9	9.9219	40'
30'	9.7419	19	9.8208	28	10.1792	8	9.9211	30'
40'	9.7438	19	9.8235	27	10.1765	8	9.9203	20'
50'	9.7457	19	9.8263	28	10.1737	9	9.9194	10'
34° 0'	9.7476	19	9.8290	27	10.1710	8	9.9186	56° 0'
10'	9.7494	18	9.8317	27	10.1683	9	9.9177	50'
20'	9.7513	19	9.8344	27	10.1656	8	9.9169	40'
30'	9.7531	18	9.8371	27	10.1629	9	9.9160	30'
40'	9.7550	19	9.8398	27	10.1602	9	9.9151	20'
50'	9.7568	18	9.8425	27	10.1575	9	9.9142	10'
35° 0'	9.7586	18	9.8452	27	10.1548	8	9.9134	55° 0'
10'	9.7604	18	9.8479	27	10.1521	9	9.9125	50'
20'	9.7622	18	9.8506	27	10.1494	9	9.9116	40'
30'	9.7640	18	9.8533	27	10.1467	9	9.9107	30'
40'	9.7657	17	9.8559	26	10.1441	9	9.9098	20'
50'	9.7675	18	9.8586	27	10.1414	9	9.9089	10'
36° 0'	9.7692	17	9.8613	27	10.1387	9	9.9080	54° 0'
	L Cos	d	L Cot	cd	L Tan	d	L Sin	Angle

Table 6   Four-place Logarithms of Trigonometric Functions

Angle	L Sin	d	L Tan	cd	L Cot	d	L Cos	
36° 0′	9.7692		9.8613		10.1387		9.9080	54° 0′
10′	9.7710	18	9.8639	26	10.1361	10	9.9070	50′
20′	9.7727	17	9.8666	27	10.1334	9	9.9061	40′
30′	9.7744	17	9.8692	26	10.1308	9	9.9052	30′
40′	9.7761	17	9.8718	26	10.1282	10	9.9042	20′
50′	9.7778	17	9.8745	27	10.1255	9	9.9033	10′
37° 0′	9.7795	17	9.8771	26	10.1229	10	9.9023	53° 0′
10′	9.7811	16	9.8797	26	10.1203	9	9.9014	50′
20′	9.7828	17	9.8824	27	10.1176	10	9.9004	40′
30′	9.7844	16	9.8850	26	10.1150	9	9.8995	30′
40′	9.7861	17	9.8876	26	10.1124	10	9.8985	20′
50′	9.7877	16	9.8902	26	10.1098	10	9.8975	10′
38° 0′	9.7893	16	9.8928	26	10.1072	10	9.8965	52° 0′
10′	9.7910	17	9.8954	26	10.1046	10	9.8955	50′
20′	9.7926	16	9.8980	26	10.1020	10	9.8945	40′
30′	9.7941	15	9.9006	26	10.0994	10	9.8935	30′
40′	9.7957	16	9.9032	26	10.0968	10	9.8925	20′
50′	9.7973	16	9.9058	26	10.0942	10	9.8915	10′
39° 0′	9.7989	16	9.9084	26	10.0916	10	9.8905	51° 0′
10′	9.8004	15	9.9110	26	10.0890	10	9.8895	50′
20′	9.8020	16	9.9135	25	10.0865	11	9.8884	40′
30′	9.8035	15	9.9161	26	10.0839	10	9.8874	30′
40′	9.8050	15	9.9187	26	10.0813	10	9.8864	20′
50′	9.8066	16	9.9212	25	10.0788	11	9.8853	10′
40° 0′	9.8081	15	9.9238	26	10.0762	10	9.8843	50° 0′
10′	9.8096	15	9.9264	26	10.0736	11	9.8832	50′
20′	9.8111	15	9.9289	25	10.0711	11	9.8821	40′
30′	9.8125	14	9.9315	26	10.0685	11	9.8810	30′
40′	9.8140	15	9.9341	26	10.0659	10	9.8800	20′
50′	9.8155	15	9.9366	25	10.0634	11	9.8789	10′
41° 0′	9.8169	14	9.9392	26	10.0608	11	9.8778	49° 0′
10′	9.8184	15	9.9417	25	10.0583	11	9.8767	50′
20′	9.8198	14	9.9443	26	10.0557	11	9.8756	40′
30′	9.8213	15	9.9468	25	10.0532	11	9.8745	30′
40′	9.8227	14	9.9494	26	10.0506	12	9.8733	20′
50′	9.8241	14	9.9519	25	10.0481	11	9.8722	10′
42° 0′	9.8255	14	9.9544	25	10.0456	11	9.8711	48° 0′
10′	9.8269	14	9.9570	26	10.0430	12	9.8699	50′
20′	9.8283	14	9.9595	25	10.0405	11	9.8688	40′
30′	9.8297	14	9.9621	26	10.0379	12	9.8676	30′
40′	9.8311	14	9.9646	25	10.0354	11	9.8665	20′
50′	9.8324	13	9.9671	25	10.0329	12	9.8653	10′
43° 0′	9.8338	14	9.9697	26	10.0303	12	9.8641	47° 0′
10′	9.8351	13	9.9722	25	10.0278	12	9.8629	50′
20′	9.8365	14	9.9747	25	10.0253	11	9.8618	40′
30′	9.8378	13	9.9772	25	10.0228	12	9.8606	30′
40′	9.8391	13	9.9798	26	10.0202	12	9.8594	20′
50′	9.8405	14	9.9823	25	10.0177	12	9.8582	10′
44° 0′	9.8418	13	9.9848	25	10.0152	13	9.8569	46° 0′
10′	9.8431	13	9.9874	26	10.0126	12	9.8557	50′
20′	9.8444	13	9.9899	25	10.0101	12	9.8545	40′
30′	9.8457	13	9.9924	25	10.0076	13	9.8532	30′
40′	9.8469	12	9.9949	25	10.0051	12	9.8520	20′
50′	9.8482	13	9.9975	26	10.0025	13	9.8507	10′
45° 0′	9.8495	13	10.0000	25	10.0000	12	9.8495	45° 0′
	L Cos	d	L Cot	cd	L Tan	d	L Sin	Angle

Summary of Formulas

Trigonometric and Inverse Trigonometric Functions

Let θ be an angle in standard position, $P(x, y)$ be any point other than the origin on the terminal side of θ, and $r = \sqrt{x^2 + y^2}$. Then:

$$\sin \theta = \frac{y}{r} \quad \cos \theta = \frac{x}{r} \quad \tan \theta = \frac{y}{x} \qquad 7$$

$$\csc \theta = \frac{r}{y} \quad \sec \theta = \frac{r}{x} \quad \cot \theta = \frac{x}{y} \qquad 8$$

$$y = \operatorname{Sin}^{-1} x \text{ if and only if } \sin y = x \text{ and } -\frac{\pi}{2} \leq y \leq \frac{\pi}{2}$$

$$y = \operatorname{Cos}^{-1} x \text{ if and only if } \cos y = x \text{ and } 0 \leq y \leq \pi$$

$$y = \operatorname{Tan}^{-1} x \text{ if and only if } \tan y = x \text{ and } -\frac{\pi}{2} < y < \frac{\pi}{2} \qquad 68\text{–}69$$

Basic Trigonometric Identities

$$\tan x = \frac{\sin x}{\cos x} \qquad \cot x = \frac{\cos x}{\sin x} \qquad \sec x = \frac{1}{\cos x}$$

$$\csc x = \frac{1}{\sin x} \qquad \tan x = \frac{1}{\cot x} \qquad \cot x = \frac{1}{\tan x}$$

$$\sin^2 x + \cos^2 x = 1 \qquad 1 + \tan^2 x = \sec^2 x \qquad 1 + \cot^2 x = \csc^2 x \qquad 86$$

$$\cos (a \pm b) = \cos a \cos b \mp \sin a \sin b$$
$$\sin (a \pm b) = \sin a \cos b \pm \cos a \sin b \qquad 94\text{–}96$$

$$\sin 2x = 2 \sin x \cos x \qquad\qquad \cos 2x = \cos^2 x - \sin^2 x$$

$$\sin \frac{x}{2} = \pm \sqrt{\frac{1 - \cos x}{2}} \qquad\qquad \cos \frac{x}{2} = \pm \sqrt{\frac{1 + \cos x}{2}} \qquad 100\text{–}101$$

$$\tan (a \pm b) = \frac{\tan a \pm \tan b}{1 \mp \tan a \tan b}$$

$$\tan 2x = \frac{2 \tan x}{1 - \tan^2 x} \qquad\qquad \tan \frac{x}{2} = \frac{\sin x}{1 + \cos x} \qquad 104\text{–}105$$

Oblique Triangles

Law of cosines $c^2 = a^2 + b^2 - 2ab \cos C$ 126

Law of sines $\dfrac{\sin A}{a} = \dfrac{\sin B}{b} = \dfrac{\sin C}{c}$ 132

Area formulas $K = \dfrac{1}{2}bc \sin A$ $\qquad K = \dfrac{1}{2}a^2 \dfrac{\sin B \sin C}{\sin A}$

$K = \sqrt{s(s - a)(s - b)(s - c)}$, where $s = \dfrac{1}{2}(a + b + c)$ 142

The equations $y = a \sin b(x - c) + d$ and $y = a \cos b(x - c) + d$

The graph of such an equation has amplitude $|a|$ and period $\dfrac{2\pi}{|b|}$. The graph can be obtained by shifting the graph of $y = a \sin bx$
(or $y = a \cos bx$)
    to the right c units if $c > 0$ and to the left $|c|$ units if $c < 0$ and
    upward d units if $d > 0$ and downward $|d|$ units if $d < 0$ 154–155, 159

Vectors

To find the sum of two vectors **u** and **v**, draw **u** and draw **v** with the initial point of **v** at the terminal point of **u**. The sum, **u** + **v**, is the vector whose tail is the initial point of **u** and whose tip is the terminal point of **v**.

To find the product $r\mathbf{v}$ of a scalar r and a vector **v**, multiply the length of **v** by $|r|$ and reverse the direction if $r < 0$.

Three properties of the norm are:
$$\|\mathbf{v}\| = 0 \text{ if and only if } \mathbf{v} = \mathbf{0}$$
$$\|r\,\mathbf{v}\| = |r|\,\|\mathbf{v}\|$$
$$\|\mathbf{u} + \mathbf{v}\| \le \|\mathbf{u}\| + \|\mathbf{v}\|$$ 192–195

If $\mathbf{u} = a\mathbf{i} + b\mathbf{j}$ and $\mathbf{v} = c\mathbf{i} + d\mathbf{j}$, then:
$$\mathbf{u} = \mathbf{v} \text{ if and only if } a = c \text{ and } b = d$$
$$\mathbf{u} + \mathbf{v} = (a + c)\mathbf{i} + (b + d)\mathbf{j}$$
$$r\mathbf{u} = (ra)\mathbf{i} + (rb)\mathbf{j} \text{ for every scalar } r$$
$$\|\mathbf{u}\| = \sqrt{a^2 + b^2}$$ 209

The dot product of **u** and **v** is given by:
$$\mathbf{u} \cdot \mathbf{v} = \|\mathbf{u}\|\,\|\mathbf{v}\| \cos \theta$$
$$= ac + bd$$ 211

The work W done by a force **F** in moving an object from A to B is given by $W = \mathbf{F} \cdot \mathbf{d}$, where $\mathbf{d} = \overrightarrow{AB}$. 217

Polar Coordinates

The formulas below enable us to convert from polar coordinates (r, θ) to rectangular coordinates (x, y) and vice versa.

Coordinate Changes	
From polar to rectangular	From rectangular to polar
$x = r \cos \theta$	$r = \pm\sqrt{x^2 + y^2}$
$y = r \sin \theta$	$\cos \theta = \dfrac{x}{r}, \sin \theta = \dfrac{y}{r}$

231

Complex Numbers

Let $z = x + yi$ and $w = u + vi$. Then:

$$z + w = (x + u) + (y + v)i$$
$$zw = (ux - vy) + (vx + uy)i$$
$$\bar{z} = x - yi$$
$$|z| = \sqrt{x^2 + y^2}$$

Furthermore:

$$z\bar{z} = |z|^2$$
$$|z|\,|w| = |zw|$$
$$\left|\frac{z}{w}\right| = \frac{|z|}{|w|} \qquad (w \neq 0)$$
$$|z + w| \leq |z| + |w|$$

247–248

If $z = r(\cos \alpha + i \sin \alpha)$ and $w = s(\cos \beta + i \sin \beta)$, then:

$$zw = rs(\cos (\alpha + \beta) + i \sin (\alpha + \beta))$$
$$\frac{z}{w} = \frac{r}{s}(\cos (\alpha - \beta) + i \sin (\alpha - \beta))$$
$$z^n = r^n(\cos n\alpha + i \sin n\alpha) \quad n \text{ an integer}$$

255–259

The nth roots of $r(\cos \theta + i \sin \theta)$ are given by:

$$r^{\frac{1}{n}}\left(\cos \frac{\theta + k \cdot 360°}{n} + i \sin \frac{\theta + k \cdot 360°}{n}\right),$$

where $k = 0, 1, 2, \ldots, n - 1$

262

Infinite Series

$$\sin x = x - \frac{x^3}{3!} + \frac{x^5}{5!} - \frac{x^7}{7!} + \cdots \qquad \cos x = 1 - \frac{x^2}{2!} + \frac{x^2}{4!} - \frac{x^6}{6!} + \cdots$$

$$e^x = 1 + x + \frac{x^2}{2!} + \frac{x^3}{3!} + \frac{x^4}{4!} + \cdots \qquad e^{ix} = \cos x + i \sin x \qquad \text{296–299}$$

$$\cosh x = 1 + \frac{x^2}{2!} + \frac{x^4}{4!} + \frac{x^6}{6!} + \cdots \qquad \sinh x = x + \frac{x^3}{3!} + \frac{x^5}{5!} + \frac{x^7}{7!} + \cdots \qquad \text{304}$$

Hyperbolic Identities

$$\cosh^2 x - \sinh^2 x = 1$$

$$\cosh 2x = \cosh^2 x + \sinh^2 x \qquad \sinh 2x = 2 \sinh x \cosh x \qquad \text{305}$$

Vectors in Space

The distance between the points $P_1(x_1, y_1, z_1)$ and $P_2(x_2, y_2, z_2)$ is given by:

$$P_1 P_2 = \sqrt{(x_2 - x_1)^2 + (y_2 - y_1)^2 + (z_2 - z_1)^2} \qquad \text{320}$$

Let $\mathbf{a} = a_1\mathbf{i} + a_2\mathbf{j} + a_3\mathbf{k}$ and $\mathbf{b} = b_1\mathbf{i} + b_2\mathbf{j} + b_3\mathbf{k}$. Then:

$$\mathbf{a} = \mathbf{b} \text{ if and only if } a_1 = b_1, a_2 = b_2, \text{ and } a_3 = b_3$$

$$\mathbf{a} + \mathbf{b} = (a_1 + b_1)\mathbf{i} + (a_2 + b_2)\mathbf{j} + (a_3 + b_3)\mathbf{k}$$

$$t\mathbf{a} = (ta_1)\mathbf{i} + (ta_2)\mathbf{j} + (ta_3)\mathbf{k}$$

$$\|\mathbf{a}\| = \sqrt{a_1^2 + a_2^2 + a_3^2} \qquad \text{323}$$

$$\mathbf{a} \cdot \mathbf{b} = \|\mathbf{a}\| \, \|\mathbf{b}\| \cos \theta = a_1 b_1 + a_2 b_2 + a_3 b_3 \qquad \text{324}$$

If line l contains P_0 and is parallel to $\mathbf{m}$, then $\mathbf{r} = \mathbf{r}_0 + t\mathbf{m}$.

If line l contains P_0 and P_1, then $\mathbf{r} = (1 - t)\mathbf{r}_0 + \mathbf{r}_1 t$.

Points P_0 and P_1 have position vectors $\mathbf{r}_0$ and $\mathbf{r}_1$, respectively. 328–329

A scalar equation of the plane with normal vector $\mathbf{n} = a\mathbf{i} + b\mathbf{j} + c\mathbf{k}$ and containing (x_0, y_0, z_0) is

$$a(x - x_0) + b(y - y_0) + c(z - z_0) = 0. \qquad \text{333}$$

The distance between $P_1(x_1, y_1, z_1)$ and the plane $ax + by + cz + d = 0$ is given by

$$D = \frac{|ax_1 + by_1 + cz_1 + d|}{\sqrt{a^2 + b^2 + c^2}}. \qquad \text{335}$$

GLOSSARY

Conic sections (p. 240): The curves formed when a plane intersects a cone. The set of all points whose distances from a fixed point, the focus, and a fixed line, the directrix, have a constant ratio, e, the eccentricity.

Conjugate of a complex number (p. 248): The conjugate of $z = x + yi$ is $\overline{z} = x - yi$.

Coordinate system in space (p. 319): Use of three mutually perpendicular axes to describe the position of points in space.

Cosecant (p. 24): $\csc \theta = \dfrac{r}{y}$, where θ is an angle in standard position, $P(x, y)$ a point on the terminal side of θ, r the distance OP, and $y \neq 0$.

Cosine (p. 24): $\cos \theta = \dfrac{x}{r}$, where θ is an angle in standard position, $P(x, y)$ is a point on the terminal side of θ, and r is the distance OP.

Cos x (p. 68): $\cos x$ with domain restricted to $0 \leq x \leq \pi$.

Cos^{-1} x (p. 68): The inverse of Cos x.

Cotangent (p. 24): $\cot \theta = \dfrac{x}{y}$, where θ is an angle in standard position, $P(x, y)$ a point on the terminal side of θ, and $y \neq 0$.

Coterminal angles (p. 3): Two angles that have coincident terminal sides when in standard position.

Cycle of a graph (p. 53): A portion of the graph of a periodic function with period p over an interval of length p.

D

De Moivre's theorem (p. 258): For every integer n, $[r (\cos \theta + i \sin \theta)]^n = r^n(\cos n\theta + i \sin n\theta)$.

Directrix (p. 240): *See* Conic sections.

Displacement (p. 198): Movement of an object from A to B, indicated by $\overrightarrow{AB}$.

Dot product (p. 211): Let θ be the angle between **u** and **v**. Then $\mathbf{u} \cdot \mathbf{v} = \|\mathbf{u}\| \|\mathbf{v}\| \cos \theta$.

E

Eccentricity (p. 240): *See* Conic sections.

Ellipse (p. 240): A conic section in which the eccentricity is between 0 and 1. (p. 242): A set of points in a plane such that, for each point of the set, the sum of its distances to two fixed points is a constant.

Equal vectors (p. 192): Vectors with the same length and direction.

Euler's formula (p. 299): $e^{ix} = \cos x + i \sin x$ where x is a real number.

Even function (p. 54): A function f such that $f(-x) = f(x)$ for all x in the domain of f.

Explicit definition (p. 282): A sequence is defined explicitly when a formula in terms of n is given for a_n.

Exponential function (p. 290): A function of the form $y = b^x$ where b is positive and nonzero. If $b = e$, then the function is called the natural exponential function.

F

Field (p. 247): An algebraic system that consists of a set F together with two binary operations that satisfy certain axioms.

Finite sequence (p. 281): A finite ordered set.

Finite series (p. 282): The sum of the terms of a finite sequence.

Focus (p. 240): *See* Conic sections.

Fourier Series (p. 306): An infinite series of the form $a_0 + a_1 \cos x + b_1 \sin x + a_2 \cos 2x + b_2 \sin 2x + a_3 \cos 3x + b_3 \sin 3x + \cdots$ where the a's and b's are constants.

Frequency of a simple harmonic motion (p. 168): Denotes oscillations per unit of time $\left(f = \dfrac{1}{T} = \dfrac{\omega}{2\pi} \right)$.

Function (p. 6): A rule that assigns to each member of a set D, the domain, a unique member of a second set R, the range.

Fundamental (p. 172): The note of lowest frequency when a string (as on a violin) is vibrated.

Fundamental period (p. 53): The least positive period, if there is one, of a periodic function.

G

Geometric sequence (p. 283): A sequence defined by $a_1 = a$ and $a_n = ra_{n-1}$ where r is a nonzero constant and $n > 1$.

Geometric series (p. 283): A series whose terms form a geometric sequence.

Great circle (p. 356): The circle determined by a plane cutting through the surface of a sphere and its center.

Ground speed (p. 198): Speed relative to the ground.

H

Heading (p. 198): Direction in which a craft is headed, or pointed.

Hero's formula (p. 142): Area $\triangle ABC = \sqrt{s(s - a)(s - b)(s - c)}$ where $s = \dfrac{1}{2}(a + b + c)$.

Hyperbola (p. 240): A conic section whose eccentricity is greater than 1.

Hyperbolic functions (p. 302): The hyperbolic functions sinh and cosh are defined by $\sinh x = \dfrac{1}{2}(e^x - e^{-x})$ and $\cosh x = \dfrac{1}{2}(e^x + e^{-x})$.

Horizontal-line test (p. 65): A test used to determine whether a function is one-to-one.

I

Identity (p. 85): An equation that is true for all values of the variable or variables for which both sides of the equation are defined.

Imaginary number (p. 246): Any number of the form $x + yi$ where $y \neq 0$.

Imaginary unit (p. 246): i, or $\sqrt{-1}$.

Infinite sequence (p. 286): A function whose domain is the set of positive integers and whose range is a subset of the real numbers.

Infinite series (p. 287): A series with infinitely many terms.

Initial point of a vector (p. 192): The tail of the arrow representing a vector.

Inverse function (p. 64): The function g such that for functions f and g, $g(y) = x$ if and only if $f(x) = y$. The inverse of f is denoted by f^{-1}. In general, a function f has an inverse if and only if the function is one-to-one; that is, if and only if $f(x_1) = f(x_2)$ implies $x_1 = x_2$.

J

Joule (p. 216): The basic unit of work and energy in the metric system; the work done by a force of 1 newton in moving an object 1 m.

K

Kilowatt-hour (p. 216): $1 \text{ kW} \cdot \text{h} = 3.6 \times 10^6 \text{ J}$.

L

Latitude (p. 357): The distance of a point from the equator, measured in degrees.

Law of cosines (p. 126): In any triangle, the square of the length of any side equals the sum of the squares of the lengths of the other two sides and the cosine of the included angle.

Law of sines (p. 132): The sines of the angles of any triangle are proportional to the lengths of the corresponding sides.

Limit of a sequence (p. 286): If the terms of a sequence eventually become and remain arbitarily close to some real number L as n becomes larger and larger, then L is the limit of the sequence.

Linear combinations of $a_1, a_2, \ldots, a_n$ (p. 194): Expressions of the form $t_1 a_1 + t_2 a_2 + \cdots + t_n a_n$, where the a's are vectors and the t's are scalars.

Linear interpolation (p. 39): A method for approximating $f(x)$ $(a < x < b)$ by a proportion involving x, a, b, $f(a)$, and $f(b)$.

Linear speed (p. 49): The rate at which the distance a point P moving in uniform circular motion travels along an arc with respect to time t.

Logarithmic function (p. 291): If $y = b^x$, then its inverse is the logarithmic function with base b. If $b = e$, the inverse is called the natural logarithmic function.

Longitude (p. 357): The angle between the meridian of a point and the prime meridian.

M

Meridians (p. 356): Great circles containing the North and South Poles.

Modulus of a complex number (p. 248): For $z = x + yi$, the modulus is $|z| = \sqrt{x^2 + y^2}$. (p. 252): The number r when a complex number is written as $r(\cos \theta + i \sin \theta)$.

Multiplication of a vector by a scalar (p. 193): For vector u and scalar k, ku is the product of the length of u by $|k|$. If $k < 0$, the direction is reversed.

N

Nautical mile (p. 34, p. 360): The length of a one-minute great-circle arc; about 1.15 land miles or 1.85 kilometers.

Newton (p. 203): Basic unit of force in the metric system; the force necessary to give a 1 kg mass an acceleration of 1 m/s^2.

Norm of a vector (p. 195): The magnitude of a vector, **u**, denoted by $\|\mathbf{u}\|$.

nth root of a complex number z (p. 262): w if $w^n = z$.

O

Oblique triangle (p. 125): A triangle having no right angle.

Odd function (p. 54): A function f such that $f(-x) = -f(x)$ for all x in the domain of f.

Orthogonal vectors (p. 213): Perpendicular vectors.

Overtones (p. 172): Tones weaker than the fundamental.

P

Parabola (p. 240): A conic section whose eccentricity is 1.

Parallelogram rule (p. 193): *See* Addition of vectors.

Parallels (p. 356): Small circles cut by planes parallel to the equator.

Period of simple harmonic motion (p. 167): The time necessary for an object moving in simple harmonic motion to complete one "round trip" in its path.

Periodic function (p. 53): A function f such that there is some fixed positive number p having the property that whenever x is in the domain of f, both $x - p$ and $x + p$ are also in the domain and $f(x - p) = f(x) = f(x + p)$ for all x in the domain. If there is at least such positive number p, then p is called the *fundamental period* of f.

Phase shift (p. 159): In the case of a sinusoid, the number of units a graph is translated left or right.

Pi (π) (p. 30): Ratio of the circumference of a circle to its diameter.

Polar axis (p. 230): The ray used as a reference in the system of polar coordinates.

Polar coordinate system (p. 229): A system for locating points in the plane by their distance and direction from a given point and given ray, respectively.

Polar coordinates (p. 230): The ordered pair (r, θ) used to locate a point in the system of polar coordinates.

Polar form of a complex number (p. 252): The polar form of a complex number $z = x + yi$ is $z = r(\cos \theta + i \sin \theta)$ for $r \geq 0$.

Pole (p. 230): The origin in the reference system for polar coordinates.

Power series (p. 295): A series of the form $a_0 + a_1 x + a_2 x^2 + a_3 x^3 + \cdots$, where the a's are constants.

Primitive root (p. 265): An nth root w, of unity, is primitive if every nth root of unity is an integral power of w.

Pure imaginary number (p. 246): A number of the form yi.

Pythagorean identities (p. 86): $\sin^2 x + \cos^2 x = 1$; $1 + \cot^2 x = \csc^2 x$; $1 + \tan^2 x = \sec^2 x$.

Q

Quadrantal angle (p. 25): An angle whose terminal side lies along an axis.

Quadrants (p. 25): The four regions into which the coordinate axes divide the plane.

R

Radian (p. 30): A unit of angle measure. The measure of a central angle of a circle whose intercepted arc is equal in length to the radius of the circle.

Recursive definition (p. 282): A sequence is defined recursively if the first term (or terms) is given and the nth term is found from preceding terms.

Reference angle (p. 25): The positive acute angle x between the x-axis and the terminal side of any angle θ.

Reference triangle (p. 25): The triangle formed by the terminal side of an angle, the x-axis, and a perpendicular from the terminal side to the x-axis.

Resultant (p. 193): If $\mathbf{w} = \mathbf{u} + \mathbf{v}$, then $\mathbf{w}$ is the resultant of $\mathbf{u}$ and $\mathbf{v}$.

Rotation (p. 3): The amount by which one side (the initial side) of an angle must be turned to coincide with the second side (the terminal side) of the angle.

S

Scalar (p. 192): A quantity having magnitude but without direction.

Scalar projection of v onto a (p. 213): The norm of $\mathbf{v_a}$, the vector projection of $\mathbf{v}$ onto $\mathbf{a}$.

Secant (p. 24): $\sec \theta = \dfrac{r}{x}$, where θ is an angle in standard position, $P(x, y)$ a point on the terminal side of θ, r the distance OP, and $x \neq 0$.

Significant digit (p. 13): Any nonzero digit or any zero that has a purpose other than indicating the position of the decimal point.

Simple harmonic motion (p. 167): The motion that occurs when an object is displaced from an equilibrium (at rest) position.

Sin x (p. 69): $\sin x$ with domain restricted to $-\dfrac{\pi}{2} \leq x \leq \dfrac{\pi}{2}$.

Sin^{-1} x (p. 69): The inverse of Sin x.

Sine (p. 24): $\sin \theta = \dfrac{y}{r}$, where θ is an angle in standard position, $P(x, y)$ a point on the terminal side of θ, and r the distance OP.

Sinusoid (p. 153): An equation or graph of an equation of the form $y = a \sin b(x - c) + d$ or $a \cos b(x - c) + d$.

Solving a triangle (p. 17): Finding the lengths of the sides and the measures of the angles of a triangle.

Spherical angle (p. 357): The angle formed by two great-circle arcs having a common endpoint.

Spherical triangle (p. 357): The triangle formed by three great-circle arcs.

Standard position of a vector (p. 208): A vector whose initial point is the origin.

T

Tan x (p. 69): tan x with domain restricted to $-\dfrac{\pi}{2} < x < \dfrac{\pi}{2}$.

Tan^{-1} x (p. 69): The inverse of Tan x.

Tangent (p. 24): $\tan \theta = \dfrac{y}{x}$ where θ is an angle in standard position, $P(x, y)$ a point on the terminal side of θ, and $x \neq 0$.

Taylor series (p. 296): A series which approximates a function to any desired degree of accuracy. Also called a power series expansion.

Tension (p. 204): In a rope or cable, the magnitude of the force it exerts.

Terminal point of a vector (p. 192): The tip of the arrow representing a vector.

Translation (p. 158): A shift of the graph of a function in which the final position of the graph is parallel to its original position.

Triangle inequality (p. 251): The property $|w + z| \leq |w| + |z|$, where w and z are complex numbers. (p. 195) Also $\|\mathbf{u} + \mathbf{v}\| \leq \|\mathbf{u}\| + \|\mathbf{v}\|$ where $\mathbf{u}$ and $\mathbf{v}$ are vectors.

Trigonometric equation (p. 108): An equation involving one or more circular or trigonometric functions.

Trigonometric series (p. 306): Any series of the form
$$a_0 + a_1 \cos x + b_1 \sin x + \cdots + a_n \cos nx + b_n \sin nx + \cdots.$$

True course (p. 198): Bearing of the vector that points in the direction in which a craft is actually traveling.

U

Uniform circular motion (p. 48): The motion of a point P moving with a constant speed in a circular path.

Unit circle (p. 43): Circle with radius 1 and center at the origin.

Unit vector (p. 210): A vector having norm 1.

V

Vector projection of v onto a (p. 213): The vector whose initial point is the initial point of **a** and whose terminal point is the projection of the terminal point of **v** onto the line containing **a**.

Vector quantity (p. 192): A quantity having magnitude and direction.

Vertical-line test (p. 65): A test used to determine if a relation is a function.

Vertical shift (p. 159): In the case of a sinusoid, the number of units a graph is translated upward or downward.

Z

Zero vector (p. 193): Vector with magnitude 0; direction not defined.

Index

Acknowledgments

Cover design by Martucci Studio
Book design by The Quarasan Group
pp. x-1   The Royal Observatory, Edinburgh
p.41   © 1988 Ralph Brunke
pp. 42-43   Jeffrey Blackman Index/Stock
pp. 81   Jerry Howard/Positive Images
pp. 84-85   H. Armstrong Roberts
pp. 121   Runk/Schoenberger/Grant Heilman Photography, Inc.
pp. 124-125 © 1988 Donald C. Johnson. All rights reserved
pp. 151   Phyllis Graber Jensen/Stock, Boston
pp. 152-153 © Jay Lurie/FPG International
pp. 187   H. Armstrong Roberts
pp. 190-191 Berl Brechner
pp. 227   H. Armstrong Roberts
pp. 228-229 © L.E. Ellsworth/FPG
pp. 273   NASA
pp. 280-281 Harold Edgerton © 1988, courtesy Palm Press, Inc.
pp. 317   Carol Lee/TSW/Click/Chicago, Ltd.
pp. 318-319 Baron Wolman

Selected Answers

Chapter 1   Trigonometric Functions

Exercises 1-1, pages 4–6

1. 60°   **3.** −45°   **5.** −120°   **7.** 480°

9. 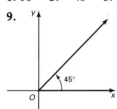   **11.**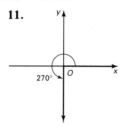

13. 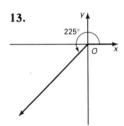   **15.**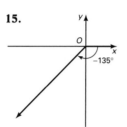

17. 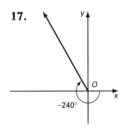   **19.**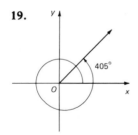

21. 270°   **23.** −135°   **25.** −144°   **27.** 600°
29. 23° 45′   **31.** 15° 25′ 48″   **33.** 24° 41′ 13″
35. 9.75°   **37.** 67.375°   **39.** 0.0075°
41–45. Answers may vary.
41. 450°; −270°   **43.** 630°; −90°   **45.** 120°;
−600°   **47.** 100°   **49.** 300°   **51.** 240°
53. 30°   **55.** 101.7°   **57.** 276.75°
59. 41° 42′ 20″   **61.** 107° 41′ 20″   **63.** 25
65. 6   **67.** 24

Exercises 1-2, pages 9–11

1. $\sin \theta = 0.8000$; $\cos \theta = 0.6000$;
$\tan \theta = 1.3333$

3. $\sin \theta = 0.2800$; $\cos \theta = 0.9600$;
$\tan \theta = 0.2917$
5. $\sin \theta = 0.4472$; $\cos \theta = 0.8944$;
$\tan \theta = 0.5000$
7. $\sin \theta = 0.8321$; $\cos \theta = 0.5547$;
$\tan \theta = 1.5000$
9. $\sin \theta = 0.8660$; $\cos \theta = 0.5000$;
$\tan \theta = 1.7321$
11. $\sin \theta = 0.4000$; $\cos \theta = 0.9165$;
$\tan \theta = 0.4364$

13. $\sin \alpha = \dfrac{3}{5}$; $\tan \alpha = \dfrac{3}{4}$; $\csc \alpha = \dfrac{5}{3}$;
$\sec \alpha = \dfrac{5}{4}$; $\cot \alpha = \dfrac{4}{3}$

15. $\sin \alpha = \dfrac{21}{29}$; $\cos \alpha = \dfrac{20}{29}$; $\csc \alpha = \dfrac{29}{21}$;
$\sec \alpha = \dfrac{29}{20}$; $\cot \alpha = \dfrac{20}{21}$

17. $\sin \alpha = \dfrac{2\sqrt{10}}{7}$; $\cos \alpha = \dfrac{3}{7}$; $\tan \alpha = \dfrac{2\sqrt{10}}{3}$;
$\csc \alpha = \dfrac{7\sqrt{10}}{20}$; $\cot \alpha = \dfrac{3\sqrt{10}}{20}$

19. $\sin \alpha = \dfrac{3}{7}$; $\cos \alpha = \dfrac{2\sqrt{10}}{7}$; $\tan \alpha = \dfrac{3\sqrt{10}}{20}$;
$\csc \alpha = \dfrac{7}{3}$; $\sec \alpha = \dfrac{7\sqrt{10}}{20}$

21. $\sin \alpha = \dfrac{1}{2}$; $\cos \alpha = \dfrac{\sqrt{3}}{2}$; $\tan \alpha = \dfrac{\sqrt{3}}{3}$;
$\sec \alpha = \dfrac{2\sqrt{3}}{3}$; $\cot \alpha = \sqrt{3}$

23. $\cos \alpha = \dfrac{\sqrt{21}}{5}$; $\tan \alpha = \dfrac{2\sqrt{21}}{21}$; $\csc \alpha = \dfrac{5}{2}$;
$\sec \alpha = \dfrac{5\sqrt{21}}{21}$; $\cot \alpha = \dfrac{\sqrt{21}}{2}$ **25.** 45°

27. 30°   **29.** $\angle B = 30°$; $a = 6\sqrt{3}$; $b = 6$
31. $\angle A = 45°$; $a = 4$; $c = 4\sqrt{2}$
33. $\angle A = 60°$; $\angle B = 30°$; $c = 2$ **35.** 65°

37. 10°   **39.** 20°   **41.** $\dfrac{\sqrt{6} - \sqrt{2}}{4}$

43. $\dfrac{\sqrt{6} + \sqrt{2}}{4}$ **45.** $\sqrt{6} - \sqrt{2}$ **47.** $\sqrt{6} + \sqrt{2}$

49. $\dfrac{15\sqrt{2} + 5\sqrt{6}}{2}$ **51.** $24 - 8\sqrt{3}$

Exercises 1-3, pages 14–16

1. 0.0310   **3.** 0.00400   **5.** 21.00
7. (a) 1.060×10^2 **(b)** 4   **9. (a)** 3.40×10^{-2}
(b) 3   **11. (a)** 1.0230×10^2 **(b)** 5
13. (a) 3.21×10^2 **(b)** 3   **15.** 0.5000
17. 0.8693   **19.** 0.2368   **21.** 1.074   **23.** 3.018
25. 0.9171   **27.** 17°   **29.** 68°   **31.** 21.6°
33. 57.6°   **35.** 70.5°   **37.** 62.1   **39.** 98.8
41. 67.7°   **43.** 65.4°; 3.82 m   **45.** 166 m
47. 2.26×10^8 m/s   **51. (a)** $\dfrac{\sqrt{5} + 1}{4}$
(b) 0.8090

Self Quiz   1-1 / 1-2 / 1-3, page 17

1. 63.4°   **2.** $10\sqrt{2}$
3. $\sin A = \dfrac{7}{25}$; $\cos A = \dfrac{24}{25}$; $\tan A = \dfrac{7}{24}$;
$\csc A = \dfrac{25}{7}$; $\sec A = \dfrac{25}{24}$; $\cot A = \dfrac{24}{7}$
4. (a) 190°   **(b)** 220°   **5.** 25°
6. (a) 6.50×10^{-3} **(b)** 3.0400×10^2
7. (a) 1.54°   **(b)** 32° 40′ 12″

Exercises 1-4, pages 20–23

1. $\angle B = 58.5°$; $a = 132$; $b = 215$
3. $\angle A = 25° \, 20′$; $b = 75.2$; $c = 83.2$
5. $\angle B = 18°$; $b = 0.15$; $c = 0.49$
7. $c = 4.38$; $\angle B = 34.8°$; $\angle A = 55.2°$
9. $a = 43.0$; $\angle B = 46.9°$; $\angle A = 43.1°$
11. $a = 3\sqrt{3}$; $\angle A = \angle B = 45°$
13. $\angle B = 43°$; $\angle C = 94°$; $b = 27$; $c = 39$
15. $\angle A = \angle B = 51.8°$; $b = 16.8$; $c = 20.8$
17. 45° 30′   **19.** 35.4°   **21.** 1960 m   **23.** 21 ft
25. 1.4 m   **27.** 30.1°   **29.** 71.7 cm; 45.7 cm
31. 25.8°   **33.** 4.0 cm   **35.** 146 m
37. $x = y \, (\cot \alpha + \cot \beta)$ **39.** 130 ft
41. $y = \dfrac{x \tan \alpha \tan \beta}{\tan \beta - \tan \alpha}$ **43.** 869 m; 3070 m
45. $FA = 1.87$ km; $FB = 1.70$ km   **47.** 12 ft

Exercises 1-5, pages 27–29

1. Quadrant I   **3.** Quadrant III
5. Quadrant IV   **7.** -0.8192 **9.** -1.0724
11. -1.6064 **13.** 0.5544
15. $\sin 240° = \dfrac{-\sqrt{3}}{2}$; $\cos 240° = \dfrac{-1}{2}$;
$\tan 240° = \sqrt{3}$; $\csc 240° = \dfrac{-2\sqrt{3}}{3}$;

$\sec 240° = -2$; $\cot 240° = \dfrac{\sqrt{3}}{3}$
17. $\sin 225° = \cos 225° = \dfrac{-\sqrt{2}}{2}$; $\tan 225° =$
$\cot 225° = 1$; $\csc 225° = \sec 225° = -\sqrt{2}$
19. $\sin(-240°) = \dfrac{\sqrt{3}}{2}$; $\cos(-240°) = \dfrac{-1}{2}$;
$\tan(-240°) = -\sqrt{3}$; $\csc(-240°) = \dfrac{2\sqrt{3}}{3}$;
$\sec(-240°) = -2$; $\cot(-240°) = -\dfrac{\sqrt{3}}{3}$
21. $\sin 480° = \dfrac{\sqrt{3}}{2}$; $\cos 480° = \dfrac{-1}{2}$; $\tan 480° =$
$-\sqrt{3}$; $\csc 480° = \dfrac{2\sqrt{3}}{3}$; $\sec 480° = -2$;
$\cot 480° = -\dfrac{\sqrt{3}}{3}$ **23. (a)** $\sin \theta = \dfrac{4}{5}$;
$\cos \theta = \dfrac{-3}{5}$ **(b)** 126.9°
25. (a) $\sin \theta = \dfrac{-15}{17}$;
$\cos \theta = \dfrac{-8}{17}$ **(b)** 241.9°   **27. (a)** $\sin \theta = \dfrac{-2}{3}$;
$\cos \theta = \dfrac{\sqrt{5}}{3}$ **(b)** 318.2°
29. (a) $\sin \theta = \dfrac{-\sqrt{6}}{3}$;
$\cos \theta = \dfrac{\sqrt{3}}{3}$ **(b)** 305.3°   **31.** $\sin \theta = \dfrac{-3}{5}$;
$\tan \theta = \dfrac{-3}{4}$ **33.** $\cos \theta = \dfrac{-2\sqrt{2}}{3}$; $\tan \theta = \dfrac{\sqrt{2}}{4}$
35. $\sin \theta = \dfrac{-\sqrt{5}}{5}$; $\cos \theta = \dfrac{-2\sqrt{5}}{5}$
37. $\sin \theta = \dfrac{\sqrt{15}}{4}$; $\tan \theta = \sqrt{15}$
39. $\sin 0° = 0$; $\cos 0° = 1$; $\tan 0° = 0$; $\cot 0°$
is undefined; $\sec 0° = 1$; $\csc 0°$ is undefined.
41. $\sin 180° = 0$; $\cos 180° = -1$; $\tan 180° = 0$;
$\cot 180°$ is undefined; $\sec 180° = -1$;
$\csc 180°$ is undefined.   **43. (a)** α
(b) $180° - \alpha$ **(c)** $180° + \alpha$ **(d)** $360° - \alpha$
45. IV; II; I   **47.** II; I; IV   **49.** 90°, 270°
51. 0°, 180°   **53.** 90°   **55.** 180°
57. 50°, 130°   **59.** 40°, 320°

Exercises 1-6, pages 32–34

1. $\dfrac{7\pi}{6}$ **3.** $\dfrac{5\pi}{3}$ **5.** $\dfrac{-\pi}{2}$ **7.** $\dfrac{\pi}{5}$ **9.** $\dfrac{7\pi}{12}$ **11.** $\dfrac{19\pi}{6}$

13. 225° **15.** 270° **17.** −120° **19.** 330°
21. 900° **23.** 390° **25.** 0.7549 **27.** 1.745
29. 45.84° **31.** 136.42° **33.** $\dfrac{-\sqrt{2}}{2}$ **35.** $\sqrt{3}$

37. 0.8415 **39.** 0.2553
41. $s = 6.00$ cm; $A = 9.00$ cm²
43. $\theta = 1.50 = 85.9°$; $A = 2700$ m²
45. $r = 3.00$ m; $A = 3.60$ m²
47. $\theta = 1.33 = 76.4°$; $s = 8.00$ cm
49. $r = 1.20$ m; $s = 1.80$ m **51.** 4706 km
53. 36 km **55.** 40,000 km, or 25,000 miles
57. 250 cm

Self Quiz 1-4 / 1-5 / 1-6, pages 34–35

1. $\sin\theta = \dfrac{-4}{5}$; $\cos\theta = \dfrac{3}{5}$; $\tan\theta = \dfrac{-4}{3}$;

$\csc\theta = \dfrac{-5}{4}$; $\sec\theta = \dfrac{5}{3}$; $\cot\theta = \dfrac{-3}{4}$

2. $\angle R = 33.5°$; $r = 28.2$; $q = 51.1$
3. (a) $\overset{\frown}{AB} = 9\pi$ (b) $A = 54\pi$
4. $\cos\theta = \dfrac{7}{25}$; $\tan\theta = \dfrac{-24}{7}$ **5.** (a) $\dfrac{1}{2}$
(b) −1

Additional Problems, pages 35–36

1. 1.38 m **3.** 15° **5.** 2100 **7.** 36π cm²
9. 1.178

11. $\sin\phi = \dfrac{21}{29}$; $\cos\phi = \dfrac{20}{29}$; $\tan\phi = \dfrac{21}{20}$;

$\csc\phi = \dfrac{29}{21}$; $\sec\phi = \dfrac{29}{20}$ **13.** −0.80

15. $3\sqrt{3}$ **17.** 40.3° and 49.7°
19. $x = d \sin\alpha \tan\beta$

Chapter 2   Circular Functions, Graphs, and Inverses

Exercises 2-1, pages 46–47

1. $\sin\dfrac{2\pi}{3} = \dfrac{\sqrt{3}}{2}$; $\cos\dfrac{2\pi}{3} = \dfrac{-1}{2}$;

$\tan\dfrac{2\pi}{3} = -\sqrt{3}$

3. $\sin\dfrac{7\pi}{6} = \dfrac{-1}{2}$; $\cos\dfrac{7\pi}{6} = \dfrac{-\sqrt{3}}{2}$;

$\tan\dfrac{7\pi}{6} = \dfrac{\sqrt{3}}{3}$

5. $\sin\dfrac{-4\pi}{3} = \dfrac{\sqrt{3}}{2}$; $\cos\dfrac{-4\pi}{3} = \dfrac{-1}{2}$;

$\tan\dfrac{-4\pi}{3} = -\sqrt{3}$

7. $\sin\dfrac{3\pi}{2} = -1$; $\cos\dfrac{3\pi}{2} = 0$; $\tan\dfrac{3\pi}{2}$ is
undefined.
9. $\sin\dfrac{11\pi}{4} = \dfrac{\sqrt{2}}{2}$; $\cos\dfrac{11\pi}{4} = \dfrac{-\sqrt{2}}{2}$;

$\tan\dfrac{11\pi}{4} = -1$

11. $\sin\dfrac{14\pi}{3} = \dfrac{\sqrt{3}}{2}$; $\cos\dfrac{14\pi}{3} = \dfrac{-1}{2}$;

$\tan\dfrac{14\pi}{3} = -\sqrt{3}$

13. $\cot\dfrac{2\pi}{3} = \dfrac{-\sqrt{3}}{3}$; $\sec\dfrac{2\pi}{3} = -2$;

$\csc\dfrac{2\pi}{3} = \dfrac{2\sqrt{3}}{3}$

15. $\cot\dfrac{7\pi}{6} = \sqrt{3}$; $\sec\dfrac{7\pi}{6} = \dfrac{-2\sqrt{3}}{3}$;

$\csc\dfrac{7\pi}{6} = -2$

17. $\cot\dfrac{-4\pi}{3} = \dfrac{-\sqrt{3}}{3}$; $\sec\dfrac{-4\pi}{3} = -2$;

$\csc\dfrac{-4\pi}{3} = \dfrac{2\sqrt{3}}{3}$

19. $\cot\dfrac{3\pi}{2} = 0$; $\sec\dfrac{3\pi}{2}$ is undefined;

$\csc\dfrac{3\pi}{2} = -1$ **21.** 0.3624 **23.** 1.557

25. 0.8314 **27.** −1.158 **29.** 0.8637
31. 1.522 **33.** 0.71 **35.** 2.25 **37.** 3.86
39. 2.67 **41.** $\cos t = 0.69$; $\tan t = 1.0$
43. $\sin t = -0.97$; $\tan t = 3.9$ **49.** 1
51. $(s + \cos(-s), \sin(-s))$

Exercises 2-2, pages 50–52

1. 48 cm/s **3.** 72 rad/min **5.** 4.8π cm/min
7. 5 ft **9.** $33\frac{1}{3}$ rpm **11.** 42 m/s
13. 36,000 π cm/min, or 113,097 cm/min

15. $(-3, 3\sqrt{3})$ **17.** $\left(\dfrac{-9}{2}, \dfrac{-9\sqrt{3}}{2}\right)$

19. $(2.93, 2.73)$ **21.** $(-2\sqrt{3}, 2)$

23. $(5\sqrt{2}, -5\sqrt{2})$ **25.** $(0, 7)$ **27.** $(1, \sqrt{3})$

29. $(-2\sqrt{3}, 2)$ **31.** $(4, -4)$

33. 10.9 miles/min   **35.** 194 rpm

37. $F = mr\omega^2$ **39.** $\dfrac{\pi}{21,600}$ rad/s; $\dfrac{\pi}{360}$ mm/s

41. 105 km/h clockwise

Exercises 2-3, pages 56–58

1. $p = 4$; odd   **3.** $p = 2$; even

5. odd

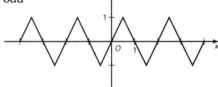

7. neither

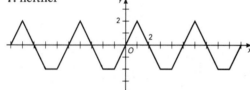

9. odd

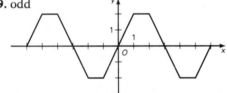

11. (a)

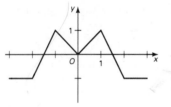

(b)

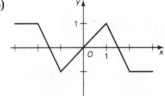

13. (a)

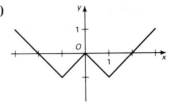

(b)

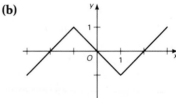

15. odd   **17.** neither   **19.** even   **21.** odd

23. neither   **25.** $x = -1; y = 1$

27. $x = 0; y = 2$ **31.** even

Self Quiz   2-1 / 2-2 / 2-3, pages 58–59

1. (a) $\sqrt{3}$ **(b)** $\dfrac{2\pi}{3}$

2. (a)

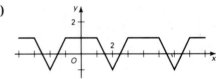

(b) even

3. 9 cm pulley, $\omega = 7000$ rad/min;
14 cm pulley, $\omega = 4500$ rad/min
4. $x = 2; y = 2$ **5.** $(-2, 2\sqrt{3})$
6. $\cos t = 0.80$; $\tan t = -0.75$

Exercises 2-4, pages 62–63

1.

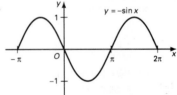

$y = -\sin x$

3.

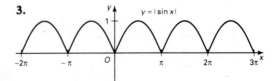

$y = |\sin x|$

5.

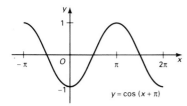

$$y = \cos(x + \pi)$$

7.

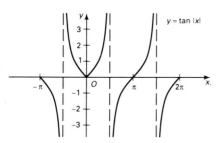

$$y = \tan |x|$$

9.

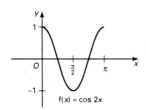

$$f(x) = \cos 2x$$

11.

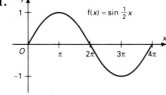

$$f(x) = \sin \frac{1}{2}x$$

13.

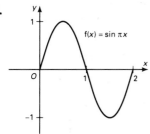

$$f(x) = \sin \pi x$$

15. odd   **17.** neither   **19.** even

21.

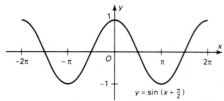

$$y = \sin\left(x + \frac{\pi}{2}\right)$$

23. $\cos x = \sin\left(x + \dfrac{\pi}{2}\right)$ and $\sin x = \cos\left(x - \dfrac{\pi}{2}\right)$

25.

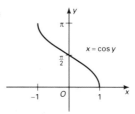

$$x = \cos y$$

27.

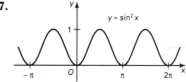

$$y = \sin^2 x$$

29.

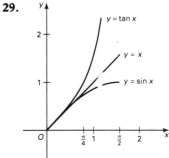

$$y = \tan x$$
$$y = x$$
$$y = \sin x$$

Exercises 2-5, pages 66–67

1. (a) domain: $-3 \le x \le 3$;
range: $-1 \le y \le 3$ **(b)** yes   **(c)** no
3. (a) domain: $-3 \le x \le 3$;
range: $-3 \le y \le 3$ **(b)** yes   **(c)** yes

5. (a)

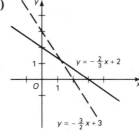

$$y = -\frac{2}{3}x + 2$$
$$y = -\frac{3}{2}x + 3$$

(b) yes   **(c)** yes

7. (a)

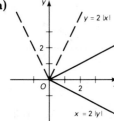

(b) no
(c) yes
9. (a) 5
(b) 5
(c) 4

13.

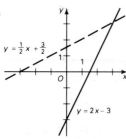

15.

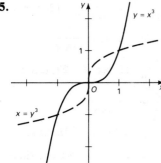

17. $\dfrac{2}{1 - x}$　**19.** $x^2 + 1$　**21.** $8 - \sqrt[3]{x}$

25. $\dfrac{b - dx}{cx - a}$　**27. (a)** $\dfrac{x(a^2 + bc) + ab + bd}{x(ac + cd) + bc + d^2}$

(b) $d = -a$

Exercises 2-6, pages 71–74

1. $\dfrac{\pi}{2}$　**3.** 0　**5.** $\dfrac{\pi}{2}$　**7.** 0　**9.** $\dfrac{3\pi}{4}$　**11.** $\dfrac{\pi}{6}$　**13.** $\dfrac{\pi}{6}$

15. $\dfrac{-\pi}{3}$　**17.** 1.3500　**19.** 0.2157

21. -1.2100　**23.** 1.3400　**25.** 0.4　**27.** 6.0

29. $\dfrac{4}{5}$　**31.** 3　**33.** $\dfrac{\sqrt{7}}{4}$　**35. (a)** domain:

$-1 \le x \le 1$; range: $0 \le y \le \pi$

(b) domain: $-1 \le x \le 1$; range: $\dfrac{-\pi}{2} \le y \le \dfrac{\pi}{2}$

37. $\dfrac{\pi}{4}$　**39.** $\dfrac{\pi}{2}$　**41.** $\sqrt{1 - u^2}$　**43.** $\dfrac{1}{u}$

45. $\dfrac{1}{\sqrt{u^2 + 1}}$　**47. (a)** 105.5°; 86.8°; 26.5°

(b) 0.7 h　**49. (a)** $\theta = \operatorname{Tan}^{-1}\left(\dfrac{840 - 3t}{2000}\right)$

(b) 37　**(c)** 22.5°; 21.9°; 19.4°

(d) $-1.7°$; The balloon has landed.

51. (a) angle of depression when $0 \le t < \dfrac{75}{11}$,

$\theta = \operatorname{Tan}^{-1}\left(\dfrac{75 - 11t}{40}\right)$; angle of elevation

when $t \ge \dfrac{75}{11}$, $\theta = \operatorname{Tan}^{-1}\left(\dfrac{11t - 75}{40}\right)$

(b) 75.6°; 86.1°

55.

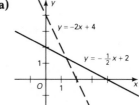

Self Quiz　2-4 / 2-5 / 2-6, page 74

1. (a)

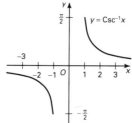

(b) yes
(c) yes

2. $\dfrac{\sqrt{5}}{3}$　**3. (a)** even　**(b)** neither

4.

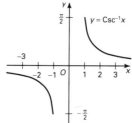

5. $\dfrac{\sqrt{1 - u^2}}{u}$

6. $\dfrac{3}{2x - 1}$

Additional Problems, pages 75–76

1. (a) $\sin\dfrac{5\pi}{6} = \dfrac{1}{2}$; $\cos\dfrac{5\pi}{6} = \dfrac{-\sqrt{3}}{2}$; $\tan\dfrac{5\pi}{6} = \dfrac{-\sqrt{3}}{3}$

(b) $\sin\dfrac{7\pi}{4} = \dfrac{-\sqrt{2}}{2}$; $\cos\dfrac{7\pi}{4} = \dfrac{\sqrt{2}}{2}$; $\tan\dfrac{7\pi}{4} = -1$

(c) $\sin 7\pi = 0$; $\cos 7\pi = -1$; $\tan 7\pi = 0$

3. (a) $\dfrac{2\pi}{3}$ **(b)** $\dfrac{\pi}{3}$ **(c)** $\dfrac{-\pi}{4}$ **5. (a)** -0.5048

(b) 0.7602 **(c)** 1.076 **7. (a)** 0.2527

(b) 2.165 **(c)** -0.9978

9. (a)

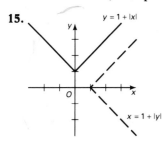

(b) odd

11. $7200\,\pi$ cm/min; 720 rpm

15.

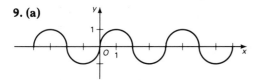

R is a function. R^{-1} is not a function.

17. $\dfrac{\sqrt{5}}{2}$ **19. (a)** $\dfrac{x - b}{m}$ **(b)** $f(x) = x$, or $f(x) = -x + b$ for some constant b.

Cumulative Review pages 82–83

Chapter 1

1. (a) $247°\ 30'$ **(b)** $145.25°$ **3.** $60°$

5. $\sqrt{2} + \sqrt{3}$

7. The fire is approximately 9.9 miles from A and 7.1 miles from B. **9. (a)** I, IV **(b)** II, IV

11. -2.112 **13. (a)** $270°$ **(b)** $210°$ **(c)** $540°$

Chapter 2

1. (a) -0.36 **(b)** -0.99

3. (a) 2π rad/min, or 1 rpm **(b)** 2π cm

5. (a) odd **(b)** even

7.

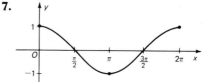

9. (a) $\dfrac{\pi}{6}$ **(b)** $\dfrac{4}{3}$ **(c)** $\dfrac{\pi}{3}$ **(d)** $\dfrac{\pi}{6}$ **11. (a)** $21°$

(b) 28 s

Chapter 3 Properties of Trigonometric Functions

Exercises 3-1, pages 87–88

1. $\dfrac{\sin^2 x}{1 - \sin^2 x}$ **3.** $\dfrac{\sin x}{1 - \sin^2 x}$ **5.** $\sin x$

7. $\dfrac{1}{\sin x}$ **9.** $\dfrac{1}{\cos^2 x}$ **11.** $\dfrac{1}{\cos^2 x}$ **13.** $1 - \cos x$

15. $\dfrac{1}{1 - \cos x}$ **17.** $\cot x$, or $\dfrac{1}{\tan x}$

19. $\cot t$, or $\dfrac{1}{\tan t}$ **21.** $\sec^2\theta$ **23.** $\tan x$

25. $\sin x$ **27.** $\sec\theta$, or $\dfrac{1}{\cos\theta}$

29. $\cot y$, or $\dfrac{1}{\tan y}$ **31.** $\sin x$ **33.** $\tan x$

35. $\cos\theta$ **37.** $\sec^4 x$ **41.** 0 **43.** 0 **45.** 0

Exercises 3-2, pages 91–93

9. no **11.** no **13.** yes **15.** no

Exercises 3-3, pages 96–99

1. $\dfrac{\sqrt{6} + \sqrt{2}}{4}$ **3.** $\dfrac{\sqrt{6} - \sqrt{2}}{4}$ **5.** $\dfrac{\sqrt{6} - \sqrt{2}}{4}$

7. $\dfrac{\sqrt{6} - \sqrt{2}}{4}$ **9.** $\dfrac{\sqrt{2} + \sqrt{6}}{4}$ **11.** $\dfrac{-\sqrt{2} - \sqrt{6}}{4}$

13. $\dfrac{63}{65}$ **15.** $\dfrac{56}{65}$ **17.** $\dfrac{-84}{85}$ **19.** $\dfrac{77}{85}$

21. $\dfrac{-3\sqrt{3} - 4}{10}$ **43.** $\dfrac{84}{85}$ **45.** $\dfrac{-204}{325}$

47. $\dfrac{\sqrt{6} - \sqrt{2}}{4}$

Self Quiz 3-1 / 3-2 / 3-3, page 99

1. $\sin x$ **2.** $\tan^2 x$ **5.** $\dfrac{-33}{65}$ **6.** $\dfrac{-56}{65}$

Exercises 3-4, pages 102–103

1. (a) $\dfrac{24}{25}$ (b) $\dfrac{-7}{25}$ **3.** (a) $\dfrac{-120}{169}$ (b) $\dfrac{-119}{169}$

5. (a) $\dfrac{3\sqrt{7}}{8}$ (b) $\dfrac{-1}{8}$ **7.** (a) $\dfrac{-4\sqrt{2}}{9}$ (b) $\dfrac{-7}{9}$

9. (a) $\dfrac{-4\sqrt{6}}{25}$ (b) $\dfrac{-23}{25}$ **11.** $\dfrac{-239}{28561}$

13. $\dfrac{-527}{625}$ **15.** $\dfrac{-79}{81}$ **17.** $\dfrac{17}{32}$ **19.** $\dfrac{\sqrt{2-\sqrt{3}}}{2}$

21. $\dfrac{\sqrt{2-\sqrt{2}}}{2}$ **23.** $\dfrac{\sqrt{2+\sqrt{3}}}{2}$ **25.** $\dfrac{\sqrt{34}}{10}$

27. $\dfrac{3}{5}$

Exercises 3-5, pages 106–108

1. $2+\sqrt{3}$ **3.** $\sqrt{3}-2$ **5.** $\dfrac{23}{9}$ **7.** $\dfrac{3}{4}$ **9.** -2

11. $\dfrac{-44}{117}$ **13.** $\dfrac{\sqrt{2}}{2}$ **15.** $\dfrac{10\sqrt{2}}{23}$

Exercises 3-6, pages 112–114

1. $135°, 315°$ **3.** $116.6°, 296.6°$
5. $60°, 120°, 240°, 300°$ **7.** $90°, 270°$
9. $30°, 90°, 150°$ **11.** $0°, 60°, 180°, 300°$

13. $\dfrac{\pi}{6}, \dfrac{5\pi}{6}, \dfrac{3\pi}{2}$ **15.** $0, \dfrac{2\pi}{3}, \pi, \dfrac{4\pi}{3}$

17. $\dfrac{3\pi}{8}, \dfrac{7\pi}{8}, \dfrac{11\pi}{8}, \dfrac{15\pi}{8}$

19. $\dfrac{\pi}{12}+k\pi, \dfrac{5\pi}{12}+k\pi, \dfrac{3\pi}{4}+k\pi$

21. $\dfrac{\pi}{6}+2k\pi, \dfrac{5\pi}{6}+2k\pi$

23. $0.72+2k\pi, 5.56+2k\pi, \pi+2k\pi$

25. $\dfrac{\pi}{3}+2k\pi, \dfrac{5\pi}{3}+2k\pi, \pi+2k\pi$

27. $\dfrac{\pi}{4}+k\pi, \dfrac{3\pi}{4}+k\pi$ **29.** $\dfrac{\pi}{6}\le x\le\dfrac{5\pi}{6}$

31. $\dfrac{\pi}{4}\le x<\dfrac{\pi}{2}$ **33.** $\dfrac{7\pi}{6}, \dfrac{3\pi}{2}, \dfrac{11\pi}{6}$

35. $\dfrac{\pi}{3}, \dfrac{2\pi}{3}, \dfrac{4\pi}{3}, \dfrac{5\pi}{3}$ **37.** $\dfrac{\pi}{3}, \dfrac{\pi}{2}, \dfrac{2\pi}{3}, \dfrac{4\pi}{3}, \dfrac{3\pi}{2}, \dfrac{5\pi}{3}$

39. $0\le x\le\dfrac{\pi}{4}$

41. $0\le x\le\dfrac{\pi}{12}; \dfrac{5\pi}{12}\le x\le\dfrac{7\pi}{12}; \dfrac{11\pi}{12}\le x\le\pi$

43. $\dfrac{\pi}{6}, \dfrac{\pi}{3}, \dfrac{7\pi}{6}, \dfrac{4\pi}{3}$ **45.** $\dfrac{2\pi}{3}, \dfrac{4\pi}{3}$ **47.** $\dfrac{\pi}{12}, \dfrac{5\pi}{12}$

49. $15°, 75°$ **51.** $15°, 75°, 90°$
53. $30°, 150°, 210°, 330°$
55. $0°, 45°, 135°, 180°, 225°, 315°$

Self Quiz   3-4 / 3-5 / 3-6, page 114

3. $\dfrac{5\sqrt{26}}{26}$ **4.** $-2-\sqrt{3}$ **5.** 3 **6.** $\dfrac{-24}{7}$

7. $\dfrac{\pi}{3}+2k\pi, \dfrac{5\pi}{3}+2k\pi, \pi+2k\pi$

8. $0, \dfrac{2\pi}{3}, \pi, \dfrac{4\pi}{3}$ **9.** $0, \dfrac{\pi}{3}, \dfrac{2\pi}{3}, \dfrac{4\pi}{3}, \dfrac{5\pi}{3}, \pi$

Additional Problems, page 116

1. (a) $\dfrac{-84}{85}$ (b) $\cos 2\theta = \dfrac{7}{25}$; $\cos\dfrac{\theta}{2} = \dfrac{3\sqrt{10}}{10}$

5. $\dfrac{\tan x-1}{\tan x+1}$ **7.** $4\tan x$ **11.** $\sin x$

13. $4\cos^3 x - 3\cos x$

Mixed Review: Chapters 1–3, pages 122–123

1. $\sin 60° = \dfrac{\sqrt{3}}{2}$; $\cos 60° = \dfrac{1}{2}$; $\tan 60° = \sqrt{3}$;

$\sec 60° = 2$; $\csc 60° = \dfrac{2\sqrt{3}}{3}$; $\cot 60° = \dfrac{\sqrt{3}}{3}$

3. 0.97 **7.** (a) 10 cm (b) 40π cm^2

9. $\dfrac{\pi}{4}+\dfrac{k\pi}{2}$ **11.** 9.0 cm **13.** $41.8°$

15. (a) $-\dfrac{\pi}{2}$ (b) $\dfrac{4}{5}$ **17.** $\cos\theta = \dfrac{-2\sqrt{13}}{13}$;

$\sin\theta = \dfrac{-3\sqrt{13}}{13}$;

$\tan\theta = \dfrac{3}{2}$; $\sec\theta = \dfrac{-\sqrt{13}}{2}$;

$\csc\theta = \dfrac{-\sqrt{13}}{3}$; $\cot\theta = \dfrac{2}{3}$; $\theta = 236.3°$

19. $5.345°$ **21.** $(\sqrt{3}, -1); d = 20\pi$

23. $\dfrac{3\pi}{8}, \dfrac{5\pi}{8}, \dfrac{11\pi}{8}, \dfrac{13\pi}{8}$ **27.** $3\sqrt{7}$ **29.** $-\sqrt{7}$

31.

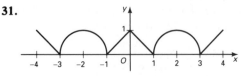

33. neither

Chapter 4   Solving Oblique Triangles

Exercises 4-1, pages 128–131

1. 7   **3.** 13   **5.** 60°   **7.** 109.6°
9. 8.7 cm, 14.0 cm   **11.** 14.5 cm, 18.4 cm
13. 60°, 120°   **15.** 82.3 km   **17.** 121.6°
19. 15 m   **21.** 63.7 ft   **23.** 274 cm
25. Pythagorean theorem
31. 40.1 m, 31.5 m   **33.** 16.8   **35.** 8.4
37. 3 or 5   **39.** 6   **41.** 13   **43.** $\sqrt{55}$
45. 8.4 m   **47.** 6.6 miles   **49.** 8

Exercises 4-2, pages 133–135

1. 25.3   **3.** 3.9   **5.** 19.5   **7.** 51.5°   **9.** 6
11. $\frac{5}{6}$ **13.** $\frac{25}{39}$ **15.** 23.1 m   **17.** 48.1 m
19. 6943 m   **25.** 5.7 m   **27.** 19.3 m
29. 280 m   **31.** 236 m
33. 242 m/s or 870 km/h

Self Quiz   4-1 / 4-2, page 136

1. 4.2   **2.** 41.5   **3.** 60°   **4.** 38.4°   **5.** 15.8 m
6. 120°

Exercises 4-3, pages 139–141

1. $b = 20.1$; $\angle A = 44.3°$; $\angle C = 93.7°$
3. $a = 23.4$; $c = 31.1$; $\angle A = 45°$
5. $\angle A = 25.5°$; $\angle B = 43.8°$; $\angle C = 110.7°$
7. $a = 8.3$; $\angle A = 22.2°$; $\angle C = 42.8°$
9. no solution
11. $\angle A = 53.1°$; $\angle B = 36.9°$; $\angle C = 90°$
13. $a = 15.7$; $\angle A = 67.9°$; $\angle B = 62.1°$;
$a = 3.6$; $\angle A = 12.1°$; $\angle B = 117.9°$
15. $b \geq c \sin B$ **17.** 17.2 m
19. 232 km, 40.4°   **21.** 55.4°   **23.** 1.5 m
25. 106 ft   **27.** 71.3°   **31.** 2.2   **33.** 3.9; 4.3

Exercises 4-4, pages 144–145

1. 210.8   **3.** 154.1   **5.** 3.98   **7.** 56.7
9. no solution   **11.** 10.1   **13.** 90.3; 20.4
15. 423.8 m²   **17.** 580.9 m²   **19.** 48
21. 282.8   **23.** 9.1 m   **25.** 46.4   **29.** $\frac{1}{2}cl \sin \phi$
31. 114,000 cm²   **33.** 18.0
35. (a) $K = \frac{1}{2}nr^2\sin\left(\dfrac{360°}{n}\right)$
(b) $K = \frac{1}{2}nr^2\sin\left(\dfrac{2\pi}{n}\right)$

Self Quiz   4-3 / 4-4, page 146

1. 12.1 square units   **2.** 249 square units
3. 166 square units   **4.** 10,214 square units
5. (a) 1   **(b)** 0   **(c)** 2   **(d)** 1   **6.** 15 cm
7. 252 km; 47.4°

Additional Problems, pages 146–147

1. $b = 28.0$, $\angle A = 35.3°$, $\angle C = 28.7°$
3. $a = 10.8$, $b = 8.1$, $\angle C = 32°$
5. $a = 61.3$, $\angle A = 61.9°$, $\angle C = 67.1°$;
$a = 19.3$, $\angle A = 16.1°$, $\angle C = 112.9°$
7. 121   **9.** 23.2   **11.** 1524; 479
13. 11,700 m   **15.** 4.68 m²   **17.** 2034 km
19. 4.8 km   **21.** $2\sqrt{a^2 - b^2 \sin^2 A}$

Chapter 5   Sinusoidal Variation

Exercises 5-1, pages 156–157

1. amplitude = 2; period = 2π; graph (c)
3. amplitude = 3; period = π; graph (b)

5.

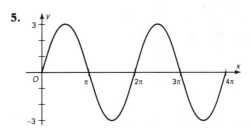

7.

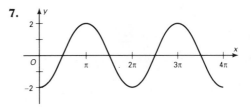

9.

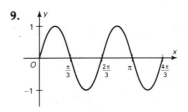

11.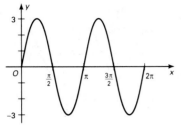

13.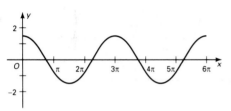

15.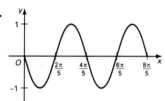

17.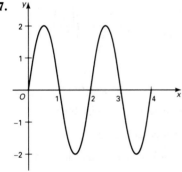

19.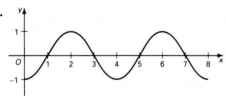

21.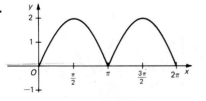

23.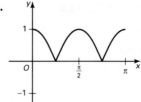

25.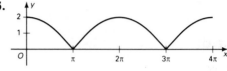

27.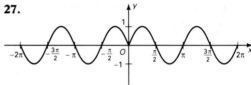

29. $y = 3 \cos x$

31. $y = 3 \sin \dfrac{2\pi x}{3}$

33. domain, $-1 \le x \le 1$;
range, $0 \le y \le 2\pi$
(See figure at right.)

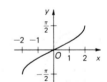

35. domain, $-2 \le x \le 2$;
range, $\dfrac{-\pi}{2} \le y \le \dfrac{\pi}{2}$

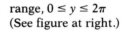

37. domain, all
real numbers;
range,
$-\pi < y < \pi$

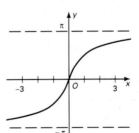

39.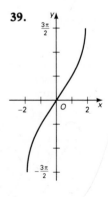

41.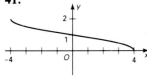

43.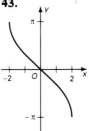

Exercises 5-2, pages 160–161

1.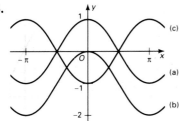

3.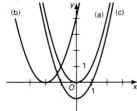

5.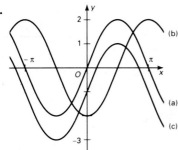

7.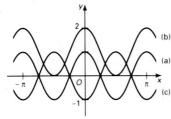

9.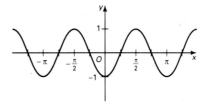

11.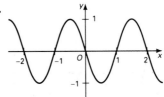

13.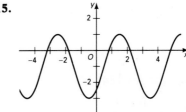

15.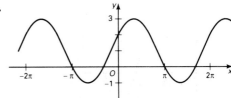

17. $\dfrac{\pi}{4} + n\dfrac{\pi}{2}$, where n is an integer

19. n, where n is an integer

21. $\pi + 2n\pi, \dfrac{5\pi}{3} + 2n\pi$, where n is an integer

23. $\dfrac{5}{6} + 4n, \dfrac{13}{6} + 4n$, where n is an integer

25. $\cos\left(2x - \dfrac{\pi}{6}\right)$ **27.** $\sin\dfrac{x}{3}$

29. amplitude $= \dfrac{1}{2}$; period $= \pi$

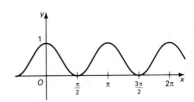

31. domain, $-1 \le x \le 1$;
range, $0 \le y \le 2$

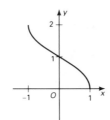

33. domain, $0 \le x \le 2$;
range, $\dfrac{-\pi}{2} \le y \le \dfrac{\pi}{2}$

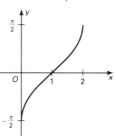

35. domain, $0 \le x \le 2$;
range, $\dfrac{-\pi}{2} \le y \le \dfrac{\pi}{2}$

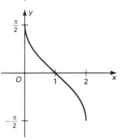

Exercises 5-3, pages 164–166

1. $\sqrt{2}\cos 55°$ **3.** $2\cos 70°$
5. $2\cos 50°$, or $2\cos 310°$
7. $2\cos 75°$, or $2\cos 285°$
9. $y = 2\sqrt{2}\cos\left(x - \dfrac{\pi}{4}\right)$

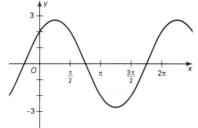

11. $y = 2\cos\left(x + \dfrac{\pi}{6}\right)$

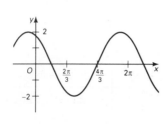

13. $y = 5\cos(x - 0.93)$

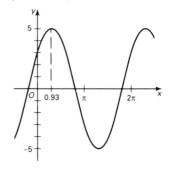

15. $y = 13\cos(x + 1.18)$

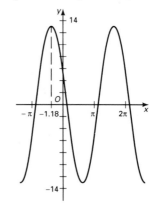

17. $\dfrac{3\pi}{4} + n\pi$, where n is an integer
19. $\dfrac{\pi}{3} + n\pi$, where n is an integer
21. $2.50 + n\pi$, where n is an integer
23. $0.39 + n\pi$, where n is an integer
25. $105°, 345°$ **27.** $11.4°, 288.6°$
29. $119.6°, 346.7°$
31. $8.8°, 114.9°, 188.8°, 294.9°$
33. $15°$ or $75°$
35. $C = \sqrt{A^2 + B^2}$; γ is such that
$\cos\gamma = \dfrac{B}{C}$ and $\sin\gamma = \dfrac{A}{C}$.
37. $15.2°$
39. $28.1°$; 55.3 N
41. $C = \sqrt{a^2 + b^2 + 2ab\sin(\beta - \alpha)}$; γ is
such that $\cos\gamma = \dfrac{a\cos\alpha + b\sin\beta}{c}$ and
$\sin\gamma = \dfrac{-a\sin\alpha + b\cos\beta}{c}$.

Self Quiz 5-1 / 5-2 / 5-3, page 166

1.

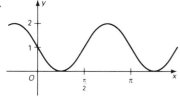

2.

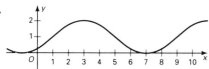

3. $2 \cos 10°$ **4.** $255°, 345°$

5. $\cos \left(2x - \dfrac{5\pi}{6} \right)$

6.

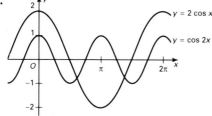

Exercises 5-4, pages 170–171

1. (a)

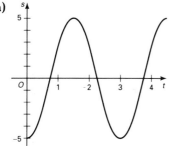

(b) $s = -5 \cos \dfrac{2\pi}{3}t$ **(c)** $\dfrac{1}{3}$ cycle per second

3. (a)

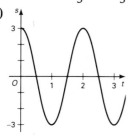

(b) $s = 3 \cos \pi t$ **(c)** $\dfrac{1}{2}$ cycle per second

5. (a)

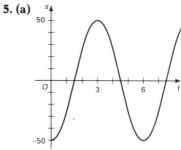

(b) $s = -50 \cos \dfrac{\pi}{3}t$ **(c)** $\dfrac{1}{6}$ cycle per second

7. (a) $s = 4 \cos \left(\pi t - \dfrac{\pi}{2} \right)$

(b)

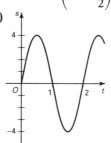

(c) $\dfrac{1}{2}$ cycle
per second

9. (a) $s = 4000 \cos \dfrac{\pi}{2550} t$

(b)

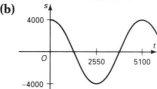

(c) $\dfrac{1}{5100}$ cycle per second $= \dfrac{1}{85}$ cycle/min

11. $y = 5 \sin \left(\dfrac{2\pi}{3}t - \dfrac{\pi}{2} \right)$

13. $y = 3 \sin \left(\pi t + \dfrac{\pi}{2} \right)$

15. $y = 50 \sin \left(\dfrac{\pi}{3}t - \dfrac{\pi}{2} \right)$ **17.** $y = 4 \sin \pi t$

19. $y = 4000 \sin \left(\dfrac{\pi t}{2550} + \dfrac{\pi}{2} \right)$

21. (a) $v = 120\sqrt{2} \sin (120\pi t + \beta)$

(b) $\dfrac{\pi}{4}$, or $\dfrac{3\pi}{4}$ **23.** 24π cm/s, or 75.4 cm/s

25. (a) $\dfrac{1}{5\pi}$ cycles per second   **(b)** 25 cm

27. $\dfrac{980}{\pi^2}$ cm, or 99.3 cm   **29.** $\dfrac{\sqrt{3}}{2}\,a$

Exercises 5-5, page 174

1.

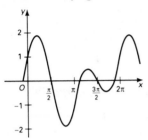

3.

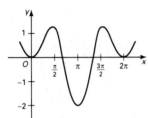

5.

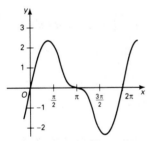

7.

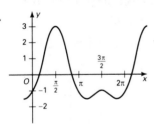

9.

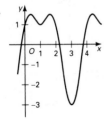

11–19. Answers may vary.

11. $\dfrac{\pi}{2}, \dfrac{7\pi}{6}, \dfrac{3\pi}{2}, \dfrac{11\pi}{6}$ **13.** $0, \dfrac{2\pi}{3}, \dfrac{4\pi}{3}$ **15.** $0, \pi$

17. $0.375, 2.767$ **19.** $2.239, 3.761$

21.

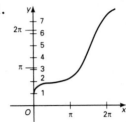

23.

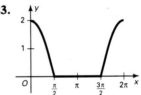

25. $y = \cos 3x + \cos x$

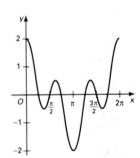

27. $y = \sin 3x - \sin x$

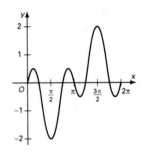

29. $y = \dfrac{1}{2}\sin 2x - \dfrac{1}{2}\sin x$

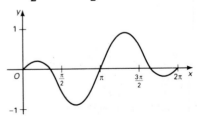

Exercises 5-6, pages 178–180

Note: In exercises 1–7, the functions are given in terms of sine.

1. $y = 2 \sin \dfrac{1}{2}(x - \pi) + 2$

3. $y = 3 \sin \dfrac{\pi}{6}(x - 4)$

5. $y = 3 \sin \dfrac{\pi}{6}(x - 1) + 1$

7. $y = 2 \cos \dfrac{1}{2}(x - 2\pi) + 2$

9. $y = 3 \cos \dfrac{\pi}{6}(x - 7)$

11. $y = 3 \cos \dfrac{\pi}{6}(x - 4) + 1$

13. $y = \dfrac{3}{2} \sin 2\left(x - \dfrac{\pi}{4}\right) + \dfrac{3}{2}$ or

$y = \dfrac{3}{2} \cos 2\left(x - \dfrac{\pi}{2}\right) + \dfrac{3}{2}$

15. $y = \dfrac{1}{2} \sin \dfrac{\pi}{4}(x + 1) - \dfrac{1}{2}$ or

$y = \dfrac{1}{2} \cos \dfrac{\pi}{4}(x - 1) - \dfrac{1}{2}$

17. $h = 6 \sin\left(\dfrac{2\pi}{15}t - \dfrac{\pi}{2}\right) + 8 =$

$6 \sin \dfrac{2\pi}{15}\left(t - \dfrac{15}{4}\right) + 8$

19. (a) $p(t) = 50 + \dfrac{1}{2}t + 2 \sin \dfrac{\pi}{4} t$

(b) $p(22) = 59$ **(c)** $p(25) = 63.9$

21. Answers may vary. If $d = 0$ at $t = 0$, then

$d = 2 \sin \dfrac{\pi}{30} t + 8$.

23. $y = \sin \dfrac{\pi x}{60} + 0.5 \sin \dfrac{\pi x}{30}$

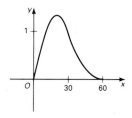

25. $y = \sin \dfrac{\pi x}{30} + \sin \dfrac{\pi x}{20}$

(See next column for diagram.)

25.

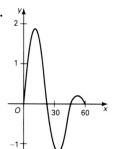

27. $h = \dfrac{\pi}{l}$;

$k = 2\pi f$

Self Quiz   5-4 / 5-5 / 5-6, page 180

1. $s = -6 \cos \dfrac{\pi}{2}t$

2. See diagram at right.

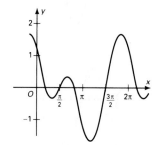

3. $\dfrac{\pi}{2}, \dfrac{\pi}{6}, \dfrac{3\pi}{2}, \dfrac{5\pi}{6}$ **4. (a)** $y = 2 \cos 2x + 2$

(b) $y = 2 \sin 2\left(x + \dfrac{\pi}{4}\right) + 2$ **5. (a)** π s

(b) $\sqrt{10}$ cm   **(c)** $2\sqrt{10}$ cm/s

6. $y = 2 \sin \dfrac{\pi}{2}(x - 2) + 1$ or

$y = 2 \cos \dfrac{\pi}{2}(x - 3) + 1$

Additional Problems, pages 181–182

1. 2 cos 20° or 2 cos 340°

3. $s = 3 \cos \pi t$ or $s = 3 \sin\left(\pi t + \dfrac{\pi}{2}\right)$

5. domain:
$-2 \le x \le 2$
range:
$\dfrac{-\pi}{2} \le y \le \dfrac{\pi}{2}$

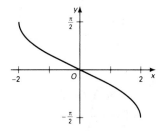

7. $y = \sin \dfrac{\pi}{2}(x - 1) + 2$ or $y =$

$-\cos \dfrac{\pi}{2}x + 2$

9.

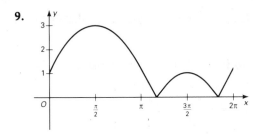

11. 2.50, 5.64   **13.** $\dfrac{\pi}{3}, \pi, \dfrac{5\pi}{3}$

15.

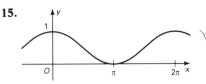

17. $s = 3 \sin 4t$; $T = \dfrac{\pi}{2}$ **19.** $\dfrac{3\pi}{4}; \dfrac{7\pi}{4}$

21. (a) π s   (b) 5 cm

Cumulative Review, pages 188–189

Chapter 3

1. $-\cos x$ **5.** (a) $-\left(\dfrac{3\sqrt{3} + 4}{10}\right)$

(b) $-\left(\dfrac{4\sqrt{3} + 3}{10}\right)$

(c) $\dfrac{48 + 25\sqrt{3}}{39}$, or $\dfrac{16}{13} + \dfrac{25\sqrt{3}}{39}$ **7.** $\sqrt{2} - 1$

9. $0, \dfrac{7\pi}{6}, \dfrac{11\pi}{6}$

Chapter 4

1. $3\sqrt{6}$ **3.** 70°   **5.** 468 cm²

Chapter 5

1. (a) period $= \dfrac{2\pi}{3}$; $a = 2$

(b)

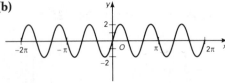

3. (a) $T = \dfrac{2\pi}{3}$; $a = 2$

(b)

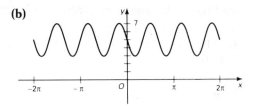

5. $10 \cos (\theta - 126.9°)$

7.

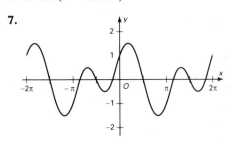

Chapter 6   Vectors in the Plane

All diagram answers for this chapter appear in the Solution Key.

Exercises 6-1, pages 195–197

9. $\mathbf{w} - 2\mathbf{v}$ **11.** $\mathbf{w} - \dfrac{1}{2}\mathbf{v}$ **13.** yes

23. Diagonals of a rhombus are perpendicular.
25. Equality holds if and only if $\mathbf{u}$ and $\mathbf{v}$ have the same direction, or $\mathbf{u} = \mathbf{0}$, or $\mathbf{v} = \mathbf{0}$.
27. (a) $\mathbf{u} = \mathbf{a} + \mathbf{b}$; $\mathbf{v} = 2\mathbf{a} + \mathbf{b}$
(b) $\mathbf{a} = -\mathbf{u} + \mathbf{v}$; $\mathbf{b} = 2\mathbf{u} - \mathbf{v}$
29. (a) $\mathbf{u} = 2\mathbf{a} - \mathbf{b}$; $\mathbf{v} = 3\mathbf{a} - \mathbf{b}$
(b) $\mathbf{a} = -\mathbf{u} + \mathbf{v}$; $\mathbf{b} = -3\mathbf{u} + 2\mathbf{v}$
33. $2\|\mathbf{u}\|^2 + 2\|\mathbf{v}\|^2 = \|\mathbf{u} + \mathbf{v}\|^2 + \|\mathbf{u} - \mathbf{v}\|^2$

Exercises 6-2, pages 201–202

1. 71.6°; 126 km   **3.** 325 km/h; 96.9°
5. 75.5 km; 304.7°   **7.** 4.3 km/h
9. 23.3°; 125 mi/h   **11.** 363 km   **13.** 277°
15. 1:54 P.M.; 316.6°

Exercises 6-3, pages 205–207

1. parallel, 110 N; perpendicular, 273 N
3. 1455 N   **5.** 3810 N; 2490 N
7. 1680 N; 13.1°   **9.** 1044 N; 16.7°
11. 9.8 N; 33.3°   **13.** 13.8 N; 29.4°   **15.** 11.3°
17. 185 N   **19.** 91 kg

Self Quiz 6-1 / 6-2 / 6-3, pages 207–208

3. $w = -3u + v; z = u - v$
4. 261°; 202 km/h **5.** 99.8°; 57 km **6.** 230 N
7. 1897 N; 2418 N

Exercises 6-4, pages 210–211

1. $3i - 4j; i + 6j; \frac{3}{5}i - \frac{4}{5}j$

3. $-2i - 2j; 4i - 6j; \frac{-\sqrt{2}}{2}i - \frac{\sqrt{2}}{2}j$

5. $8i - 15j; 6i + 5j; \frac{8}{17}i - \frac{15}{17}j$

7. $2i + 2\sqrt{3}\,j; 4\sqrt{3}j; \frac{1}{2}i + \frac{\sqrt{3}}{2}j$ **9.** $11i$

11. $\frac{5}{2}i - j$ **13.** $s = -1; t = 1$

15. $s = 6; t = 7$ **17.** $s = 2; t = -1$

Exercises 6-5, pages 215–216

1. 0 **3.** 0 **5.** 3 **7.** $p^2 - q^2$ **9.** (a) **v** and **t**;
(b) **w** and **t**, **v** and **w** **11.** 38 **13.** 51
15. 60° **17.** 26.6° **19.** 121.3°
21. (a) $\frac{-\sqrt{2}}{2}i + \frac{\sqrt{2}}{2}j$; (b) $\frac{\sqrt{2}}{2}i + \frac{\sqrt{2}}{2}j$

23. (a) $\frac{4}{5}i - \frac{3}{5}j$; (b) $\frac{3}{5}i + \frac{4}{5}j$

25. (a) $\frac{x}{\sqrt{x^2 + y^2}}i + \frac{y}{\sqrt{x^2 + y^2}}j$;

(b) $\frac{y}{\sqrt{x^2 + y^2}}i - \frac{x}{\sqrt{x^2 + y^2}}j$ **27.** $\frac{1}{2}i + \frac{1}{2}j$;

$\|v_a\| = \frac{\sqrt{2}}{2}$ **29.** $\frac{4}{25}i - \frac{3}{25}j; \|v_a\| = \frac{1}{5}$

31. $v = \frac{1}{2}a - \frac{1}{2}b$ **33.** $v = \frac{5}{17}a + \frac{14}{17}b$

Exercises 6-6, pages 218–219

1. 10 J **3.** (a) 70 J (b) 70 J **5.** (a) 42 J
(b) 42 J **7.** 8,820,000 J = 2.45 kW · h
9. 1.12×10^4 J = 0.00311 kW · h
11. 3.99×10^7 J = 11.1 kW · h **13.** 566 J
15. 6.19×10^6 J = 1.72 kW · h

Self Quiz 6-4 / 6-5 / 6-6, page 220

1. $4i + 9j$ **2.** $9i + j$ **3.** $\frac{3}{5}i + \frac{4}{5}j$
4. $s = -3; r = 1$ **5.** -2 **6.** 59°

7. $v_a = \frac{6}{25}i + \frac{8}{25}j; \|v_a\| = \frac{2}{5}$ **8.** 2.04 kW · h
9. 4599 J

Additional Problems, pages 221–222

1. $-3u + v$ **3.** $-3u + 2v$
5. $r = \frac{-17}{5}; s = \frac{14}{5}$
7. $\|v - u\|^2 = \|u\|^2 + \|v\|^2 - 2\,u \cdot v$
9. $\frac{17\sqrt{353}}{353}i - \frac{8\sqrt{353}}{353}j$ **11.** 59.2

13. $\frac{-10}{169}i + \frac{24}{169}j$ **15.** 0.68 kW · h

17. $\frac{7}{2}a - \frac{1}{2}b$ **21.** 28 J

Chapter 7 Complex Numbers

Exercises 7-1, pages 232–233

1. $(\sqrt{3}, -1)$ **3.** $(-\sqrt{3}, -1)$ **5.** $(\sqrt{3}, -1)$
7. $(-\sqrt{3}, 1)$ **9.** $(-\sqrt{2}, \sqrt{2})$ **11.** $(2, -2\sqrt{3})$
13. $(2, 30°)$ **15.** $(4, 120°)$ **17.** $(2, 225°)$
19. $(2\sqrt{2}, 30°)$
21. $r = -2 \csc \theta$, or $r \sin \theta = -2$
23. $r = -2 \sec \theta$, or $r \cos \theta = -2$
25. $\theta = 45°$ **27.** $r = 2$
29. $\theta = -\frac{\pi}{6}$, or $\theta = 150°$
31. $r(\cos \theta + \sin \theta) = 4$
33. The circle of radius 1 centered at the
origin
35. The line $x = 2$
37. The line $y = -x$
39. The circle $x^2 + (y - 1)^2 = 1$
41. The circle $\left(x - \frac{1}{2}\right)^2 + \left(y - \frac{1}{2}\right)^2 = \frac{1}{2}$
43. The hyperbola $xy = 1$
47. $a^2 = r^2 + r_0^2 - 2rr_0 \cos(\theta_0 - \theta)$

Exercises 7-2, pages 238–239

1. 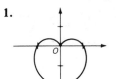 **3.**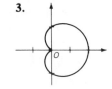

5.

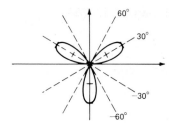

7.

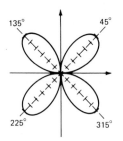

9.

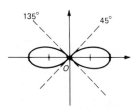

11.

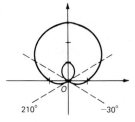

13.

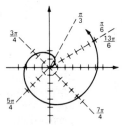

15.

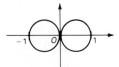

17.

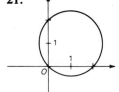

19.

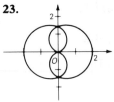

21.

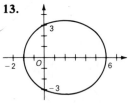

23.

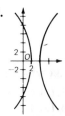

Exercises 7-3, pages 244–245

1. $e = 1$; parabola   **3.** $e = \frac{1}{2}$; ellipse

5. $e = 3$; hyperbola   **7.** $e = \frac{2}{3}$; ellipse

9. $e = 1$; parabola

11.

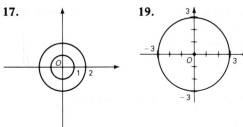

13.

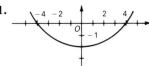

15.

17.

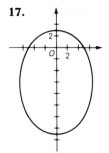

19.

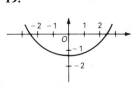

21. $(r \cos \theta)^2 - (r \sin \theta)^2 = 1$; $x^2 - y^2 = 1$; hyperbola

23. Alternate solutions use the law of cosines. **25.** $\dfrac{(x + 2)^2}{16} + \dfrac{y^2}{12} = 1$

Self Quiz   7-1 / 7-2 / 7-3, page 245

1. Answers may vary. **(a)** $\left(10, \dfrac{5\pi}{6}\right)$

(b) $\left(2\sqrt{2}, -\dfrac{\pi}{4}\right)$ **2. (a)** $(2\sqrt{3}, -2)$; **(b)** $(0, -6)$

3. $r^2 = -\sec 2\theta$, or $r^2 = -\dfrac{1}{\cos 2\theta}$

4. The circle $(x + 4)^2 + (y - 3)^2 = 25$

5.

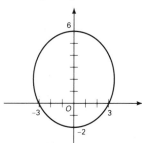

6. $e = \frac{1}{2}$; ellipse   **7.** $e = 1$; parabola

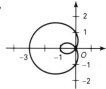
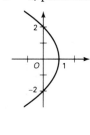

Exercises 7-4, pages 248–250

1. (a) 4 **(b)** $4 - 2i$ **(c)** $1 + 2i$
3. (a) $3 + 9i$ **(b)** $-20 + 15i$ **(c)** $\frac{4}{5} + \frac{3}{5}i$
5. (a) -2 **(b)** 4 **(c)** $-\dfrac{1}{2} - \dfrac{\sqrt{3}}{2}i$ **7. (a)** $2 - i$
(b) $\sqrt{5}$ **(c)** $\dfrac{2}{5} - \dfrac{1}{5}i$ **9. (a)** $\dfrac{\sqrt{3}}{2} + \dfrac{1}{2}i$ **(b)** 1
(c) $\dfrac{\sqrt{3}}{2} + \dfrac{1}{2}i$ **11.** $-2i$; $2 + 2i$
13. $-2 - 2\sqrt{3}\,i$; 8 **15.** $\dfrac{-7}{2} - \dfrac{3}{2}i$ **17.** $\dfrac{8}{5} - \dfrac{1}{5}i$
19. (a) i; **(b)** -1; **(c)** 1 **21.** i **23. (a)** $-i$
(b) -1 **(c)** i **(d)** 1 **25.** $-i$ **27.** $-i$

Exercises 7-5, pages 253–254

1.

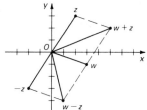

3.

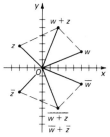

5.

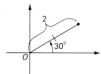

7.

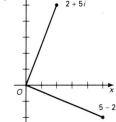

9. $2\sqrt{2}\,(\cos 45° - i \sin 45°)$ or $2\sqrt{2}\,(\cos 315° + i \sin 315°)$
11. $3(\cos 90° + i \sin 90°)$
13. $2\sqrt{7}\,(\cos 220.9° + i \sin 220.9°)$ or $2\sqrt{7}(\cos 139.1° - i \sin 139.1°)$
15. $-6\sqrt{3} + 6i$ **17.** $7.660 - 6.428i$

19. (a)

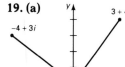

(b)

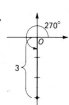

(c)

(d) The vectors representing z and iz are perpendicular.

Self Quiz 7-4 / 7-5, page 254

1. (a) $8 (\cos 300° + i \sin 300°)$;
(b) $3\sqrt{2} (\cos 225° + i \sin 225°)$
2. (a) $- i$ **(b)** i **(c)** 0 **3. (a)** 4 **(b)** $4 - 2i$
(c) $\dfrac{1}{5} - \dfrac{2}{5}i$ **(d)** $-2 - 2i$
4. $s = r$; $\alpha = 360° - \theta$, or $\alpha = -\theta$

Exercises 7-6, page 257

1. $2\sqrt{2} (\cos 225° + i \sin 225°)$
3. $8 (\cos 180° + i \sin 180°)$

5. (a) $\dfrac{1}{2} (\cos 120° - i \sin 120°)$

(b) $-\dfrac{1}{4} - \dfrac{\sqrt{3}}{4} i$

7. (a) $\dfrac{\sqrt{2}}{2} (\cos 135° - i \sin 135°)$

(b) $-\dfrac{1}{2} - \dfrac{1}{2} i$ **9.** $\dfrac{\sqrt{2}}{2} (\cos 45° + i \sin 45°)$

11. $\dfrac{1}{2} (\cos 60° - i \sin 60°)$

13. (a) $2 (\cos 190° + i \sin 190°)$
(b) $2 (\cos 10° + i \sin 10°)$

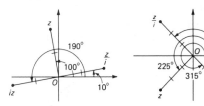

Ex. 13   Ex. 15

15. (a) $2\sqrt{2} (\cos 315° + i \sin 315°)$
(b) $2\sqrt{2} (\cos 135° + i \sin 135°)$
19. $\bar{z} = r (\cos \theta - i \sin \theta)$
21. $\dfrac{1}{z} = \dfrac{1}{r} (\cos \theta - i \sin \theta)$
23. $(\bar{z})^2 = r^2 (\cos 2\theta - i \sin 2\theta)$
25. (a) Im $zw = \sin (\theta + \phi) = 0$; $\theta + \phi = n\pi$, n an integer

(b) Re $zw = \cos (\theta + \phi) = 0$; $\theta + \phi = \dfrac{n\pi}{2}$, n an odd integer

Exercises 7-7, pages 260–261

1. $64 + 0i$ **3.** $-1 + 0i$ **5.** $-\dfrac{1}{16} + 0i$

7. $-8 + 0i$ **9.** $0 + 32i$ **11.** $0 + 9i$
13. $0 + \dfrac{1}{4}i$ **15.** $-1 + 0i$ **17.** $\dfrac{\sqrt{3}}{4} + \dfrac{1}{4}i$

Exercises 7–8, pages 264–265

1. $w_1 = -\dfrac{1}{2} + \dfrac{\sqrt{3}}{2}i$; $w_2 = -\dfrac{1}{2} - \dfrac{\sqrt{3}}{2}i$;
$w_3 = 1$; $w = w_1$; $w_2 = w^2$; $w_3 = w^3$
3. $w_1 = 0.3091 + 0.9511i$;
$w_2 = -0.8090 + 0.5878i$;
$w_3 = -0.8090 - 0.5878i$;
$w_4 = 0.3090 - 0.9511i$; $w_5 = 1$; $w = w_1$;
$w_2 = w^2$; $w_3 = w^3$; $w_4 = w^4$; $w_5 = w^5$
5. $w_1 = \dfrac{\sqrt{3}}{2} + \dfrac{1}{2}i$; $w_2 = -\dfrac{\sqrt{3}}{2} + \dfrac{1}{2}i$;
$w_3 = -i$

7. $w_1 = 2^{\frac{1}{10}}(\cos 45° + i \sin 45°)$;
$w_2 = 2^{\frac{1}{10}}(\cos 117° + i \sin 117°)$;
$w_3 = 2^{\frac{1}{10}}(\cos 189° + i \sin 189°)$;
$w_4 = 2^{\frac{1}{10}}(\cos 261° + i \sin 261°)$;
$w_5 = 2^{\frac{1}{10}}(\cos 333° + i \sin 333°)$

9. $w_1 = 2^{\frac{1}{6}}(\cos 25° + i \sin 25°)$;
$w_2 = 2^{\frac{1}{6}}(\cos 85° + i \sin 85°)$;
$w_3 = 2^{\frac{1}{6}}(\cos 145° + i \sin 145°)$;
$w_4 = 2^{\frac{1}{6}}(\cos 205° + i \sin 205°)$;
$w_5 = 2^{\frac{1}{6}}(\cos 265° + i \sin 265°)$;
$w_6 = 2^{\frac{1}{6}}(\cos 325° + i \sin 325°)$

11. $\dfrac{\sqrt{6}}{2} + \dfrac{\sqrt{2}}{2}i$; $-\dfrac{\sqrt{2}}{2} + \dfrac{\sqrt{6}}{2}i$;
$\dfrac{-\sqrt{6}}{2} - \dfrac{\sqrt{2}}{2}i$; $\dfrac{\sqrt{2}}{2} - \dfrac{\sqrt{6}}{2}i$
13. $\sqrt{2} (\cos 45° + i \sin 45°)$;
$\sqrt{2} (\cos 165° + i \sin 165°)$;
$\sqrt{2} (\cos 285° + i \sin 285°)$
19. $-i$; $-2 + i$

23. $\cos \dfrac{\pi}{6} + i \sin \dfrac{\pi}{6}$; $\cos \dfrac{5\pi}{6} + i \sin \dfrac{5\pi}{6}$;

$\cos \dfrac{7\pi}{6} + i \sin \dfrac{7\pi}{6}$; $\cos \dfrac{11\pi}{6} + i \sin \dfrac{11\pi}{6}$

Self Quiz   7-6 / 7-7 / 7-8, page 266

1. $n = 6$; $n = 12$; any $n = 6k$
2. $\sqrt{2} + \sqrt{2}i$, 2 (cos 45° + i sin 45°);
$-\sqrt{2} + \sqrt{2}i$, 2 (cos 135° + i sin 135°);
$-\sqrt{2} - \sqrt{2}i$, 2 (cos 225° + i sin 225°);
$\sqrt{2} - \sqrt{2}i$, 2 (cos 315° + i sin 315°)
3. $-8 + 8\sqrt{3}\,i$
4. $-6 + 6\sqrt{3}\,i$; 12 (cos 120° + i sin 120°)
5. $\dfrac{3\sqrt{3}}{2} - \dfrac{3}{2}i$; 3 (cos 30° − i sin 30°)
6. $w_1 = 2$ (cos 30° + i sin 30°)
$w_2 = 2$ (cos 102° + i sin 102°)
$w_3 = 2$ (cos 174° + i sin 174°)
$w_4 = 2$ (cos 246° + i sin 246°)
$w_5 = 2$ (cos 318° + i sin 318°)

Additional Problems, pages 266–267

1. (a) $25 = x^2 + y^2$ **(b)** $r = 5$
3. (a) $0 = x^2 - 2x + y^2$ **(b)** $2 \cos \theta = r$
5. The circle $(x - 2)^2 + (y - 2)^2 = (2\sqrt{2})^2$

7.

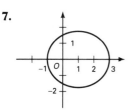

ellipse

9. (a) $\sqrt{2}$ (cos 45° + i sin 45°)   **(c)** $n = 3k$
11. cos 0° + i sin 0° = 1;

$\cos 120° + i \sin 120° = -\dfrac{1}{2} + \dfrac{\sqrt{3}}{2}\,i$;

$\cos 240° + i \sin 240° = -\dfrac{1}{2} - \dfrac{\sqrt{3}}{2}\,i$

13. $z^{-1} = \dfrac{1}{2}(\cos(-30°) + i \sin(-30°))$;

$z^0 = (\cos 0° + i \sin 0°)$;
$z^1 = 2(\cos 30° + i \sin 30°)$;
$z^2 = 4(\cos 60° + i \sin 60°)$;
$z^3 = 8(\cos 90° + i \sin 90°)$
See next column for diagram.

13.

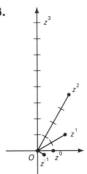

17. (a) $x^2 - y^2 = 9$;
(b) $2xy = 6$; $xy = 3$;
hyperbola

Cumulative Review   pages 274–275

Chapter 6

1. See Solution Key for diagrams.
3. 621 N   **5.** 10
7. (a) $\dfrac{\sqrt{10}}{10}\mathbf{i} - \dfrac{3\sqrt{10}}{10}\mathbf{j}$ or $-\dfrac{\sqrt{10}}{10}\mathbf{i} + \dfrac{3\sqrt{10}}{10}\mathbf{j}$
(b) $\dfrac{\sqrt{10}}{10}\mathbf{i} + \dfrac{3\sqrt{10}}{10}\mathbf{j}$ or $-\dfrac{\sqrt{10}}{10}\mathbf{i} - \dfrac{3\sqrt{10}}{10}\mathbf{j}$
9. 135°   **11.** 6212 J

Chapter 7

1. $\left(-\dfrac{7}{2}, \dfrac{7\sqrt{3}}{2}\right)$ **3.** The line $y = x + 1$
5. parabola (See figure at right.)
7. (a) $-\dfrac{5}{13} - \dfrac{12}{13}i$

(b) 1

(c) $-\dfrac{5}{13} - \dfrac{12}{13}i$

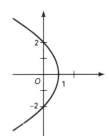

11. $-117.2 + 43.6\,i$
13. $0.9239 + 0.3827\,i$
$-0.3827 + 0.9239\,i$
$-0.9239 - 0.3827\,i$
$0.3827 - 0.9239\,i$

Mixed Review: Chapters 1–7, pages 276–279

1. (a) $\dfrac{5\pi}{6}$ **(b)** $-35°\,20'$

3. $y = 4 \cos\left(\dfrac{2\pi}{3}t + \pi\right)$; $\dfrac{1}{3}$ **5.** 256
7. 167° 54′ 46″   **9.** 8.4   **13.** 1.025 π **15.** 18.9

17. $\dfrac{-\sqrt{2}+\sqrt{3}}{2}$

19.

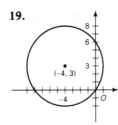

$(-4, 3)$

21. (a) $2\sqrt{2} - i$ (b) 3 (c) $\dfrac{2\sqrt{2}}{9} - \dfrac{1}{9}i$

23. $3\sqrt{5}$ (cos 206.6° + i sin 206.6°) 25. 59.2

27. $s = 2$; $t = \dfrac{5}{3}$

29. (a) cos (−10°) + i sin (−10°)
   (b) cos 170° + i sin 170°

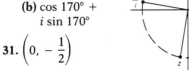

31. $\left(0, -\dfrac{1}{2}\right)$

33. 0.8704; −0.4924; −1.767 35. $\csc^2 \theta$

37. $\sin \theta = \dfrac{20}{29}$; $\cos \theta = \dfrac{-21}{29}$; $\tan \theta = \dfrac{-20}{21}$;

   $\cot \theta = \dfrac{-21}{20}$; $\sec \theta = \dfrac{-29}{21}$

45.

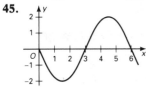

47. $0 + i$; $\dfrac{-\sqrt{3}}{2} + \dfrac{-1}{2}i$; $\dfrac{\sqrt{3}}{2} - \dfrac{1}{2}i$

49. $\dfrac{-\sqrt{2}-\sqrt{6}}{4}$; $\dfrac{-\sqrt{6}+\sqrt{2}}{4}$; $\dfrac{\sqrt{3}+3}{3-\sqrt{3}}$

51. There are no such values.
53. Translate the graph $y = \cos x$ the left π units. Then reflect the result in the x-axis. Translate the result 1 unit up.

55.

57. (a) i (b) −1 59. −53.7 − 34.9i

63. $\dfrac{3\pi}{4} + 2n\pi$, $\dfrac{7\pi}{4} + 2n\pi$;

Chapter 8   Infinite Series

Exercises 8-1, pages 284–286

1. 18, 6, 2, $\dfrac{2}{3}$, $\dfrac{2}{9}$ 3. −7, −4, −1, 2, 5

5. 1, 0, 1, 0, 1 7. 1, 1, 2, 3, 5

9. 10, 20, 40, 80, 160 11. $\dfrac{1}{3}$, 1, 3, 9, 27

13. −160 15. $\dfrac{3}{32}$ 17. 48

19. $\dfrac{1}{4} + \dfrac{1}{9} + \dfrac{1}{16} + \dfrac{1}{25} + \dfrac{1}{36}$

21. $\dfrac{2}{1} + \dfrac{3}{4} + \dfrac{4}{9} + \dfrac{5}{16} + \dfrac{6}{25}$ 23. −124

25. 89.99991 27. −85 29. 2.2222

31. $a_n = 18\left(\dfrac{1}{3}\right)^{n-1}$

33. $a_n = -7 + 3(n-1) = 3n - 10$

35. $a_n = \dfrac{1-(-1)^n}{2}$ 37. $\dfrac{-5}{16}$ 39. $\pm \dfrac{\sqrt{3}}{3}$

41. (a) $1 - r^n$

Exercises 8-2, pages 288–290

1. 1 3. 3 5. no limit 7. −1 9. $\dfrac{\pi}{4}$ 11. 3

13. 24 15. $\dfrac{1000}{27}$ 17. $\dfrac{1}{4}$ 19. $\tan^2 x$ 21. $\dfrac{7}{9}$

23. $\dfrac{2}{11}$ 25. 5, $\dfrac{5}{6}$, $\dfrac{5}{36}$, $\dfrac{5}{216}$ 27. $x = \dfrac{1}{2}$

29. (b) 1
31. $S_{200} = 3.13659269$, $S_{500} = 3.13959267$; agreement to one decimal place; slowly

Exercises 8-3, page 294

1. $\dfrac{1}{2}$ 3. 7 5. 5 7. 64 9. 2.5052

11. 2.6309 13. 1.7918 15. $-\dfrac{1}{2}$ 17. $-\dfrac{3}{2}$

19. $\dfrac{1}{e} - 1$ 21. 3 23. ± 3 25. 2.7167

27. 1.3956 29. (a) e^2 (b) 6.192 (c) 7.267

31. (a) $e^{\frac{1}{2}}$ or $\sqrt{e}$ (b) 1.629 (c) 1.649

Self Quiz   8-1 / 8-2 / 8-3, pages 294–295

1. $-\dfrac{5}{64}$ **2.** $\dfrac{-7}{8}$ **3. (a)** 6   **(b)** $e^{\frac{1}{3}}$ **4.** -1

5. 25   **6.** $\dfrac{7}{33}$ **7.** $\dfrac{1}{2}, \dfrac{2}{3}, \dfrac{3}{4}, \dfrac{4}{5}, \dfrac{5}{6}$; $S_n = \dfrac{n}{n+1}$

8. 7

Exercises 8-4, pages 297–298

1. $1 + x^2 + x^4 + x^8 + \cdots + x^{2(n-1)} + x^{2n} + \cdots$

3. $2 - 2x + 2x^2 - 2x^3 + \cdots + 2\,(-x)^{n-1} + 2\,(-x)^n + \cdots$

5. $4 + 2x + x^2 + \dfrac{1}{2}x^3 + \cdots + 4\left(\dfrac{x}{2}\right)^{n-1} +$

$4\left(\dfrac{x}{2}\right)^n + \cdots$ **7. (a)** 0.9998   **(b)** 0.4969

9. (a) 0.908   **(b)** 0.909

(c) They are very close; In fact $\sin x =$

$\sin (\pi - x)$.   **13.** $x^2 - \dfrac{x^6}{3!} + \dfrac{x^{10}}{5!} - \dfrac{x^{14}}{7!} + \cdots$

15. $1 - x + \dfrac{x^2}{2!} - \dfrac{x^3}{3!} + \cdots$

17. $1 + x^2 + \dfrac{x^4}{2!} + \dfrac{x^6}{3!} + \dfrac{x^8}{4!} + \cdots$

Exercises 8-5, pages 300–301

1. $\dfrac{1}{2} + \dfrac{\sqrt{3}}{2}i$ **3.** $\dfrac{-\sqrt{2}}{2} - \dfrac{\sqrt{2}}{2}i$ **5.** $-e^2$

7. $-e^{-2}$ **9.** $\dfrac{3\sqrt{2}}{2} - \dfrac{3\sqrt{2}}{2}i$ **11.** $\dfrac{1}{3}i$

13. $b = 2k\pi$ where k is an integer

15. $a = \ln 5$ **21. (a)** $1 - x^2 + x^4 - x^6 + \cdots$;

$i\,(x - x^3 + x^5 - x^7 + \cdots)$

(b) $\dfrac{1}{1 + x^2}$; $i\left(\dfrac{x}{1 + x^2}\right)$

Exercises 8-6, pages 305–306

1. (a) 1   **(b)** 0   **5.** 2.129   **7.** 3.756   **9.** $\pm\dfrac{3}{4}$

11. $x + \dfrac{x^3}{3!} + \dfrac{x^5}{5!} + \dfrac{x^7}{7!} + \cdots$

15. $\cosh ix = \cos x$

Exercises 8-7, pages 309–310

1.

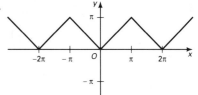

3.

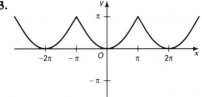

5.   **7.**

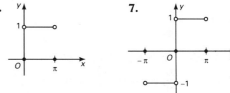

9. $11 - \dfrac{11}{4} + \dfrac{11}{9} - \dfrac{11}{16} + \cdots$

Self Quiz   8-4 / 8-5 / 8-6 / 8-7, pages 310–311

2. 0.8415   **3.** $\dfrac{13}{12}$ **5.** $\dfrac{-1}{2e} + \dfrac{\sqrt{3}}{2e}i$

6. $3 + 3x + 3x^2 + 3x^3 + \cdots$

7. $1 - \dfrac{1}{3^3} + \dfrac{1}{5^3} - \cdots = 1 - \dfrac{1}{27} + \dfrac{1}{125} - \cdots$

8. $\dfrac{\pi}{2} = 2 - \dfrac{2}{3} + \dfrac{2}{5} - \cdots$

Additional Problems, pages 311–312

1. 48   **3.** 0.5833

5. $S_n = \ln\left(\dfrac{1}{2} \cdot \dfrac{4}{3} \cdot \dfrac{9}{8} \cdot \cdots \cdot \dfrac{n^2}{n^2 - 1}\right) =$

$\ln\left(\dfrac{n}{n+1}\right)$ **9.** $\dfrac{\pi^2}{8} = \dfrac{3}{2} - \dfrac{3}{8} + \dfrac{3}{18} - \dfrac{3}{32} + \cdots$

11. (a) 1   **(b)** -1; 1

Chapter 9   Vectors in Space

Exercises 9-1, pages 321–322

1. (a)

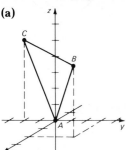

(b) *ABC* is isosceles

3. (a)

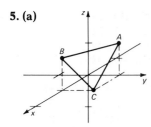

(b) *ABC* is neither

5. (a)

(b) *ABC* is isosceles (equilateral)

7. $x^2 + y^2 + z^2 - 25 = 0$
9. $x^2 + y^2 + z^2 - 4x + 2y + 4z = 0$
11. center $(1, 0, 0)$; radius $= 1$
13. center $(2, -3, -1)$; radius $= 3$
15. $(5, 0, 3), (5, 0, -1), (2, 0, -1), (2, 2, -1),$
$(2, 2, 3), (5, 2, 3)$ **17.** $(0, 0, 4)$
19. $x - y = 0$
21. $x^2 + y^2 + z^2 - 6x + 8z = 0$
23. $x^2 + y^2 + z^2 - 4x \pm 4y - 6z + 13 = 0$
25. $(1, 0, 0)$; 2
27. $100x^2 + 36y^2 + 100z^2 - 225 = 0$

Exercises 9-2, pages 325–326

1. (a) $3\mathbf{i} + 3\mathbf{k}; \mathbf{i} - 8\mathbf{j} + 5\mathbf{k}$
(b) $\|\mathbf{a}\| = 3\sqrt{2}; \|\mathbf{b}\| = 3; \|\mathbf{a} + \mathbf{b}\| = 3;$
$\|\mathbf{a} - \mathbf{b}\| = 3\sqrt{5}$ **3. (a)** $3\mathbf{i} + \mathbf{j} + 2\mathbf{k}$
(b) $-3\mathbf{a} + \mathbf{b} + 6\mathbf{c}$ **5. (a)** $2\mathbf{i} + 6\mathbf{j} + \mathbf{k}$

(b) $-2\mathbf{a} - \mathbf{b} + 5\mathbf{c}$ **7. (a)** **a, c; c, d (b) a, d**
9. $90°$ **11.** $109.5°$ **13.** $\frac{8}{9}\mathbf{a} + \frac{13}{9}\mathbf{b} - \frac{1}{9}\mathbf{c}$
15. $r = 3\mathbf{i} - \mathbf{j} + 2\mathbf{k}$ **17.** $2\mathbf{i} - \mathbf{j} + 3\mathbf{k}$

Self Quiz   9-1 / 9-2, page 327

1. right
2. $x^2 + y^2 + z^2 + 10x - 6y - 2 = 0$
3. $x^2 + y^2 + z^2 - 8x - 14y + 8z = 0$
4. $13\mathbf{i} - 7\mathbf{j} + 9\mathbf{k}; \|\mathbf{a}\| = \sqrt{14}; \|\mathbf{b}\| = \sqrt{11};$
$\|\mathbf{a} + \mathbf{b}\| = 3$ **5.** $115.2°$ **6.** $r = \mathbf{i} + 5\mathbf{j} + 4\mathbf{k}$

Exercises 9-3, pages 331–332

1. $r = (1 + t)\mathbf{i} + (2 - t)\mathbf{j} + 3t\mathbf{k}$
3. $r = 2t\mathbf{i} + 3t\mathbf{j} - t\mathbf{k}$ **5.** $(5, 4, 0)$ **7.** $2\sqrt{5}$
9. $\dfrac{\sqrt{114}}{3}$
11. $r = t\mathbf{i} + (-1 + 2t)\mathbf{j} + (3 - t)\mathbf{k}$
15. $(4, 2, -2), (-2, -1, 1)$ **17.** $(1, 2, 1)$
19. $r = (1 + t)\mathbf{i} + (2 + t)\mathbf{j} + t\mathbf{k}$

Exercises 9-4, pages 335–337

1. (a) $2x + y + 2z - 4 = 0$
(b)

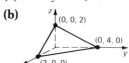

3. (a) $x + 2z - 10 = 0$
(b)

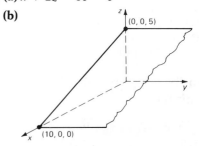

5. $r = (3 + 2t)\mathbf{i} + (2 + t)\mathbf{j} + (1 - 2t)\mathbf{k}$
7. $r = 3t\mathbf{i} + (-2 + 2t)\mathbf{j} + (2 - 6t)\mathbf{k}$ **9.** 4
11. $\dfrac{20}{7}$ **13.** $2x - y + 1 = 0$ **15.** $\dfrac{17}{6}$
17. $(0, -6, 4)$
21. $r = (2 - t)\mathbf{i} + (1 + 2t)\mathbf{j} + t\mathbf{k}$
23. (a) one   **(b)** infinitely many
(c) infinitely many   **(d)** one   **(e)** none
(f) none

Self Quiz 9-3 / 9-4, page 337

1. $(-1 + 6t)\,\mathbf{i} + (3 + t)\,\mathbf{j} + (-2 - 5t)\,\mathbf{k}$

2. $\sqrt{11}$ **3.** $\left(\dfrac{26}{5}, \dfrac{23}{5}, 0\right)$ **4.** $3x - y + 8z = 53$

5. $4x + 3y - 6z + 3 = 0$

6. $(5 + 3t)\,\mathbf{i} + (2 - 4t)\,\mathbf{j} + (-1 + t)\,\mathbf{k} = \mathbf{r}$

Additional Problems, page 338

1. 3 **3.** $\mathbf{v} = \mathbf{a} + 2\,\mathbf{b} - 3\,\mathbf{c}$ **5.** $60°$

7. $x^2 + y^2 + z^2 + 2x - 4z = 116$ **9.** $\dfrac{8}{13}$

11. $r = (2 - t)\,\mathbf{i} + (-3 - 4t)\,\mathbf{j} + (1 + 6t)\,\mathbf{k}$

13. $3x - 8y + 9z + 32 = 0$

Appendix   Spherical Trigonometry

Exercises, page 359

1. $\angle N = 16° \, 40'$ **3.** $\angle N = 82° \, 35'$

5. $\angle B = 113° \, 20'$ **7.** $a = 105° \, 30'$

9. $b = 79.1°$ **11.** $\angle A = 123.3°$ **13.** $30° \, N$

Exercises, page 363

15. (a) 2040 nautical miles
(b) 7440 nautical miles
(c) 3360 nautical miles
17. $5° \, 22' \, W; 119° \, 45'$
19. $54° \, 18' \, N; 37° \, 42' \, W$
21. (a) 4961 nautical miles **(b)** $291° \, 16'$
(c) $220° \, 35'$ **23. (a)** 6458 nautical miles
(b) $240° \, 16'$ **(c)** $235° \, 55'$
25. (a) 4962 nautical miles **(b)** $229° \, 57'$
(c) $218° \, 18'$

Appendix   Computation with Logarithms

Exercises, pages 366–367

1. $\log 127 + \log 42$ **3.** $15 \log 17$
5. $\log 59.7 - \log 23.2$

7. $\log 14.1 + 2 \log 2.5 - \log 3.2$

9. $\dfrac{2}{3}\log 10.2 + \dfrac{1}{3}\log 5.1 - \log 4.1 - 3 \log 6.7$

11. 0.7177 **13.** 1.7559 **15.** 1.3566
17. 3.1895 **19.** $9.5567 - 10$
21. $6.9729 - 10$ **23.** 453 **25.** 359
27. 0.0305 **29.** 3.565 **31.** 0.03533
33. 132,400 **35.** 49.4 **37.** 30.4

41. 1.45 **45.** 2,300,000
47. $a = 13.6; b = 31.4; \angle C = 59°$
49. $\angle C = 24° \, 50'; \angle A = 45° \, 10'; a = 96.6$

Extra Practice

Chapter 1, pages 368–369

1. 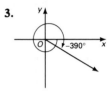 **3.**

5. 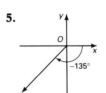 **7.**

9. (a) 0.8000; 0.6000 **(b)** $53°$
11. (a) 0.9428; 0.3333 **(b)** $71°$
13. (a) 0.7454; 0.6667 **(b)** $48°$
15. (a) 0.8944; 0.4472 **(b)** $63°$
17. 0.4014; 0.9159 **19.** 0.9075; 0.4200
21. $76° \, 10'$ or $76.17°$ **23.** $57° \, 10'$ or $57.17°$

25. $\sin \theta = \dfrac{5}{13}; \cos \theta = \dfrac{12}{13}; \tan \theta = \dfrac{5}{12};$
$\csc \theta = \dfrac{13}{5}; \sec \theta = \dfrac{13}{12}; \cot \theta = \dfrac{12}{5}$

27. $\sin \theta = \dfrac{21}{29}; \cos \theta = \dfrac{20}{29}; \tan \theta = \dfrac{21}{20};$
$\csc \theta = \dfrac{29}{21}; \sec \theta = \dfrac{29}{20}; \cot \theta = \dfrac{20}{21}$

29. $\sin \theta = \cos \theta = \dfrac{\sqrt{2}}{2}; \csc \theta = \sec \theta = \sqrt{2};$
$\tan \theta = \cot \theta = 1$

31. $\sin \theta = \dfrac{2\sqrt{5}}{5}; \cos \theta = \dfrac{\sqrt{5}}{5}; \csc \theta = \dfrac{\sqrt{5}}{2};$
$\sec \theta = \sqrt{5}; \tan \theta = 2; \cot \theta = \dfrac{1}{2}$

33. $a = 101; \angle A = 57°; \angle B = 33°$
35. $c = 41; \angle A = 59°; \angle B = 31°$
37. $\angle A = 54°; a = 99.1; c = 122.5$
39. $c = 30; \angle A = 36.9°; \angle B = 53.1°$

41. $\sin 225° = \dfrac{-\sqrt{2}}{2}$; $\cos 225° = \dfrac{-\sqrt{2}}{2}$;

$\tan 225° = 1$; $\csc 225° = -\sqrt{2}$; $\sec 225° = -\sqrt{2}$; $\cot 225° = 1$

43. $\sin 270° = \csc 270° = -1$; $\cos 270° = 0$; $\sec 270°$ and $\tan 270°$ are undefined; $\cot 270° = 0$ **45.** 0.3640 **47.** -0.9833

49. 10.39 **51.** $\dfrac{4}{3}\pi$ **53.** $\dfrac{1}{9}\pi$ **55.** $75°$

57. $-15°$ **59.** $360°$

Chapter 2, pages 369–370

1. $\dfrac{-\sqrt{3}}{2}$; $\dfrac{-1}{2}$; $\sqrt{3}$; -2; $\dfrac{-2\sqrt{3}}{3}$; $\dfrac{\sqrt{3}}{3}$

3. $\dfrac{1}{2}$; $\dfrac{\sqrt{3}}{2}$; $\dfrac{\sqrt{3}}{3}$; $\dfrac{2\sqrt{3}}{2}$; 2; $\sqrt{3}$

5. $\dfrac{-\sqrt{3}}{2}$; $\dfrac{1}{2}$; $-\sqrt{3}$; 2; $\dfrac{-2\sqrt{3}}{3}$; $\dfrac{-\sqrt{3}}{3}$

7. $\dfrac{-\sqrt{3}}{2}$; $\dfrac{1}{2}$; $-\sqrt{3}$; 2; $\dfrac{-2\sqrt{3}}{3}$; $\dfrac{-\sqrt{3}}{3}$ **9.** $(3, 0)$

11. $(-2, 0)$ **13.** $(4, 0)$ **15.** $(4, 0)$

17.

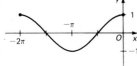

19.

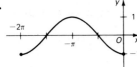

21.

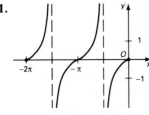

23.

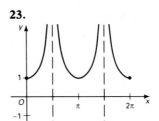

25. $45°$, or $\dfrac{\pi}{4}$

27. $30°$, or $\dfrac{\pi}{6}$

29. 0

31. 0

33. $\dfrac{\pi}{2}$

Chapter 3, pages 370–371

1. $\dfrac{1 - 2\sin^2 x}{\sin^2 x}$ **3.** $\dfrac{\cos^2 x \sqrt{1 - \cos^2 x}}{1 - \cos^2 x}$

5. $\dfrac{\cos^3 x}{\cos^2 x - 1}$ **7.** $2\tan x$ **17.** $\dfrac{-63}{65}$

19. $\dfrac{-56}{65}$ **21.** $\dfrac{16}{65}$ **23.** $\dfrac{-56}{65}$ **25. (a)** $\dfrac{120}{169}$

(b) $\dfrac{-119}{169}$ **27. (a)** $\dfrac{-24}{25}$ **(b)** $\dfrac{-7}{25}$ **29.** $\dfrac{-\sqrt{34}}{12}$

31. $\dfrac{1}{7}$ **33.** $\dfrac{\pi}{6}, \dfrac{5\pi}{6}, \dfrac{3\pi}{2}; \dfrac{\pi}{2}$

35. $0; \pi; \dfrac{\pi}{4}, \dfrac{3\pi}{4}, \dfrac{5\pi}{4}, \dfrac{7\pi}{4}$ **37.** $\dfrac{k\pi}{2}$ **39.** $\dfrac{\pi}{2} + k\pi$

Chapter 4, pages 371–372

1. $48.2°$ **3.** 4.58 **5.** 15.39 **7.** 49.34; 34.87
9. $a = 4.6$; $c = 6.3$ **11.** $a = 23.2$; $b = 15.2$
13. $b = 25.5$; $c = 21.8$
15. $a = 12.0$; $b = 15.7$
17. no solution; $b \sin A = 12.6 > a$
19. $23.2°$ or $156.8°$

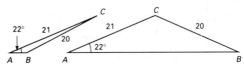

21. $90°$

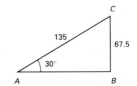

23. no solution; $a = 6 < b \sin 55° 40' = 7.43$
25. 47.8 square units   **27.** 236.9 square units
29. 3.4 square units   **31.** 27.5 square units

Chapter 5, pages 372–374

1.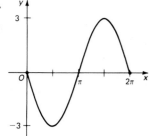

3.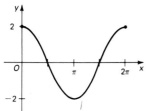

5.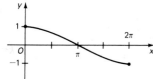

7.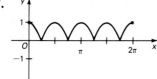

9.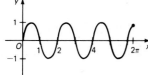

11.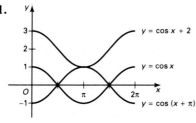

13.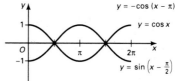

15.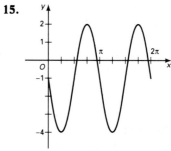

17.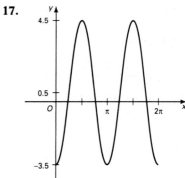

19. $\sqrt{2} \cos (\theta - 45°)$ **21.** $4 \cos (\theta + 49°)$
23. $5 \cos (\theta - 127°)$ **25.** $\dfrac{5}{2} \cos (\theta - 37°)$

27.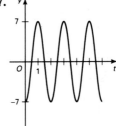

29.

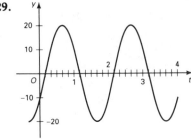

31. $y = 7 \cos (\pi t \pm \pi)$

33. $y = 20 \cos \left(\pi t - \dfrac{2\pi}{3} \right)$

35.

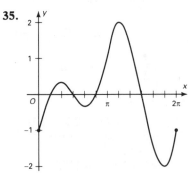

37.

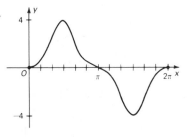

39.

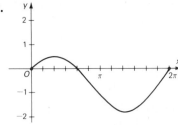

41.

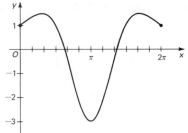

Chapter 6, pages 374–376

1.

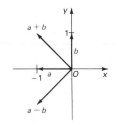

$\|\mathbf{a}\| = 1$; $\|\mathbf{b}\| = 1$; $\|\mathbf{a} + \mathbf{b}\| = \sqrt{2}$;
$\|\mathbf{a} - \mathbf{b}\| = \sqrt{2}$

3.

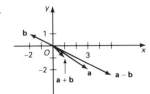

$\|\mathbf{a}\| = \sqrt{13}$; $\|\mathbf{b}\| = \sqrt{5}$;
$\|\mathbf{a} + \mathbf{b}\| = \sqrt{2}$; $\|\mathbf{a} - \mathbf{b}\| = \sqrt{34}$
5. $\mathbf{i} + 13\mathbf{j}$ **7.** $\mathbf{i} + \mathbf{j}$ **9.** 206 km; 346°
11. 74 km; 236°   **13.** 329 km/h; 175°
15. 315 km; bearing of A from B, 99°;
bearing of B from A, 279°
17. perpendicular 1842 N; parallel 670 N
19. 2695 N
21. 2585 N on wire at 50°; 1919 N on wire at
30°   **23.** 14.3 N at 19.4°   **25.** -4; no; no

27. -20; no; yes   **29.** $\dfrac{-1}{5}\mathbf{a} + \dfrac{3}{5}\mathbf{b}$

31. $\dfrac{11}{25}\mathbf{a} + \dfrac{2}{25}\mathbf{b}$ **33.** 1 J   **35. (a)** -16 J

(b) -16 J   **37.** 1,102,500 J or 1.1025×10^6 J
39. 2,572,500 J or 2.5725×10^6 J

Chapter 7, pages 376–378

1. $3 = r \sin \theta$ **3.** $r = 3$

5.

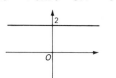

7.

9.

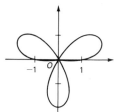

11.

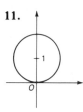

13.

15.

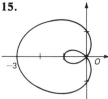

17. parabola   **19.** hyperbola

21.

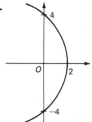

23.

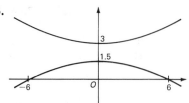

25. (a) $5 + 0i$ (b) $7 + i$ (c) $\dfrac{1}{2} + \dfrac{1}{2}i$

27. (a) $5 + i$ (b) $12 + 0i$ (c) $0 - \dfrac{2}{3}i$

29. (a) $3 + i$ (b) $\sqrt{10}$ (c) $\dfrac{3}{10} + \dfrac{1}{10}i$

(d) $8 - 6i$ **31.** (a) $\dfrac{4}{5} - \dfrac{3}{5}i$ (b) 1

(c) $\dfrac{4}{3} - \dfrac{3}{5}i$ (d) $\dfrac{7}{25} + \dfrac{24}{25}i$

37. $\sqrt{2}\,(\cos 135° + i \sin 135°)$
39. $2\sqrt{3}\,(\cos 330° + i \sin 330°)$
41. $4\,(\cos 240° + i \sin 240°)$
43. $2\sqrt{2}\,(\cos 345° + i \sin 345°)$
45. $3\sqrt{2}\,(\cos 90° + i \sin 90°)$

47. $\dfrac{\sqrt{2}}{2}\,(\cos 285° + i \sin 285°)$

49. $\dfrac{\sqrt{2}}{3}\,(\cos 30° + i \sin 30°)$

51. (a) $\cos 60° + i \sin 60°$ (b) $\dfrac{1}{2} + i\,\dfrac{\sqrt{3}}{2}$

53. (a) $\dfrac{\sqrt{2}}{2}\,(\cos 45° + i \sin 45°)$ (b) $\dfrac{1}{2} + \dfrac{1}{2}i$

55. $\dfrac{-27\sqrt{2}}{2} + i\,\dfrac{27\sqrt{2}}{2}$ **57.** $\dfrac{1}{32} + 0i$

59. $\dfrac{-1}{4} + 0i$ **61.** $-4 + 0i$

63. $1.1 + 0.3i;\ -0.8 + 0.8i;\ -0.3 - 1.1i$
65. $0.5 + 1.0\,i;\ -0.8 + 0.8\,i;\ -1.0 - 0.5\,i;$
$0.2 - 1.1\,i;\ 1.1 - 0.2\,i$
67. $\cos 67.5° + i \sin 67.5°;\ \cos 157.5° +$
$i \sin 157.5°;\ \cos 247.5° + i \sin 247.5°;$
$\cos 337.5° + i \sin 337.5°$

69. $2^{\frac{1}{8}}\,(\cos 78.75° + i \sin 78.75°);$

$2^{\frac{1}{8}}\,(\cos 168.75° + i \sin 168.75°);$

$2^{\frac{1}{8}}\,(\cos 258.75° + i \sin 258.75°);$

$2^{\frac{1}{8}}\,(\cos 348.75° + i \sin 348.75°)$

Chapter 8, page 379

1. (a) 0.8677 (b) 0.7104
(c) $\sin \dfrac{\pi}{3} = 0.8660;\ \sin \dfrac{\pi}{4} = 0.7071$
3. (a) 0.805 (b) -0.809
(c) $\sin 2.2 = 0.8085;\ \sin(-0.94) = -0.8076$
5. 1.543 **7.** -1 **9.** $\dfrac{\sqrt{2}}{2} + i\,\dfrac{\sqrt{2}}{2}$

Chapter 9, pages 380–381

1.

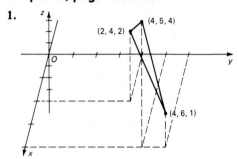

3. $x^2 + y^2 + z^2 - 9 = 0$

5. $x^2 + y^2 + z^2 - 4x - 4z - 17 = 0$

7. $(0, 0, 0)$; 4   **9.** $(0, -1, 2)$; 2

11. (a) $4\mathbf{i} - 14\mathbf{j} + 8\mathbf{k}$; $2\mathbf{i} + 8\mathbf{j} - 6\mathbf{k}$

(b) $\|\mathbf{a}\| = \sqrt{6}$; $\|\mathbf{b}\| = \sqrt{13}$;

$\|\mathbf{a} + \mathbf{b}\| = \sqrt{3}$; $\|\mathbf{a} - \mathbf{b}\| = \sqrt{35}$ **13.** $61.9°$

15. $118.1°$ **17.** $\dfrac{2}{9}\mathbf{a} + \dfrac{7}{9}\mathbf{b} - \dfrac{1}{9}\mathbf{c}$

19. $l{:}r = (1 + t)\mathbf{i} + t\mathbf{j} + (1 - 2t)\mathbf{k}$

21. $l{:}r = (1 + t)\mathbf{i} + t\mathbf{j} + (3 + t)\mathbf{k}$

23. $(1, 0, 1)$ **25.** $\dfrac{\sqrt{222}}{3}$

27. $3x + y + 2z - 7 = 0$

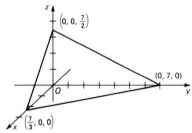

29. $\mathbf{r} = (2 + 3t)\mathbf{i} + (2 - t)\mathbf{j} + 2t\,\mathbf{k}$

31. $\dfrac{\sqrt{14}}{7}$ **33.** $x + 2y - 3z = 0$ **35.** $\dfrac{5\sqrt{14}}{7}$

Algebra and Geometry Review Exercises

The Pythagorean Theorem and Radicals, page 382

1. 17.0   **3.** 36.5   **5.** 82.5   **7.** $5\sqrt{3}$ **9.** $3\sqrt{5}$

11. $4\sqrt{6}$ **13.** $\sqrt{2}$ **15.** $\dfrac{\sqrt{2}}{4}$

Parallel Lines, page 383

1. $90°$ **3.** $110°$ **5.** $40°$ **7.** $55°$

Special Right Triangles, page 384

1. 14   **3.** 68   **5.** $\dfrac{17\sqrt{3}}{2}$ **7.** $20\sqrt{3}$ **9.** $2\sqrt{6}$

Graphing Equations, page 385

1.

3.

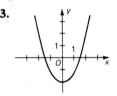

5.

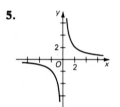

7.

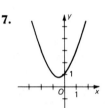

9.

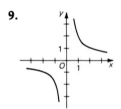

Simplifying Rational Expressions, page 385

1. $\dfrac{x}{x + 5}$ $(x \neq -2, -5)$ **3.** $\dfrac{2z - 3}{3}$ $(z \neq -1)$

5. $\dfrac{4x}{x + 1}$ $(x \neq 1, -1)$ **7.** $\dfrac{x + 2}{x + 3}$ $(x \neq -3, 1)$

9. $(a + 2)(a - 1)$ $(a \neq 2, -1)$

Solving Equations, page 386

1. -1 or -4 **3.** $\dfrac{2}{3}$ or $\dfrac{3}{2}$ **5.** no solution

7. 14 or -4 **9.** $\dfrac{2}{3}$ **11.** $3 \pm \sqrt{11}$

13. 10 or -2 **15.** $-4 \pm \sqrt{17}$

17. $-3 \pm 2\sqrt{10}$

Area of a Polygon, page 387

1. 108   **3.** 169   **5.** 35   **7.** $\dfrac{363\sqrt{3}}{2}$ **9.** $\dfrac{81\sqrt{3}}{4}$

Conic Sections, page 388

1.

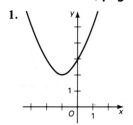

$y = (x + 1)^2 + 2$; parabola

3.

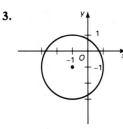

$(x + 1)^2 + (y + 1)^2 = 4$; circle

5.

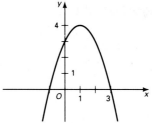

$y = -(x - 1)^2 + 4$; parabola

7.

$\dfrac{x^2}{4} - \dfrac{y^2}{25} = 1$; hyperbola

Systems of Equations, page 389

1. $\left(\dfrac{20}{3}, 6\right)$ **3.** $(1, 1)$ **5.** $(-4, -3)$

7. $\left(\dfrac{2}{5}, \dfrac{8}{5}, 0\right)$ **9.** $(3, -2, 0)$ **11.** $(4, 2, 3)$